Soils
and
Soil Fertility

McGRAW-HILL PUBLICATIONS
IN THE AGRICULTURAL SCIENCES

JOHN R. CAMPBELL
Department of Dairy Husbandry
University of Missouri
Consulting Editor in Animal Science

CARL HALL
College of Engineering
Washington State University
Consulting Editor in Agricultural Engineering

LAWRENCE H. SMITH
Department of Agronomy
University of Minnesota
Consulting Editor in Plant Science

ADRIANCE AND BRISON   Propagation of Horticultural Plants
BROWN   Farm Electrification
CAMPBELL AND LASLEY   The Science of Providing Milk for Man
CAMPBELL AND MARSHALL   The Science of Animals that Serve Mankind
CHRISTOPHER   Introductory Horticulture
CRAFTS AND ROBBINS   Weed Control
CRUESS   Commercial Fruit and Vegetable Products
DICKSON   Diseases of Field Crops
ECKLES, COMBS, AND MACY   Milk and Milk Products
EDMONDS, SENN, ANDREWS, AND HALFACRE   Fundamentals of Horticulture
JONES   Farm Gas Engines and Tractors
KIPPS   The Production of Field Crops
KOHNKE   Soil Physics
KOHNKE AND BERTRAND   Soil Conservation
KRIDER AND CARROLL   Swine Production
LASSEY   Planning in Rural Environments
LAURIE AND RIES   Floriculture: Fundamentals and Practices
LAURIE, KIPLINGER, AND NELSON   Commercial Flower Forcing
MAYNARD AND LOOSLI   Animal Nutrition
METCALF, FLINT, AND METCALF   Destructive and Useful Insects
MUZIK   Weed Biology and Control
RICE, ANDREWS, WARWICK, AND LEGATES   Breeding and Improvement of Farm Animals
SMITH AND WILKES   Farm Machinery and Equipment
THOMPSON AND TROEH   Soils and Soil Fertility
THOMPSON AND KELLY   Vegetable Crops
THORNE   Principles of Nematology
TRESHOW   Environment and Plant Response
WALKER   Plant Pathology

FOURTH EDITION

SOILS AND SOIL FERTILITY

LOUIS M. THOMPSON
Associate Dean of Agriculture
Professor of Agronomy
Iowa State University

FREDERICK R. TROEH
Professor of Agronomy
Iowa State University

McGRAW-HILL BOOK COMPANY

New York St. Louis San Francisco Auckland Bogotá Düsseldorf
Johannesburg London Madrid Mexico Montreal New Delhi
Panama Paris São Paulo Singapore Sydney Tokyo Toronto

Library of Congress Cataloging in Publication Data

Thompson, Louis Milton.
  Soils and soil fertility.

  Includes bibliographical references and index.
  1. Soil science. 2. Soil fertility. I. Troeh,
Frederick R., Joint author. II. Title.
S591.T47 1978   631.4   77-21404
ISBN 0-07-064411-X

SOILS AND SOIL FERTILITY

5 6 7 8 9 10 11 **KPKP** 8 9 8 7 6 5 4 3 2

This book was set in Times Roman by Cobb/Dunlop
Publisher Services Incorporated.
The editor was C. Robert Zappa and the production
supervisor was Milton J. Heiberg.
Kingsport Press, Inc. was printer and binder.

Contents

Preface

Soils and Soil Fertility is intended for use as a text for the introductory course in soils for students in agriculture and related sciences. It originated at Texas A. & M. in 1937 as a mimeographed supplement prepared by the senior author for use with available textbooks. The first two versions in book form were published by William C. Brown of Dubuque, Iowa. After that, two editions by the senior author were published by McGraw-Hill Book Company. The third and fourth editions by McGraw-Hill were revised by the junior author with chapter-by-chapter consultation with the senior author. Students' comments and reviews by competent colleagues provided helpful guidance in the revisions.

This fourth edition is an extensive revision incorporating significant changes in every chapter. Appropriate new materials have been integrated into the text to supplement proven materials from earlier editions. The aim has been to provide an up-to-date text that is an easily readable source of accurate information. The result represents over 30 years of teaching experience in Texas A. & M. and in Iowa State University by the senior author, and 15 years of teaching experience at the University of Idaho and Iowa State University by the junior author. The junior author was also able to draw on over 10 years of soil survey experience in Idaho, California, New York, Uruguay, and Iowa.

Most students in agriculture take only one course in soils. This book has therefore been made comprehensive enough to stand by itself for students who will have only the first course in soils. This broad coverage also makes the book well suited to serve as a background for the more advanced courses in soils. It is complete enough to serve as a reference for soil fertility and other advanced courses or to be used as a refresher course for those who need to review the field of soil science.

An ideal preparation for the study of soils would include courses in both inorganic and organic chemistry, geology, and biology. Few students begin their study of soils with such a comprehensive background, however, so the authors have made an effort to supply essentials from these related sciences. Widely understood terms have been chosen wherever possible so that persons from any field of interest can read and understand the contents of this book. Necessary technical terms are explained at the point where they are first used. Even so, experience has shown that a course in college chemistry is very helpful in the study of soils and should be a prerequisite for the introductory soils course.

Special effort was made in both the third and fourth editions to broaden the book to match the varied backgrounds and interests of today's students. The chapter that was once called Crop Rotations and Soil Fertility has been changed into a chapter on Land Use and Soil Management that includes urban as well as rural land use. Similarly, the former chapter on Principles and Practices of Liming has been replaced by a chapter on Amending the Soil that includes not only liming but also acidification of alkaline soils and the making of artificial soils for greenhouse and other uses. A new chapter entitled Water Management, dealing with weather, irrigation, and drainage, was added to the third edition. Another new chapter dealing with Soil Pollution has been added to the fourth edition to cover an area that is both current and pertinent.

An attempt has been made throughout the book to make the material understandable as well as accurate. The approach has been to treat the material as the first exposure of the reader to the subject. Concepts are developed gradually, and new terms are defined or explained when they are introduced in the text. The organization both within and among the chapters has been carefully considered to give continuity to the material. The chapters are arranged in a sequence that permits the later chapters to build on the previous chapters. Students will therefore benefit from reading the book through from the beginning rather than being asked to skip around for assigned reading. Chapter 1, for example, is entitled Soil because it contains an overall viewpoint including the development of a number of basic concepts that are needed for understanding the succeeding chapters. Chapter 6, Soil Mineralogy, Chapter 9, Fertilizers, and Chapter 16, Soil Classification and Survey, also contain much background material for the succeeding chapters.

Earlier editions used both the metric and the English systems of measurement according to prevailing usage in the United States. The fourth edition uses metric units throughout with only a few references to the English system of units. Sections have been rewritten, data recalculated, and adjustments made wherever needed to express the material smoothly in the metric system. An appendix has been added to provide conversion factors between the two systems.

The literature on soils has proliferated greatly in recent years. The authors have had access to excellent libraries and other sources of information. References cited in the text and other particularly relevant sources are listed at the end of each chapter, but these items represent only a fraction of the literature that was consulted.

Many people have assisted in the development of this book. Special recognition goes to John Schafer for having read every chapter and made important contributions to each one. Other colleagues who contributed significantly to the present content of the book include Minoru Amemiya, C. A. Black, L. R. Frederick, J. R. George, D. A. Gier, Don Kirkham, Murray Milford, Larry Miller, W. C. Moldenhauer, John Pesek, W. H. Pierre, Don Post, F. F. Riecken, W. H. Scholtes, A. D. Scott, C. H. Sherwood, J. A. Stritzel, Steve Thien, J. R. Webb, and L. V. Withee. Special acknowledgment is hereby given to Margaret Thompson for having typed the manuscripts of the earlier editions and to Miriam Troeh for typing the material for the third and fourth editions.

Louis M. Thompson
Frederick R. Troeh

Chapter 1

Soil

A thin layer of soil covers most of the earth's land surface. This layer, varying from a few centimeters to 2 or 3 meters in thickness, might appear insignificant relative to the bulk of the earth. Yet it is in this thin layer of soil that the plant and animal kingdoms meet the mineral world and establish a dynamic relationship. Plants obtain water and essential nutrients from the soil. Animals depend on plants for their lives. Plant and animal residues find their way back to the soil and are decomposed by the teeming microbial population living there. Life is vital to soil and soil is vital to life.

Humanity's contact with soil is so universal that each person has his or her own concept of the nature of soil. To an engineer it may be a construction material or the foundation material for a building. To the farmer it is a medium for growing crops. A child may use it to make mud pies, but then it becomes dirt to be washed from hands and clothes. To all of us, soil is the source from which springs our food, clothing, and shelter. Our existence depends on soil.

Liebig and other early chemists considered soil as a storehouse for plant nutrients. Early geologists concluded that soil is weathered rock. These early concepts are not wrong, but neither are they complete. The origin of the word *soil* illustrates another facet of its character. It comes from the Latin word *solum,* which means *floor.* The French word *sol* and the Spanish word *suelo* are still used to mean either soil or floor.

Any definition assigned to such a complex material as soil depends greatly on the viewpoint of the person formulating the definition. An edaphologist, considering soils in relation to their use as media for growing plants, may define soil as a *mixture of mineral and organic matter that is capable of supporting plant life.* A pedologist, studying soil as a distinct entity, can define it as the *natural product formed from weathered rock by the action of climate and living organisms.* The concept of life is vital in these definitions. In one, the soil supports life; in the other, life helps to form the soil. Both viewpoints are correct. Soil supports living organisms and its characteristics are partly determined by the action of the living organisms.

Beneath the soil, a mass of loose mineral matter lacking the influence of living things is frequently encountered above bedrock. Such material is properly termed

Figure 1-1 A characteristic common to nearly all soils is the darkening of the upper part of the soil by the accumulation of organic matter in and near the surface. *(Soil Conservation Service, USDA.)*

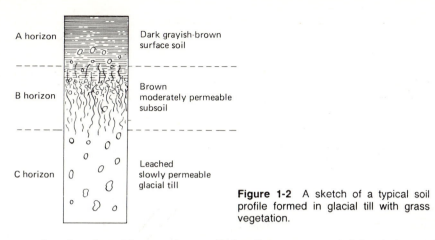

Figure 1-2 A sketch of a typical soil profile formed in glacial till with grass vegetation.

weathered rock and is sometimes called *soil parent material,* but it is not the same as soil. The word *regolith* is used as an inclusive term for all the loose material above bedrock.

THE SOIL PROFILE

Soils develop distinct layers at varying depths below the land surface (Figure 1-1). A vertical section of the soil to expose the layering is called a *profile.* The upper layer is usually higher in organic matter and darker in color than the layers below. This upper layer is called the *A horizon,* or topsoil.

The middle part of the profile usually contains more clay and has a brighter color than the topsoil. This layer is called the *B horizon,* or subsoil. The A and B horizons together are referred to as the *solum,* or true soil.

The *C horizon,* commonly referred to as the *soil parent material,* occurs beneath the solum and extends downward to bedrock. It may be thick, thin, or even absent. The soil profile includes the A and B horizons and at least the upper portion of the C horizon if a C is present (Figure 1-2).

How the Soil Profile Develops

The material in the A and B horizons was once part of the C horizon but has undergone many changes. The changes that produce A and B horizons are referred to as *soil development.*

The upper layer became an A horizon as it accumulated organic matter from roots and plant residues. Other changes resulted from forces that cause physical disintegration and chemical decomposition. The A horizon is the portion of the soil most exposed to the weathering action of the sun, rain, wind, and ice and to the action of living things. The more easily decomposed materials weather away first, thus leaving the more resistant minerals concentrated in the topsoil. The more mature the soil, that is, the more weathering that has taken place in the soil, the more concentrated resistant materials become in the topsoil. This fact is of great significance to plant growth. The more resistant minerals remaining behind in strongly weathered soils may decompose too slowly to be a good source of plant nutrients. The more fertile soils are less weathered and still contain minerals that decompose and release new plant nutrients each year.

Weathering processes break down some of the mineral particles in the soil to smaller sizes. Part of the sand is reduced to silt size, and part of the silt is reduced to clay. Some constituents become soluble and are removed from the soil by the leaching action of water. The A horizon is the most leached portion of the soil.

It would be reasonable to anticipate that the clay concentration of the A horizon would gradually increase as weathering action reduces the sizes of mineral particles. There are, however, opposing processes. Some of the clay may weather into soluble materials that are leached from the soil. Another, generally more important, process is the downward movement of solid clay particles in percolating water. The net result is that the percentage of clay in the A horizon is likely to remain about the same for thousands of years. However, clay movement continues even after clay formation slows down; many old soils have lost most of the clay from their A horizons.

The clay content of B horizons increases with time. Part of the increase is caused by clay from the A horizon being deposited in the B horizon, and part of it comes from the weathering of silt and sand in the B horizon to form clay. Thus the B horizon may accumulate a much higher concentration of clay than either the A or the C horizon of the same soil. Soils with large differences in the clay contents of their A and B horizons are said to be *strongly differentiated*.

In summary, two of the most widespread soil characteristics the world over are the accumulation of organic matter in the A horizon and the accumulation of clay in the B horizon. These characteristics must be considered as norms that fail to occur only when there is some unusual overriding factor.

COMPOSITION OF SOIL

The soil inherits mineral matter from its parent rock and organic matter from its living organisms. These materials constitute the solid portion of the soil and form its skeleton. Voids known as *pore spaces* occur between the solid particles. The pore space usually constitutes about half of the volume of the A horizon and somewhat less than half the volume of the B and C horizons. Water and air share the pore space in variable proportions. The smaller pores generally contain water, and the larger ones contain air. The shape and continuity of the larger pores determine to a large extent how well the soil is aerated.

It is desirable for the water that enters the soil to continue moving downward through the profile until the pore space contains about two-thirds water and one-third air. Soils with high clay contents, especially those low in organic matter, may hold so much water that they are poorly aerated. Sandy soils may let the water pass through too rapidly and not retain enough to support plant growth through a dry period. Desirable proportions of air, water, and solids are illustrated in Figure 1-3.

Mineral Components of Soils

The mineral particles in soils range in size from submicroscopic clay to stones several meters across. The pieces over 76 mm in diameter are referred to as *stones*, and those

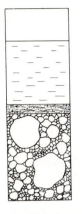

Air

Water

Organic matter

Mineral matter

Figure 1-3 The relative volumes of soil components in a typical A horizon at the maximum desirable water content.

smaller than stones but over 2 mm in diameter are called *gravel*. Stones and gravel are inert in terms of supporting plant growth but can affect physical properties such as permeability and erodibility and can be limiting factors for tillage operations.

Mineral particles smaller than 2 mm in diameter are divided into sand, silt, and clay. *Sand* particles, according to U.S. Department of Agriculture (USDA) standards, are between 0.05 and 2 mm in size and can easily be seen by the unaided eye. Sand gives soil a gritty feel. *Silt* includes the particles from 0.002 to 0.05 mm in diameter; only the coarse silt particles are visible to the eye. When silty material is rubbed between the thumb and fingers, it has a smooth feel like flour or talcum powder. Those particles smaller than 0.002 mm in diameter are called *clay*. Clay feels gritty and hard when dry but becomes sticky and plastic when wet. Certain types of clay swell considerably by absorbing water and then shrink and cause the soil to crack open when it dries. Clay is much more active chemically than sand and silt. Available plant nutrients are stored by the clay and the soil organic matter.

The elemental composition of the mineral matter is over half oxygen by weight. In fact, oxygen and hydroxyl ions are the only anions (negatively charged ions) in the most abundant soil minerals. The principal cations (positively charged ions) in their usual order of abundance are silicon, aluminum, iron, potassium, calcium, magnesium, and sodium. All the other elements found in the mineral portion usually total less than 5 percent of its weight. The most abundant of all minerals are called *silicates* or *aluminosilicates* because of the predominance of oxygen, silicon, and aluminum in their composition. Another important group, known as the *oxide minerals,* is dominated by iron, aluminum, and silicon oxides. The oxide minerals are most abundant in highly weathered materials in tropical areas. Minerals are discussed more fully in Chapter 6.

Organic Components of Soils

Organic matter constitutes between 1 and 6 percent of the topsoil weight of most upland soils. Topsoils with less than 1 percent organic matter are mostly limited to desert areas. At the other extreme, the organic matter contents of soils in low wet areas may be as high as 90 percent or more. Soils with more than 12 to 18 percent organic carbon are called *organic soils* (see Chapter 5 for more details).

Soil organic matter is partly alive and partly dead. Living plant roots are usually excluded, but both fresh and partly decomposed residues of plants and animals are included, along with the tissue of living and dead microorganisms.

It should be obvious that the soil organic matter contains some of every element that is essential to plant growth. These elements are released and made available to plants as the organic matter decomposes, but not all at the same time and rate. Almost all the nitrogen utilized by growing plants must come via an organic source (unless supplied by people) because the primary minerals contain only traces of nitrogen. Decomposing organic matter is also an important source of phosphorus and sulfur. Furthermore, the decomposition of organic matter produces acids and other substances that cause soil minerals to decompose and release plant nutrients.

Supplying plant nutrients is one of the important functions of organic matter in soil. Another is to help bind the mineral particles into aggregate units providing an open structure with adequate pore space for good aeration. Structure is especially important in soils having moderate to high clay contents.

THOUSANDS OF DIFFERENT SOILS

Parent material, climate, living organisms, topography, and time are known as the *factors of soil formation.* A soil can be considered as the result of the integrated action of these five factors. There are thousands of different soils because different kinds and degrees of the soil-forming factors can be combined in thousands of different ways. Each combination produces a different soil with its own unique properties. Fortunately, many soil properties can be predicted from a study of the five soil-forming factors as they pertain to individual soils.

Parent Material

Soils inherit hundreds of different minerals from their parent materials. These minerals have a wide variety of chemical compositions and a wide range of weathering rates. The individual particles differ greatly in size and shape. The arrangement and the amount of consolidation or binding together of the individual particles also vary widely. Thus a particular soil may originate from a parent material that is distinct, at least in some measure, from all other soil parent materials.

Climate

The weathering forces that attack the parent materials are as varied as the materials. The complexity of climate is seldom overstated. It is true that precipitation and temperature are the basic components, but neither of these is adequately described by citing averages. Maxima, minima, and seasonal patterns are vital characteristics of both temperature and precipitation. Type and intensity of precipitation are also significant. The coordination of the temperature and precipitation patterns must be considered. Does the soil become thoroughly dry at some time during the year? If so, it will differ from another soil that remains moist. Does it freeze? Is the precipitation intense enough to cause runoff and erosion? Is there ever enough water present at any one time to percolate through the entire solum into the C horizon and rock layers beneath? The answer to this last question is one of the most important features of the climate. Even a temporary excess of water over the amount the soil can absorb leaches materials from the solum.

Living Organisms

Plant and animal life of both macroscopic and microscopic forms take part in soil formation, working with climatic forces to alter the parent material and help make it into soil. Generally the most obvious variable in this factor is the vegetation. An easily noted difference between grassland soils and forested soils is in the distribution of organic matter within the soil. The organic-matter concentration in a grassland soil is highest near the surface and declines gradually with depth through the solum. This pattern closely follows the distribution of grass roots because much of the organic matter comes from the annual death and regrowth of grass roots.

A horizon

B horizon

Figure 1-4   A soil formed under forest vegetation showing a lighter color in the lower part of the A horizon than in either the upper A horizon or the B horizon. This may be contrasted with the grassland soil shown in Figure 1-1. *(Courtesy of Wells Andrews.)*

The top part of a forested soil may be as dark colored and contain as much organic matter as the corresponding part of a grassland soil, but the lower part of the A horizon under forest is commonly light colored and low in organic matter (Figure 1-4). Tree roots live for a long time and therefore add little organic matter to the soil year by year.

Trees deposit leaves and twigs as litter on the soil surface. This litter decomposes gradually. Part of it becomes mixed into the upper layer of soil and helps to produce the thin dark-colored upper A horizon. The decomposition processes produce organic acids that leach down through the soil and tend to make forested soils more acid than grassland soils. Intense acid leaching often produces more distinct concentrations of clay in the B horizons of forested soils than in grassland soils.

There are, of course, many different kinds of grasses and trees. Each kind has its own influence on soil formation. Evergreen trees, for example, usually produce a more acid leachate than do deciduous trees. Various types of grasses, bushes, and other plants produce different kinds and amounts of soil organic matter. Each soil reflects the vegetation under which it formed.

Vegetation is a very important variable in determining the type of soil that forms in a particular place, but it is not a completely independent variable. The climate of an area has much to do with the type of vegetation that will grow there. Also, the nutrient-supplying capacity the soil inherits from its parent material helps determine the kind of vegetation and influences the amount of growth produced on the soil. Soil and vegetation are each influenced by the other and by the action of climate and other factors.

Topography

Another two-way relationship exists between soil and topography. Some soils erode easily and readily permit wide valleys to form. Other soils resist erosion and may cause hills and steep slopes. In turn, the steepness of slope influences how fast the soil erodes. Rapid erosion keeps soils shallow and young. Slower erosion allows the formation of deeper, more strongly differentiated soils as time passes.

Soils with strong differentiation of A and B horizons usually occur in nearly level areas where there is little or no erosion. Water tends to accumulate in these soils and make them wetter than the sloping soils. Level soils tend to have poor drainage both externally and internally. The wetness often results in a lush growth of water-loving plants. Poor aeration in the wet soils causes organic matter to accumulate and produce dark colors.

Soils on steep slopes are drier, are usually shallower, and have lighter, often redder, colors than soils on gentle slopes. These characteristics are especially prevalent in soils that face toward the early afternoon sun. South- (in the Northern Hemisphere) and west-facing soils lose water first by runoff and second by rapid evaporation. Vegetative growth is sparser, and decomposition is more rapid than elsewhere. The soil organic matter contents are therefore relatively low on the slopes with the warmest soils.

Time

Soil formation is a slow but continuing process. The soils change as the years, centuries, and millennia pass. The nature of the soil today at any one place therefore depends partly on how long it has been exposed to weathering. Young soils are similar to their parent materials, whereas older soils become strongly differentiated. Too much differentiation is unfavorable because the clay in the B horizon of an old soil makes it too hard and poorly aerated for plant roots to penetrate easily.

Declining fertility is another unfortunate characteristic usually associated with old soils. Fertility, however, can be improved by the proper use of fertilizers.

NAMING SOILS

It has become standard practice to give each soil the name of a town, school, church, stream, or other geographic feature located near the area where the soil is first identified. Soil names such as *Amarillo* and *Fargo* automatically infer soils of northwestern Texas and North Dakota, respectively. The same name is given to all soil areas that have profile characteristics falling within its defined limits. The unit thus formed is known as a soil *series.* Theoretically, any two areas that combine the same type and degree of each of the five soil-forming factors will have the same soil series. This principle applies theoretically whether the two areas occur near each other or on separate continents. In practice, a soil series name is usually used only in one country because of political and organizational problems. Certain broad units of soil classification, however, are in almost worldwide use.

Some range of characteristics must be allowed within each soil series because differences can be found between any two soil profiles. But these ranges must be carefully defined. The limits must be narrow enough to permit interpretations to be made for agricultural, engineering, and other uses, yet broad enough to permit significant areas of soil to be called by the same name.

Soil series names are given to areas, not points. Surface features such as slope and roughness are soil characteristics to be considered in naming a soil. A soil profile by itself does not qualify because it essentially represents a point on the landscape having a vertical dimension but lacking area and volume. The *pedon* is the three-dimensional soil unit defined to have all the characteristics of a soil. It corresponds to an individual plant or animal as the smallest individual unit of the soil series. Arbitrarily, each pedon covers an area of 1 m^2 except in certain special circumstances where it must be larger to fully represent the soil.

Classifying Soils

There are thousands of soil series—too many for any one person to be acquainted with all of them. More inclusive soil names are needed to group soils into a smaller number of classes. Several systems of classification are used for this purpose in

Table 1–1   Generalized Descriptions of the Orders in Soil Taxonomy

Order	Description
Alfisols	Soils with medium to light-colored A horizons and with significant clay accumulations in their B horizons. Most Alfisols formed under forest vegetation.
Aridisols	Soils of arid regions. Aridisols are light-colored and most are alkaline in reaction.
Entisols	Very young soils that have little or no horizon differentiation.
Histosols	Soils dominated by organic materials. Histosols form in wet and/or cold conditions.
Inceptisols	Soils in an early stage of development that lack significant clay accumulations. Most Inceptisols formed under forest vegetation in humid climates.
Mollisols	Soils with thick dark-colored A horizons. Most Mollisols formed under grass vegetation in temperate climates.
Oxisols	Very highly weathered soils. Most Oxisols occur in the tropics and have low natural fertility.
Spodosols	Very strongly leached soils of cool humid areas. Spodosols have bright colors, high acidity, and low fertility.
Ultisols	Strongly weathered soils formed in warm humid regions under forest vegetation. Ultisols are redder and less fertile than Alfisols.
Vertisols	Soils high in clay that form deep cracks at least 1-cm wide during dry seasons.

various places. Both the new and the old soil classification systems of the USDA are explained in Chapter 16.

One of the best-known higher levels of soil classification is called a *soil order*. Most soils can be fitted into one or another of 10 soil orders. Each soil order includes many different soil series that have several important characteristics in common. For example, the soil in Figure 1-1 is a Mollisol because it has a thick, dark-colored A horizon characteristic of soils formed under grass vegetation in subhumid temperate climates. The soil in Figure 1-4 is an Alfisol, formed under forest vegetation in a humid, temperate climate. These and other soil names will be used in various parts of this book. Generalized descriptions of the 10 orders in the new USDA soil classification system are given in Table 1-1. More detailed descriptions are given in Chapter 16.

CONTRIBUTIONS OF SOIL TO PLANT GROWTH

Soil provides higher plants with many things essential to their growth. Principal among these are mechanical support, plant nutrients, water, and oxygen for root respiration.

Plants produce sugar and give off oxygen in the process of photosynthesis. This complex reaction may be shown as follows:

$$\text{Light energy} + 6CO_2 + 6H_2O \rightarrow C_6H_{12}O_6 + 6O_2$$

An important aspect of photosynthesis is that it takes place only when and where light energy strikes plant tissue containing chlorophyl or some similar substance. Plants also carry on respiration and utilize some of the energy stored in the sugar produced by photosynthesis by the following generalized reaction:

$$C_6H_{12}O_6 + 6O_2 \rightarrow 6CO_2 + 6H_2O + energy$$

Respiration is carried on in all living tissue. The required supply of oxygen is no problem for aboveground plant parts bathed in an atmosphere that is over 20 percent oxygen. Roots, however, respiring belowground may deplete the soil air of its oxygen content if the soil is excessively wet. Too much water in the soil not only reduces the amount of air in the pore space but also restricts its circulation with the atmosphere. Under such conditions the soil air becomes depleted of oxygen and enriched with carbon dioxide. Microorganisms in the soil compete with the plant roots for oxygen for their own respiration as they decompose organic materials in the soil. The oxygen supply in a poorly drained soil may be so low that the life processes are drastically altered. Plants fail to develop normally, and the microbial population of the soil changes. A soil may be rich in plant nutrients and filled with water but produce a poor crop because of lack of oxygen in the soil air.

Root respiration is necessary for the uptake of plant nutrients and the absorption of water. Plants may suffer from a deficiency of an element that is actually present in adequate quantities in the soil if the oxygen supply is depleted. It is also possible for plants to wilt with their roots implanted in a soil saturated with water (Kramer, 1949). A seeming contradiction to this principle arises when plants are grown in water cultures. Oxygen is still vital to the plant roots, but it reaches them the same way it reaches a fish—by dissolving in the water. The water of the cultures is changed frequently or air is bubbled through it to replenish the oxygen supply. Oxygen limitations are circumvented by certain plants such as rice that are equipped to absorb oxygen for root respiration through special cell structures at the surface of the water, but most plants require that oxygen be present in their growth media.

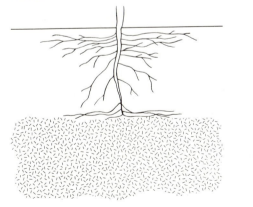

A horizon

B horizon
(saturated with water)

Figure 1-5 Roots spread sideways rather than enter airless soil.

Much, but not all, of the water that enters the soil ordinarily becomes available for plant growth. Excess water may percolate through the soil and eventually reach the local water table after the soil has absorbed all it can hold. Some water may be lost by evaporation at the surface, especially that contained in the top 8 or 10 cm of soil. The remaining water includes some that is unavailable to growing plants. Clay and organic matter hold the unavailable part too tightly for the plant to withdraw it.

Mechanical Support

The fact that soil provides mechanical support for plants is rather obvious. The possibility of a problem arising in this regard is less obvious but nonetheless true. A problem of inadequate support can arise from some organism in the soil attacking and weakening the plant roots. Inadequate support can also result from physical soil conditions. For example, the dominant soil at College Station, Tex., has a B horizon with an extremely high clay content. Most of the pore spaces in the B horizon are so small that they hold water and not air most of the time. Thus the B horizon is poorly aerated in spite of its relatively high percentage of total pore space. Plant roots therefore penetrate the B horizon only to a very limited extent and are principally anchored in the sandy A horizon, as illustrated in Figure 1-5. Trees growing in such soil are occasionally uprooted by windstorms because of their inadequate mechanical support.

Plant Nutrients

Sixteen elements are recognized as being essential to plant growth. Three of these, carbon, hydrogen, and oxygen, are supplied by water and air (carbon dioxide). The remaining 13 elements are considered to be *plant nutrients* and can be grouped into six macronutrients, those needed in large amounts by plants, and seven micronutrients needed only in trace amounts. The macronutrients can be further classified according to whether growing plants obtain them mostly from the mineral matter or from the decomposing organic matter in the soil:

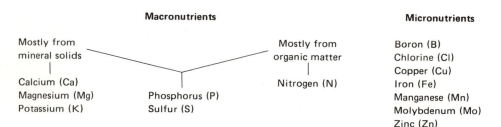

Macronutrients		Micronutrients
Mostly from mineral solids	Mostly from organic matter	Boron (B)
		Chlorine (Cl)
		Copper (Cu)
Calcium (Ca)	Nitrogen (N)	Iron (Fe)
Magnesium (Mg) Phosphorus (P)		Manganese (Mn)
Potassium (K) Sulfur (S)		Molybdenum (Mo)
		Zinc (Zn)

Table 1-2 shows the average percentages of the eight most abundant elements contained in a large number of plants (excluding carbon, hydrogen, and oxygen). Plants contain larger amounts of sodium and chlorine than they do of some macronutrients. Sodium and chlorine are absorbed in large amounts by plants because of their abundance in soils. Sodium is considered nonessential for plant growth. Chlorine is listed as a micronutrient even though it is absorbed in macronutrient quantities. Three different quantities may be distinguished for any plant nutrient—

Table 1-2 Average Chemical Composition of a Large Number of Plants

Analysis	N	K	Ca	Cl	Na	Mg	S	P
Percent	1.52	1.47	0.77	0.73	0.37	0.30	0.26	0.22

the amount essential, the amount beneficial, and the amount absorbed. For chlorine the amount essential is so small that it was long thought to be nonessential; the amount beneficial is considerably larger but not nearly as large as the amount usually absorbed. Sodium may also be beneficial either for its own sake or as a partial substitute for potassium. Sodium and chlorine are useful in another sense, too. Animals have relatively high requirements for both sodium and chlorine and derive most of their needs for these elements by feeding on plants.

Plants also contain appreciable amounts of silicon and aluminum, but these elements are not considered essential. They are absorbed by the plants according to their availability in the soil. It has long been known that soluble aluminum in some acid soils is absorbed in such large amounts that it becomes detrimental or even toxic to many plants (Pierre, Pohlman, and McIlvaine, 1932). Similarly, under certain conditions plants may absorb such large amounts of nitrogen (a macronutrient essential to both plants and animals) that it becomes toxic to animals that eat the plants. Plants absorb some of any element that happens to be available, not because they need all of them but merely because they are available. Over 50 different elements have been identified in plant tissue even though most of them are not needed by the plants.

Nutrient Availability

Only a small portion of each nutrient present in soil is available to plants. Most of it is locked up so firmly in the mineral and organic matter that it is unavailable until decomposition takes place. Such decomposition occurs slowly over a long period, and nutrients are gradually released.

Plants absorb elements from the soil in ionic forms. For example, calcium, magnesium, and potassium are absorbed as Ca^{++}, Mg^{++}, and K^+ ions. Nitrogen is absorbed as NH_4^+ or NO_3^-. Phosphorus is absorbed as $H_2PO_4^-$ or HPO_4^{--}, and sulfur is absorbed as SO_4^{--}. Plant roots either absorb an equal number of positive and negative charges or make a simple exchange of one ion for another. Living plants can absorb ions even when the ion concentration in the plant exceeds that in the soil water if adequate oxygen is present for respiration to provide the required energy.

A portion of the available plant nutrients is present as ions dissolved in the soil water. Dissolved ions are the most readily available supply of plant nutrients stored in the soil, but they are usually only a small fraction of the needs of a crop for a whole season. Most of the cations (positively charged ions) that will be absorbed during the season are stored on the surfaces of clay and humus (organic matter) particles. Ions from these storage sites are exchanged with ions from the soil water in a very important process that will be considered in detail in Chapter 6. A much smaller quantity of anions (negatively charged ions) may be similarly stored, but the most significant quantity of anions becomes available from structural portions of organic materials that decompose during the season. Slightly soluble mineral particles that dissolve during the season are another source of both cations and anions.

The slow breakdown of soil minerals into smaller particles also yields soluble materials that serve as plant nutrients.

Even though the soil water is usually a very dilute solution, it is almost impossible to completely deplete its supply of plant nutrients. Growing plants remove some of the ions from solution, but others move from the mineral and organic materials to take their places. The plant may suffer from a deficiency of one or more nutrients, however, if the rate of replenishment for that nutrient is too slow. Farmers, gardeners, and other plant growers often find it advantageous to apply fertilizer to supplement the nutrients released by the soil when they want rapid plant growth.

REFERENCES

Alderfer, R. B., 1973, The Lawn and Its Soil, *Plants Gard.* **29**(1):9–12.

Bridges, E. M., 1970, *World Soils,* Cambridge Univ. Press, 89 p.

Carson, E. W. (ed.), 1974, *The Plant Root and Its Environment,* Univ. Press of Virginia, Charlottesville, 714 p.

Grimes, D. W., R. J. Miller, and P. L. Wiley, 1975, Cotton and Corn Root Development in Two Field Soils of Different Strength Characteristics, *Agron. J.* **67**:519–523.

Holt, D. F., D. R. Timmons, W. B. Voorhees, and C. A. Van Doren, 1964, Importance of Stored Soil Moisture to the Growth of Corn in the Dry to Moist Subhumid Climate Zone, *Agron. J.* **56**:82–85.

Johnson, W. M., 1963, The Pedon and the Polypedon, *Soil Sci. Soc. Am. Proc.* **27**:212–215.

Kramer, Paul J., 1949, *Plant and Soil Water Relationships,* McGraw-Hill, New York.

Limbrey, Susan, 1975, *Soil Science and Archeology,* Academic Press, New York, 384 p.

Meyer, B. S., and D. B. Anderson, 1952, *Plant Physiology,* Van Nostrand, New York.

Olsen, S. R., and W. D. Kemper, 1968, Movement of Nutrients to Plant Roots, *Advan. Agron.* **20**:91–151.

Pierre, W. H., G. G. Pohlman, and T. C. McIlvaine, 1932, Soluble Aluminum Studies. I. The Concentration of Aluminum in the Displaced Soil Solution of Naturally Acid Soils, *Soil Sci.* **34**:145–160.

Riecken, F. F., 1963, Some Aspects of Soil Classification in Farming, *Soil Sci.* **96**:49–61.

Shourbagy, N. E., and A. Wallace, 1965, Sodium Accumulation and Sodium Responses of Five Varieties of Barley, *Agron. J.* **57**:449–450.

Simonson, R. W., 1957, What Soils Are, in *Soil,* 1957 Yearbook of Agriculture, U.S. Government Printing Office, pp. 17–31.

Skogley, C. R., 1973, Lawns Need Fertilizer and Lime, *Plants Gard.* **29**(1):13–16.

Soil Survey Staff, 1975, The Soil that We Classify, Chap. 1 in *Soil Taxonomy,* U.S. Government Printing Office, Washington, D.C. pp. 1–5.

Sopher, C. D., and R. J. McCracken, 1973, Relationships between Soil Properties, Management Practices, and Corn Yields on South Atlantic Coastal Plain Soils, *Agron. J.* **65**:595–599.

Chapter 2

Soil Formation

Soil develops from parent material[1] by processes of soil formation different from the processes of rock weathering or disintegration that produced the parent material. The process of forming soil from a hard rock (for example, granite) can therefore be divided into two fairly distinct stages—rock weathering and soil formation.

Physical processes are dominant during the formation of parent material by disintegration of rock. The crystals of various minerals contained in the original rock (primary minerals) are separated from one another by several processes. Ice formation, root penetration and expansion, and physical movements fracture rocks. The swelling and shrinking effects of wetting and drying and expansion and contraction caused by temperature changes also tend to disintegrate rock. Chemical processes may participate to the extent of removal of cementing agents between the rock crystals, but the dominant action in rock weathering is physical in nature. The product is an unconsolidated mass of internally unaltered grains that becomes the soil parent material.

[1]Some persons argue that the term *parent material* is a misnomer for this material on the basis that the true parent material has been altered and is now soil. Such logic causes a useful term to be discarded. The term *parent material* will be used herein for the C horizon, especially where it is presumably similar to the material from which the soil formed. This material is readily convertible into soil and possibly will become soil at some future time.

Disintegrated rock material that remains in place is called *residuum*. If it is transported to another site, it becomes a sedimentary deposit. Either way, the loose rock material can serve as the parent material of a soil when it is exposed to the soil-forming processes. Sedimentary deposits include and in some places consist largely of material eroded from other soils. Even material from former soils will be altered when a new soil is formed in it and should be considered as parent material rather than soil until it becomes part of a new solum.

Soil formation is mainly a biochemical weathering process in contrast with the dominantly physical weathering processes involved in parent material formation. The material is located near enough to the surface that the influences of living organisms and of chemical weathering alter it. Even the individual mineral grains gradually change, and new minerals are formed. These new minerals are called *secondary minerals* to distinguish them from the *primary minerals* inherited from the original rock. It should be emphasized that the transformation of minerals is a very slow process. Mineral transformation acts first on primary minerals from the parent material helping to convert it into soil. Subsequently the material that is already soil is further transformed and gradually forms a more strongly developed, more mature soil. Soils are dynamic and continually changing; their natures are not static.

The soil-forming processes develop a distinctive set of horizons in each soil. The changes required to form a soil are largely humus formation coupled with leaching and eluviation. Leaching was discussed in Chapter 1 as the removal of materials from the soil by solution processes. *Eluviation* refers to the movement of materials from the A horizon to the B horizon either mechanically or chemically. The A horizon is referred to as the *zone of eluviation* because it has lost the materials. The B horizon receives them and is called the *zone of illuviation*. The actual materials that move from the A to the B horizon may be clay, humus, or certain iron and aluminum compounds.

The presence of organic matter in the soil facilitates both leaching and eluviation. The acids produced by the decomposition of organic matter hasten the breakdown of minerals. A considerable portion of the breakdown products are soluble cations and anions that are leached from the soil. Moreover, some of the organic compounds form complexes with otherwise insoluble compounds of iron and aluminum and cause them to move downward in the profile of strongly acid soils.

SOIL HORIZON NOMENCLATURE

The ABC soil horizons introduced in Chapter 1 are the master horizons that serve as a skeleton for the more complete system shown in Figure 2-1. Two more master layers are added where they occur; O represents an overlying organic layer, and R represents the underlying hard bedrock. The master horizons and layers may then be subdivided as shown in the figure. It must be realized that any one soil contains some but not all of these horizons. The particular set of horizons present in the soil at a particular place are a result of the combination of soil-forming factors of that place. The O layers and the A2 horizon, for example, are most common in forested

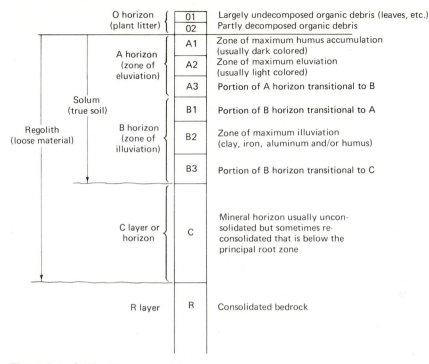

Figure 2-1 Soil horizon nomenclature.

soils. A young soil may contain A and C horizons but not yet have formed any B horizon. An eroded soil may consist entirely of B and C horizons because its A horizon has been lost. Soils forming from hard rock often lack distinct C horizons. Thus, no one horizon is universal, but every soil contains some horizons.

In some places a soil forms in two or more parent materials with a portion of its profile in each. A common example is where windblown material overlies some other material. Suppose the division is in the middle of the B3 horizon. The upper portion of the B3 would be designated B31 to show that it is the uppermost part of the B3. The lower portion would be designated IIB32 showing by the II that it is the second kind of parent material in the profile and by the 2 that it is the second portion of the B3 encountered. The C horizon and the R layer would also have Roman numeral prefixes, II if they were from the second source or type of material, III if from the third, etc. Thus the "genetic" soil horizons and the original parent material differences are both identified.

Sometimes it is desirable to indicate some other features of the soil horizon. This can be done by adding a small letter after the horizon designations already indicated. For instance, plowed layers are designated as Ap horizons. Several features are so described. A few of the more common abbreviations are *b* for buried soil horizon, *ca* for accumulation of calcium carbonate, *g* for gleying,[1] and *sa* for accumulation of soluble salts. Some designations are limited to particular parts of the soil profile.

[1]Gleying refers to the gray soil colors that result from reducing conditions in waterlogged soils.

For example, *h* stands for an accumulation of illuvial humus, *ir* for illuvial iron, and *t* for illuvial clay. These illuvial notations necessarily are appendages of B horizons because they are the zones of illuviation.

SOIL PARENT MATERIALS

The mineral matter inherited from rocks is referred to as *soil parent material* because it is the principal ingredient from which most soils are formed. The exceptions are organic soils. The principal parent materials of organic soils are decomposing plant parts. Actually, all soils have both organic and inorganic parent materials, but whichever dominates the soil's characteristics is referred to as the soil parent material.

Two of the most important properties of soil parent material are its texture and its mineral composition. Both of these properties are carried over as characteristics of the soil formed from the parent material, though they are altered somewhat as the soil ages. Such alteration generally reduces particle size by weathering action. The fine particles, however, are most subject to removal from the A horizon by erosion, eluviation, and leaching. A sandy, gravelly, or stony soil may result from the gradual removal of much of the finer material.

Texture

Soil texture is a means of describing the particle sizes present in a soil and is discussed more fully in Chapter 3. Texture has much to do with the passage of air, water, and roots through soil. Sandy materials usually cause little restriction on movement through soil, whereas clayey materials often delay or prevent movement of air, water, and roots. A favorable arrangement of the clay particles into a good soil structure helps to alleviate the problem but may not eliminate it. Another property related to soil texture is the amount of water that can be stored in the soil. Clay soils can hold much more water than can sandy soils.

The permeability and the water-storage aspects of texture combine to cause soils developing in materials with a high clay content to have shallower sola[1] than those developing in coarser-textured materials under comparable climatic and topographic conditions. Extremes of this effect occur where A horizons several feet thick form in stony materials containing little clay. More clay would reduce water penetration. Extremely stony soils may contain B horizons at great depths. Figure 2-2 relates soil depth and horizon differentiation to the clay content of parent materials.

Mineral Composition

The mineral content of parent materials is of particular significance in determining the fertility level of young soils. The presence of a high proportion of calcium-magnesium minerals and a high content of total potassium is associated with a youthful and fertile soil. As the materials undergo long periods of weathering, there is a decrease in calcium-magnesium minerals, and some of the potassium minerals

[1]Sola is the plural of solum.

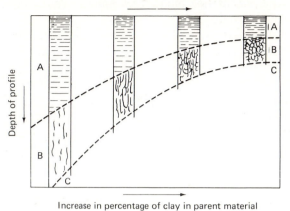

Increase in degree of differentiation

Depth of profile

Increase in percentage of clay in parent material

Figure 2-2 The effect of texture of parent material on profile development. (*After Smith, Allaway, and Riecken,* 1950.)

also disappear. The older and more strongly weathered materials are a poor source of plant nutrients, and they support only those species of plants having low fertility requirements. The organic-matter content of the soil generally increases with the fertility level and is highest in the most fertile soils (except when the climate or the type of vegetation changes).

Changes occur as the parent material becomes soil, but the nature of the parent material has much to do with the nature of the soil, especially of young soils. The influence of parent material gradually declines as the soil becomes older and more strongly developed. In some old soils of tropical regions such great changes have occurred in the soil material that it is difficult to determine the nature of the original rocks.

Rocks

The first rocks were igneous in origin; that is, they formed by the cooling of a molten mass. Materials weathered from these early rocks were moved elsewhere by water and wind erosion resulting in the formation of deposits of loose, transported materials. These deposits differed from the source rocks in that the mineral grains in them were not cemented together—that is, they were unconsolidated. Wind and water deposit layer upon layer of unconsolidated materials during the formation of sedimentary rocks. When such deposits form above water level, they may have vegetation growing on them during the depositional process and function as soils. These vegetated deposits constitute the *current deposits* class of soil parent material.

Eventually the environment under which materials are transported and deposited changes enough that deposition ceases. The materials then constitute a geologic formation and are referred to as *soft sedimentary rock*[1] or as a *soft rock formation.*

[1] *Rock* is used here in a broad sense that includes an extensive geologic formation whether hard or soft, consolidated or unconsolidated.

Hard sedimentary rock is formed by cementation of soft rock material into a new consolidated rock such as sandstone or shale.

Older rock layers may be buried by new layers of sedimentary rock and/or lava rock and thus be subjected to heat and pressure. Such burial may lead to gradual changes in the rocks forming *metamorphic rocks.* Metamorphism does not include conditions hot enough to melt the rock because that would produce a new igneous rock when the mass cooled again.

Any of the aforementioned rocks may be exposed at the surface of the earth by processes of erosion that remove the overlying material. Their influences on soil formation are such that soil parent materials may be divided into four broad classes:

1. Residuum from hard rocks
2. Soft rock formations
3. Current deposits
4. Organic materials

The course of soil development is somewhat distinct for each of these classes. One very basic difference is that the first two classes are subject to erosion, and the soils formed from them gradually sink into the landscape. The last two classes receive deposition, either mineral matter or organic, and therefore gradually rise to a higher position in the landscape. The four broad classes are divided into more specific classes in the section that follows. The occurrence of several of these classes of soil parent materials in the United States is illustrated in Figure 2-3.

Residuum

Disintegration of hard rock with the material remaining at the site produces residuum. The residual material accumulates at a rate dependent upon the rock and the weathering forces. Usually the rate is slow, and soil formation often acts as rapidly as disintegration, thereby producing a soil without any C horizon. The B horizon, or even the A horizon, then rests directly on bedrock. The texture of residual material, and therefore of residual soil, is closely related to the grain size of the parent rock. A fine-grained rock tends to produce a clay soil, and a coarse-grained rock tends to produce a sandy soil.

Forming soil from hard rock is usually a slow process. Residual soils are generally shallower than those formed from the other three classes of parent material, especially while the soils are young.

The nature of soil depends so much on the nature of its parent rock that a discussion of the most important hard rocks is included here. Such rocks may be igneous, sedimentary, or metamorphic in terms of their mode of origin.

Igneous Rocks   Igneous rocks are those formed by solidification of a molten material. If the molten mass flows out on the surface, it is called *lava.* Lava cools rapidly and produces a fine-grained rock. Slow cooling deep within the earth produces a coarse-grained rock. A very simple classification of the most abundant igneous rocks based on grain size and presence or absence of quartz is shown in Table 2-1.

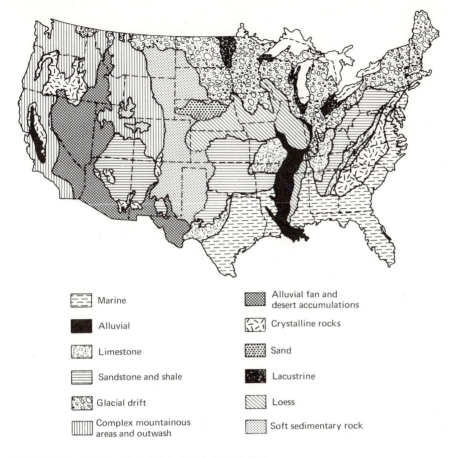

Marine

Alluvial

Limestone

Sandstone and shale

Glacial drift

Complex mountainous
areas and outwash

Alluvial fan and
desert accumulations

Crystalline rocks

Sand

Lacustrine

Loess

Soft sedimentary rock

Figure 2-3 Soil parent materials in the United States.

Several other igneous rocks are recognized by geologists, but these are mostly gradations between those shown. The two most common igneous rocks are basalt and granite.

Basalt is the most common "lava rock" and is the principal hard rock underlying the ocean basins. Its crystals are so small that when basalt weathers it produces a large percentage of clay. Much of the undecomposed rock is either bedrock or loose stones. Soils formed from basalt are high in clay, may be stony, and frequently contain no C horizons.

Granite, being a coarse-grained rock, falls apart into its separate crystals as it weathers. It produces a much sandier soil than basalt, one that may contain fine gravel or large rock outcrops but seldom small stones. A significant portion of the sand and gravel is quartz and is very resistant to weathering. Granite usually produces a deeper soil than basalt, including a C horizon that grades gradually into unaltered rock.

Igneous rocks are the direct source for the parent material of about 10 percent of the earth's soil area. The remaining 90 percent of the land area has a mantle of

Table 2–1 A Simple Classification of Igneous Rocks

	Quartz crystals present	Quartz crystals absent
Fine-grained rock	Rhyolite	Basalt
Coarse-grained rock	Granite	Gabbro

sedimentary or metamorphic rocks. Indirectly, however, igneous rocks are the source material for all mineral matter in soils because the material in sedimentary and metamorphic rocks once came from older igneous rocks.

Sedimentary Rocks   Most hard sedimentary rocks result from the cementation of materials deposited by wind, water, ice, or gravity. The cementing agent in coarse-grained rocks such as sandstone and conglomerate is a chemical precipitate of iron, aluminum, or silicon compounds, of calcium carbonate, or of some combination of these materials. The strength of cementation and the degree of resistance to weathering vary widely.

Fine-grained rocks such as shale do not require cementing agents. Clay particles will cling together and produce shale when submitted to the pressure of burial beneath other rocks. Shale usually weathers fairly easily when again exposed at the earth's surface. The weathering resistance of shales can, of course, be augmented by a degree of cementation with such materials as calcium carbonate.

Limestone is a noteworthy sedimentary rock formed by chemical precipitation in shallow lakes and seas. Precipitation of the cementing agent and principal component, calcium carbonate, is triggered by the action of plants such as blue-green algae controlling the carbon dioxide content of the water. The secondary components, sand, silt, and clay, are very important because the calcium carbonate is dissolved and removed by weathering as soil forms in the material that is left behind. The resulting soil often contains quartz stones, called *chert,* that have formed in the limestone by recrystallization of silicon dioxide from the sand, silt, and clay. The removal of calcium carbonate during soil formation is often so complete that the soils formed are acid in reaction and require liming to achieve good crop production.

Metamorphic Rocks   Metamorphic rocks are formed under heat and pressure intense enough to change the nature of their minerals but not hot enough to melt the material. Melting would produce an igneous rock. There are many variations of metamorphic rocks because metamorphism can vary considerably in intensity and can act on any kind of rock.

Certain metamorphic rocks are distinctive in nature because of the nature of the preexisting rock. Sandstone may be changed to quartzite, shale to slate, and limestone to marble. Most other metamorphic rocks can be classified as either schist or gneiss. Schist forms where metamorphism is "low-grade," that is, not too severe. Schists contain large amounts of fine-grained flaky minerals like mica. "High-grade" metamorphism produces gneiss. The mineral grains in gneiss are somewhat elongated and larger than those in schist. The elongated grains are so arranged as to give gneiss a banded appearance that helps to distinguish it from otherwise similar coarse-grained igneous rocks.

The weathering rates of metamorphic rocks are equal to or slower than the weathering rates of the preexisting rock. Gneiss and quartzite weather much like coarse-grained igneous rocks. Marble weathers like limestone by solution processes. Slate and schist weather faster than gneiss and quartzite but slower than marble.

Soft Rock Formations

Most sedimentary deposits are initially unconsolidated and are included in soft rock formations unless they have been consolidated into hard sedimentary rocks. The initial stages of soil formation are much more rapid in these materials than in residuum because disintegration is not required. The material already qualifies as C horizon. It is permeable to air, water, and roots as soon as it is exposed at the surface of the earth. The processes of chemical and biological weathering that convert it into true soil begin to take effect immediately rather than having to wait for the material to be loosened by physical weathering.

Many soft rock materials at one time belonged to the current deposits group. These two groups are distinguished on the basis of whether they are presently gaining material by depositional processes (current deposits) or are losing material by erosional processes (soft rock formations). If both processes occur, the one that is dominant over a period of years determines the classification. The important difference is whether the surface horizons are being gradually thickened by deposition or thinned by erosion. Soil profile differentiation is generally more pronounced, and a more distinct B horizon is formed under gradual erosion than under deposition.

The most important soft rock formations are glacial deposits, wind deposits (especially loess), and old water deposits now exposed to erosion. The nature of the water deposits will be considered under Current Deposits.

Glacial Deposits The last million years of geologic history are known as the Pleistocene era or the age of glaciers. Great areas of glacial ice covered major parts of the land at four different times or stages. Some of the largest ice sheets occurred in North America moving from centers in Canada and reaching as far south as southern Illinois (Figure 2-4). Similar events occurred in Europe and Asia. Glacial ice even crossed the English Channel and covered a portion of the British Isles.

Every glacier has an area of accumulation and an area of attrition. The accumulation process occurs wherever snowfall exceeds melting. Accumulation continues until the ice is thick enough to move toward an area of attrition. Eventually it reaches an area where melting exceeds snowfall, and from there on the amount of ice gradually diminishes until finally its outer edge or front is reached. The glacier front moves forward, backward, or remains stationary, according to the balance between snowfall, glacier movement, and melting.

Glaciers pick up soil and rock materials in their areas of accumulation. Central Canada, for example, has relatively shallow soils because the glaciers scoured away everything loose only a few thousand years ago. Much of this material was ground

Figure 2-4 A map of the area covered by continental glaciation in North America and the principal centers of origin of the glaciers.

to smaller sizes as it was transported by the ice. In the area of attrition the ice thickness decreases and the glacier can no longer transport all its accumulated rock and soil debris. A portion of this unsorted material is then left on the landscape as a deposit called *glacial till.*

Much of the glacial deposition occurs near the front of the glacier when the last of the ice melts. Where the glacier front remains nearly stationary or moves back and forth over the same area for many years, it leaves an accumulation of unsorted materials of all sizes in a rough topography along its margin. The topography consists of a series of small hills called a *terminal moraine.* At other times the glacier melts back rapidly for several years in succession, leaving a relatively smooth deposit of glacial till called *ground moraine.* When the glaciers retreated from central United States, they left wide bands of ground moraine between narrower bands of terminal moraine. Wet soils occur in many of these areas where the glaciers left small basins.

Great torrents of meltwater pour forth from the glacier front regardless of whether the glacier is advancing or retreating. Sand, silt, clay, and even sizable stones are carried out in front of the glacier. The water sorts this material, depositing the stones first, then the gravel and sand in succession as it gradually spreads out and slows down. Some of the silt may finally be deposited similarly, or it may be carried into a glacial lake and deposited along with the clay. If no lake is formed, the finer particles may be deposited on a river floodplain or carried to the ocean. The material deposited by the flowing water is both sorted and stratified. It is called *glacial outwash* or *stratified drift.* A more general term including both glacial till and glacial outwash is *glacial drift.*

Wind Deposits   Wind picks up material from any source area where vegetation is scarce. Deserts, beaches, river floodplains, and glacial outwash are common sources. Wind can pick up particles the size of fine sand or smaller. The wind sorts the material, leaving large particles behind, carrying the fine sand a short distance, and the silt a long distance. The sand is often deposited in dunes near beaches and bottomlands. The Sand Hills of Nebraska are an example of a large region of windblown sand deposits. Dune sand usually supports scant vegetation because the sand holds little water for plant growth. If the vegetation is too scant, the sand may continue to blow and move the dune gradually downwind.

Most windblown silt and clay are spread over large areas as loess deposits. Loess from a single source area may blanket a region several tens or even hundreds of kilometers downwind. Some deposits, such as those east of the Missouri and Mississippi Rivers, are over 30 m thick near their source areas but become gradually thinner with distance until they are no longer detectable. The dominant particle size shifts gradually from very fine sand or coarse silt near the source to fine silt farther away, and the clay content increases as the distance from the source increases. Leaching during deposition is most effective in the thinner part of each deposit, and consequently the calcium carbonate content of the loess decreases with distance. The great significance of this fact to plant growth and cropping management will become apparent in later chapters.

Loess deposits have a very high silt content and generally lack distinguishable layers except in the area near the source where the deposit is thickest. They have the interesting property of being able to stand vertically but being highly erodible on slopes. Highway cuts made in loess deposits cause much less difficulty if the banks are made vertical than if they are sloped (Figure 2-5).

Gullies form easily in loess unless waterways are well protected with a good vegetative cover or some mechanical device. The gullies develop a characteristic U shape with vertical banks and nearly level bottoms. Gullies frequently erode to the bottom of the loess deposit.

The source areas of the large loess deposits in China are mostly from deserts. Most of the world's loess deposits, however, have river valleys as their source areas. The largest of these deposits relate to glacial times when the rivers were swollen with silt-laden meltwater from the glaciers. Broad floodplains were formed with little vegetation on them. During dry periods between floods, the wind whipped across the floodplains picking up silt and carrying it to the growing loess deposit. Deposition may still be proceeding at present but at a much slower rate in the absence of

Figure 2-5 A vertical roadbank is more stable than a sloping one in loess.

glaciers. Most loess deposits are now being eroded faster than they are receiving deposition.

Volcanic eruptions occasionally spew out large quantities of ash, and the wind spreads the material like a blanket over a considerable area. The deposit is called *volcanic ash* and is usually thin but extensive. Soils formed from volcanic ash generally have high clay contents.

Water Deposits   Any of the alluvial materials discussed in the following section (Current Deposits) can eventually be separated from their source of deposition and become subject to erosional processes instead. A *terrace* is an example of such an occurrence. It was originally an alluvial deposit on a floodplain, but it is now high enough above the stream level to escape flooding (Figure 2-6). The land may have risen, or erosional processes may have permitted the stream to cut down to a lower level. The terrace is left as a remnant of the old floodplain. Usually the soils forming on terraces are highly desirable for cultivation. They have been removed from the hazard of floods but most of them are still young enough to be well supplied with plant nutrients. The topography is nearly level at first but may be dissected later by erosion.

Lacustrine deposits are composed of materials carried into lakes by streams flowing in from the surrounding uplands. Normally the sand and gravel carried into a lake are deposited in a delta near the point of entrance. The main part of the lake therefore receives only silt and clay sediments. The fine textured soils that form in low topographic positions often have poor drainage even long after the lake is drained.

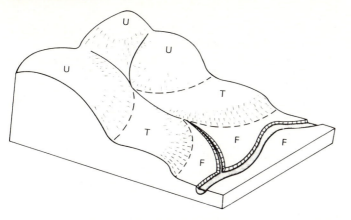

Figure 2-6 A landscape diagram showing a terrace (T) representing an old floodplain above the current floodplain (F) and below a hilly upland (U).

Marine materials are those which have been eroded from continental areas, washed into seas or oceans (salt water), and later lifted above sea level. Most of the area in the cotton belt of the United States was at one time under the water of the Gulf of Mexico. As the gulf gradually receded, layer after layer of sediments was exposed to soil-forming processes. The layers of materials vary from very acid quartz sands to neutral clays. Very good and very poor soils often occur within a few miles of each other in regions of marine materials. The layers of marine materials are exposed as bands of material a few meters to many kilometers in width, approximately parallel to the original shoreline of the gulf.

Where the marine material accumulated slowly as erosional debris from continental areas, it often included a large amount of skeletal material of aquatic animals. Bones from the skeletons are high in both calcium and phosphorus minerals that contribute to soil fertility. The calcareous marine materials of the cotton belt have developed into the most productive upland soils of the region.

The topography near the Gulf of Mexico and away from the rivers is quite level for many kilometers. Near river outlets there is much swampy land. The farther the material is from the gulf, the older it is, the higher the relief (elevation difference between hills and valleys), and the more rolling the topography. The same topographic sequence occurs along many other coasts including much of the Atlantic coast of the southeastern United States.

Current Deposits

Depositional processes tend to remain active in the same areas over lengthy periods of time. Materials are brought in and laid down slowly and intermittently, but gradually a deposit is built. Vegetation grows on such deposits and soil forms in them. Generally the soil profiles are weakly differentiated because the action of the soil-forming processes gradually shifts to a new position as the surface rises. Only rapidly produced changes can keep pace with the depositional processes. The most distinctive of these relatively rapid changes is the accumulation of organic matter that produces an A1 horizon. Many soils formed in current deposits therefore have thick A1 horizons over C horizons that include buried A1 horizons.

The loess deposits described among the soft rock materials were good examples of the current deposits class as long as the rate of deposition exceeded the rate of erosion. Certain areas fit the current deposits class even now where loess is still being deposited, but most of the large loess deposits are now undergoing a net loss by erosion.

The most varied and widespread current deposits are those transported by water. They reflect the variable conditions of water transport from rushing streams to tranquil lakes. Variations from one time to another are shown by the stratified nature of most water deposits. Successive layers differ in particle size and in orientation of the particles. The chemical and mineralogical nature of the materials may also vary. Usually a high degree of sorting takes place so that particles of one size range are deposited in a particular area while other sizes are deposited elsewhere.

Alluvium   Alluvium is material deposited by streams. The streams may be large or small, swift or slow. The bulk of alluvial material is transported during flood stage when the water comes pouring from the mountains and hills. *Floodplains* are inundated by water that moves much slower after it leaves its channel. Therefore most of the material carried by the water is deposited on the floodplains, first the coarser particles and then progressively finer ones. Most of the stones and pebbles are left in the steep headwater areas or in the main channel of the stream. Sand is deposited in intermediate areas, either near the stream banks, in secondary channels, or in the upper reaches of the floodplains. Silt is deposited farther downstream or farther to the side. Clay is deposited in the slackwater areas where the water moves very slowly. The whole action is changeable as the currents come and go. Material of one texture is often deposited on top of material of another texture, thus forming a stratified deposit. The periodic addition of material on the floodplains usually maintains a high fertility level. The Nile River Valley in Egypt is an example of a floodplain that has remained fertile through thousands of years of agricultural use.

Alluvial fans are another form of stream deposit that occurs where a smaller stream enters the valley of a larger stream. The openness and flatness of the larger valley spreads the water and reduces its velocity so that it deposits sediment in a fan-shaped area. The coarser material is deposited in the upper portion of the alluvial fan and the finer particles are carried down to the toe of the fan or perhaps into the larger stream.

Deltas are somewhat like alluvial fans but form where a stream enters a lake or sea. The material being transported is then dumped in a roughly triangular area named after the Greek letter Δ (delta). A portion of the deposit is above water level and can support plant growth. The deltas of some large rivers are hundreds of kilometers across. Deltas generally have a high fertility level, and some of them are very productive agricultural land. Deltas have high water tables unless the level of the lake or sea drops relative to the land surface. As in all current deposits, the soils formed in deltas remain perpetually young as long as the deposition continues.

Colluvium   Colluvium is material transported by gravity. It is usually intermixed with alluvium, but since gravity can move big particles as easily as small ones, colluvium is unsorted. It commonly occupies small areas above floodplains or terraces but below hillsides. Larger deposits of colluvium occur in mountainous areas where steep slopes may produce large landslides. Most colluvium, however, is the result of gradual movement down a slope just barely steep enough to cause the movement when the material is weakened by being saturated with water. The movement commonly is slow enough and smooth enough that the soil forming in the colluvium remains intact and the vegetation growing on it is undisturbed. Such soils often become thicker than other soils of the area as they move into flatter areas where the creeping movement becomes still slower.

Organic Materials

Marshes and swamps occupy low, wet areas where organic matter accumulates because lack of oxygen slows decomposition. The plant parts retain their form for a long time, then gradually fall apart. The oldest materials are on the bottom with progressively younger, fresher material toward the top. Admixtures or even layers of mineral matter may be washed or blown into the deposit, keeping it from being pure organic matter.

The organic materials that accumulate in swamps and marshes are called *peat*. There are several kinds of peat such as woody peat, fibrous peat, disintegrated peat, and sphagnum moss peat. Peats are too wet for agricultural use but may be made highly productive by drainage. Considerable change occurs in the material when drainage occurs. Removal of water causes peat to shrink, and the entrance of air causes its rate of decomposition to be greatly accelerated. The surface level may drop several centimeters the first year and more slowly thereafter. Oxidation causes chemical changes that are generally helpful to fertility. The oxidized material is called *muck* to distinguish it from the unoxidized peat from which it is formed. Muck generally contains a higher percentage of mineral matter than peat because part of the organic material has been lost by oxidation.

CLIMATE

Climatic factors are involved in converting rock into parent material and in converting parent material into soil. Climate has a direct influence on the nature of soil through temperature and precipitation and an indirect influence through vegetation and topography. These influences are so strong that the climatic pattern of the world is the main control factor on the broad soil pattern of the world.

The effects of rainfall and temperature on the chemical processes of weathering are of considerable importance in the processes of soil formation. Chemical weathering is most severe where high rainfall is coupled with high temperatures. Most of the primary minerals decompose, and their components either recombine to form secondary minerals or are leached from the soil in drainage water. The resulting soil bears little resemblance to its parent material. The effects are less drastic but nonetheless real in cooler and drier climates.

The amount of water percolating through the solum greatly affects the process of soil formation. The presence of water is essential for both chemical and biological weathering. Water is also the transporting agent involved in leaching and in eluviation. Probably no other agent has such wide-ranging influence on the soil-forming process.

The amount of water entering a soil (infiltrating) depends on both climate and topography. Precipitation zones for the United States are shown in Figure 2-7, but these must be modified on a local basis for microclimatic effects and for run-on and runoff. Most of the water that infiltrates leaves the soil by transpiration and evaporation but not before it has participated in the process of soil formation. Large amounts of precipitation coming when the temperature is only slightly above freezing produce a strong leaching effect because of low evapotranspiration. A high relative humidity also increases leaching by reducing evapotranspiration.

Effects of Climate on Soils in the United States

The effects of precipitation on soil formation are well represented in the soils of the United States. Soil formation occurs rather slowly in the arid regions of the western United States because there is a deficiency of water for both chemical reactions and biological activity. The soils are relatively shallow because the water does not penetrate very deeply. They are light colored because little organic matter accumulates under the sparse vegetation. The soils are neutral or even alkaline in reaction because the soluble salts are not thoroughly leached from the sola. An accumulation of calcium carbonate is commonly deposited at the bottom of the solum constituting a Cca horizon.

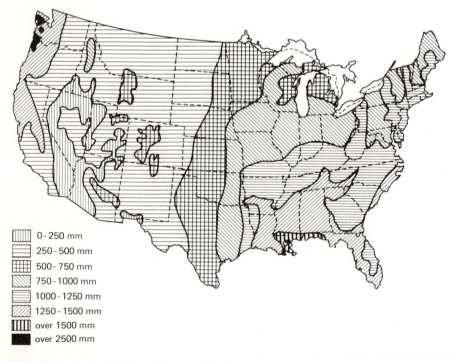

0-250 mm
250-500 mm
500-750 mm
750-1000 mm
1000-1250 mm
1250-1500 mm
over 1500 mm
over 2500 mm

Figure 2-7 Generalized map of annual precipitation.

The soils in the more humid grasslands of the central United States are subjected to more intense weathering action than in the arid areas. These soils are deeper, darker colored, and more thoroughly leached than the drier soils. Eluviation moves clay to the B horizons more rapidly in the humid grasslands than in the drier areas. The accumulation of calcium carbonate is moved to a greater depth as the precipitation increases.

Calcium carbonate is leached completely out of the sola of the eastern half of the United States. The distinction between soils with and without Cca horizons was proposed by one of America's great soil scientists, C. F. Marbut, as the line that should divide all soils into two broad classes. Marbut named the soils with Cca horizons *Pedocals* and those without them *Pedalfers.* His dividing line fell within the 500- to 750-mm precipitation zone in Figure 2-6. The distinction between Pedocals and Pedalfers is not used much now because it is difficult to apply to soils formed in parent materials lacking calcium carbonate.

The soils of the eastern United States are dramatically different from the Pedocals partly because the climate is humid and partly because the vegetation changes from grass to forest. The differences caused by vegetation will be discussed in the section on living organisms. Actually, some effects of vegetation are indirect climatic effects because climate is an important control factor on the natural vegetation of the area.

Temperature is another important feature of climate. The effects of temperature can be illustrated by comparing the soils of the northern and southern United States. Soils in the northern states have gray and black colors in their A horizons and brown colors in their B horizons. The soils farther south have progressively brighter and redder colors. This brightening trend continues in soils of tropical regions where still redder colors prevail. The reason is that the material dominating the soil color changes from humus in cool regions to iron compounds in warm regions. Decomposition is slow in cool regions, and organic materials therefore accumulate as humus. Organic materials decompose rapidly in warm regions and leave the iron compounds dominating the soil color. Other chemical reactions are also accelerated in warm regions so that mineral weathering is faster where temperatures are warmer.

LIVING ORGANISMS

Soil teems with life. Vegetation is far from the only factor included in living organisms. A gram of soil supports a population of millions or even billions of bacteria and other microorganisms. Earthworms, burrowing insects, moles, and other animals that live in the soil are potent weathering and mixing agents. Their action prevents some soils from developing strong differences between their A and B horizons. Some soils have been so thoroughly mixed that the dark color usually associated with A horizons extends through the entire solum.

Apart from the mixing action mentioned above, two general functions of living organisms are involved in soil formation: higher plants produce organic matter from inorganic materials; animals and microorganisms decompose organic materials for their own food and energy.

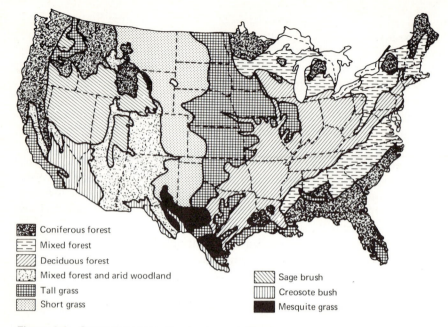

Coniferous forest
Mixed forest
Deciduous forest
Mixed forest and arid woodland   Sage brush
Tall grass   Creosote bush
Short grass   Mesquite grass

Figure 2-8   Generalized map of natural vegetation.

The two most important general groups of higher plants are grasses and trees. Trees are often subdivided into conifers and hardwoods and grasses into sod-forming grasses, tall bunchgrasses, and short bunchgrasses. There are also other types of vegetation such as brush, cacti, and a multitude of herbaceous plants. Figure 2-8 shows the dominant occurrence of several types of natural vegetation in the United States. Each type of plant has its own subtle influence on soil properties, but the most dramatic differences are between grasses and trees. Some of these differences are illustrated in Figure 2-9.

Much of the organic matter produced by grasses dies and becomes part of the soil each year. Organic matter is added to the soil throughout the root zone, but the concentration is greatest near the surface and declines with depth. Aboveground parts that fall on the soil may decompose on the surface or become mixed into the upper part of the soil. Grass roots also add more organic matter to the upper layers than to lower layers because root density declines with depth. Many roots die annually so a large percentage of the organic matter produced by grass roots becomes soil organic matter each year. It has been estimated that even perennial grasses produce whole new root systems every 3 years or less (Weaver and Zink, 1946).

Most grassland soils are less acid than forest soils because grasses thrive in drier climates where leaching is less intense than in forested areas. However, the soil under grass is less acid than the soil under forest even where they occur in the same vicinity so climate must not be the only reason for the difference in acidity. Another important factor is that grasses are generally more effective than trees at offsetting leaching

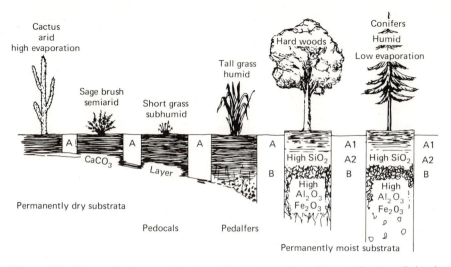

Figure 2-9 Schematic diagram showing relation of climate and vegetation to soil development.

by recycling soluble materials. Recycling occurs when ions are absorbed by roots and translocated up through the plant. Ions are returned to the soil surface when the plant dies and are leached back into the soil to complete the cycle. The recycling process is especially significant for cations such as calcium, magnesium, potassium, and sodium that are gradually leached from the soil. These cations are known as *bases* because they tend to give a more basic (less acid) reaction to the soil. The more bases a plant utilizes, the more it offsets the leaching process. Grasses are generally heavy feeders on bases.

A humid climate favors forest vegetation. Forests, in turn, exert a tremendous influence on the kind of soil developed—an influence that adds to the leaching effect of the humid climate. Most of the organic matter from trees is deposited on the soil in the form of fallen leaves, twigs, and eventually the entire tree trunk and its branches. The litter layer thus formed produces organic acids as it decomposes, and increases the mineral weathering and leaching effects of the water percolating through the soil. Moreover, forest vegetation generally has a lower requirement for such bases as calcium and magnesium and therefore returns less of these elements to the soil surface than does grass. There are also differences among trees in their demands for bases; a beech-maple forest, for example, produces a litter layer with a higher content of bases than that of a pine forest. In general, deciduous trees are heavier feeders on bases than coniferous trees but utilize less bases than grasses.

Trees have more large, long-lived roots and fewer small, short-lived roots than grass. Even if the total weight of tree roots is large, the amount of organic matter added to the soil each year is smaller than that contributed by grass roots. Most of the organic matter of forested soils comes from the litter layer and is concentrated in the upper few inches of the A horizon. The lower part of the A horizon is generally light colored because it is highly leached and is low in organic matter. This layer is designated A2 to separate it from the darker colored A1 horizon. A thick A1 horizon is usually associated with a grassland soil, and a thin A1 over an A2 horizon usually indicates a forested soil.

The A horizons developed under forests are usually acid. Their organic matter is more likely to be eluviated than is the organic matter developed under grass. Furthermore, certain compounds of iron and aluminum are more soluble and leach much faster in the acid conditions common under forest. The B horizons of forested soils may therefore have illuvial deposits of organic matter and/or iron and aluminum compounds in addition to or even in place of illuvial clay deposits. The extreme examples of this are the Spodosols. A Spodosol is shown in Figure 2-10.

Organic matter added to the soil is decomposed sooner or later by various *microorganisms* (also known as *microbes*). The three most important groups of microbes are *bacteria, fungi,* and *actinomycetes.* These and others will be discussed more fully in Chapter 5, but a few salient points are pertinent here. Representatives of each of these groups of organisms are found in most soils, but the proportions and types vary. On the basis of end products of decomposition it matters little which group is dominant; the end products are water, carbon dioxide, and simple inorganic

Figure 2-10   The profile of a Spodosol showing O, A1, A2, and B2 horizons. The upper part of the B2 is darker because it contains illuvial humus (Bh); the remainder of the B2 is colored by illuvial iron (Bir).

compounds. However, multitudes of intermediate organic products of decomposition are found in the soil, and these vary according to the soil conditions and the kind of organism responsible for the decomposition. Also, the microbes themselves are part of the soil organic matter, and the types of microbes present influence soil structure and other soil properties.

In general, fungi are more tolerant of acidity than bacteria and actinomycetes. Fungi predominate in most forest soils because these are usually acid in reaction. The nitrogen economy of the soil is also related to vegetation and microbes. Nitrogen constitutes a very small percentage of the woody growth produced by trees and eventually deposited on or in the soil. It also constitutes a smaller proportion of the weight of fungi than it does of bacteria and actinomycetes. Most of the nitrogen transformations (see Chapters 5 and 10) are performed by bacteria and therefore proceed more slowly in acid forest soils than in neutral grassland soils. The nitrogen supply is an important reason why crops often require less fertilizer on grassland soils than they require for the same yield on forest soils.

Both bacteria and actinomycetes are favored by an abundance of calcium and the nearly neutral pH that usually accompanies it. These organisms tend to be abundant in grassland soils. Bacteria are smaller and more numerous than actinomycetes and fungi. A gram of soil contains millions or even billions of bacteria. Among these multitudes are organisms that can live even in waterlogged soils. Some bacteria perform specific processes such as utilizing nitrogen gas from the atmosphere or transforming ammonium nitrogen into nitrate nitrogen.

Actinomycetes are less numerous than bacteria and smaller than fungi. More interest has been shown in actinomycetes in recent decades than formerly because some of them have been found to produce substances that inhibit bacterial growth. They have provided humanity with such important medicines as penicillin and streptomycin. Actinomycetes in the soil function in about the same activities as bacteria but often at a slower rate.

The relation between soil and living organisms is two-way. The organisms influence the nature of the soil and the soil influences the nature of the organisms. Trees not only tend to make soil acid but also grow better than grasses on acid soils. In forest-grassland transition areas it is common to find grasses growing on the finer-textured soils and trees on the sandier soils. The finer-textured soils are usually higher in bases and provide a favorable nutrient supply for grasses. The sandier soils are usually deeper and provide a more adequate root zone for trees.

TOPOGRAPHY

Anyone who has worked with soils in a hilly area knows that contrasting soils occur on different parts of a hill. The effect is very striking, and many people have concluded that their soils must be more variable than those in almost any other part of the world. In a general way, it may be said that the broad soil patterns of the world are controlled mostly by climate and the local soil patterns are controlled mostly by topography.

Topography influences soil indirectly in many ways. The direct variables are moisture and temperature relations, biological activity, soil movement, and the movement of water both on and in the soil.

Not all the water that strikes the soil surface infiltrates where it falls. Some runs off to a stream or enters the soil elsewhere. The steeper the slope, the more runoff is likely. Steeper soils are usually drier soils with less leaching, less plant growth, lower organic-matter contents, and shallower sola similar to those found in drier climates. The solum thickness of soils on slopes is also reduced by erosion. The surface soil is gradually eroded away, the upper part of the B horizon is eluviated and converted into A horizon, and fresh parent material is converted into soil at the bottom of the solum. The processes of soil formation occur most rapidly where the soil is shallow because the B and C horizons have less overlying material to insulate them from the forces of weathering.

Soils on slopes are generally well drained and well aerated. Increased desiccation and dehydration of iron compounds results in redder soil colors (or browner colors if the organic matter content is higher), especially in the B horizons. The effects of dryness are most pronounced on slopes facing the sun during the warmer part of the day. In the western United States the soils on steep south- and west-facing slopes are notably warmer, shallower, lighter colored, and brighter colored than those on north-facing slopes.

Figure 2-11 illustrates the differences in soil horizon thickness in various topographic positions. These differences are caused in part by the moisture and temperature effects that have already been discussed. Another related factor that can be equally important is soil movement. Soils on slopes may move downslope under the force of gravity, either suddenly as in a landslide or slowly as in *soil creep.* Soil creep is less dramatic but much more widespread than are landslides. Both are dependent upon other factors such as texture and water content of the soil in addition to steepness of slope. The effects of soil creep are like those of erosion in that soil is moved from higher areas to lower areas and is moved fastest where the slopes are steepest. As shown in Figure 2-11, soil thickness is usually least in the convex area where the hilltop meets the slope because both soil and water move away from this area faster than they move to it from above. Conversely, the soils in footslope positions (where slopes are steeper above and flatter below) are thickened because

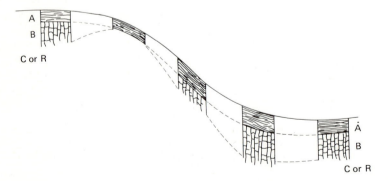

Figure 2-11 A schematic diagram of a toposequence showing the effect of slope and topographic position on the thickness and development of A and B horizons. Each separate profile may represent a different soil series.

soil and water move to these positions faster than they move away. The movement occurring by soil creep may be so gradual that the A and B horizons remain intact and normal soil development continues as the soil moves downslope.

It should be understood that most soil changes are gradual and that transitional soils occur between the contrasting profiles. Soil is a natural continuum across the landscape reflecting the gradations between topographic positions as well as the specific positions used for illustrations.

Toposequences and Catenas

A group of soils such as that shown in Figure 2-11 is called a *toposequence* because of its relation to topography. Where the parent material is similar throughout the toposequence, the group of soils is called a *catena,* from the Latin term meaning *chain.* Wind and water may have sorted the materials leaving more sand in the higher-lying soils and more clay in the lower-lying soils. A distinctly different deposit of alluvial parent material is not included in the same catena, though it may be a part of the toposequence.

Large differences in degree of profile differentiation commonly occur within a catena. As a general rule, the soils on the upper, drier slopes have less profile differentiation and are less weathered than those on the more moist lower slopes or on the flatter hilltops.

Three catenas developed in loess are illustrated in Figure 2-12. One catena formed under grass, one under forest, and one under mixed vegetation. The grass-

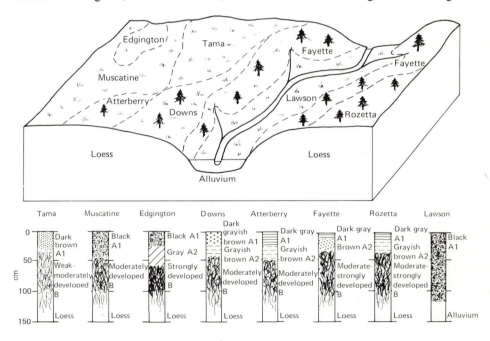

Figure 2-12   Grassland and forested soils formed in deep loess in northern Illinois and adjoining states.

land catena includes Tama soils in rolling areas, Muscatine soils in nearly level areas, and Edgington soils in depressions. The degree of development increases progressively from weak to moderate in the well-drained Tama soils through the moderately developed, imperfectly drained Muscatine soils to the strongly developed, poorly drained Edgington soils.

The Fayette and Rozetta soils shown in Figure 2-12 formed under forest vegetation. The Fayette soils occur in the steeper, better drained areas; the Rozetta soils are in nearly level imperfectly drained areas. The forested soils have light-colored A2 horizons beneath thin A1 horizons whereas the entire A horizon in the Tama and Muscatine soils is dark enough to qualify as A1. The B2 horizons of the forested soils contain more clay and have stronger structure than their grassland counterparts.

Grass and trees are intermixed in some borderline areas. The Downs and Atterberry soils occur in such areas, see Figure 2-12. These soils with transitional vegetation have properties that are transitional between the properties of the forested and the grassland soils. The term *biosequence* is used to describe the situation. Tama-Downs-Fayette constitute a biosequence of well-drained soils, and Muscatine-Atterberry-Rozetta form a biosequence of imperfectly drained soils. The differences among soils in a biosequence are attributed to the biologic factor of soil formation because the other factors (such as climate, parent material, and topography) are similar for all of the soils in the biosequence.

Catenas of the southeastern United States include soils with red, yellow, and gray colors. The gray member occurs on flats where the drainage is poor. The red soils are generally the ones with the best drainage, and the yellow soils generally have intermediate drainage. The color differences are caused by differences in the degree of oxidation and hydration of iron compounds as explained in Chapter 3.

Influence of Water Tables

Some soils are strongly influenced by the presence of a water table within a meter or two of the soil surface. Water moves upward through the soil by capillary action (discussed in Chapter 4). The balance between upward and downward movement of water strongly influences the nature of the soil. Eluviation is ineffective if the water table is near the surface but may be accelerated if the water table is at a depth of about 100 to 150 cm.

Several contrasting soils form above water tables in arid regions. The differences result partly from the type of parent material, but the most important factor is the depth to the water table. The possibilities are best illustrated by the series of terraces shown in Figure 2-13. The soils in the lowest positions are kept so wet by the water from below that the oxygen supply is very limited. Organic matter accumulates because it cannot decompose. The lowest soil may well qualify as an organic soil; one 25 to 50 cm above the water table may be a black mineral soil with a high organic-matter content.

The middle soil in Figure 2-13 is high enough for the surface to become dry. Water moves upward from the water table and evaporates at the surface. The dissolved salts accumulate at the surface as the water evaporates and form a white crust. Plant growth is reduced if the salt accumulation becomes too great.

Precipitation and consequent downward movement of water usually equal or exceed upward movement if the water table is more than a meter below the soil surface. Very deep water tables affect the soils only indirectly through a possible increase in plant growth resulting from the underground water supply. Water tables at about a meter, however, can produce sodium-affected soils. Enough water reaches the surface by upward movement to keep returning highly soluble sodium salts but not to build up a salt concentration. The result is a strongly alkaline soil that is unfavorable for plant growth. Another effect of the sodium is to cause clay particles to move more readily from the A to the B horizon and produce a strongly differentiated solum.

Soils of humid regions rarely have the salt crusts and sodium effects indicated in Figure 2-13. The precipitation in humid regions is usually adequate to prevent salts from accumulating. Instead, the rate of weathering is increased, and the moderately wet soils accumulate more clay in their B horizons than their well-drained neighbors. Peat spots and wet black soils, however, are so dominated by wetness that they have similar characteristics whether the climate is arid or humid.

TIME

Soils are referred to as *young, mature,* or *old* according to their degree of profile development. An unconsolidated material can be considered a young soil when it serves as a source of water, oxygen, and nutrients for plant growth and begins to show that it has been influenced by living organisms. Probably the only signs of profile differentiation in such a young soil would be the leaching from the solum of any soluble salts that may have been present in the parent material and a slight darkening of the upper part of the soil as an A1 horizon begins to form.

A few years suffice to form a young soil if the material is already unconsolidated. It must be realized, however, that such soil is merely entering its youthful stage and that it will be several hundreds or thousands of years old before it becomes a mature soil. In fact, some soils are kept perpetually young by constant deposition, rapid erosion, or perhaps by a parent material that is extremely resistant to weathering.

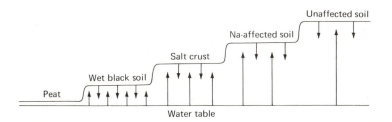

Figure 2-13 Soils developing in the presence of a water table in an arid region. The number of arrows indicates the relative amounts of water movement upward and downward in each soil.

Materials that resist weathering produce infertile soils, but the other types of young soils have good fertility.

Mature soils have reached a state of profile development which appears to be in balance with the environment. Further development is quite slow. Their condition is not one of *equilibrium,* though that term was once popular. It comes closer to a *steady state* with additions by the conversion of parent material into soil approximately equaling soil losses by leaching, erosion, or other means.

A mature soil has an A horizon showing definite eluviation and a B horizon with some type of illuvial accumulation, usually including an increased clay content. It may or may not have a C horizon. The A and B horizons continue developing during maturity, but the changes are slower than during youth. The soil will remain in the mature stage for a very long time. Differences among mature soil profiles are great enough that they are often classified as *minimal, medial,* or *maximal.* A medial profile, for example, might contain 1.5 times as high a percentage of clay in its B horizon as in its A horizon.

The fertility of mature soils would be much poorer if it were not maintained by the weathering of primary materials. Some nutrient losses by leaching, erosion, and plant removal are inevitable but are offset by mineral weathering. The nutrient release rate slows with time as the more readily weatherable minerals are consumed.

An old soil is less suitable as a medium for plant growth than it was during maturity. Its fertility has declined because most of its weatherable minerals are gone. The plant nutrients released by mineral weathering have been mostly leached away. The B horizon has accumulated so much clay (or iron compounds in some soils) that it seriously restricts the movement of water, air, and roots. It might well be said that such a soil is worn out, not by man but by nature.

Fortunately, geologic activities such as rising or sinking of the land, glaciation, etc., act to interrupt soil development at one stage or another. Many soils never reach old age; those that do become old and infertile are eventually either removed by erosion or covered by deposition. Soil formation then begins again in the new surficial material.

REFERENCES

Baxter, F. P., and F. D. Hole, 1967, Ant Pedoturbation in a Prairie Soil, *Soil Sci. Soc. Am. Proc.* **31:**425–428.

Bidwell, O. W., and F. D. Hole, 1965, Man as a Factor of Soil Formation, *Soil Sci.* **99:**65–72.

Buol, S. W., F. D. Hole, and R. J. McCracken, 1973, *Soil Genesis and Classification,* Iowa State Univ. Press, 360 p.

Foss, J. E., and R. H. Rust, 1968, Soil Genesis Study of a Lithologic Discontinuity in Glacial Drift in Western Wisconsin, *Soil Sci. Soc. Am. Proc.* **32:**393–398.

Gagarina, E. I., and V. P. Tsyplenkov, 1974, Use of the Micromorphological Method for Simulating Present-day Soil Formation, *Sov. Soil Sci.* **6:**233–240.

Gile, L. H., F. F. Peterson, and R. B. Grossman, 1966, Morphological and Genetic Sequences of Carbonate Accumulation in Desert Soils, *Soil Sci.* **101:**347–360.

Goldich, S. S., 1938, A Study in Rock-weathering, *J. Geol.* **46:**17–58.

Goss, D. W., S. J. Smith, and B. A. Stewart, 1973, Movement of Added Clay Through Calcareous Materials, *Geoderma* **9**:97–103.

Heil, R. D., and G. J. Buntley, 1965, A Comparison of the Characteristics of the Ped Faces and Ped Interiors in the B Horizon of a Chestnut Soil, *Soil Sci. Soc. Am. Proc.* **29**: 583–587.

Huggett, R. J., 1975, Soil Landscape Systems: A Model of Soil Genesis, *Geoderma* **13**:1–22.

Jenny, Hans, 1941, *Factors of Soil Formation,* McGraw-Hill, New York.

Jha, P. P., and M. G. Cline, 1963, Morphology and Genesis of a Sol Brun Acide with Fragipan in Uniform Silty Material, *Soil Sci. Am. Proc.* **27**:339–344.

Khangarot, A. S., and L. P. Wilding, 1973, Biogenic Opal in Wisconsin and Illinoian-Age Terraces of East-Central Ohio, *J. Indian Soc. Soil. Sci.* **21**:505–508.

McKeague, J. A., 1965, A Laboratory Study of Gleying, *Can. J. Soil Sci.* **45**(2):199–206.

Malo, D. D., and B. K. Worcester, 1975, Soil Fertility and Crop Responses at Selected Landscape Positions. *Agron. J.* **67**:397–401.

Riecken, F. F., 1965, Present Soil-forming Factors and Processes in Temperate Regions, *Soil Sci.* **99**:58–64.

Ruhe, R. V., and P. H. Walker, 1968, Hillslope Models and Soil Formation. II. Closed Systems, *Trans. Int. Congr. Soil Sci., 9th,* **4**:561–568.

Simonson, R. W., 1959, Outline of a Generalized Theory of Soil Genesis, *Soil Sci. Soc. Am. Proc.* **23**:152–156.

Simonson, R. W., F. F. Riecken, and G. D. Smith, 1952, *Understanding Iowa Soils,* Wm. C. Brown Company Publishers, Dubuque, Iowa.

Smith, G. D., 1942, *Illinois Loess: Variations in Its Properties and Distribution,* Illinois Agric. Exp. Sta. Bull. 490, pp. 139–184.

Smith, G. D., W. H. Allaway, and F. F. Riecken, 1950, Prairie Soils of the Upper Mississippi Valley, *Adv. Agron.* **2**:157–205.

Soil Survey Staff, 1951, Parent Materials of Soils, in *Soil Survey Manual,* USDA Handbook 18, U.S. Government Printing Office, pp. 147–154.

Stephens, C. G., 1965, Climate as a Factor of Soil Formation through the Quaternary, *Soil Sci.* **99**:9–14.

Stuart, D. M., and R. M. Dixon, 1973, Water Movement and Caliche Formation in Layered Arid and Semiarid Soils, *Soil Sci. Soc. Am. Proc.* **37**:323–324.

Weaver, J. E., and Ellen Zink, 1946, Length of Life of Roots of Ten Species of Perennial Range and Pasture Grasses, *Plant Physiol.* **21**:201–217.

Chapter 3

Physical Properties of Soils

Physical properties are those which can be evaluated by visual inspection or by feel. They can be measured against some kind of scale such as size, strength, or intensity. Each soil has its own set of physical properties depending on the natures of its component parts, the relative amounts of each component present, and the way they fit together.

Soils are composed of solids, liquids, and gases mixed together in variable proportions. The relative amounts of air and water present depend greatly on how tightly the solid particles are packed together. The packing of small particles tends to be quite different from the packing of large particles. Both the soil texture (an evaluation of the sizes of the particles) and the soil structure (the way the particles stick together) influence the amount of pore space the soil contains and the way that pore space is distributed. Depth, texture, structure, porosity, and consistency are all important physical properties of soils. These five properties plus color and temperature are emphasized in this chapter.

The physical properties have direct significance because the depth of the root zone and the air and water relations within it are determined largely by the physical makeup of the soil horizons. Furthermore, they have additional indirect significance because many chemical and biological aspects of soil fertility can be inferred from physical properties.

SOIL DEPTH

Deep soils provide a more adequate root zone and greater capacities to store water and plant nutrients than shallow soils have. Logic indicates and experiments verify that deep soils are more productive than similar but shallower soils. The differences are largest when some kind of stress affects the plants. For example, plants can endure a longer drought when they grow on soils having a higher available water-holding capacity.

Many construction problems result from shallow soils. Roadbuilding costs increase greatly when rock must be blasted out of the way. Blasting may also be required to remove bedrock from the route of a pipeline or a drainage tile. Home builders must often either omit basements or go to additional expense where the soil is shallow.

Horizon thicknesses can be considered as subdivisions of soil depth. The properties that distinguish one horizon from another influence the suitability of the horizons for various possible uses. The A1 horizon is usually the most favorable for plant growth. Experiments conducted with various thicknesses of A1 horizons have shown more growth where the A1 is thickest. Other horizons might be significant for other reasons. For example, a thick B2 horizon high in clay might be useful for making an impervious core in the dam when a pond is built.

The depth to various soil features is also important. A layer with low permeability will restrict water movement more frequently if it occurs near the surface rather than deeper in the profile. Similar significance can be assigned to the position of layers that are exceptionally high or low in essential plant nutrients or in ions that may be toxic to either plants or animals.

SOIL TEXTURE

Soil texture refers to the percentage by weight of each of the three mineral fractions, *sand, silt,* and *clay.* These fractions are defined in terms of the diameter in millimeters of the particles (nonspherical particles are considered to have an equivalent diameter somewhere between their maximum and minimum dimensions). The sand fraction may be further subdivided into narrower-size groups designated as *soil separates.* Table 3-1 gives the size groups used by the USDA along with a memory aid showing their similarity to the United States monetary system.

Particles larger than 2 mm in diameter are excluded from soil texture determinations. Stones and gravel may influence the use and management of land because of tillage difficulties, but these larger particles make little or no contribution to such basic soil properties as water-holding capacity and the capacity to store and supply plant nutrients. Their presence is noted for use and management purposes by adding an adjective to texture names such as *gravelly sand* or *stony clay.*

Naming Soil Textures

The words *sand, silt, clay,* and *loam* are used to name soil textures. Loam refers to a mixture of sand, silt, and clay that exhibits the *properties* of each fraction about

Table 3–1 Size Limits of Soil Separates*

Fraction	Soil separate	Size, mm	"Memory aid"
Sand	Very coarse sand	2.0 –1	$2.00–$1.00
	Coarse sand	1.0 –0.5	1.00– 0.50
	Medium sand	0.5 –0.25	0.50– 0.25
	Fine sand	0.25–0.10	0.25– 0.10
	Very fine sand	0.10–0.05	0.10– 0.05
Silt	Silt	0.05–0.002	
Clay	Clay	Below 0.002	

* These size limits are used by the U.S. Department of Agriculture. Certain other organizations use different size limits for soil separates. For example, the International Society of Soil Science divides the separates as follows: coarse sand, 2.0 to 0.2 mm; fine sand, 0.2 to 0.02 mm; silt, 0.02 to 0.002 mm; and clay, below 0.002 mm.

equally. It contains less clay than sand and silt because clay properties are strongly expressed relative to the amount of clay present.

Each soil texture name specifies that the weight percent of each fraction falls within certain defined limits. The texture names corresponding to specified percentages of sand, silt, and clay can be determined from the triangle shown in Figure 3-1. The triangle is divided into 12 areas containing all possible proportions of sand, silt, and clay. The numbers on the three scales are angled to indicate the slopes of the lines to which they apply. Thus, the intersection of lines above the y in sandy clay loam represents 30 percent clay (along a horizontal line), 10 percent silt (along a line parallel to the left side of the triangle), and 60 percent sand (along a line parallel to the right side of the triangle). Lines may be similarly traced on the triangle to show that the following mixtures have the indicated names:

60% sand, 25% silt, and 15% clay = sandy loam
25% sand, 45% silt, and 30% clay = clay loam
28% sand, 54% silt, and 18% clay = silt loam

Sometimes the point representing the texture of a soil sample happens to fall exactly on the line between two texture names. It is customary to use the name of the finer fraction when this happens. For example, a sample containing 40 percent clay, 30 percent silt, and 30 percent sand is called clay rather than clay loam.

The adjectives *very coarse, coarse, fine,* and *very fine* are used to identify the dominant sand size in texture names containing the words *sand* or *sandy.* Thus, a sample containing 10 percent clay, 20 percent silt, and 70 percent sand, most of which is coarse sand, is named *coarse sandy loam.* Medium sand is implied when there is no adjective.

Particle-size Analysis

The determination of the percentages of the soil separates present in a sample is called *particle-size analysis.* A complete particle-size analysis involves sample prepa-

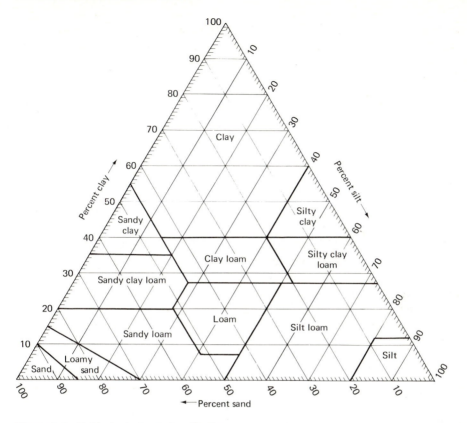

Figure 3-1 Guide for textural classification.

ration followed by two kinds of procedures. The sand separates are determined by sieving, and the silt and clay separates are determined by their rates of settling in water.

The sample preparation consists of removing materials that are not to be considered in the analysis. Particles larger than 2 mm in diameter are removed by sieving. Precise results require that any organic matter present be removed, usually by oxidizing it with hydrogen peroxide. Organic matter binds the clay particles together causing them to be measured as groups. Calcium carbonate has a similar effect and may be removed by treating samples that contain it with acid. Finally, the samples are dried to remove water and then weighed.

A *sieve analysis* is made with a "nest" of sieves placed one above the other. The top sieve has 2-mm holes; below it are sieves with successively smaller holes, 1 mm, 0.5 mm, 0.25 mm, 0.10 mm, and 0.05 mm. A soil sample is subdivided into its various sand separates by placing it on the top sieve and shaking the assembly until each sand particle is caught on a sieve through which it cannot pass. The silt and clay are caught in a pan placed beneath the bottom sieve. The sand separates are then weighed to determine the amount of each present. The combined weight of silt plus clay can also be determined, but these two fractions cannot be separated by sieving. Even if a sieve were made with 0.002-mm openings, it would be extremely difficult to get the clay to pass through it.

The silt and clay can be determined by either the *pipette* or the *hydrometer* methods of analysis. Both of these methods are based on the principle that larger particles fall faster through water than do smaller particles. This principle is quantified in Stokes' law, which in its simplest form can be expressed as:

$$v = kD^2$$

where v = velocity
k = constant that depends on water temperature and density of soil particles
D = diameter of soil particles

Strictly speaking, Stokes' law applies only to spherical particles of about the size of very fine sand, silt, or coarse clay. It cannot be used to distinguish between various sand sizes, but this is done by sieves. For the finer particles, Stokes' law permits the determination of the size of a spherical particle that would settle at the same rate as the particle being evaluated.

The percentages of silt and of clay are determined by treating a weighed soil sample with a sodium compound (usually sodium polyphosphate or sodium oxalate) and dispersing it by means of a shaker or an electric mixer. Alternatively, ultrasonic vibration can be used to disperse the soil particles. Dispersion breaks soil aggregates apart so the particles act individually in the analysis. The suspension of dispersed soil is transferred to a tall liter cylinder and shaken end over end a few times, then placed upright. The particles fall through the water, and the sand soon settles from the upper few centimeters, leaving behind silt and clay. The amount of silt plus clay present is then determined either by withdrawing a sample with a pipette or by measuring the density of the suspension with a hydrometer. Then the settling is allowed to continue for a few hours to eliminate the silt, and the measuring process

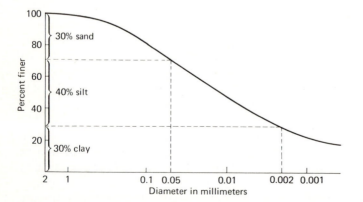

Figure 3-2 The particle-size distribution curve of a clay loam soil containing 30 percent sand, 40 percent silt, and 30 percent clay.

is repeated to determine how much clay is present. Additional readings are made if it is desired to subdivide the silt and clay fractions.

The percentages of sand, silt, and clay can be calculated from the data obtained in the preceding analyses. These can be expressed in a soil texture name. Another means of representing the data is in a particle-size distribution curve such as that shown in Figure 3-2. The curve shows the percentage of mineral matter composed of particles smaller than any specified size.

Texture by Feel

It is often necessary to determine soil texture quickly or in some place where a laboratory determination is not feasible. Soil mappers and others who work with soil learn to estimate texture by feeling the soil between the thumb and fingers. The technique used can be divided into two parts. First, the amount of clay is estimated by how hard the sample is when dry, how much water it can absorb, and its degree of stickiness and plasticity when nearly saturated with water. Plasticity is probably the best guide. It is evaluated by attempting to squeeze the soil into a thin ribbon as it is pressed with a rolling motion between thumb and forefinger. An alternative method is to try to roll the soil so that it resembles a thin wire.

A moist clay soil can readily be formed into a long, durable ribbon or wire (Figure 3-3). A clay loam will form a ribbon or a wire but not as long or durable

Figure 3-3 Soil texture can be estimated by forming moist soil into ribbons or wires about 3 mm thick. The wires and the long ribbon in the hand indicate a higher clay content than the short ribbons on the right.

as that formed by a clay. Some loam soils can be formed into short ribbons but others will not form any ribbon.

The second step is to decide if the sample is sandy or silty. The sample should be wetter for this test than for the clay test. Sand gives the wet soil a gritty feel; silt gives it a smooth, floury feel. Thus a soil that forms a moderately good ribbon would be called a silty clay loam if it had a very smooth feel, a sandy clay loam if it had a very gritty feel, or simply a clay loam if the smooth and the gritty materials were approximately equal. A practiced hand can identify by feel any of the textures shown in the texture triangle in Figure 3-1.

Importance of Soil Texture

Each type of particle present makes its contribution to the nature of the soil as a whole. Clay works with organic matter to store water and plant nutrients. The finer particles also help to bind soil particles together into structural aggregates. The largest particles serve as a skeleton to the soil and support most of the weight borne by the soil and help to make the soil permeable and well aerated. Coarse sandy soils usually are able to support heavy loads with little compaction.

Sandy soils are usually quite permeable to air, water, and roots, but they have two important limitations. One is their relatively low water-holding capacities; the second is that they are poor storehouses for plant nutrients. They must receive frequent additions of water and plant nutrients in order to be highly productive. The presence of a high percentage of organic matter would help to compensate for low clay contents, but most sandy soils are low in organic matter. These limitations of sandy soils can, of course, be overcome if both fertilizer and irrigation water are available, but the costs are large. One hazard is the loss of applied fertilizer by leaching if too much fertilizer and water are applied.

The limited capacity of sandy soil to store water and plant nutrients is related to the relatively small surface area of its soil particles. Surface area per gram is inversely proportional to particle diameter. This relation is illustrated in Figure 3-4.

Clay not only has a large surface area, it is also electrically charged. The charge gives clay the capacity to hold plant nutrients on its surface in forms available to plants, but sand lacks this capacity. The loss of plant nutrients by leaching from clay is therefore very small compared with the losses that might occur if the same amount of nutrients were present in a sandy soil.

Clays hold much more water than sands because they have a large surface area to be covered with water. An amount of water that would leach through a sandy soil might not wet a clay soil deep enough to cause leaching. Dissolved nutrients are lost from the soil only when water penetrates beyond the reach of plant roots and becomes drainage water.

Soils containing too much clay may have high water-holding capacities but inadequate aeration. It may seem surprising, but a high content of organic matter

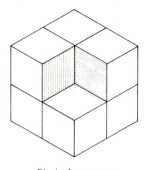

Blocks 1 cm across

A 1-cm block sliced vertically
and horizontally into blocks
0.2 cm across

Figure 3-4 Each 1-cm block has 6 cm² of surface. The block sliced into smaller blocks 0.2 cm across results in 125 pieces with 30 cm² of total surface exposed. Slicing the block into still smaller pieces 0.001 cm across produces 1 billion clay-sized particles with 6,000 cm² of surface area.

is as helpful for overcoming the problem of too much water in a clay soil as it is for the problem of too little water in a sandy soil. Organic matter helps to hold the clay particles together in clusters that have air space between them.

One important problem with clay soils is their stickiness. When wet, they stick to plows, boots, or whatever they contact. Many motorists can testify that clays are slippery when wet. Clays become hard when dry, and so their optimum water content for any kind of work is in midrange—when they are dry enough to lose their stickiness but moist enough to avoid being hard. Even so, plowing and other tillage operations require considerably more power in clay soils than in sandy or loamy soils. Soils high in clay are therefore known as "heavy" soils and sandy soils are called "light" soils. As will be seen later, the actual weight of soil solids is usually greater in the sandy soil than in the clay soil.

Loam and silt loam soils are highly desirable for most uses. They have enough clay to store adequate amounts of water and plant nutrients for optimum plant growth but not so much clay as to cause poor aeration or to make working with them difficult. They contain enough silt to gradually form more clay (to replace that lost by eluviation and erosion) and to release fresh plant nutrients by weathering. A soil containing between 7 and 27 percent clay and approximately equal amounts of silt and sand has a *loam* texture. Loam soils that contain several percent organic matter are very good for most uses. This is not to say that all other soils are poor. For example, the *Houston clay* soils of Texas have over 50 percent clay but are highly productive. These soils have a high content of organic matter helping to give them a good granular structure that provides adequate aeration.

Texture names are assumed to apply to the topsoil unless stated otherwise. They apply, of course, to the entire profiles of uniform soils, but most soils have some variations in texture. In many soils there is enough difference to give different texture names to the horizons. Texture variations can be highly important. B horizons with high clay contents may seriously restrict air, water, and root penetration. In general, any considerable abrupt change in texture causes some retardation of water movement. This retardation may be favorable if, for example, it occurs low in the profile and retains additional water in the soil long enough for it to be utilized by plants.

PROFILE DIFFERENTIATION

Most soils have higher percentages of clay in their B horizons than in their A horizons. The differences in clay content increase as the soils become older and soil formation takes its course. Figure 3-5 illustrates in a schematic way the clay distribution in a soil at three different stages of profile differentiation.

Part of the increased clay content in the B horizon was eluviated from the A horizon currently above it. Part of the clay may have come from material that has since been removed by erosion. Part of it may have been formed in the B horizon itself by mineral weathering. Clay formation in the A horizon often equals clay eluviation during the young to mature stages of soil development. Many mature soils therefore have about equal clay percentages in their A and C horizons. B horizons accumulate clay from weathering in addition to clay they originally had.

Sand-size particles of quartz and other resistant minerals accumulate in the A horizon. Eventually the more weatherable minerals are depleted enough to reduce the rate of clay formation. But, clay eluviation continues and some of the clay is weathered into soluble materials and leached from the soil. The result is a decline in the clay content of the A horizons. How long this takes depends greatly on the intensity of weathering. The prevalence of sandy loam A horizons in the southeastern United States indicates soils that have been subjected to intensive weathering for a long time.

A strongly differentiated soil handicaps soil management and crop production. The A horizon has lost much of its fertility because mineral weathering has become too slow to provide a good supply of plant nutrients. Water passes very slowly through the B horizon because it has a high clay content. Root penetration is limited by the strength and the poor aeration of the B horizon. Most of the root development

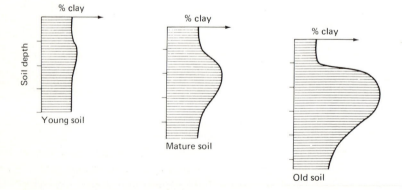

Figure 3-5 A schematic representation of the way soil profile differentiation increases with time. A minimal amount of soil erosion is indicated by the gradual lowering of the soil surface.

then occurs in the A horizon. Plant growth may suffer from too much water during wet seasons and too little water after a week or two of drought. Such a soil can also be a problem for engineering uses because the A horizon becomes saturated easily. Saturation reduces its weight-bearing strength. A roadbed underlain by wet soil may break apart. Wet soil may fail to support the weight of a building.

The advanced stages of profile differentiation are delayed when either geologic erosion or deposition is active. These processes cause fresh material to be incorporated into the soil at either the bottom or the top of the profile.

SOIL STRUCTURE

Two soils with the same texture may have distinctly different physical properties because the soil particles are arranged in different ways. The arrangement of the individual particles into larger units is called *soil structure*. It results from the tendency of the finer soil particles, especially clay and humus, to stick together.

Units of soil structure are called *peds*. Peds are natural groups of primary particles (sand, silt, and clay) that occur and persist within the soil. Their natural occurrence and persistence distinguish them from clods. The latter are caused by disturbances such as plowing.

Peds occur in several different forms and may be coarse, medium, or fine in size and weak, moderate, or strong in their degree of development. Close observation is required to detect weak peds, whereas strong peds are easily visible and readily separable from each other. Peds are completely absent from some soil materials, especially those low in clay. The presence of peds is especially significant in soils high in clay because it improves the permeability. Air, water, and roots move through the spaces between peds much more readily than through the peds. A soil with structure is much more permeable than it would be without structure.

Types of soil structure are classified as follows and are illustrated in Figure 3-6:

1 Structureless
   a *Single grain.* Each soil particle is independent of all others. This condition is typical of very sandy materials.
   b *Massive.* The entire soil mass clings together with no definite lines of weakness. This condition is typical of soil parent materials that contain significant amounts of clay.
2 Structured
   a *Granular.* The primary soil particles are grouped into roughly spherical peds. There is space between the granules because they do not fit tightly together. This is the most desirable soil structure for growing plants because of the openness between peds. Water can be held within the granules, and yet air can circulate between them. Granular structures occur mostly in A1 horizons because the peds in the lower horizons are usually squeezed more tightly together by the weight of the overlying soil. A particularly porous form of granular structure is sometimes called *crumb* structure.

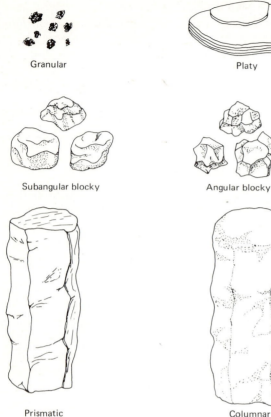

Figure 3-6 Types of soil structure.

b *Platy.* Platy soil structure has peds with horizontal dimensions greater than vertical dimensions. It is most common either at the soil surface or in an A2 horizon. Platy structure is often associated with lateral movement of water through the soil.

c *Blocky.* Blocky peds are approximately equal in their vertical and horizontal dimensions. They are distinguished from granules because the blocks fit together like a three-dimensional jigsaw puzzle. Two types of blocky structure are distinguished—*angular blocky* for peds with mostly sharp corners and *subangular blocky* for peds with mostly rounded corners. Angular blocks are common in B horizons, and subangular blocks are likely to be found in either A or B horizons.

d *Prismatic.* Prismatic peds are taller than they are wide. While occasionally found in A horizons, prismatic structure is most common in B horizons of well-developed soils. The prisms in the B horizons of some very strongly developed soils have rounded tops owing to having lost their top corners by eluviation. This special case is known as *columnar structure,* and usually results from either old age or a high sodium content in the soil solution.

3 Structure destroyed

    a *Puddled.* When soils, especially those high in clay, are trampled or plowed while too wet, they become puddled or "run together." Structure is destroyed, the larger pores collapse, and the soil is left in an undesirable condition.

How Structure Develops

Soil parent material is presumably structureless, either massive or single grain, depending on its texture. Structure develops as weathering agents convert the parent material into soil. Wetting and drying causes swelling and shrinking that shift soil particles back and forth. Planes of weakness gradually develop into ped faces. Freezing and thawing help develop structure in soils. Plant roots penetrate the mass, cause further shifting and loosening, and ultimately decay, leaving pore spaces in the soil. Animal activity may loosen the soil and break it into distinct groups of particles. Plants, animals, and microbes produce cementing agents that help to stabilize the peds. All these forces continue working throughout the existence of the soil with the result that soil structure commonly becomes more distinct with time. B horizon peds become more obvious because they often are coated with illuvial clay.

Importance of Structure in Topsoils

Topsoil structure receives more attention than that of subsoil because it is more observable and more amenable to treatment. The structure of the surface soil is important for (1) aeration, (2) water permeability and its relation to runoff, (3) the degree of resistance the soil has to erosion, and (4) formation of a good seedbed to initiate plant growth.

A granular structure is ideal for air and water relations. The space between granules is important for aeration and for water percolation. All of the water entering the soil must pass through the upper part so the permeability of the A horizon is most significant. Rainfall that cannot enter the soil becomes runoff and may cause erosion. The size and strength of the peds has a marked effect on the velocity of runoff water required to cause erosion. Bonding between peds helps prevent erosion unless the peds merge and form a crust that reduces infiltration and increases runoff.

A good seedbed provides a suitable environment for seedlings to become established. There should be enough aggregates contacting the seed to provide it with moisture. There should be enough pore space to provide oxygen for the seed to sprout and the seedling to grow. One of the main purposes of tillage is to create a good seedbed even if the soil has less desirable structure. Unfortunately, stirring the soil by tillage has much the same effect as opening the draft on a furnace. The resulting oxidation decomposes some of the binding materials that help to stabilize soil structure. Even so, the preparation of a good seedbed usually requires some tillage. The resulting increase in oxidation is not all bad since it leads to an increased supply of available plant nutrients from the decomposing organic matter. Important aspects of this effect are discussed in Chapters 5 and 10.

Importance of Structure in Subsoils

The significance of structure in B horizons depends partly on their textures. B horizon structure, or the lack thereof, seldom causes any problem if the texture is sandy. A compact B horizon with a high clay content is another matter. Water passes through it very slowly, and the A horizon above may be saturated with water for a time. Continuing rains can keep the lower A saturated with water for long periods of time each year and leach it so much that an A2 forms regardless of the type of vegetation.

Root penetration may be limited by lack of oxygen for root respiration. A lack of oxygen may occur in the lower part of the A horizon, but it persists longer when it occurs in the B horizon. High clay content and lack of large pores permit some B horizons to absorb and hold so much water that they remain saturated for days or weeks after the rain stops. A desirable soil structure provides large enough pores between the peds that water and air movement is never seriously restricted. Some farmers try to create such conditions in their subsoils by deep plowing or chiseling. A few soils have brittle layers that can be shattered to improve drainage and aeration, but most experiences with subsoil cultivation have been disappointing. The high cost of such practice is not repaid by sufficiently high increase in yields. Some good is done, of course, and yields are improved temporarily following the deep tillage, but the improvement usually is short-lived.

Deep-rooted perennials, particularly alfalfa, help to open up channels for air and water movement in heavy clay subsoils. As the large roots decay, they leave temporary channels between the usually close-fitting aggregates.

One of the most important lessons in the study of soils is gaining an appreciation of the limitations placed on soil productivity by a dense and poorly aerated subsoil. The appraised value of any soil should be based to some extent on the physical properties of the subsoil. The topsoil can be altered much more easily and more economically than can the subsoil.

Aggregate Stability

The most important features of soil structure are (1) the arrangement of soil particles into *aggregates* (groups of soil particles that may become peds or parts of peds if they are stable and enduring) of desirable shape and size to provide an adequate amount of large pores with connecting channels between them, and (2) the stability of the aggregates when exposed to water. Water-stable aggregates maintain the ability of the soil to absorb water and help it to resist erosion. Weak aggregates will slake and disperse in water or fall apart when struck by raindrops. Dispersed silt and clay then move into the pores of the surface soil and clog them. The permeability to air and water declines when the pores clog, and much water runs off that would otherwise infiltrate. Furthermore, dispersed soil is easily carried away in the runoff water.

The weakness of unstable aggregates in a surface soil is particularly apparent when rain falls on a freshly planted seedbed. Rain is important to provide moisture for germination and seedling growth, but falling raindrops break up many aggregates and the detached particles plug soil pores. Continual pounding by raindrops produces a puddled surface that dries into a hard crust.

The stability of aggregates is related to the soil texture, the kind of clay, the kinds of ions associated with the clay, the kind and amount of organic matter present, and the nature of the microbial population. Some clays are inherently stickier than others. Some types of clay expand like an accordion as they absorb water. Expansion and contraction of such clay shift and crack the soil mass in ways that may either create or break apart soil peds. Hydrogen and calcium ions associated with clay promote aggregation, whereas the presence of a high proportion of sodium ions causes dispersion. Microbes produce many different kinds of organic compounds, some of which help to hold soil aggregates together. The mycelial growth of fungi appears to have a binding effect on soil. Furthermore, it has been demonstrated that certain species of fungi are much more effective than others in promoting aggregate stability (Downs et al., 1955).

Organic materials mixed with soil have an immediate effect of preventing the soil from forming a crust. The organic materials keep the soil loose and improve aeration. But the mixing of organic matter with the soil has little influence on aggregate stability until some decomposition occurs. Certain products of decomposition have a binding effect on soil particles. These products are thought to be complex chainlike organic compounds that bridge across between the soil particles. The naturally occurring organic compounds which have this binding effect are subject to decomposition and must therefore be replaced from time to time by new additions of organic matter.

Aggregate stability can be measured by determining the percentage of soil aggregates of a specified size remaining on a sieve after repeated dipping in a container of water. The wet aggregate stability declines rapidly in soils planted to a clean-tilled crop. The wet stability increases while the soil is in sod crops like alfalfa

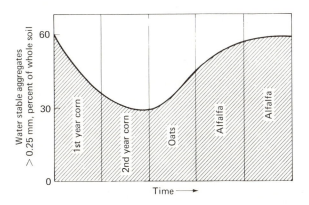

Figure 3-7 Schematic diagram showing effect of time and crop on soil aggregate stability. (*Browning et al.*, 1948.)

and bromegrass. The graph in Figure 3-7 illustrates the changes in aggregate stability that can be anticipated during a 5-year rotation of corn, oats, and alfalfa. When the same crop is grown 2 or more years in succession, the aggregate stability begins to level out at an equilibrium level for that crop. Thus second-year corn and second-year alfalfa cause smaller changes than the first years caused. Third and later successive years of the same crop result in only small changes in stability. The particular level at which equilibrium is reached under continuous cropping depends upon the crop, the method and timing of tillage, and the yields attained as well as on such fixed characteristics as climate and soil texture. In general, intensive tillage lowers aggregate stability; close-growing crops help to improve aggregate stability.

Aggregate stability is more closely related to runoff and soil erosion than to crop yields. The aggregate stability in most soils can decline somewhat without influence on yields if the fertility is maintained by the use of fertilizers and if water is not a limiting factor. Excessive erosion will eventually lower the productive potential of many soils, but this usually takes a long time. Faster effects show on some soils which become puddled when wet and crusted when dry. Poor seedling emergence leads to poor stands and reduced production. Another effect of declining aggregate stability that occurs in certain soils with restricted drainage is increased waterlogging caused by reduced permeability. Waterlogging, too, reduces production.

It is worth noting that all the effects of aggregate stability on plant growth are indirect in nature. These effects, furthermore, are most likely to become significant in soils with high clay contents.

Probable Mechanisms of Aggregate Stability

Mineralogical, chemical, and biological factors are all involved in aggregate stability. The importance of any one mechanism varies greatly from one soil to another. Some combination of the mechanisms explained in the following paragraphs occurs in any soil that has structure, but there are many possible combinations.

Sand and silt have little or no tendency to form aggregates except when they happen to be included in groups of particles bound together by clay and organic matter. The sand and silt will therefore be ignored in the discussion that follows.

The nature of clay minerals will be discussed more fully in Chapter 6, but some features must be mentioned here as background for understanding aggregation. The clay fraction can be divided into three groups of minerals for this purpose: (1) Clay-sized particles of quartz, feldspars, and other minerals that are relatively inert and need not be considered here (these minerals are usually considered to be more typical of the sand and silt fractions), (2) hydrous oxides of iron, aluminum, and other metals that are important constituents of many tropical soils, and (3) the silicate clay minerals such as montmorillonite, kaolinite, and other similar minerals, including some micas. The silicate clay minerals are the most universal of the three groups and are commonly assumed when clay is mentioned.

1   Chemical bonding in soil aggregates

Chemical bonding is possible because most clay particles have electric charges. Particles of silicate clay minerals approximate six-sided plates. Their broad

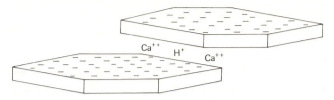

Figure 3-8 An expanded representation of a pair of overlapping clay plates held together by *cation linkage*.

surfaces have negative charges that serve as storage sites for cations. Some of these negative-charge sites are also involved in chemical bonding. These negative charges must, of course, be balanced by positive charges. Three types of positive charges are significant in the chemical bonding of soil aggregates:

a *Cation linkage* as shown in Figure 3-8 results from positively charged cations being attracted to the negatively charged surfaces of two clay particles and holding them together. The strength of the bond depends on the size and charge of the cation. Hydrogen ions (very small) and calcium ions (double charges) favor aggregation. Sodium ions (large hydrated ions with single charge) force the particles farther apart.

b The *cardhouse effect* represented in Figure 3-9 results from the presence of positive charges at the edges of clay plates (both positive and negative charges occur in edge positions). This charge situation permits the edge of one plate to be attracted to the surface of the next. The resulting three-dimensional arrangement has abundant small pore spaces favorable for storing water but is fragile and easy to break down if the soil is worked.

c *Plinthite* (formerly called laterite) contains hydrous oxides of iron and aluminum along with silicate clay minerals. The hydrous oxides have positive charges rather than negative. The combination of the two types of clay produces very stable aggregation by bonding the two together. True plinth-

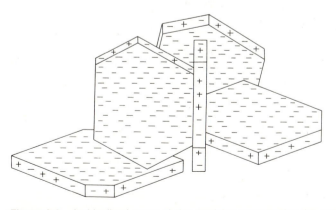

Figure 3-9 An idealized group of clay particles representing the *cardhouse effect* in soil structure wherein positive charges on edge positions are attracted by negative charges on the broad surfaces.

ite can be cut into blocks that can be dried in the sun to form hard bricks. The drying locks the chemical bonds together irreversibly.

2 Organic bonding in soil aggregates

Many types of organic compounds occur in soils. Detailed treatment of this subject must be left for Chapter 5. For the present, these materials may be considered as long chains of carbon atoms with frequent side branches and several types of reactive groups that can ionize to produce sites with either positive or negative charges. Chang and Anderson (1968) showed that the effectiveness of organic materials for flocculating clay depends on the number of reactive groups in the organic matter. Usually the negatively charged sites are the most abundant but perhaps not the most important in soil structure. The organic compounds may act in several ways to bind soil particles into aggregates.

a *Chemical linkage* may occur between the positively charged sites on an organic molecule and negatively charged sites on clay particles (or, less commonly, between negative organic charges and positive mineral charges). Linkage of the same large molecule to two or more clay particles helps to stabilize the aggregate.

b *Cation linkage* may help to hold together negatively charged sites on both the organic matter and the mineral matter.

c *Filamentous* (threadlike) mycelial growth of fungi and actinomycetes can produce a tangled mass that can wrap around soil particles and tie them together.

d *Gelatinous exudates* from plant roots and from bacteria and other microorganisms may engulf portions of soil particles or fill the corners between particles and act as cementing agents to glue the particles together.

Aggregates of clay particles combine with sand and silt particles into a soil structure such as that shown in Figure 3-10. Coatings of organic matter and iron and aluminum oxides cover most of the particle surfaces.

Artificial Soil Conditioners

In 1952 an artificial soil conditioner sold under the trade name of Krilium was announced. Other similar products soon followed with the anticipation that agriculture might be revolutionized. Actually, the materials proved to be too expensive for general agricultural use, but they are useful as a research tool. It is possible to study the effect of soil structure on plant growth through the use of soil conditioners. Previous methods of improving soil structure also improved fertility, and the structural effects could not be evaluated separately.

Soil conditioners are polymerized organic compounds with a high resistance to decomposition. They have many negatively charged sites whereby they can bind soil particles together through the same chemical linkage and cation linkage mechanisms as the natural organic materials in the soil. In use, the soil conditioner must be thoroughly mixed with the soil, and a favorable soil structure created by tillage or other means. The soil conditioner will then stabilize the aggregates and make them last for years.

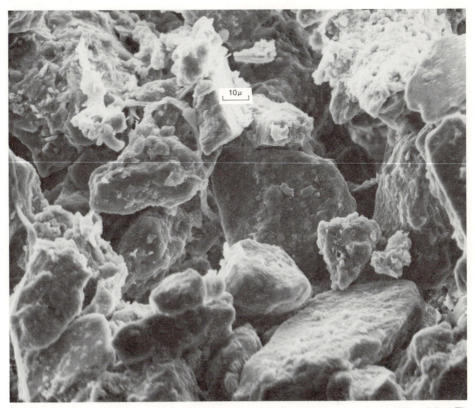

Figure 3-10 A scanning electron micrograph showing a natural arrangement of soil particles. The line in the upper center is 10 microns (0.01 mm) long. *(Courtesy of Steve Thien.)*

SOIL POROSITY

A very significant part of the process of changing rock into soil is the loosening of the material so there are pore spaces in the mass. These pore spaces are sometimes referred to as empty spaces, but this is an error. They contain air and water. Without pore space the soil would not be a suitable medium for plant growth.

The relative volumes of each of the three states of matter present in a soil can be illustrated by a diagram such as Figure 3-11. Usually the solid matter in the A horizon occupies roughly half of the volume; the remainder is pore space. The solids are mostly mineral matter, but a small portion such as that shown in the figure are organic (in organic soils this portion is much larger and the mineral portion is smaller). The pore space is divided between air and water in proportions that vary with the wetness or dryness of the soil. In general, the larger pores contain air unless the soil is completely waterlogged, and the smallest pores contain water unless the soil is thoroughly dried. Air and water move in and out of intermediate-sized pores whenever the water content of the soil changes.

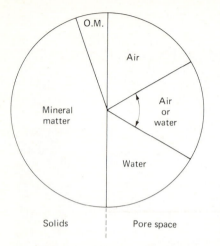

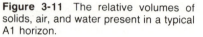

Figure 3-11 The relative volumes of solids, air, and water present in a typical A1 horizon.

Sandy soils usually have less pore space than finer-textured soils, but they are almost invariably well aerated (unless there is a subsurface limitation on water movement). Their good aeration results because most of their pore spaces are large enough to permit at least a part of the water that enters to drain out. Air circulates through the drained pores and reaches all but a few isolated pore spaces. Of course, this easy loss of water from a sandy soil means that it has a rather low capacity for storing water for plant growth.

Clay loams and clays generally have high total pore space, but they hold large amounts of water even when there are no subsurface restrictions on water percolation. Their pore spaces are numerous but tiny. Unless a good structure is present, most of the pore spaces in clay soils may be no thicker than the water films that can be held in place around each soil particle. Even those larger pores that are present may be isolated from circulating air because the connecting passages are all tiny pores that are filled with water much of the time.

Capillary and Aeration Porosity

From the previous discussion it should be understood that aeration is provided by the larger pore spaces in the soil if these larger pore spaces are adequately interconnected. The connecting passageways need to be at least 0.01 mm in diameter at their smallest points for water to drain from them at field capacity. Such pore space is called *aeration porosity* in contrast with the *capillary porosity* that may be filled with water. Figure 3-12 shows how these two types of porosity vary with depth in two soils. Minden is a silt loam soil with slight profile differentiation. The Minden soil is well aerated through the solum (about 1 m), but the aeration porosity declines in the C horizon. Edina is a more strongly developed soil. It has a silt loam A horizon, but the B horizon has a high clay content. The total porosity of the Edina B horizon is adequate, but the aeration porosity is very low. This B horizon has the gray subsoil color usually associated with poor aeration.

Total pore space is a poor measure of aeration because the water content is variable. The finer-textured soils generally have the highest percentage of total pore

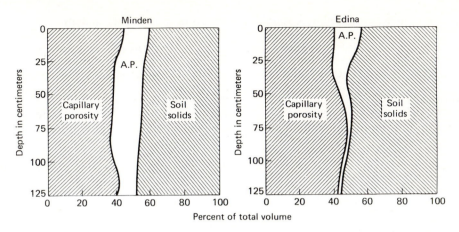

Figure 3-12 The capillary and aeration porosity in a slightly differentiated Minden profile and in a strongly differentiated Edina profile. (*Ulrich*, 1949.)

space, but they also hold the most water and usually require the most care to maintain adequate aeration. Some clay B horizons have nearly 60 percent total pore space and yet are so poorly aerated that they seriously inhibit root penetration. The oxygen supply is likely to be inadequate for good root development whenever the soil holds so much water that it has less than 9 or 10 percent air space at field capacity (Patt, Carmeli, and Zafrir, 1966).

The usual method of measuring aeration porosity involves allowing a saturated soil to drain under a specified "tension." The volume of water that drains from the soil is taken as the volume of the aeration pore space. The amount of tension used is usually equivalent to a column of water between 40 and 60 cm long. This is equivalent to the tension that exists in a layer of soil located 40 to 60 cm above a water table. This amount of tension causes water to drain from the larger pore spaces.

The tension method of evaluating porosity may be the best available but it has its limitations. The water tension to use is necessarily arbitrary, but it influences the results. Also, some pores may be drained and yet remain isolated from other drained pores. As a consequence, the oxygen in these pores may soon become depleted, and a condition of poor aeration may exist in spite of their presence. Various techniques have been devised to measure air circulation and oxygen diffusion rates in soils rather than aeration pore space, but there are problems with these approaches also. None of these techniques has become a standard procedure.

Effect of Cultivation on Porosity

Cultivation loosens the soil temporarily, but the cultivated soil gradually packs down and usually becomes denser than adjacent uncultivated soils of the same type. Cultivation generally causes a reduction in organic-matter content, compaction by heavy machinery, and exposure of the soil to the direct impact of falling raindrops. These factors tend to break down the larger aggregates and reduce the sizes of many

Table 3-2   Changes in Pore Space Resulting from Cultivation of Several Soils

| | Total pore space, %* | | Aeration pore space, %[†] | |
Soil type	Uncultivated	Cultivated	Uncultivated	Cultivated
Webster loam	65.6	57.0	19.3	7.9
Carrington silt loam	57.3	57.3	9.5	10.3
Ida silt	55.0	48.8	8.3	4.8
Marshall silt loam	59.2	59.2	7.1	9.1
Grundy silt loam	61.5	54.7	7.2	5.4
Edina silt loam	60.8	48.8	9.0	7.1

* Calculated from bulk-density values assuming a particle density of 2.65 g/cm^3 in all soils.
[†] Pore space drained under a tension of 40 cm of water.
Source: Anderson and Browning, 1949.

of the pores. Furthermore, material dispersed by the raindrops filters down into some of the pores below.

The data in Table 3-2 show changes in pore space resulting from cultivation in six different soils. For each soil type in the table a sample of cultivated soil was collected just a few meters away from the same soil type under natural conditions of grass cover. Pore space determinations were then made on the paired samples. Four of the six soils had less total pore space and less aeration pore space in the cultivated samples. The other two soils, Carrington and Marshall, showed slight increases in aeration pore space and no change in total pore space under cultivation. The two increases were considerably smaller and less significant than the large reductions in aeration pore space shown by some of the soils.

Practices that improve soil structure also help to improve aeration. The use of grass-legume meadows in a rotation helps to loosen the soil and to develop aeration porosity. Plowing at the right moisture level also loosens the soil, but this effect may disappear within a few weeks. Too much tillage speeds the decomposition of organic matter and reduces aggregate stability and porosity. The incorporation of large amounts of crop residues into the soil after each crop helps to maintain the organic-matter content of the soil and thereby favors both fertility and aeration.

Measuring Soil Porosity

Data such as those shown in Table 3-2 and represented in Figure 3-12 are generally obtained in two ways. The aeration porosity is measured by the tension method already described. Total porosity is calculated from data on bulk density and particle density.

Bulk Density   The bulk density is the weight of the soil solids per unit volume of total soil. The pore space is a part of the volume of soil measured for bulk density, but the soil is oven-dried to drive the water out of the pores before the soil is weighed.

Direct measurement of bulk density requires both field and laboratory work. It is now possible to measure soil density in the field by means of gamma-ray attenuation. A radioactive source of γ-rays is lowered into a hole along with a detector. The source and the detector are arranged one above the other so that the γ-rays may reach the detector by passing along a curved path through the soil. Soil particles tend to absorb the γ-rays. The denser the soil, the more the γ-rays are absorbed and the smaller the count will be at the detector. The count can therefore be taken as an indication of soil bulk density. Allowance must be made in the interpretation for the water content of the soil, since water is excluded from the bulk density. Gamma-ray attenuation is therefore usually used along with a neutron probe to determine soil moisture content. The operation of the neutron probe is described in Chapter 4.

Bulk-density data are necessarily expressed in units of weight and volume. These can be any properly specified units, but the usual units are grams per cubic centimeter (g/cm^3). An older term, *volume weight,* was long used for this same property but is seldom seen in recent literature. Volume weight is determined by dividing the weight of a given volume of oven-dry soil by the weight of the same volume of water. Volume weight is therefore a number without any units. This number is the same as the numerical part of the bulk density expressed in grams per cubic centimeter.

The bulk density of the A horizons of mineral soils is usually between 1.0 and 1.6 g/cm^3 (that of organic soils is lower and can be as low as 0.1 g/cm^3 in sphagnum moss peats). The variation results mostly from differences in total pore space. As a general rule, the finer-textured soils have more pore space and lower bulk densities than sandy soils. Of course, the bulk density of any one soil varies according to its degree of compaction. Packing a soil decreases its pore space and increases its weight per unit volume. Overburden weight tends to compact the lower horizons and give them higher bulk densities than the A1 horizons. A bulk density given without specifying the horizon is usually for an A1 or an Ap horizon.

Organic matter decreases bulk density in two ways. First, organic matter is much lighter in weight than a corresponding volume of mineral matter; second, organic matter gives increased aggregate stability to a soil. The latter is by far the most important of the two effects in most soils, but both act to give lower bulk

Table 3–3   Bulk Densities and Organic-matter Contents for Several Cultivated and Uncultivated Soils

Soil type	Bulk density g/cm^3		Organic matter, %	
	Uncultivated	Cultivated	Uncultivated	Cultivated
Webster loam	0.91	1.14	10.9	8.4
Carrington silt loam	1.13	1.13	7.8	7.2
Ida silt	1.19	1.36	5.2	1.7
Marshall silt loam	1.08	1.08	5.5	3.8
Grundy silt loam	1.02	1.20	8.5	5.4
Edina silt loam	1.04	1.36	5.9	4.2

Source: Anderson and Browning, 1949.

densities in soils with higher organic-matter contents. The bulk density usually increases when cultivation causes a loss in organic matter from a soil. Some of these changes are shown in Table 3-3. Cultivation reduced the organic-matter contents in all six of these soils and increased the bulk densities of most of them. The uncultivated Webster loam in this table illustrates how organic matter can reduce the bulk density even below the usual 1.0 to 1.6 g/cm^3 range of mineral soils.

The bulk density is useful for calculating the weight of various masses of soil. For example, the weight in kilograms of a cubic meter of soil with a bulk density of 1.2 g/cm^3 is:

$$1.2 \text{ g/cm}^3 \times \frac{(100 \text{ cm/m})^3}{1,000 \text{ g/kg}} = 1,200 \text{ kg/m}^3$$

The weight of soil with a bulk density of 1.1 g/cm^3 filling a planter box 1 m long, 20 cm wide, and 15 cm deep is:

$$1.1 \text{ g/cm}^3 \times \frac{100 \text{ cm} \times 20 \text{ cm} \times 15 \text{ cm}}{1,000 \text{ g/kg}} = 33 \text{ kg}$$

Weights of other soil volumes can be calculated similarly if the bulk density is known. A reasonable estimate of weight can often be made from an assumed bulk density of 1.0 g/cm^3 for loose soil, 1.3 g/cm^3 for undisturbed soil, or 1.6 g/cm^3 for compacted soil. The weight of soil needed to pot a plant or to fill a wheelbarrow or a truck can be calculated by multiplying the bulk density by the volume.

The soil mass of most interest to a farmer is often the plow layer of the field. The hectare-furrow slice is used as a reference unit for calculating the number of kilograms of soil amendments, fertilizers, or other chemicals needed for a field.

The weight of a hectare-furrow slice depends on its depth as well as its bulk density. Common plow depths range from 15 to 20 cm. Combining the greater depth with a low bulk density of 1.0 g/cm^3 gives:

$$1.0 \text{ g/cm}^3 \times \frac{20 \text{ cm} \times (100 \text{ cm/m})^2 \times 10,000 \text{ m}^2/\text{ha}}{1,000 \text{ g/kg}} = 2 \text{ million kg/ha}$$

The same result can be obtained by assuming a shallower depth but a larger bulk density, such as 15 cm and 1.33 g/cm^3. A hectare-furrow slice is, therefore, often assumed to weigh 2 million kg as a reasonable estimate. As it turns out, pounds per acre (the comparable unit in the English system) is so nearly the same as kilograms per hectare that the two units are often considered equal and an acre-furrow slice is taken as weighing 2 million lb. Where a more accurate conversion is needed, the factor 1.0 kg/ha = 0.89 lb/ac may be used. Actually, the 2 million value is an approximation that is often in error by 10 or 20 percent in either system.

In spite of being an approximation, the 2 million kg/ha or lb/ac value is widely used for converting laboratory data to field use. Soil analyses usually are reported in either parts per million (ppm) or milliequivalents per 100 grams (meq/100 g). Multiplication by 2 converts ppm to pp2m (parts per 2 million) for use on either a kilogram per hectare or a pound per acre basis. Similarly, multiplying 20 times an equivalent weight converts meq/100 g to pp2m (the meq/100 g unit will be explained in Chapter 8). The use of ppm data can be illustrated by a soil containing 8 ppm of available phosphorus. This is equivalent to 16 pp2m or 16 kg/hectare-furrow slice of a representative soil.

Particle Density   Particle density as well as bulk density must be known to calculate total pore space. Particle density is the average density of the soil particles. The oven-dry weight of the soil is divided by the volume of the soil solids *excluding pore space.* The units used are almost always grams per cubic centimeter.

Much of the older literature uses the term *specific gravity* as a measure of the density of the soil particles. Like volume weight, it is determined by dividing by the weight of an equivalent volume of water and has no units. The specific gravity is numerically equal to the particle density in grams per cubic centimeter.

Particle density can be measured by means of a small bottle known as a *pycnometer.* The pycnometer is weighed once full of water and once with a known weight of soil plus enough water to fill the bottle. Thus the amount of water displaced by the soil can be determined, and the particle density (actually the specific gravity) can be calculated.

Particle densities of mineral soils vary much less than bulk densities. Most are between 2.6 and 2.7 g/cm^3. The usual practice is to assume that the particle density of a mineral soil is 2.65 g/cm^3 unless very precise data are sought. Very careful work is required to measure the particle density and obtain a result more accurate than this assumed value.

Particle density is not altered by differences in particle size nor by change in pore space. It can be regarded as an average of the densities of the solid materials that are in the soil. It is a reflection of the densities of the most abundant minerals in soils and rocks as shown in Table 3-4. The soils most likely to have particle densities markedly different from 2.65 g/cm^3 are the organic soils (considerably lower) and some tropical soils high in hydrous oxides of iron where the particle densities can be quite high.

Table 3–4   Densities of Common Minerals

Mineral	Density, g/cm^3
Quartz	2.65
Feldspars	
Orthoclase	2.56
Plagioclase	2.60–2.76
Micas	2.76–3.0
Silicate clay-minerals	2.00–2.7
Hydrous oxides of Fe and Al	2.40–4.3

Percent Pore Space   The percent pore space in soils is calculated from data on bulk density and particle density. Usually the bulk density is determined by weighing a sample of measured volume, and the particle density is assumed to be 2.65 g/cm^3. Given these two values, the percent of the soil volume occupied by solids is:

$$\% \text{ solid} = \frac{\text{bulk density}}{\text{particle density}} \times 100$$

and

$$\% \text{ pore space} = 100 - \% \text{ solid}$$

If preferred, these two equations can be combined into one:

$$\% \text{ pore space} = 100 - \frac{\text{bulk density}}{\text{particle density}} \times 100$$

or, usually

$$\% \text{ pore space} = \left(1 - \frac{\text{bulk density}}{2.65 \text{ g/cm}^3}\right) \times 100$$

A soil with a bulk density of 1.2 g/cm^3 and a particle density of 2.65 g/cm^3 is:

$$\frac{1.2}{2.65} \times 100 = 45\% \text{ solid}$$

and

$$\left(1 - \frac{1.2}{2.65}\right) \times 100 = 55\% \text{ pore space}$$

It should be noted that increasing bulk density signifies an increased percent solid and a corresponding decrease in percent pore space. A simple exercise in calculations involving bulk density involves verifying part of or all the figures presented in Table 3-5. The values in any of the four columns can be calculated from the values in the other columns. Percent solid, for example, can be calculated from density in kilograms per cubic meter (kg/m^3) as follows:

$$\% \text{ solid} = \frac{\text{kg/m}^3}{2.65 \times 1,000} \times 100$$

Soil strength also varies with bulk density. Engineers often specify that a soil must be compacted to a certain bulk density so that it can support a road, a building,

Table 3–5   Relation of Bulk Density to Percent Solid and Percent Pore Space for Soils with Particle Density Equal to 2.65 g/cm^3

Bulk density			
g/cm^3	kg/m^3	% solid	% pore space
1.0	1,000	38	62
1.1	1,100	42	58
1.2	1,200	45	55
1.3	1,300	49	51
1.4	1,400	53	47
1.5	1,500	57	43
1.6	1,600	60	40

or some other structure. Excess compaction, however, is unfavorable for plant growth. Root growth and water penetration are likely to be slowed significantly by soils with bulk densities of 1.5 to 1.6 g/cm^3. Root growth is usually stopped by soil layers with bulk densities of 1.7 to 1.9 g/cm^3.

Compact layers resist root penetration because of increased soil strength, reduced oxygen supply, and accumulation of carbon dioxide.

A soil crust is a common form of compact, excessively strong soil. Crusts result from the pounding action of raindrops falling on bare soil—often the soil of a freshly prepared seedbed. Many soil crusts are too strong for seedlings to break through. The seedlings are deflected downward again and die.

SOIL CONSISTENCE

The cohesiveness holding soil particles together in clumps or chunks is called *soil consistence.* Depending on water content, consistence may be expressed in terms of hardness, firmness, plasticity, or stickiness. A complete description of soil consistence requires that the soil be tested in the dry, moist, and wet conditions so it can be evaluated on all of these scales. The scales usually are used on a qualitative or semiquantitative basis rather than fully quantitatively.

A dry soil may be loose, soft, slightly hard, hard, very hard, or extremely hard. These terms not only describe how hard it is to break a small clod, they also indicate how much resistance there is to root penetration and how much power is required to cultivate or excavate the soil. In general, increasing the clay content of a soil will make it harder because clay particles stick to each other and to sand and silt particles as well. Other factors influencing hardness include type of clay, packing of the soil particles, and several chemical and biological relationships.

The consistence of moist soil is described as loose, very friable, friable, firm, very firm, or extremely firm. The moist consistence is determined with the soil's pore space about half filled with water. Less pressure is required to fracture a moist soil than is required by the same soil in the dry condition. The moist soil tends to deform under pressure, then break into small aggregates. These aggregates will usually cohere again when pressed together. Friable soils crush readily, but firm soils require distinct pressure to break.

Both plasticity and stickiness are evaluated with the soil nearly saturated with water. *Plasticity* is the ability to take and hold a new shape when pressure is applied and removed. A nonplastic soil will fall apart when confining pressure is removed. Plastic soils may be classed as slightly plastic, plastic, or very plastic depending on the pressure required to deform the wet soil. Increasing plasticity is correlated with increasing clay content.

Stickiness is a measure of the tendency of wet soil to adhere to other objects. Very sandy soils are nonsticky. Increasing clay content makes soils slightly sticky, sticky, or very sticky. Stickiness disappears when the water content drops below field capacity or rises above saturation. Soils high in clay are sticky and plastic over a wider range of water content and exhibit stronger degrees of stickiness and plasticity than do sandy soils. Some soils are so sticky and plastic that they prevent machinery and vehicles from working in them during wet periods. Clay soils may stick to tires until wheels will no longer turn.

SOIL COLOR

Color is one of the most noticeable characteristics of a soil. Color is important because it is related to organic-matter content, climate, soil drainage, and soil mineralogy.

The natural color of most minerals is white or light gray though there are some blacks, reds, and other colors. The A2 horizon may be very nearly the color of its constituent minerals, but the colors in the remainder of the solum are strongly influenced by humus and by iron compounds. These two materials coat the soil particles and determine the observable color much like a coat of paint determines the color of whatever it covers. Organic materials become very finely divided and almost black in color by the time they have been subjected to microbial action and converted into humus. Humus coats the soil particles so thoroughly that only about 5 percent organic matter is required to give the soil a black or nearly black color.

Iron compounds appear as coatings or stains on the surfaces of mineral particles. The iron does not have to come from external sources—it is present as a component of some of the unweathered minerals. Most of these minerals are dark-colored, but the iron inside these mineral structures does not show its own color until it is released by weathering action. After weathering, the iron exists in one of the oxide forms shown in Table 3-6. Like humus, these materials are finely divided and are efficient coloring agents. The amount of water associated with the hydrated ferric oxide is variable and is therefore indicated by an x. The least-hydrated forms are nearer the reddish color of hematite than the more hydrated forms. A species of limonite called *goethite* is sometimes represented as $2Fe_2O_3 \cdot 3H_2O$.

The particular iron oxide present in a soil depends on the amount of water and oxygen in the soil. Well-drained soils are normally well-enough aerated for the iron to be in the oxidized or ferric state. In somewhat wetter conditions the iron will still

Table 3–6   Iron Compounds that Influence Soil Color

Compound (mineral)	Chemical formula	Color
Ferric oxide (hematite)	Fe_2O_3	Red
Hydrated ferric oxide (limonite)	$Fe_2O_3 \cdot xH_2O$	Yellowish brown
Ferrous oxide	FeO	Bluish gray

be in the ferric state but in the hydrated form. In very wet soils there is a deficiency of oxygen, and the iron is in the ferrous state.

In cool climates, such as those prevailing in the northern United States, humus normally dominates the color of the A1 horizons. If the humus content is low, the color of the A1 is a shade of grayish brown. If it is high (more than about 5 percent), the color is probably black when moist and dark gray or dark-grayish brown when dry. The humus content is usually low in warm climates because of rapid decomposition; the black colors occur only in wet soils (or in a few other special conditions). Enough iron is in evidence to impart a somewhat reddish or yellowish-brown color to most of the soils in warm climates.

B2 horizons generally contain less humus than the A1 horizons of the same soils and are therefore more likely to show the influence of iron coloration. A mixed color is quite common in the B2 horizons of well-drained soils of the northern United States wherein specks of black humus and red iron produce a brown color.

The iron compounds in wetter soils accumulate in spots called *mottles* that consist of spots of one color on a background of another color. Mottled conditions exist where the oxidation state varies and both ferrous and ferric iron are present. The relative solubilities of these two forms of iron help concentrate the iron in rust mottles. The iron moves in the ferrous form because it is many times as soluble as the ferric form. The lower solubility of the ferric iron results in a precipitate where the dissolved iron encounters oxygen. Gray mottles form where the iron is removed leaving the base minerals and organic matter to control the color. The precipitated iron forms rust mottles that are commonly a few millimeters across but sometimes develop into continuous layers in the soil. A water table at a constant depth may form a rust layer, but a fluctuating water table will distribute mottles through much of the solum.

Rust mottles are common in imperfectly drained soils but not in very poorly drained soils. Oxygen is scarce in waterlogged soils, and iron is present in the ferrous form. The medium-to-dark-gray color of ferrous iron is often tinted slightly bluish or greenish. Such waterlogged material is sometimes called "blue clay" in common parlance even though some of it is nearly devoid of clay.

The shades of brown, red, yellow, gray, and black already mentioned include by far the majority of the colors found in soils. A few other colors such as tinges of olive and purple also occur but are unusual. Most of the unusual colors and some of the usual colors are inherited from the parent material rather than resulting from processes of soil formation.

The broad climatic areas in the world are clearly shown in the soil colors. Allowance must be made for the effects of specific parent materials, poor soil drainage on flatlands, vegetation, and age of soils. What remains is still a well-defined climatic pattern. The soil colors become progressively lighter in drier areas and redder in warmer climates.

Arid regions such as much of the western United States have soils with gray, grayish-brown, and reddish-brown colors. The arid soils are broadly classified as Aridisols (systems of classifying soils are explained in Chapter 16). The colors are strongly influenced by the natural color of the minerals because the contents of humus and iron oxides are low in arid climates. The central part of the United States has more precipitation, more humus in the soils, and darker soil colors. This trend continues into the humid grasslands of Minnesota and Iowa, where the dominant topsoil colors are nearly black and the soils are classified as Mollisols.

Forested soils generally occur in climates sufficiently humid that color changes caused by precipitation differences are small but temperature differences remain important. Many forested soils are classified as either Alfisols or Ultisols in soil taxonomy. The same soils were called Gray-Brown Podzolics or Red-Yellow Podzolics in the 1938 System of Soil Classification. The Gray-Brown Podzolics were named for the dominant soil colors in cool-humid regions such as the northeastern United States. Alfisols and Ultisols contain A2 horizons that are often very nearly the basic light-gray color of the soil minerals. Usually there is a very high concentration of quartz sand in the A horizons of Ultisols. Quartz generally has a milky white color, but this may be tinted by other minerals and by small amounts of iron oxides and humus even in the A2 horizons. The red and yellow colors are actually better described as reddish brown and yellowish brown in most of these soils. They are generally brightest in the B horizons. There is a tendency for the colors to be redder in the higher, better-drained positions and yellower in the wetter topographic positions, but this relation is far from perfect.

Soils of humid tropical areas can be considered as an extension of the Ultisols into warmer climates. The higher temperatures accelerate the rate of weathering and thereby increase the content of iron oxides in the soils (also of aluminum and titanium oxides, but these have neutral grayish colors and are therefore not so noticeable). Organic-matter contents are as low in many tropical soils as in the Aridisols (generally less than 1 percent) but for a different reason. Little organic matter is produced in the deserts. Organic matter production is high in the tropics, but so is the rate of decomposition of both organic matter and soil minerals. Iron and aluminum oxides accumulate in the soils and they are therefore classified as Oxisols. Oxisols have the reddest colors of any soils in the world, and these red colors often are relatively uniform throughout their deep profiles.

The preceding discussion of soil colors has necessarily been very general, and there are important exceptions to these generalities. For example, Argentina is noted for its large areas of extremely flat land known as *pampas*. These areas have poorly drained soils with gray colors even where the climate would lead one to anticipate Ultisols. Another exception occurs in certain soils that contain large amounts of calcium carbonate. Much of the iron in these soils is dark-gray ferrous carbonate.

This plus a small amount of humus produces black soils in the midst of red and yellow soils.

Describing Soil Color

There was a time when soil scientists invented colorful terms such as "tawny brown" or "mousy gray" to describe soil colors. The results were not very precise until an orderly system utilizing Munsell color notations was introduced. The Munsell system describes colors in terms of three variables—hue, value, and chroma.

Hue   Hue refers to the dominant wavelength of light reflected by an object and is defined in terms of five cardinal colors and mixtures of these colors: blue, green, yellow, red, and purple. These can be arranged in a circle with gray in the center, as shown in Figure 3-13. Gradations in hue are noted with numbers followed by letter abbreviations of one or two of the cardinal colors. The subdivisions shown in Figure 3-13 are those commonly used in describing soil colors.

Chroma   Chroma is represented by radial distance from the center of Figure 3-13. A pure color reflecting only one wavelength of light would have a chroma of about 20. Actual soil colors can be duplicated by mixing different amounts of pure hue with neutral gray colors. Grays have zero chroma and are designated N for neutral because they have no hue. That is, they reflect equal percentages of each wavelength of light. Mixing a pigment of some particular hue with gray causes its wavelength to be reflected more strongly than others. The more pigment used, the purer the reflected light becomes. Chroma is a measure of the degree of color

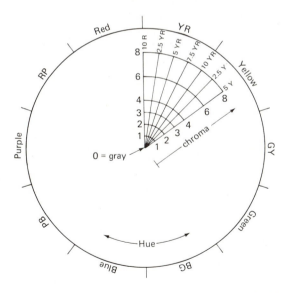

Figure 3-13   The hues used to define colors in the Munsell color system. The graduated sector shows the hues and chromas most commonly used in describing soil colors.

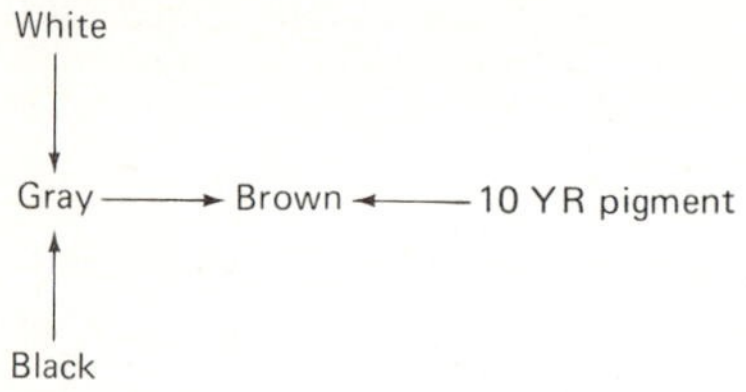

White

Gray ————→ Brown ◄———— 10 Y R pigment

Figure 3-14 A brown color can be formed by mixing black, white, and 10YR pigment.

Black

saturation or purity. Soil chromas seldom exceed 8 on a scale from 0 to 20. High chromas give an impression of being "bright" or "colorful."

Value   Value is a measure of the lightness or darkness of the color. It is given a numerical value equal to the square root of the percentage of incident light reflected by the sample being described. It may be helpful to consider a color as being formed by mixing black and white to produce the proper gray tone and then adding pigment of the appropriate hue, as indicated in Figure 3-14. The amount of white needed can be considered as a measure of the color value. The value scale ranges from 0 to 10, with 0 representing pure black and 10 representing pure white.

A Munsell color notation combines hue, value, and chroma in a standard symbol such as 10YR 5/3 (hue, value/chroma). The notation is determined by comparing the soil color with standard color "chips" arranged in a color book. Each page in the book represents one hue with chromas increasing from the left to the right side of the page and values increasing from the bottom to the top of the page. The 10YR hue is the most common one in soils of temperate regions.

Color names are assigned to the various colors that occur in soils and defined in terms of the appropriate Munsell designations. The usual practice in describing soils is to give both the color name and the Munsell notation for each significantly different color in the soil. Thus, a mottled B2 horizon might be described as having a brown (10YR 5/3) color with yellowish-brown (10YR 5/6) mottles.

SOIL TEMPERATURE

Temperatures vary from day to day and season to season, but this does not detract from the significance of temperature as a soil property. The influence of soil temperature on plant growth is considerable. This influence can be observed every spring in climates where the winters are cold. There is a period of several days or a few weeks of warm weather before most plants show much evidence of growth. The lawn, for example, does not have to be mowed during this period. It is not growing much because the soil is cold even though the air is warm. It takes time to warm the soil in the spring and to cool it in the fall.

Several factors influence soil temperature including the angle of the sun's rays, any cover on the soil, the color of the soil, the water content, and the depth and time of measurement. The angle of the sun's rays relates not only to the latitude and climate of the area but also to the direction the slope faces. The more squarely the sun's rays strike the soil surface, the warmer its temperature will be. Tropical areas are warm because the sun crosses directly overhead. Polar regions are cold because the rays always reach them obliquely. Soils on south-facing slopes in the Northern Hemisphere and north-facing slopes in the Southern Hemisphere are several degrees warmer than the soils that face away from the sun. The warmth of slopes facing the sun is a significant factor to be considered in choosing sites for temperature-sensitive crops such as many fruits, vegetables, and flowers.

Growing plants, plant residues, or other cover on the soil provide shade and insulation that reduce the warming effect of the sun. The insulation provided also reduces radiation cooling at night so that temperature variations are reduced in both directions. Soil color also influences temperature extremes because dark colors tend to absorb and release heat more rapidly than light colors.

The usual rule that dark-colored objects absorb more heat and become warmer than light-colored objects does not necessarily hold in soils. Darker-colored soils are generally higher in organic matter than lighter-colored soils. The higher-organic-matter contents usually occur in wet soils. The wetness may keep the dark-colored soils cooler than the others.

Wet soils tend to be cooler than dry soils, especially in the spring. Evaporation helps cool the moist soil by dissipating heat. And, the high heat capacity of water reduces the temperature change from the heat that is absorbed by the soil. Finally, when the soil becomes warm it dries rapidly because high temperatures accelerate both evaporation and transpiration.

The upper layers of soil help to insulate the lower layers so temperature fluctuations decrease with depth (Figure 3-15). Below 6 m the yearly variation in tempera-

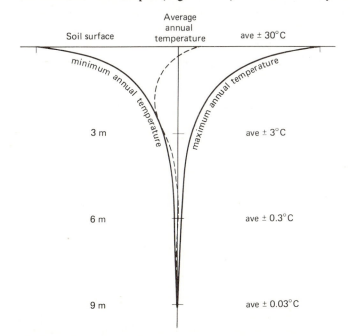

Figure 3-15 Soil temperature variability in a temperate-region soil. The dashed line shows the soil temperature at various depths on a warm spring day.

ture is less than 1 degree. The temperature in caverns remains constant throughout the year because of such insulation. Slow heat penetration also creates a time lag that increases with depth. The maximum temperature at a depth below 3 m occurs months after the maximum temperature at the surface.

People can influence soil temperature in at least two ways. One way is to drain a wet soil to eliminate excess cooling. The other way is to vary the amount of cover (such as plant residues) left on the soil surface. One argument against leaving residues on the surface to control erosion is that the residues delay warming up of the soil in the spring.

Average Soil Temperature

The temperature in a cavern or at the bottom of a hole 6 m or more deep is the average soil temperature for the site regardless of what time of year it is measured. The average soil temperature is normally between 1 and 5°C higher than the average air temperature above the soil because soil absorbs the sun's heat much more readily than air does. The 1 to 5°C variation depends greatly on the amount of cooling resulting from evaporation and transpiration and the amount of cover present on the soil at various times. For example, a blanket of snow insulates the soil and keeps it from getting as cold as bare soil in the same vicinity. On the other hand, plant residues covering the soil in the spring intercept the sun's rays and keep the soil from warming up as fast as it otherwise would.

Permafrost naturally occurs in polar areas where average annual temperatures are below 0°C. The soil may thaw each summer, but the substratum below does not. It is possible to grow some crops above a permafrost layer because the days are long and the air temperature becomes reasonably warm. Building construction can be a problem, though. The continual downward penetration of heat from the building eventually melts the permafrost and the building settles unevenly.

Microbial Activity

Microbial activity increases with increasing temperature within the limits of cold soil at one end of the scale and hot dry soil at the other end. Practically no biological activity takes place below freezing, so the temperature range of interest begins at 0°C. The upper limit is less definite and depends on the heat tolerance of the microbes and the water content of the soil. Temperatures of some surface soils reach 50°C or above during the summer, but the higher temperatures usually are associated with dry soil.

Microbial response to increased soil temperature generally follows a logarithmic curve such as the solid line shown in Figure 3-16 up to some maximum temperature. Some organisms reach their peak activity at about 35°C. Good examples of this group are the nitrifying bacteria that oxidize ammonium ions to nitrates. The activities of other microbes peak at different temperatures. Thompson (1950) showed that carbon dioxide evolution was greater and decomposition of organic phosphorus compounds was more rapid at 40°C than at lower temperatures.

While a higher rate of organic-matter decomposition is a good index of microbial activity, it may not indicate maximum numbers of microorganisms. Some studies have indicated that maximum numbers of organisms occur between 30 and 35°C. The higher rates of decomposition above 35°C are associated with higher rates

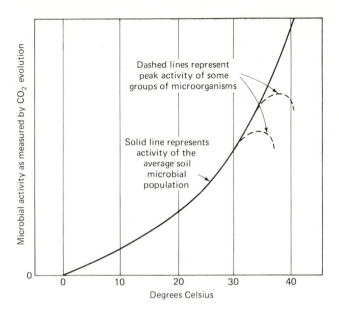

Figure 3-16 A schematic diagram showing the relation of CO_2 evolution to soil temperature. The peak in the biological range is about 50°C. CO_2 evolution continues to increase to a peak at about 70°C (Thompson, 1950), but the decomposition is probably more chemical than biological at temperatures above 50°C.

of respiration. Carbon dioxide evolution continues to increase at still higher temperatures up to about 70°C, but the decomposition is probably more chemical than biological at temperatures above 50°C.

Slow microbial activity at soil temperatures below 10°C can cause plant nutrient deficiencies in the early spring and sometimes in late fall. This is particularly true of nitrogen because practically all the available nitrogen comes from the decomposition of organic matter (unless nitrogen fertilizer is applied). Phosphorus and sulfur availability are similarly affected but to a lesser extent.

Seed Germination

The most favorable temperature for seed germination depends on the species of plant. Oats, alfalfa, and bromegrass germinate when the daily soil temperature in the top 5 cm averages about 10°C. Corn germinates when the average daily temperature of the top 5 cm of soil is about 15°C. Cotton needs a seedbed temperature of 20°C or a little above.

Soil temperature is of particular concern in growing corn. If corn is planted too soon, the seed may rot before germinating or the young plants may suffer frost damage in March in the southern United States or in May in the northern United States. Nevertheless, it is generally desirable to plant the corn early to escape the hot and dry part of the summer in the southern states and to escape frost damage in the early fall in the northern states.

Two factors are especially important in determining when the soil temperature will be warm enough for germination. One is the air temperature. Air temperature can be observed but not controlled by the farmer. The other factor, drainage, can sometimes be controlled. Water retards the warming of wet soils because it has a heat capacity of 1.0 cal/g, whereas that of oven-dry soil is only about 0.2 cal/g. Some retardation from excess water is almost certain to occur in any soil high in clay because of the high water-holding capacity of clay.

Temperatures at later stages of growth may be either too warm or too cool for optimum growth. Allmaras, Burrows, and Larson (1964) found that the optimum temperature for corn roots was near 27°C at a depth of 10 cm. The more the temperature deviated from the optimum, the lower the yield was. Rykbost et al. (1975) piped warm water through soil to raise its temperature and obtained yield increases ranging from 19 to 54 percent for several different agronomic and vegetable crops. The greatest effects occurred in the spring. Figure 3-17 illustrates some of their results.

Not only the size but also the shape of a root system can be influenced by soil temperature. Mosher and Miller (1971) found that corn radicles grew vertically downward in warm soil (32°C at 3 cm depth) but grew horizontally in cool soil.

Root Growth

Like other biological activities, root growth is slow at low temperatures. The relation is much like that for microbial growth including the effect of excess wetness on soil temperatures. Low temperatures and slow growth can be critical for seedlings because they cannot extract nutrients from any sizable volume of soil until their root systems develop. This handicap to a fast start can be partially overcome by the use of "starter" fertilizer. Starter fertilizer must be carefully placed to be within reach of the early plant roots and yet not so close as to give a salt concentration high enough to be toxic to either the seed or the seedling (Chapter 9).

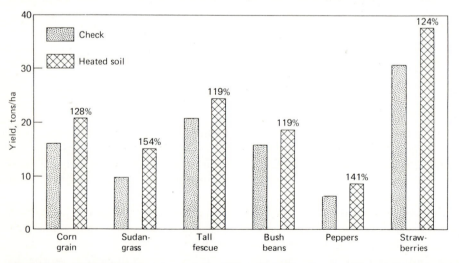

Figure 3-17 Yields of six crops grown in the Willamette Valley, Oregon with and without additional heat to raise the soil temperature from 0 to 100 cm an average of about 10°C. (*Data of Rykbost, Boersma, Mack, and Schmisseur, 1975.*)

Effect of Temperature on Soil Properties

Several relations between temperature and other soil properties have already been explained and need only be summarized here. Microbial action and mineral weathering are both more rapid in warmer climates. Soils of warmer climates contain less organic matter and more iron and aluminum oxides than those of cooler climates, other factors being equal. The accumulation of iron oxides causes brighter colors in the soils of warmer climates (reddish and yellowish browns rather than the grayish browns and blacks of the soils of cooler climates).

REFERENCES

Allmaras, R. R., W. C. Burrows, and W. E. Larson, 1964, Early Growth of Corn as Affected by Soil Temperature, *Soil Sci. Soc. Am. Proc.* **28**:271–275.

Anderson, M. A., and G. M. Browning, 1949, Some Physical and Chemical Properties of Six Virgin and Six Cultivated Iowa Soils, *Soil Sci. Soc. Am. Proc.* **14**:370–374.

Anderson, W. B., and W. D. Kemper, 1964, Corn Growth as Affected by Aggregate Stability, Soil Temperature, and Soil Moisture, *Agron. J.* **56**:453–456.

Baier, R. E., E. G. Shafrin, and W. A. Zisman, 1968, Adhesion: Mechanisms that Assist or Impede It, *Science* **162**:1360–1368.

Barley, K. P., 1963, Influence of Soil Strength on Growth of Roots, *Soil Sci.* **96**:175–180.

Bouyoucos, G. J., 1951, A Recalibration of the Hydrometer Method of Making Mechanical Analysis of Soils, *Agron. J.* **43**:434–438.

Cagauan, B., and B. Uehara, 1965, Soil Anisotropy and Its Relation to Aggregate Stability, *Soil Sci. Soc. Am. Proc.* **29**:198–200.

Chang, C. W., and J. U. Anderson, 1968, Flocculation of Clays and Soils by Organic Compounds, *Soil Sci. Soc. Am. Proc.* **32**:23–27.

DeVries, J., 1969, *In Situ* Determination of Physical Properties of the Surface Layer of Field Soils, *Soil Sci. Soc. Am. Proc.* **33**:349–353.

Downs, S. C., T. M. McCalla, and F. A. Haskins, 1955, *Stachybotrys atra,* an Effective Aggregator of Peorian Loess, *Soil Sci. Soc. Am. Proc.* **19**:179–181.

Edwards, A. P., and J. M. Bremner, 1967, Microaggregate Formation in Soils, *J. Soil Sci.* **18**:64–73.

Edwards, W. M., J. B. Fehrenbacher, and J. P. Vavra, 1964, The Effect of Discrete Ped Density on Corn Root Penetration in a Planosol, *Soil Sci. Soc. Am. Proc.* **28**:560–564.

Genrich, D. A., and J. M. Bremner, 1972, A Reevaluation of the Ultrasonic-Vibration Method of Dispersing Soils, *Soil Sci. Soc. Am. Proc.* **36**:944–947.

Grable, A. R., and E. G. Siemer, 1968, Effects of Bulk Density, Aggregate Size, and Soil Water Suction on Oxygen Diffusion, Redox Potentials, and Elongation of Corn Roots, *Soil Sci. Soc. Am. Proc.* **32**:180–186.

Harris, R. F., G. Chesters, O. N. Allen, and O. J. Attoe, 1964, Mechanisms Involved in Soil Aggregate Stabilization by Fungi and Bacteria, *Soil Sci. Soc. Am. Proc.* **28**:529–532.

Hedrick, R. M., and D. T. Mowry, 1952, Effect of Synthetic Polyelectrolytes on Aggregation, Aeration, and Water Relationships of Soils, *Soil Sci.* **73**:427–441.

Holder, C. B., and K. W. Brown, 1974, Evaluation of Simulated Seedling Emergence through Rainfall-Induced Soil Crusts, *Soil Sci. Soc. Am. Proc.* **38**:705–710.

Jackson, M. L., 1963, Aluminum Bonding in Soils: A Unifying Principle in Soil Science, *Soil Sci. Soc. Am. Proc.* **27**:1–10.

Jackson, R. D., 1963, Porosity and Soil-water Diffusivity Relations, *Soil Sci. Soc. Am. Proc.* **27**:123–126.

King, F. H., 1914, *Physics of Agriculture,* 6th ed., published by Mrs. F. H. King, Madison, Wis.

Martin, W. P., J. P. Martin, and J. D. DeMent, 1954, Soil Aggregation: Microbiological and Synthetic Conditioner Effects, *Proc. Soil Microbiological Conf.,* Purdue Univ.

Moody, J. E., J. N. Jones, Jr., and J. H. Lillard, 1963, Influence of Straw Mulch on Soil Moisture, Soil Temperature, and the Growth of Corn, *Soil Sci. Soc. Am. Proc.* **27**: 700–703.

Mosher, P. N., and M. H. Miller, 1971, Soil Temperature and the Geotropic Response of Corn Roots, *1970 Progress Report,* Dept. of Soil Science, Univ. of Guelph, Guelph, Ontario, Canada, pp. 49–50.

Nelson, W. E., G. S. Rahi, and L. Z. Reeves, 1975, Yield Potential of Soybean as Related to Soil Compaction Induced by Farm Traffic, *Agron. J.* **67**:769–772.

Olmstead, L. B., L. T. Alexander, and H. E. Middleton, 1930, A Pipette Method of Mechanical Analysis of Soils Based on Improved Dispersion Procedure, *USDA Tech. Bull.* 170.

Patt, J., D. Carmeli, and I. Zafrir, 1966, Influence of Soil Physical Conditions on Root Development and on Productivity of Citrus Trees, *Soil Sci.* **102**:82–84.

Russell, E. W., 1950, *Soil Conditions and Plant Growth,* 8th ed., Longmans, New York.

Rykbost, K. A., L. Boersma, H. J. Mack, and W. E. Schmisseur, 1975, Yield Response to Soil Warming, *Agron. J.* **67**:733–745.

Scott, T. W., and A. E. Erickson, 1964, Effect of Aeration and Mechanical Impedance on the Root Development of Alfalfa, Sugar Beets and Tomatoes, *Agron. J.* **56**:575–576.

Swanson, C. L. W., and J. B. Peterson, 1942, The Use of Micrometric and Other Methods for the Evaluation of Soil Structure, *Soil Sci.* **53**:173–185.

Taylor, H. M., G. M. Roberson, and J. J. Parker, Jr., 1966, Soil Strength—Root Penetration Relations for Medium- to Coarse-textured Soil Materials, *Soil Sci.* **102**:18–22.

Thompson, L. M., 1950, *The Mineralization of Organic Phosphorus, Nitrogen, and Carbon in Virgin and Cultivated Soils,* doctoral dissertation, Iowa State University, Ames.

Ulrich, Rudolph, 1949, *Some Physical and Chemical Properties of Planosol and Wiesenboden Soil Series as Related to Loess Thickness and Distribution,* doctoral dissertation, Iowa State University, Ames.

Voorhees, W. B., D. A. Farrell, and W. E. Larson, 1975, Soil Strength and Aeration Effects on Root Elongation, *Soil Sci. Soc. Am. Proc.* **39**:948–953.

White, E. M., 1966, Subsoil Structure Genesis: Theoretical Considerations, *Soil Sci.* **101**: 135–141.

Williamson, R. E., 1964, The Effect of Root Aeration on Plant Growth, *Soil Sci. Soc. Am. Proc.* **28**:86–90.

Willis, W. O., 1955, Freezing and Thawing, and Wetting and Drying in Soils Treated with Organic Chemicals, *Soil Sci. Soc. Am. Proc.* **19**:263–267.

Wittsell, L. E., and J. A. Hobbs, 1965, Soil Compaction Effects on Field Plant Growth, *Agron. J.* **57**:534–537.

Chapter 4

Soil Water

The presence of adequate water in the soil is vital to plant growth, not only because plants need water for their physiological processes but also because the water contains nutrients in solution. Water comes from rain and other precipitation, but rain would do plants little good if the soil could not store water for plant use between rains. The capacity of a soil to store water depends upon its depth, texture, structure, and other properties.

The amount of water that a soil can store in a form available for plant use is known as the *available water-holding capacity* of the soil. Most soils can store between 5 and 25 cm of available water in their sola (measured as rainfall before it enters the soil). The difference between a high and a low available water-holding capacity is often decisive in determining crop yield (or any other index of plant growth). Soil fertility also can and does limit plant growth but is more amenable to treatment than is the water-holding capacity. Management of the available soil water becomes increasingly important when higher yields are sought with large fertilizer applications.

WATER MOLECULES

The two hydrogens in a water molecule are not located together as the formula H_2O might imply nor are they on opposite sides of the oxygen as the form HOH suggests. Rather, the outer surface of the oxygen has four pairs of electrons centered

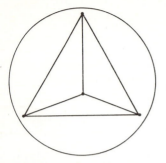

Figure 4-1 The four points of a tetrahedron inscribed in a sphere are 105° apart from each other when measured around the surface of the sphere. The hydrogens in water molecules give positive charges to two of these points whereas the other two points have net negative charges.

near the four points of a tetrahedron (Figure 4-1). The hydrogen atoms are covalent bonded to two of these tetrahedral points and give net positive charge to these two points. The other two tetrahedral points have net negative charge.

The positive and negative charge sites cause water molecules to cohere to each other and to adhere to many other materials. This cohesive tendency lets water molecules join together in chains and sheets that give water a higher viscosity and greater surface tension than it would otherwise have. The adhesive tendency makes it possible for soil particles to hold layers of water next to their surfaces. Also, small amounts of water occupying narrow spaces between particles can adhere to both sides and help hold the particles together.

SUBDIVISIONS OF SOIL WATER

Figure 4-2 illustrates some significant subdivisions in the film of water of the maximum thickness likely to occur next to a soil particle. The differences shown actually pertain to how tightly the water is held and not to any change in the nature of the water. The boundaries between the different types of water are chosen to reflect whether the water can be removed by downward percolation, plant growth, or evaporation.

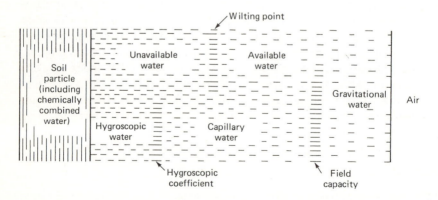

Figure 4-2 A schematic diagram showing the forms of soil water present shortly after a rain. The closer-spaced lines near the particle indicate greater tension. The soil particle is not to scale; it should be thicker than the water film.

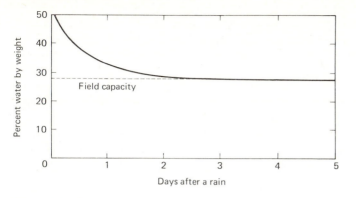

Figure 4-3 The percent water contained in a loam A1 horizon of a well-drained soil during the first few days after a rain if water is lost only by drainage.

Field Capacity

The moisture content of the soil when downward movement of water has virtually ceased is called *field capacity*. This condition usually exists in a well-drained soil about two or three days after a rain (Figure 4-3). Field capacity is quantitatively evaluated as the water present expressed as a percentage of the oven-dry weight of the soil. The water content of the soil remains at field capacity until water is removed by plant growth or evaporation or until the water content is temporarily raised by another rain or other source of water. The concept can be demonstrated by placing dry soil in a container and adding water to one side of the soil (a glass tube is ideal for observing what happens). Water movement from the moist to the dry soil is rapid at first but soon slows and eventually stops with the moist part near field capacity.

Field capacity is closely related to soil texture and is influenced by the organic-matter content, types of minerals present, and soil structure. The characteristics of adjoining layers are also important because more water will move downward if the layer below has a greater attraction for water than the layer above. Clay layers with small pores delay downward water movement by their low permeabilities, but they also exert strong tensions that pull water from adjoining loamy layers. On the other hand, a loamy layer above a sandy layer will hold more water at field capacity than will the same texture in a uniform soil.

The effects of structure and of the characteristics of adjoining layers make field capacity difficult to duplicate in the laboratory. The usual approach is to wet a soil sample and then apply a pressure of ⅓ atm to expel the excess water. The force required is illustrated more dramatically by an older approximation of field capacity known as the *moisture equivalent*. The moisture equivalent is determined by excessively wetting soil samples and centrifuging them at a force of 1,000 times gravity. The centrifuge spins off the excess water and leaves the soil near field capacity.

Wilting Point

The wilting point is the lower end of the soil water content useful to plants. About half of the water contained in a soil at field capacity is held so tightly that it is useless to plants. The wilting point is reached when the rate the plant absorbs water from the soil becomes so slow that the plant wilts and cannot recover. The plant may still obtain some water but not fast enough to meet its needs. Plants sometimes wilt temporarily at higher water contents on hot, dry days but recover at night or when water is again supplied. The wilting point is taken as the water content of the soil when the plants lose the ability to recover. It, like field capacity, is expressed as a percentage of the oven-dry weight of the soil.

The wilting point varies slightly according to the particular plant being grown and the atmospheric conditions that influence the rate of transpiration. This variation is usually not large in terms of percent water present, and it can be minimized by specifying the conditions for making the test. Dwarf sunflowers are usually used as the test plant. Under the usual conditions the upper part of the soil dries out first, but the plant does not permanently wilt until most of its root zone has been dried to the wilting point.

Determining the wilting point by growing plants is a slow process. The centrifuge approach used for the moisture equivalent will not spin off enough water to reach the wilting point. Other laboratory procedures have therefore been developed. One standard method of simulating a wide range of conditions including the wilting point has been the application of air pressure to one side of a soil sample to force water out the other side. The wilting point is taken as the water held against 15 atm pressure. A newer procedure proposed by Nelson (1975) uses hectorite clay to absorb water from a soil sample and dry it to the wilting point.

Hygroscopic Coefficient

The hygroscopic coefficient is the percentage of water remaining in an air-dry soil. Evaporation can remove more water from the soil than can growing plants. There is, however, a small amount of water that cannot be removed from the soil under field conditions, not even by prolonged exposure to the air without any rainfall. For precise determination of the hygroscopic coefficient, the soil is exposed to an atmosphere with a 98.2 percent relative humidity during the air-drying process. Under normal field conditions the soil will always contain at least as much water as the hygroscopic coefficient.

Oven-dry Soil

Oven-dry soil is the basic reference weight for all soil moisture calculations. It can be determined by drying the soil in an oven at a temperature between 105 and 110°C. Temperatures a few degrees above the boiling point of water hasten the drying process and allow for the increased boiling point of water caused by the presence of dissolved salts. The sample is left in the oven for several hours or longer, depending on its size. There must be enough air circulation through the oven to allow the water vapor to escape.

Forms of Soil Water

Even oven-dry soil will lose some water if heated to a still higher temperature. Most of this water probably comes from hydroxyl groups contained in the mineral and organic solids. Such *chemically combined* water is considered to be a part of the soil solids rather than of the liquid phase. Removing the chemically combined water by heating to high temperatures alters the nature of the mineral and organic materials present.

The water contents at the field capacity, wilting point, and the hygroscopic coefficient can be used to define several forms of soil water as follows (and as shown in Figure 4-2):

Hygroscopic water = hygroscopic coefficient minus oven-dry weight
Capillary water = field capacity minus hygroscopic coefficient
Unavailable water = wilting point minus oven-dry weight
Available water = field capacity minus wilting point
Gravitational water (if present) = percent water present minus field capacity

The names of the various forms of soil water suggest some of their characteristics. Gravitational water is water that temporarily occupies aeration pore space but will drain down to a lower depth if there is drier soil below. Gravity is the most obvious factor in this movement even though the soil moisture tension of the drier soil is usually the stronger factor. Capillary water is held in the small pores or "capillaries" in much the same manner that water can be held in a thin glass tube. Hygroscopic water not only is retained by air-dried soil, it can be absorbed again from the air by oven-dried soil. Available water is the water that can be utilized by plants, and unavailable water is the portion that is held too tightly to sustain plant life. It should be noted that available and unavailable water on the one hand and capillary and hygroscopic water on the other hand are two different ways of subdividing the same soil water.

PERCENT WATER CALCULATIONS

Amounts of soil water are most commonly expressed as weight percentages. *The reference weight for calculating percent water is always the oven-dry weight of the soil.* A constant base is essential for calculations, and the oven-dry weight can be determined more accurately and consistently than any of the moist-soil weights. The weight of the soil solids (oven-dry weight) is taken as 100 percent, and the weight of water present is considered as an additional percentage. Example:

Weight of moist soil	75 g
Weight of oven-dry soil	60 g
Weight of water present	15 g

$$\text{Percent water present} = \frac{15 \text{ g}}{60 \text{ g}} \times 100 = 25\%$$

Division by the oven-dry weight of 60 g causes the percent water to be 25 percent rather than 20 percent (the result of dividing by 75 g in this example). One consequence of this method of expression is that calculations of oven-dry weight from moist weight and percent water data must also be by division. If the oven-dry weight in the preceding example were unknown, it could be calculated from the moist weight and percent water:

$$75 \text{ g} \times \frac{100\%}{100\% + 25\%} = \frac{75 \text{ g}}{1.25} = 60 \text{ g}$$

This procedure works for any water content at which both the soil weight and the percent water present are known.

Each form of soil water can be calculated directly from the appropriate soil weights. The calculation takes this form:

$$\text{Percent of any type of water} = \frac{\text{weight 1} - \text{weight 2}}{\text{oven-dry weight}} \times 100$$

where weight 1 represents the weight of the soil including the specified type of water and weight 2 represents the weight of the soil after that type of water has been removed. *The denominator is always the oven-dry weight.*

A soil having the following weights will serve as an example of some simple soil moisture calculations:

Weight at field capacity	26.0 g
Weight at wilting point	23.2 g
Air-dry weight	21.6 g
Oven-dry weight	20.0 g

From these data it can be calculated that:

$$\text{Percent water at field capacity} = \frac{26.0 \text{ g} - 20.0 \text{ g}}{20.0 \text{ g}} \times 100 = 30\%$$

$$\text{Percent water at wilting point} = \frac{23.2 \text{ g} - 20.0 \text{ g}}{20.0 \text{ g}} \times 100 = 16\%$$

$$\text{Percent available water} = \frac{26.0 \text{ g} - 23.2 \text{ g}}{20.0 \text{ g}} \times 100 = 14\%$$

The consistent use of the oven-dry weight in the denominator makes the system additive so that:

Available water (14%) + wilting point (16%) = field capacity (30%)

Students often suggest another system involving division by the weight of the sample including the water. This system is not additive and should not be used.

Other water percentages that can be calculated from the above example are hygroscopic water equals 8 percent, unavailable water equals 16 percent, and capillary water equals 22 percent. These percentages would not change if the units were in some other form such as kilograms or tons because the units cancel out in each computation.

SOIL WATER EXPRESSED IN CENTIMETERS

Rainfall amounts are expressed in depth units, and it is frequently convenient to express soil water in a similar manner. The weight percentages calculated in the previous section can be changed to volume and depth percentages by using the bulk density as a conversion factor:

$$\text{Percent water by volume} = \% \text{ water by weight} \times \frac{\text{soil bulk density}}{\text{density of water}}$$

For example, a soil with a bulk density of 1.2 g/cm^3 and a water content of 30 percent by weight contains 36 percent water by volume:

$$30\% \text{ H}_2\text{O by weight} \times \frac{1.2 \text{ g/cm}^3}{1.0 \text{ g/cm}^3} = 36\% \text{ H}_2\text{O by volume}$$

The percent water by depth is the same as the percent water by volume so the above equation serves for both. The reason for this equality is that the water must be distributed across the same area as the soil so the area can be canceled from the volume, leaving only depth. Actually, the water depth is more often expressed as a fraction than as a percentage. The example showing 36 percent water by volume is equivalent to 0.36 cm of water per cm of soil. The total depth of water in a uniform soil can be determined by multiplying the water fraction by the soil depth. Assuming that the example soil is 120 cm deep and that it has a uniform water content throughout:

$$120 \text{ cm of soil} \times 0.36 \text{ cm/cm} = 43.2 \text{ cm of water}$$

The corresponding information for nonuniform soils is obtained by considering each layer separately and adding up the total.

The same procedure applies to all forms of water so that capillary water, hygroscopic water, available water, and unavailable water can be expressed in centimeters as well as total water. In practice, the forms most often expressed in centimeters are the total water and the available water-holding capacity.

SOIL MOISTURE POTENTIAL

Several forces act upon the water in soil with a combined effect known as the *soil moisture potential.* The most important components of the moisture potential are the gravitational potential, capillary potential, and the osmotic potential (Figure 4-4). Any entity such as a plant root that removes water from the soil must overcome

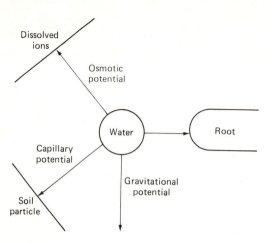

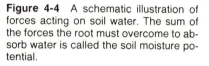

Figure 4-4 A schematic illustration of forces acting on soil water. The sum of the forces the root must overcome to absorb water is called the soil moisture potential.

the soil moisture potential. Removal of water increases the moisture potential until the soil exerts as much force as the root does and removal ceases.

The *gravitational potential* is measured relative to an arbitrary reference level. The value of the gravitational potential can be expressed in distance units as the number of centimeters a particular mass of water is above or below the reference level. This value could be considered as a pressure with positive values representing locations above the reference level. Moisture potentials, however, are taken as negative pressures so their positive values represent locations below the reference level. The soil surface is often taken as the reference level so that the gravitational potentials in the soil profile will have positive values equal to the depths of whatever positions are being considered. If only one depth is being considered, the reference level may be taken at that depth so the gravitational potential will be zero.

The *capillary potential,* sometimes called the *matric potential* or *soil moisture tension,* results from the attraction between the soil solids and water. Hydrogen bonding causes a layer of water molecules to adhere to the oxygen ions that constitute the surfaces of most soil minerals. Additional layers of water are held by the cohesion of water molecules for each other. The inner layers are held rigidly in a structure similar to that of ice, but each succeeding layer is held less tightly than the layers nearer the soil particle. The amount of water that can be held in such films is limited to the thickness where the attraction on the outermost layer is equal to the force of gravity.

Additional capillary water is held where two or more soil particles are close enough together for the water to create a bridge between them. Large pores have their corners filled with water by this bridging effect with surface tension helping to hold the water in place. Small pores are often completely filled with water. Also, certain minerals such as the montmorillonite clay described in Chapter 6 have the capacity to absorb water into their internal structures. Montmorillonite swells as it takes in water between its layers, then shrinks as it dries. Shrinkage produces notable cracks in dry soils containing high percentages of montmorillonite.

The capillary potential at any particular water content and the amount of capillary water a soil can hold are closely related to the surface area of the soil particles. The smaller particles have more surface area relative to their weight and volume; soils high in clay therefore hold more water than those low in clay. Of course, the types of minerals present, the content of organic matter, and the structural arrangement of the soil particles also influence the water-holding capacity. Organic matter and montmorillonite both have large surface areas and high capacities for holding water. Soil structure influences water-holding capacity by controlling the size and shape of pore spaces between particles. The smaller these spaces are, the more tightly water is held in them.

The *osmotic potential* results from the materials dissolved in the soil water. The soil solution is very dilute in most soils so the osmotic potential is commonly very small. But some soils in arid regions contain large concentrations of soluble salts and are called *saline soils*. The osmotic potential in some saline soils is higher than the gravitational and capillary potentials. Removal of water from the soil increases the osmotic concentration in the remaining water and thereby increases the osmotic potential.

Plants have an osmotic potential of their own resulting from dissolved solutes. The solute concentration in a plant root is normally high enough to exceed the soil moisture potential of a moist soil. Water moves into the plant roots in response to the difference in potentials. Both the capillary potential and the osmotic potential increase in the soil next to the roots as the soil becomes drier. The increased soil moisture potential near the roots causes water from other parts of the soil to move toward the roots. Water from a short distance away can thus be drawn to a root when there is a gradient in moisture potential. Effective water movement to roots is limited to about 5 cm or less by dissipation of the moisture potential gradient.

Moisture Potential Units

Moisture potentials can be measured and expressed in several different kinds of units. A parallel can be drawn between moisture potential and suction or negative pressure so that any units suitable for expressing pressure can be used for potentials as well. Atmospheres or bars are the most commonly used units. Other types of units that can be used are heights of a column of water or of mercury or weights per unit area. The variety of units used generates a need for conversion factors to relate various measurements to each other. The following are all equivalent to one atmosphere:

$$
\begin{aligned}
1 \text{ atm} &= 1.013 \text{ bars} \\
&= 1.033 \text{ kg/cm}^2 \\
&= 1{,}033 \text{ cm of water} \\
&= 76 \text{ cm of mercury} \\
&= 33.9 \text{ ft of water} \\
&= 14.7 \text{ lb/in.}^2
\end{aligned}
$$

A possible ambiguity arises because centimeters of water can be used to measure either the amount of water present or the soil moisture potential. The size of the numbers usually identifies which kind of measurement is intended since amounts of water are likely to be a few centimeters or tens of centimeters but soil moisture potentials are usually equal to hundreds or thousands of centimeters of water. Nevertheless, both the possibility of confusion and the large numbers involved when centimeters of water are used make it advisable under most circumstances to express soil moisture potential in atmospheres or bars.

As mentioned earlier, the subdivisions of soil water illustrated in Figure 4-2 are based on how tightly the water is held. These subdivisions are, therefore, closely related to the soil moisture potential and frequently are expressed in atmospheres. The following approximate values are often used:

$$
\begin{aligned}
\text{Field capacity} &= \frac{1}{3} \text{ atm} \\
\text{Wilting point} &= 15 \text{ atm} \\
\text{Hygroscopic coefficient} &= 30 \text{ atm}
\end{aligned}
$$

The precise soil moisture potentials for the above subdivisions depend on circumstances that vary from one site to another and one time to another. For example, field capacity can be near $\frac{1}{10}$ atm if the soil has a layer that restricts water percolation, or it can be near ½ atm in deep, uniform well-drained soils with small pores. The 15-atm figure for the wilting point must be considered to be plus or minus a few atmospheres depending on the plant and its environmental conditions (light, temperature, and humidity). The variability of field conditions should be recognized even though the standardized values of ⅓, 15, and 30 atm are usually used for field capacity, wilting point, and the hygroscopic coefficient.

MEASURING SOIL WATER

It is often important to measure either the amount of water present in a soil or the soil moisture potential. Several different approaches are available for making these measurements.

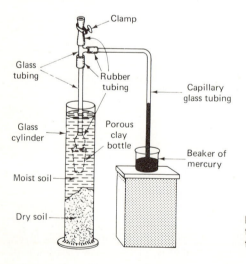

Figure 4-5 A simple mercury tensiometer. The clay bottle and glass tubing are filled with water.

Figure 4-5 illustrates a simple device for measuring soil moisture potentials near field capacity. This simple tensiometer is useful for demonstrating the increased soil moisture potential resulting from a decrease in the water content of the soil. The water content of the soil surrounding the clay bottle is reduced as water moves down into the dry soil below. The clay bottle is full of water, as is the glass tubing leading to the beaker of mercury. The reduced water content in the soil near the clay bottle causes a tension (a capillary potential) to develop, and water is pulled out through the pores of the clay bottle. Water is pulled from the glass tubing to replace that lost to the soil. Consequently, mercury is pulled up to replace the water as the tension increases. The soil surrounding the clay bottle is considered to be at field capacity when the mercury has risen to a height of 25 cm. An equivalent column of water would be 340 cm tall (25 cm X 13.6, the specific gravity of mercury).

The device shown in Figure 4-5 is suitable for classroom demonstrations and some laboratory use, but it is poorly suited for field use. Another type of tensiometer adapted to field use is shown in Figure 4-6. These tensiometers have porous clay cups connected to metal tubing filled with water. Vacuum gages replace the mercury used in the other type of tensiometer.

Moisture Meters

Tensiometers such as those shown in Figures 4-5 and 4-6 are limited to measuring soil moisture tensions of less than 1 atm because even a tiny air bubble will expand and fill the instrument as the absolute pressure (air pressure minus the tension being measured) approaches zero. Professor G. J. Bouyoucos of Michigan State University developed a moisture meter that overcomes this limitation by measuring the elec-

Figure 4-6   Three tensiometers installed at various depths to monitor the moisture potential in the soil profile. The potential is indicated by a vacuum gage rather than a mercury tensiometer.

trical resistance of a gypsum block (Figure 4-7). The gypsum block is buried in the soil and gains or loses water according to the changes in soil moisture potential of the surrounding soil.

Electrical resistance increases when the block dries out and decreases when it becomes wet. Unfortunately, the salt content of the water also influences electrical resistance so the gypsum block measurements read too wet in saline soils. Like tensiometers, the blocks respond to moisture potentials rather than to the amount of water. The meter is calibrated to read 100 percent at ⅓ atm and 0 percent at 15 atm. It thus divides the available water range into a percentage scale indicating the availability, though not really the amount, of soil water.

Even though the moisture meter is not highly accurate by some standards, it can readily detect major water use by plants at whatever depths the gypsum blocks have been buried. Standard practice is to bury the blocks at two or three chosen critical depths with their wires extending aboveground so that a portable meter can be connected to them whenever readings are desired. Such a practice is well adapted to use in experimental plots and in the irrigated fields of commercial growers. A moisture meter can even be made to turn on an irrigation system automatically when a certain soil moisture potential is reached.

A device that measures soil moisture potential by sensing heat dissipation is described by Phene, Rawlins, and Hoffman (1971). This device uses a porous block buried in the soil in a manner similar to that for the Bouyoucos moisture meter. The water content of the block is controlled by the moisture potential in the soil. This water content is measured by electrically heating the center of the porous block and measuring how fast the heat is dissipated. Faster heat dissipation indicates more

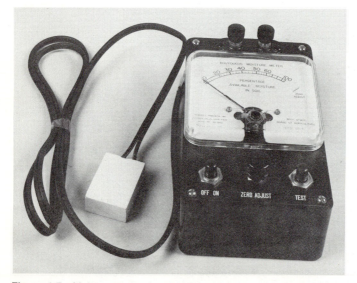

Figure 4-7 Moisture meter developed by Prof. G. J. Bouyoucos of Michigan State University. The white gypsum block is buried at the depth where the moisture potential is to be monitored.

moisture in the block and lower soil moisture potential. This device is reported to be accurate within ±0.2 atm in the 0 to 2 atm range of soil moisture potential. The accuracy decreases progressively to about ±1 atm at a soil moisture potential of 10 atm. This degree of accuracy is considerably better than that of the Bouyoucos moisture meter. This device can be connected to a recorder or to an irrigation system control if desired.

MEASURING THE PERCENT WATER IN THE SOIL

The amount of water present in the soil can be determined by weighing a sample before and after oven drying. This is basically a laboratory procedure, and it involves disturbing the soil. The development of the neutron probe in recent years has made it possible to make such measurements in situ. A neutron-emitting radioactive material is lowered into a hole along with a suitable detector. High-velocity neutrons are directed into the soil away from the detector, but their velocities are reduced and their paths deflected by hydrogen nucleii. A certain percentage of the deflected neutrons eventually reach the detector (Figure 4-8). The probability of any one neutron reaching the detector is proportional to the amount of hydrogen present in the soil. The frequency or count of neutrons reaching the detector can therefore be taken as a measure of the water content of the soil because water is the most important source of hydrogen in the soil. Corrections can be made when necessary for the hydrogen content of organic matter and other soil components.

The neutron probe is often accompanied by a gamma-ray probe that is used to measure bulk density as described in Chapter 3. The two probes can be attached to the same counting equipment. The gamma-ray probe also can be used to measure the soil water content if the bulk density is known (Topp, 1970).

RELATION BETWEEN PERCENT WATER AND CAPILLARY POTENTIAL

The amount of water present at a particular soil moisture tension varies considerably from one soil to another. Soils with large amounts of clay and organic matter hold more water than soils with small amounts of these constituents. Another important

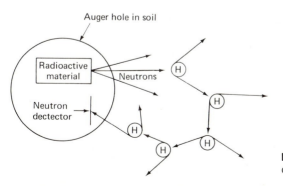

Figure 4-8 The neutron-probe method of measuring soil water content.

factor in this relation is whether the particular water content is reached by adding water to a dry soil or by removing water from a wet soil. Thus a soil has two curves relating percent water to soil moisture potential, a wetting curve and a drying curve, such as those shown in Figure 4-9. The difference between the two curves, known as *hysteresis,* is caused by the variability in the sizes of the pore passageways in the soil.

During the wetting process, movement through pores is delayed until the capillary potential is lower than the capillary pull of the widest parts of the passages. During the drying process, water tends to be held in the pores until the capillary potential exceeds the capillary pull of the narrowest parts of the passages. Thus, considerably more water can be present in the soil at a particular capillary potential during drying than at the same potential during wetting of the soil. Hysteresis represents pores that have not yet filled at that potential on the wetting curve and that have not yet emptied at the same potential on the drying curve.

The wetting and drying curves shown in Figure 4-9 represent one particular soil. The curves for other soils may lie either above or below those shown. The drying curve is usually more important than the wetting curve because wetting is usually fairly rapid but drying extends over a longer time.

WATER MOVEMENT IN SOILS

Soil water moves in both the liquid and the vapor phases. Liquid water movement is controlled by the soil moisture potential and is much more important when the soil is at a high moisture content than at the lower levels of soil moisture. The water in a relatively dry soil is held so tightly by the soil particles that liquid movement is almost nonexistent. The amount of vapor movement occurring between different parts of the soil under such conditions depends on temperature conditions. The presence of a temperature gradient causes a vapor pressure gradient that leads to the movement of water in the vapor phase from the warmer parts of the soil through

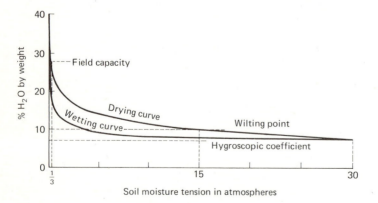

Figure 4-9   The relation between percent water and capillary potential in a Greenville loam soil in Utah. (*From data of Russell,* 1939.) The difference between the wetting and drying curves is known as *hysteresis.*

the air space in the soil pores to the cooler parts of the soil. The more open air space there is in the soil pores and the better connected the pores are, the more readily this type of water movement occurs. Water loss by evaporation occurs where the pores are so interconnected that air circulation to the soil surface occurs. The effective depth of evaporation losses is usually only 5 or 10 cm but is greatly increased where there are large cracks in the soil. Evaporation losses may extend to depths of 50 or 100 cm where the cracks extend to such depths.

When a heavy rain falls on a relatively dry soil, some of the water infiltrates and some runs off. The infiltration rate depends greatly on the soil and its physical condition and is often about 1 or 2 cm/h. The infiltration rate is usually highest shortly after the rain begins to fall and gradually declines as loose soil particles and swelling clay tend to block the soil pores and close any cracks that may have existed. The infiltrating water may nearly saturate a thin layer of soil near the surface. Below this a relatively wet zone develops with a water content somewhere between field capacity and saturation. The water must pass through this zone to reach the drier zone below. Relatively rapid water movement requires that the soil contain some gravitational water; that is, it must be above field capacity. A wetting profile such as that shown in the middle drawing of Figure 4-10 develops in the soil during the period of water infiltration.

The force of gravity obviously is a factor in the downward movement of water. The importance of gravity is greatest in saturated flow (when the pores are full of water) because the capillary potential is zero then. Capillary potential exists only when there are air spaces and air-water interfaces.

Unsaturated flow takes place when the soil contains air as well as moving water. The capillary potential is almost always a much greater factor than the gravitational potential in unsaturated flow. One evidence of the relative importance of these two factors is that water moves from wet to dry soil almost as fast in a horizontal or even in an upward direction as it moves in a downward direction. After the rain stops, the downward movement of water in the soil continues until the upper part of the soil is reduced to field capacity or until some restrictive layer is reached that prevents

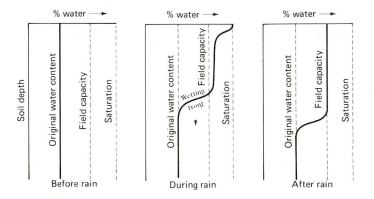

Figure 4-10 The water distribution in a soil before, during, and after a rain. The center figure illustrates a wetting profile.

further downward movement. By the time field capacity is reached, the water films around the soil particles have been reduced to a fraction of a millimeter in thickness, and only the small pores remain full of water.

CONTRASTING SOIL LAYERS

Water movement patterns are altered where there are contrasting soil layers. A clay layer with smaller pores than the soil above slows water movement. Part of the water that percolates down to such a layer is pulled into the layer by its strong capillary potential. The rate is limited by the low permeability of the clay layer, but movement continues and some water enters the clay. The rest of the water accumulates above the clay layer and produces a temporary water table. It is possible to saturate the soil all the way to the surface if water continues to infiltrate faster than it can pass through some layer in or beneath the soil profile. The restrictive layer can be a clay layer such as a heavy B horizon, or it can be some other dense layer such as bedrock or perhaps a plow sole caused by the compacting effect of tillage equipment.

Sand layers, lenses, and streaks also influence water movement but not always in the manner that might be anticipated. The relatively fast permeability of sand is often inoperative because the pore spaces in the sand have too little capillary potential to draw water from moist soil. Water moving down through a finer-textured soil stops when it reaches a sand layer. The water accumulates just above the sand until the soil there is nearly saturated and the moisture potential approaches zero. When the water finally enters the sand, it moves rapidly for a time and drains part of the excess water accumulated in the soil above. Even then the soil above the sand layer retains more water than it otherwise would and has a higher field capacity than normal because of the low capillary potential of the sand.

Sand lenses and streaks behave much like sand layers except that water may be able to go around and leave them as relatively dry areas in the soil. The same is true of other coarse materials such as chopped plant remains incorporated into the soil. Large pores are unable to pull water from smaller pores unless the soil is nearly saturated.

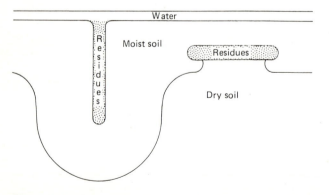

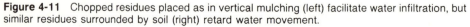

Figure 4-11 Chopped residues placed as in vertical mulching (left) facilitate water infiltration, but similar residues surrounded by soil (right) retard water movement.

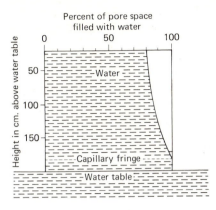

Figure 4-12 A typical distribution of water and air in the pore space of a soil above a water table. The thickness of the capillary fringe depends on the size of the largest pores (see Figure 4-13). The amount of air space at any height depends on the amount of pore space large enough to be drained at that capillary potential.

Coarse materials are good conductors of water only when they make direct contact with a water supply. A sand or gravel streak that reaches the soil surface will greatly facilitate water infiltration whereas a similar streak covered by a few centimeters of soil will hinder water movement. Plant residues have a short-term effect much like gravel and can be used to improve infiltration by means of *vertical mulching*. This practice involves placing chopped organic materials in a narrow slot opened by a tillage implement. The slot must reach to the soil surface so rain or irrigation water can flow into it (Figure 4-11).

Capillary Rise

The conditions shown in Figure 4-12 are the result of capillary rise. The attraction soil particles have for water (capillary potential) can lift and hold water against the pull of gravity. The amount of capillary rise is inversely proportional to the size of the pores, as shown in Figure 4-13.

Variable pore sizes and discontinuities normally limit capillary rise to a distance of a meter or so above a water table. The proportions of water and air contained in the soil vary with height above the water table, as shown in Figure 4-12. The reason for this variation is that the capillary potential required to hold the water is equal to the height above the water table. The water drains from the pores that are too large to exert the required potential.

Figure 4-12 shows a zone of saturated soil known as the *capillary fringe* above the water table. Even the largest pores present in this zone are able to hold water

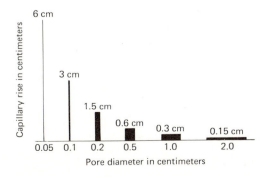

Figure 4-13 The maximum possible capillary rise in soil pores from 0.05 to 2 cm in diameter as calculated from the following formula: maximum capillary rise X pore diameter = 0.3 cm^2 (approximately).

against gravity. The true level of the water table is shown only where there is a large-enough hole to eliminate the effect of capillary rise.

Permeability

The rate at which water can pass through a soil material is known as its *permeability* or *hydraulic conductivity.* [1] The infiltration rate mentioned earlier may be considered to be the permeability at the soil surface. The permeability most commonly referred to is that of the least permeable layer in the soil profile. Soil permeabilities of 1 or 2 cm/h are common and may be considered as moderate rates. Permeabilities of more than 2.5 cm/h are rapid, and those less than 0.5 cm/h are slow. The appropriateness of these particular values depends in part upon the climate of the area where the soils occur because much of the significance of permeability is related to the likelihood of water running off during a rainstorm. Little or no runoff will occur from soils having permeabilities as high or higher than the rate of precipitation. Runoff may be low at first and increase greatly later in soils that have restrictive layers in their profiles. The infiltration rate drops from that of the soil surface to the permeability of the restrictive layer when the soil above the layer becomes saturated.

Soil permeability depends on the number, size, and continuity of pores. Continuous large pores carry much more than their share of the water so their abundance is important. Tillage that loosens the soil creates many large pores and temporarily increases the permeability. But, oversize pores created by tillage collapse when the soil is wet and cause the permeability to decrease with time. Expanding clay closes cracks and reduces pore sizes as the soil becomes wet. Also, raindrops pounding the soil surface break down soil structure, detach soil particles, and produce a crust that reduces the infiltration rate. Some of the detached soil particles move downward and reduce the permeability by plugging pores. Another factor that can reduce the infiltration rate is the increased air pressure if too much air is trapped in the soil (Linden and Dixon, 1973). Thus, the combined effect of several factors causes permeability to decrease with time.

Soil materials with low permeabilities usually have high clay contents, low percentage of pore space, or poor soil structure. The lowest permeability in the soil profile most commonly occurs either in the B2 horizon because of its high clay content or at the soil surface where the soil structure may have been damaged by traffic or by raindrop impact.

Capillary Movement

Capillary movement of water in soil occurs in response to differences in the soil moisture potential. Water moves from areas of low moisture potential to areas of higher moisture potential. In uniform soils this movement will be from areas where the water content is higher to areas where the water content is lower. In nonuniform

[1] Engineers use the term *hydraulic conductivity* because permeability has sometimes been used to mean percolation rate—the rate water level drops after an auger hole is filled with water. Permeability as used in this book is synonymous with hydraulic conductivity.

soils, however, the water movement may be to an area of equal or even higher water content than the area losing water so long as the water is moving in the direction of increasing soil moisture potential. Such movement ceases when the soil moisture potentials become nearly equal. Most of the water that moves by capillarity is actually gravitational water; capillary water is held too tightly to move very far or very fast.

Capillary action can theoretically lift water tens of meters above the water table if the soil pores are small enough. Each atmosphere of soil moisture tension is equivalent to 1,033 cm of water. Practically, however, such action is seldom effective to heights of more than 1 to 3 m. Continuously small pores would be required to raise the water farther than that, and such pores have very low permeability. Where capillary movement does reach heights of more than 3 m, it is too slow to be significant. In fact, evaporation and transpiration may dry the surface soil to the wilting point even where there is a water table at a depth of only 1 m.

Capillary movement is generally more significant in silty soil than it is in either sandy or clayey soil. Capillary movement is relatively rapid in sandy soil, but it is limited to short distances; in clayey soil the distance is much greater, but the rate of movement becomes very slow unless the clay is very well aggregated. Strongly aggregated clay may have large enough pores between the aggregates for capillary movement similar to that in a sandy soil. Such movement is limited in distance because it depends on the larger pores.

A magnified view of capillary movement of water through a soil reveals that the movement is not smooth and steady but rather progresses by a series of small jumps. A reason for this is the same variability in the pore sizes that causes hysteresis. The narrower parts of the passages produce a high capillary potential that pulls water into these parts rapidly. The forward progress of the wetting front into drier soil slows or halts, however, as the water approaches a wide part of the passageway. The soil moisture potential must drop to the level that can be exerted by the wider part before the water will pass through that part and into the next narrow section.

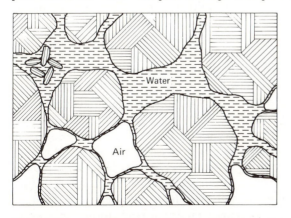

Figure 4-14 A greatly enlarged schematic cross section showing wet soil (above) and dry soil in equilibrium. The curvature of the air-water interface is constant throughout the system indicating that the capillary potential is equal throughout.

The variable capillary potential resulting from variable pore size helps to explain how soil at field capacity can exist in close proximity to relatively dry soil without water movement from one to the other. The water has stopped moving because its potential is too high for it to pass into the adjacent wider parts of the soil pores. Much larger pores are filled with water on one side of this front than on the other side (Figure 4-14).

Plant roots absorb water from soil because they are able to exert a greater potential than the soil is exerting. When the plant root removes water from one sector of the soil, the soil moisture potential in that sector is increased. This increased potential causes water to move by capillarity from adjoining parts of the soil into the part that has been partially depleted of its water. Water movement toward the plant roots is significant for a distance of a few millimeters in the soil. It can, of course, occur over much greater distances than this, but not at a rate adequate to meet the needs of a growing plant. Soil water movement is slow and roots grow rapidly, with the result that the roots grow toward the water rather than the water moving to the roots over long distances. A soil wet to field capacity to a depth of 1 m will illustrate the point. The plant obtains water from as deep as 1 m because the root system grows to such depths. It would not be possible for the roots to obtain water from the entire 1-m depth by growing to a depth of only 20 cm. The rate of capillary movement would be too slow to supply their needs after the water was depleted in their immediate vicinity.

PLANT USE OF SOIL WATER

The hygroscopic water and approximately a third of the capillary water are held too tightly for plant use. About two-thirds of the capillary water is available to plants. Gravitational water is also available, but most of it drains from the soil before the plants utilize it. In fact, drainage of gravitational water is essential for aeration.

There has been much speculation as to the relative availability of water near field capacity as compared with that near the wilting point. Data can be found to suggest that water is equally available throughout the range from field capacity to the wilting point. Other data, however, made with high fertility and irrigation indicate that higher yields are obtained where the water content is maintained near field capacity. It seems reasonable to suppose that the availability of water declines gradually as the work required to withdraw it from the soil increases. The decline must be most rapid near the wilting point as the soil moisture potential approaches the potential that plants can exert. Small changes in water content and soil moisture potential near the wilting point therefore have more drastic effects than equivalent changes near field capacity. Plants absorb water rather easily when the soil is near field capacity and only a small amount of plant energy is used in its absorption.

Capillary potential is the principal force that plants must overcome to obtain water from most soils. Saline soils (soils high in soluble salts) are an important exception. Dissolved salts create an osmotic potential that tends to hold the water where the salts are. This potential can be high enough to make plants wilt when the soil is at field capacity. More commonly the osmotic potential is not high enough to cause wilting by itself even in a saline soil, but it does add to the force from capillary potential that plants must overcome.

FACTORS AFFECTING THE AVAILABLE WATER-HOLDING CAPACITY OF SOILS

The available water-holding capacity is a soil characteristic of great importance. Where plants are grown with irrigation, the amount and frequency of water application are determined largely by the available water-holding capacity of the soil. Where plants are grown without irrigation, the available water-holding capacity determines how long a dry period the plants can tolerate. Good estimates of the available water-holding capacity of soil can be made by considering the soil texture, type of clay, structure, organic-matter content, and the thickness and sequence of layers in the soil profile.

Soil *texture* has a very considerable effect on the available water-holding capacity because the water is held as films on the surfaces of soil particles and in the small pore spaces between the particles, as shown in Figure 4-15. Fine-textured soils have high water-holding capacities because their many small particles have a large total surface area and there is a large total volume of small pore spaces located between the particles. Finer-textured soils therefore hold more water than coarser-textured soils. This relation holds true for water held at the wilting point as well as for that held at field capacity. In fact, the wilting point increases so much in clay soils that they hold less available water than do many soils containing less clay. Approximate values for field capacity, wilting point, and available water-holding capacity for soils of various textures can be read from Figure 4-16. Values read from Figure 4-16 must be considered as only rough approximations of the water-holding capacities of the various textures, since they are based only on texture and neglect several other important variables. As shown in the figure, most loamy soils have an available

Figure 4-15   Soil water stored in films around the particles and filling the smaller pores of a soil at field capacity.

water-holding capacity of about 0.17 cm per cm of soil depth. Extremely sandy soils, however, often have available water-holding capacities of less than 0.1 cm per cm of soil. Soils with low water-storage capacities are droughty; that is, they must receive rain or irrigation more frequently than the finer-textured soils in order to be productive.

A soil with a capacity to store 0.17 cm of available water per cm to a depth of 1.5 m can store about half of the water needed to grow a crop of corn on a fertile soil in the corn belt. If 25 cm of this water is in storage in the top 1.5 m of soil at planting time, then a typical crop of corn will require an additional 25 cm of water in the form of precipitation or irrigation during the growing season. This water must soak deep enough into the soil to avoid evaporation losses that occur from the upper several centimeters of soil after each rain. A few large rains are therefore more efficient in supplying water for plant use than are several smaller rains that add up to the same total precipitation.

It is frequently important to know not only how much water a soil can store but also at what depths the water is stored. For instance, a soil with about 1 m of sandy material overlying a heavy subsoil may prove to be too droughty for grasses and some field crops and yet provide adequate water for trees to grow quite satisfactorily. The trees probably need more water than the grasses, but trees have enough roots extending into the lower part of the soil to utilize water from the lower depths. Thus the soil is not droughty to trees, but it is droughty to shallow-rooted plants and to annual crops that may not have time to grow deep roots before they need the water from the lower part of the soil. In fact, the soil with the deep sandy layer may lose less water by runoff and evaporation and therefore make more water available to trees than a finer-textured soil would.

Another problem of managing sandy soils occurs because many of them are low in nitrogen. Drought damage is most common on soils low in available nitrogen because roots must extend themselves to considerable depths to obtain water that has drained deeply in the sandy soil. The plant cannot produce the protein needed

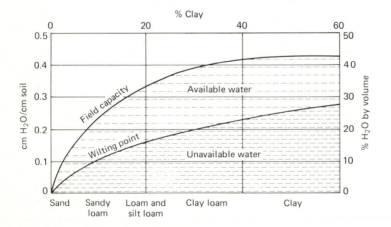

Figure 4-16 The water-holding capacities of typical soils of various textures.

for rapid extension of its root system if it runs out of nitrogen. Therefore the plant starves for water simply because its roots cannot extend themselves to reach the available water.

Soil *structure* influences the available water-holding capacity by the way in which the soil particles are arranged. Particles that are packed tightly together have room for only a thin film of water around them. Much of this water is held so tightly that it is unavailable to plants. A looser arrangement such as that of the cardhouse structure described in Chapter 3 is better. The soil particles should be arranged openly enough to include a large percentage of pore space and so that all sizes of pores are present. Such an arrangement provides for good aeration and for the storage of adequate available water for plant growth.

Loose sandy soils have such large pore spaces that they hold much air and little water. Compacting a sandy soil may rearrange the particles in such a way that there will be more fine capillaries; the available water-holding capacity is thereby increased. This increase comes at the expense of the aeration pore space and is therefore desirable only in sandy soils. Compaction of medium- and fine-textured soils has a detrimental effect on plant growth.

The *type of clay* present in a soil influences the available water-holding capacity indirectly as well as directly. The indirect effect is through the influence of clay type on soil structure. The direct effect is through the ability of certain types of clay to expand and absorb water internally. Clay types are discussed in Chapter 6.

Organic matter can hold a weight of water in excess of its own weight. Many organic soils have field capacities of 100 to 150 percent water, and some, such as sphagnum moss peat, hold as much as 500 or 600 percent water by weight. Of course, this is possible only because these materials have very low bulk densities and high porosities.

Organic matter can influence the water-holding capacity of a sandy soil. One percent organic matter in a soil at field capacity will hold about 1.5 percent water on a soil-volume basis. One percent organic matter therefore would account for about one-tenth of the water-holding capacity of a sandy soil holding 15 percent water by volume. Many soils have lost 1 percent or more of organic matter under cultivation. The amounts of water that they now store may have been seriously reduced not only because of the water-holding capacity of the organic matter but also because of changes in soil structure and water infiltration rate.

It should not be concluded that water-holding capacities can be readily increased simply by adding organic matter to the soil. In the first place, the influence of organic matter on water-holding capacities is less in finer-textured soils than it is in sandy soils, probably because much of the organic matter is so closely associated with the clay particles that the same film of water envelopes both. Secondly, the plow layer of a hectare of soil weighs about 2 million kg, and 1 percent of this represents 20,000 kg of dry weight. An equivalent moist weight of organic matter in the form of manure would be about 80,000 kg. Even if that much manure or other organic matter were applied to a soil, much of it would decompose the first year. On the other hand, it may be feasible to improve soil structure and the water infiltration rate by the addition of reasonable amounts of organic matter.

One type of material that shows promise for improving the water-holding capacity of sandy soils is a copolymer of starch and acrylonitrile called "super slurper." Super slurper is reported to absorb up to 1,400 times its weight of distilled water or perhaps 100 times its weight of water containing dissolved minerals (Agricultural Research Service, 1976). Preliminary reports indicate that super slurper added to sand can be very helpful in storing water for plant growth. Super slurper can also be used to make watery materials such as mud, sewage sludge, or liquid manure easier to handle by converting them to solids.

The *sequence of layers* in the soil profile can have considerable influence on the water-holding capacity of the soil, not only because of the different water-holding capacities of the various textures, etc., but also because of their influence on water movement. There is generally a delay in the downward movement of water wherever there is a drastic textural change in the soil profile. This delay is sometimes long enough to constitute an increase in the water-holding capacity of the upper layer. Such water is loosely held and is available to plants unless there is an aeration problem.

AERATION PORE SPACE IN SOILS

A loam soil in good structural condition contains about 50 percent pore space by volume. It can be estimated from Figure 4-16 that two-thirds of the pore space is filled with water when the loam soil is at field capacity. The soil volume is then one-half solid, one-third water, and one-sixth air. These proportions change as water is used until at the wilting point the soil volume is approximately one-half solid, one-sixth water, and one-third air. The soil has thus provided about 0.17 cm of available water per centimeter of depth for plant growth and maintained a satisfactory amount of aeration at all times.

The proportion of pore space filled with water tends to increase with increasing clay contents and with soil compaction. It is entirely possible for the 40 percent or so of water that a compacted clay soil may hold at field capacity to entirely fill its pore space. A small amount of air might remain in isolated pockets, but it would be completely surrounded by water and cut off from replenishment of its oxygen supply. Even this small amount of air might be removed by being dissolved in the soil water if the condition continued for an extended period of time.

It may be recognized, then, why some soils with clay subsoils have apparently no root penetration into their subsoils. These strongly differentiated soils have no air space remaining in their B2 horizons when they are at field capacity. Since no roots penetrate these subsoils, there is no way to withdraw water from them and reduce their water contents below field capacity. Even when crops are growing on these soils, roots do not grow into the wet subsoils because of the lack of oxygen. It is as though the soil were only as deep as its A horizon.

The partial filling of the aeration pore space by capillary rise (as shown in Figure 4-12) is of great practical significance to farmers with wet land. This effect occurs whether the water table is natural or artificial, permanent or temporary. Aeration is reduced whenever it occurs.

Wetness problems often cause slow crop growth in the spring. The wet soils are cold as well as poorly aerated. The coldness results in part from the high heat capacity of the wet soil and in part from the cooling effect of evaporation. Total water loss is relatively slow on these soils because the coldness reduces plant growth and transpiration.

THE ABSORPTION OF PLANT NUTRIENTS

Reduced water contents not only make water less available to plants, they also reduce the availability of certain plant nutrients. Phosphorus is a good example. The solubility of most phosphorus compounds is so low that the plow layer of a hectare of soil usually contains only a few tens of grams of dissolved phosphorus. As this small amount of phosphorus is utilized by the growing plants, more is dissolved from the soil solids. However, the rate of solution may not be fast enough to satisfy the needs of the plants. A higher water content helps because the amount of dissolved phosphorus and the rate of solution are roughly proportional to the amount of water present.

The available plant nutrients are usually more concentrated in the A horizon than in the B horizon of a soil. Also, most plant roots are concentrated in the A horizon. Therefore, the plants utilize water more rapidly from the A horizon than they do from the B horizon. The A horizon dries out much more quickly than the B horizon, owing to both plant absorption and evaporation of water.

A natural question that arises from the above considerations is whether the plants can continue to withdraw nutrients from the A horizon after its water content has been depleted to the wilting point. It has been shown by several workers that this is possible as long as adequate water is still being supplied by the subsoil to keep the plant turgid. Figure 4-17 shows an experiment illustrating this phenomenon. The

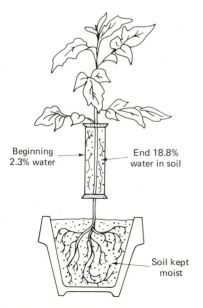

Beginning 2.3% water

End 18.8% water in soil

Soil kept moist

Figure 4-17 Sketch of a tomato plant that moved water from the moist soil in the pot to the dry soil around its stem.

tomato plant was grown for 3½ months with its root system in a pot of moist soil and with its stem encased in apparently dry soil. At the end of the experiment the originally dry soil had increased in water content from 2.3 percent to 18.8 percent. Furthermore, roots were found to be growing in this originally dry soil. Water had been transferred from the soil in the pot through the tomato stem to the soil surrounding the stem. Surely the roots in this area had been able to contact and absorb some plant nutrients. This experiment illustrates well the fact that water moves in accordance with potential gradients. Normally the movement is from the soil to the plant because the plant exerts a greater potential than does the soil. However, when the potential gradient is in the opposite direction, water can move from the plant to the soil.

It is not expected that a large amount of water will move from one part of the soil to another through the plant roots under normal field conditions. Water supply was never limiting in the case of the tomato plant used in the experiment. Furthermore, the soil wrapped around its stem was so dry that it had a very high soil moisture potential. Some plant nutrients were probably absorbed from this originally dry soil, but the amount must have been much less than it would have been had the soil been maintained in a moist condition.

RELATION OF FERTILITY TO WATER UTILIZATION

Water requirements and transpiration ratios (grams of water required to produce a gram of dry matter) depend on factors that affect transpiration (wind, temperature, and humidity) and on the level of productivity attained. Studies have shown that plants grown on fertile soil produce more dry matter per gram of water than those grown on soils of lower fertility. The addition of a needed fertilizer normally increases production even though there is no increase in the water supply. This kind of experience has led some farmers to consider fertilizer as a partial substitute for water. Such a conclusion can lead to disastrous results. Fertilizer may improve the efficiency of use of water but it is not a substitute.

The water-holding capacity of a deep soil may carry crops through one drier-than-normal year but fail to last for two dry years. Experimental results in 1953 and 1954 in southern Iowa will illustrate this point. The data are shown in Table 4-1. The soil was at field capacity when corn was planted in 1953. However, the last rain the corn received came when the crop was only waist high. The growing season

Table 4-1 Corn Yields from Plots with and without Nitrogen during the 1953—1954 Dry Seasons in Southern Iowa

	1953 yield	1954 yield
No nitrogen fertilizer (dried soil to 1.2 m in 1953)	12 q/ha	12 q/ha
135 kg of fertilizer N/ha (dried soil to 2 m in 1953)	38 q/ha	0

Source: Nicholson, Pesek, and Shrader.

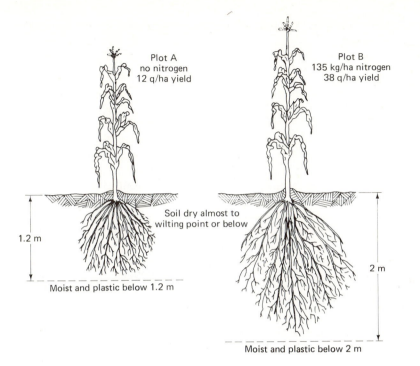

Plot A
no nitrogen
12 q/ha yield

Plot B
135 kg/ha nitrogen
38 q/ha yield

Soil dry almost to
wilting point or below

1.2 m

Moist and plastic below 1.2 m

2 m

Moist and plastic below 2 m

Figure 4-18 The relation of water removed from the soil and the yield of corn, as related also to nitrogen treatment in an experiment in Iowa. *(Unpublished data of Nicholson, Pesek, and Shrader.)*

precipitation was 28 cm below normal. Plot *A* with no nitrogen fertilizer produced only 12 q (quintals) of corn. In doing so, it reduced the water content to the wilting point in the top 1.2 m of soil. Plot *B* had received 135 kg of nitrogen fertilizer per hectare and produced 38 q/ha of corn. Its more vigorous plants had developed deeper roots and dried the soil to a depth of 2 m (Figure 4-18).

The same treatments were continued in 1954 but with different results. Winter and spring precipitation had replenished the water in the unfertilized soil but left a dry zone in the lower part of the soil of the fertilized plot. The growing season was dry again in 1954, and the unfertilized plot produced the same as in 1953. The corn on the fertilized plot grew vigorously at first but ran out of water in midseason and then withered and died. The fertilizer helped to make maximum use of the available water in 1953, but it did not substitute for water.

Prudent farmers are aware of the subsoil moisture status of their soils as well as that of the topsoil. A reasonable estimate of the soil water content can be based on the amount and timing of precipitation. Any doubts can be resolved by sampling at various depths with a soil auger. Many farmers adjust their planting and fertilizer rates in accordance with the water content of the soil at planting time. The rates are reduced when the water reserves are low so that the probability of a complete crop failure is reduced. The rates are raised to aim for a higher yield when the water reserves are adequate. A good supply of available nitrogen in the soil is particularly valuable in promoting root development so that the crop can utilize water stored in the lower part of the soil.

REFERENCES

Adams, J. E., and R. J. Hanks, 1964, Evaporation from Soil Shrinkage Cracks, *Soil Sci. Soc. Am. Proc.* **28**:281–284.

Agricultural Research Services, 1976, An Honor for Super Slurper, *Agric. Res.* **24**(7):12–13.

Amemiya, M., 1965, The Influence of Aggregate Size on Soil Moisture Content—Capillary Conductivity Relations, *Soil Sci. Soc. Am. Proc.* **29**:744–748.

Cary, J. W., 1966, Soil Moisture Transport due to Thermal Gradients: Practical Aspects, *Soil Sci. Soc. Am. Proc.* **30**:428–433.

Chaudhary, T. N., V. K. Bhatnager, and S. S. Prihar, 1975, Corn Yield and Nutrient Uptake as Affected by Water-table Depth and Soil Submergence, *Agron. J.* **67**:745–749.

Denmead, O. T., and R. H. Shaw, 1962, Availability of Soil Water to Plants as Affected by Soil Moisture Content and Meteorological Conditions, *Agron. J.* **54**:385–390.

Dixon, R. M., and A. E. Peterson, 1972, Tiny Soil Channels Determine Water Infiltration, *Crops Soils* **24**(5):11–12.

Gardner, W. H., 1962, How Water Moves in the Soil. Part II—In the Field, *Crops Soils* **15**(2):9–11.

Grover, B. L., G. A. Cahoon, and C. W. Hotchkiss, 1964, Moisture Relations of Soil Inclusions of a Texture Different from the Surrounding Soil, *Soil Sci. Soc. Am. Proc.* **28**:692–695.

Herkelrath, W. N., and E. E. Miller, 1976, High Performance Gamma System for Soil Columns, *Soil Sci. Soc. Am. J.* **40**:331–332.

Holmes, J. W., 1966, Influence of Bulk Density of the Soil on Neutron Moisture Meter Calibration, *Soil Sci.* **102**:355–360.

Horton, J. H., and R. H. Hawkins, 1965, Flow Path of Rain from the Soil Surface to the Water Table, *Soil Sci.* **100**:377–383.

Kemper, W. D., D. E. L. Maasland, and L. K. Porter, 1964, Mobility of Water Adjacent to Mineral Surfaces, *Soil Sci. Soc. Am. Proc.* **28**:164–167.

Kohl, R. A., and J. J. Kolar, 1976, Soil Water Uptake by Alfalfa, *Agron. J.* **68**:536–538.

Linden, D. R., and R. M. Dixon, 1973, Infiltration and Water Table Effects of Soil Air Pressure under Border Irrigation, *Soil Sci. Soc. Am. Proc.* **37**:94–98.

Miller, D. E., 1967, Available Water in Soil as Influenced by Extraction of Soil Water by Plants, *Agron. J.* **59**:420–423.

Miller, R. F., and K. W. Ratzlaff, 1965, Chemistry of Soil Profiles Indicates Recurring Patterns and Modes of Moisture Migration, *Soil Sci. Soc. Am. Proc.* **29**:263–266.

Nelson, R. E., 1975, Estimation of Fifteen-bar Percentage by Desorption of Soil on Hectorite, *Soil Sci.* **119**:269–272.

Nicholson, R. P., John Pesek, and W. D. Shrader. Unpublished data, Iowa Agricultural Experiment Station, Ames.

Phene, C. J., S. L. Rawlins, and G. J. Hoffman, 1971, Measuring Soil Matric Potential *in situ* by Sensing Heat Dissipation within a Porous Body: II. Experimental Results, *Soil Sci. Soc. Am. Proc.* **35**:225–229.

Reginato, R. J., and C. H. M. van Bavel, 1964, Soil Water Measurement with Gamma Attenuation, *Soil Sci. Soc. Am. Proc.* **28**:721–724.

Rubin, J., R. Steinhardt, and P. Reiniger, 1964, Soil Water Relations during Rain Infiltration: II. Moisture Content Profiles during Rains of Low Intensities, *Soil Sci. Soc. Am. Proc.* **28**:1–5.

Russell, M. B., 1939, Soil Moisture Sorption Curves for Four Iowa Soils, *Soil Sci. Soc. Am. Proc.* **4**:51–54.

Smith, D. D., 1954, Fertility Increases Efficiency of Soil Moisture, *Better Crops Plant Food* **38**(6):11.

Sykes, D. J., and W. E. Loomis, 1967, Plant and Soil Factors in Permanent Wilting Percentages and Field Capacity Storage, *Soil Sci.* **104**:163–173.

Topp, G. C., 1970, Soil Water Content from Gamma Ray Attenuation: A Comparison of Ionization Chamber and Scintillation Detectors. *Can. J. Soil Sci.* **50**:439–447.

Volk, G. M., 1947, Significance of Moisture Translocation from Soil Zones of Low Moisture Tension to Zones of High Moisture Tension by Plant Roots, *J. Am. Soc. Agron.* **33**: 93–107.

Chapter 5

Soil Organic Matter

Organic materials, living and dead, exert a profound influence on almost every facet of the nature of soil. Many significant effects of organic matter are discussed in the chapters dealing with soil formation, physical properties of soils, and the various plant nutrients. This chapter deals more specifically with the composition and properties of soil organic matter.

Organic materials of plant and animal origin incorporated into the soil provide the parent material of soil organic matter. There is great variability in the nature of the organic material added to the soil, yet there is a surprising similarity in the soil humus (the relatively stable portion of soil organic matter) of different soils. The humus from any soil is a complex mixture of organic compounds that defy detailed chemical analysis. Nevertheless, much has been learned about soil organic matter, and it is possible to characterize it in general terms.

The dynamic nature of soil organic matter can scarcely be overemphasized. Plant and animal residues undergo extensive alteration in the soil before they become humus. Various types of microorganisms attack the residues and decompose their constituents. The residues serve as a source of nutrients and energy for the life processes of the microorganisms. Readily decomposed organic compounds are utilized rapidly and would soon disappear if they were not replenished by fresh residues from time to time. The more resistant organic compounds are altered in nature by the microorganisms but persist for long periods as a part of the soil humus.

BIOLOGICAL ACTIVITY IN SOIL

Life within the soil exceeds life above the soil in terms of numbers of living organisms and total metabolic activity. The teeming soil population includes plant and animal life ranging in size from bacteria and other microscopic forms to large burrowing animals and the roots of mighty trees. The bacteria alone sometimes number billions per gram of soil. An average soil contains about 2 or 3 kg/m^2 of living plants and animals or 20,000 to 30,000 kg/ha. Much of this living mass is composed of microscopic life forms that can accelerate to very high metabolic rates when conditions are favorable.

The various biotic forms living in and on the soil are mutually interdependent. Green plants use nutrients from the soil and carbon dioxide from the air to produce plant tissue. The plant tissue serves as food for animal life both on and in the soil. Plant and animal waste and remains return to the soil where they are processed by insects, earthworms, fungi, bacteria, and other living things. Eventually the cycle is completed and the nutrients and carbon dioxide are again available for plant growth. The soil would run out of nutrients and become sterile if the cycle were not completed. Plants would stop growing and all life would cease.

Decomposition is a specialty of soil organisms. Almost anything will decompose sooner or later in the soil whether it be a plant or animal waste, a fertilizer or soil amendment, a pesticide such as DDT or 2,4-D, or even a metal object. Insects and earthworms often begin the process by chewing up the material, digesting part of it, and leaving fragments behind. Various microbial forms decompose the remaining fragments, the wastes, and eventually the dead bodies of the insects, earthworms, and other microbes. A material must be extremely resistant to avoid decomposition under the attack of all the enzymes, oxidizing and reducing conditions, and other chemical and physical processes that occur in the soil.

Decomposition is only one of many functions performed by plant and animal life in the soil. Removal of water from the soil is another. Plant use of soil water greatly reduces the intensity of leaching and thereby helps to maintain soil fertility. Even nutrients that are carried downward by leaching may be absorbed by plant roots and moved upward inside the plant.

Biological activity has much to do with soil structure, pore space, and permeability. Plant roots force their way through the soil and shift particles around as they grow. Earthworms, insects, and rodents all rearrange the soil with their tunneling activities. Even bacteria may move clay particles and help produce soil aggregates. Both large and small living things produce exudates and residues that help to bind clay and silt particles into structural units.

AMOUNTS OF SOIL ORGANIC MATTER

The amount of organic matter present in a soil is normally expressed as a percentage of the oven-dry weight of the soil. This percentage may be determined by burning out the organic matter in a furnace (a procedure which may also drive off some chemically bound water) or by the use of a chemical oxidizing agent such as hydrogen peroxide (a procedure which may not remove all the organic matter). Alternative procedures include determinations of the amounts of organic carbon or nitrogen present and multiplying by suitable factors.

Most soils contain between 1 and 6 percent organic matter. This represents between 20,000 and 120,000 kg of organic matter in an average hectare-furrow slice (about 2 million kg of soil). The organic matter in all the rest of the solum is usually about equal in total amount to that in the furrow slice. Fresh residues usually represent less than one-tenth of the weight of soil organic matter, and the rest is humus. Some writers consider the difference too small to make any distinction between humus and soil organic matter, whereas others consider organic matter to be a broader term including both decomposable and resistant organic materials.

ORGANIC MATERIALS IN THE SOIL

Soil contains living and dead organic materials of plant and animal origin ranging in size from submicroscopic particles to the largest tree roots. The composition can be classified in several different ways including the mode of origin and the chemical composition of the materials. The following groups are considered significant from the point of view of mode of origin:

1 Living macroorganisms
2 Dead but identifiable remains of macroorganisms
3 Living microbes
4 Finely divided nonliving organic materials

Estimated average amounts of these materials are given in Table 5-1.

Living Macroorganisms

Living macroorganisms are entities in themselves and are usually not considered as a part of the soil. Rather, they constitute a soil-forming factor that exerts a potent influence on the nature of the soil. They are, of course, the source of all soil organic matter and thereby indirectly influence many soil properties. These living entities also have other, more direct ways of influencing the soil. Plant roots, for example, shift and move soil particles as they penetrate the soil and grow in size. This rearrangement changes the sizes and shapes of soil peds. Later when the roots die and decompose, a channel is left in the soil that serves as aeration pore space for a time. Root channels are formed over and over again unless the soil becomes too dense and strong for roots to penetrate. The roots in many horizons are concentrated between the peds because the ped interiors are hard to penetrate. Plants differ in the ability of their roots to penetrate hard soil and in the number, size, and arrangement of pores created by their roots. Alfalfa is particularly noted for its ability to open channels in subsoils.

Roots and other plant parts growing belowground constitute most of the mass of living macroorganisms in most soils, but animal life should not be overlooked. The effect of earthworms is a well-known animal action. Small mounds at the surface of a soil are only an indication of what earthworms have done within the soil. A large population of earthworms ingests and excretes many tons of soil per hectare in a

Table 5-1   Estimated Average Contents of Organic Materials per Hectare in a Soil Profile Formed under Grass in a Subhumid Temperate Region

	kg/ha	Number of organisms per hectare
Living macroorganisms:		
Plant roots	15,000	
Insects	1,000	20,000,000
Earthworms	500	1,000,000
Nematodes	50	200,000,000
Crustaceans	40	400,000
Snails, slugs	20	10,000
Rodents, snakes	20	200
Dead but identifiable remains of macroorganisms	4,000	
Living microbes:		
Bacteria	3,000	2×10^{18}
Fungi	3,000	2×10^{14}
Actinomycetes	1,500	5×10^{16}
Protozoa	100	5×10^{12}
Algae	100	1×10^{10}
Finely divided nonliving organic materials	150,000	

year's time. They leave the soil material in a strongly aggregated granular condition
that, along with the channels they create, gives the soil a greatly increased permeability to air, water, and roots. Earthworms cause considerable soil mixing, sometimes
producing 50 cm or more of well-mixed black topsoil. Other animals also mix the
soil in which they live. It has been estimated that certain soil areas have been
completely reworked by ants during a period of about 1,000 years (Baxter and Hole,
1967). Still other organisms that mix soil are moles, badgers, prairie dogs, and
certain burrowing insects. In some wet soils, vertical tubules of mixed soil may be
attributed to crayfish. Even trees are credited with mixing the top 50 cm of certain
soils every few hundred years by tree throw. The roots of fallen trees carry large
quantities of soil into the air, and it later falls into a well-mixed heap that gradually
is leveled as the years pass.

Living macroorganisms are active participants in decomposition processes in
and out of the soil. Much coarse plant material is chewed up by ants and other insects
before microbial decomposition takes place. The residues left by the macroorganisms
include gummy materials that help glue soil particles together.

Dead but Identifiable Remains

Remains of plants and animals are the input materials from which humus is formed.
These materials are as variable as the plant and animal parts from which they come.
The amount present in the soil varies from time to time according to the season of
the year when plants mature and die or shed surplus parts. Seldom, however, would
identifiable remains constitute more than 10 percent of the mass of the soil organic

matter. Some changes in dead materials take place very rapidly, such as the leaching of soluble ions. Decomposition begins almost immediately and builds up to a maximum rate within a few days or weeks if conditions are favorable. Full decomposition, however, takes months or years and even extends to many centuries in some soils that are so wet that oxygen is virtually excluded.

Laboratory determinations of soil organic matter usually include little of the identifiable remains of living things. Most of these materials are removed by sieving during the process of sample preparation.

Microbes

The microbial population of the soil includes large numbers of microanimals and far larger numbers of microplants. The latter are sometimes referred to as *microflora* and the entire assemblage as *microorganisms* or *microbes*. The number of varieties of microbes occurring in soil is much too large for any detailed consideration here, and so only a broad classification will be presented.

Most soils contain representatives of each of the major groups discussed in the following paragraphs. The relative importance of the different groups varies somewhat according to the soil properties. Bacteria generally are the most numerous microbes, but fungi often predominate on a weight basis.

Bacteria   Bacteria (Figure 5-1) are the smallest and simplest forms of plant life. They are single-celled organisms most of which are either spherical or rod

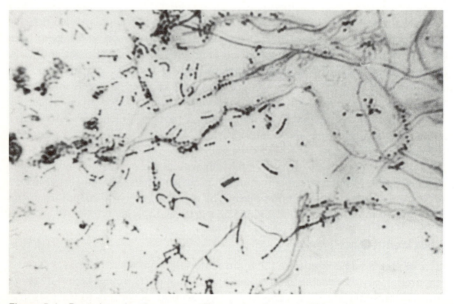

Figure 5-1   Bacteria and actinomycetes. The chains of spherical bacteria are cocci (some bacteria are rod- or spiral-shaped). The long gray lines are actinomycetes. *(Photomicrograph courtesy of Don Gier, Iowa State Univ.)*

shaped (a few types have spiral shapes). Bacteria occur in fantastic numbers. A fertile soil well supplied with an energy source (decomposable plant tissue) may contain a billion or more bacteria in a single gram of soil (Clark, 1954). Some bacteria perform specific functions such as oxidizing ammonium nitrogen to nitrate nitrogen. Others act as a part of the general process of decomposing organic materials.

Bacteria are divided into *aerobic* and *anaerobic* groups according to whether they require oxygen from the air. Anaerobic organisms obtain oxygen by reducing compounds such as sugars or ions such as NO_3^- or SO_4^{--}. Facultative anaerobes utilize atmospheric oxygen when it is available but can also function anaerobically. Anaerobic conditions result from poor aeration when most of the pore space is filled with water. Aerobic conditions are required for the root growth of most plants. The microbes involved in plant nutrient transformations also generally grow best under aerobic conditions. It is possible, however, to have both conditions at once, with anaerobic conditions inside peds while the ped exteriors are aerobic.

Bacteria are further classified as *heterotrophic* or *autotrophic*. Heterotrophic bacteria obtain their energy by oxidizing organic matter, while autotrophic bacteria obtain energy by oxidizing inorganic materials. The bacteria that oxidize ammonium ions to nitrite ions and those that oxidize nitrite ions to nitrate are autotrophic.

There are so many types of bacteria that some can live in almost any environment capable of supporting life. However, bacteria are most numerous and vigorous in neutral or slightly alkaline soils where calcium is abundant. Acid conditions generally reduce the bacterial population.

Actinomycetes Actinomycetes (Figure 5-1) are threadlike organisms that appear to be single celled. The threads are about the diameter of bacteria (and of coarse clay) and are often branched and tangled so that these microbes are difficult to count. Certainly the numbers are large, on the order of millions per gram of soil where the conditions are favorable. One important favorable factor is an abundance of calcium giving a neutral or slightly alkaline reaction. The pleasant odor of freshly plowed ground comes from actinomycetes in the soil.

Actinomycetes are much larger and less numerous than bacteria. They are smaller in diameter than fungi, and cell walls are not seen in their hyphae (the long, threadlike parts). Some of the hyphae terminate in groups of spores that somewhat resemble bacteria. There are many different types of actinomycetes, and some of them have familiar names such as *Streptomyces* because their relatives have provided us with antibiotics such as streptomycin and aureomycin. Among their numbers are organisms that can perform most of the functions performed by any of the other soil microbes.

Fungi Fungi (Figure 5-2) range in size from microscopic (but larger than either bacteria or actinomycetes) to such large and readily visible specimens as mushrooms. The smaller sizes are naturally most numerous, about a million of them occurring in a gram of soil. They often contribute more weight to the soil organic matter than any other type of microbe (Clark, 1954), especially in acid soils. Some types of fungi are known for their contributions to medicine. Penicillium is an

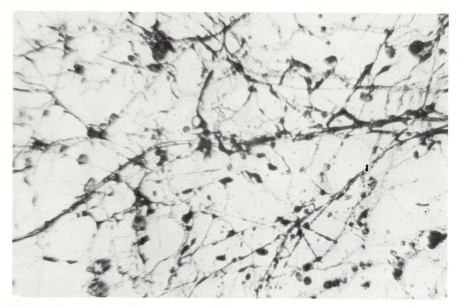

Figure 5-2   Fungi grown on a microscope slide in contact with the soil. The long lines are hyphae, and the larger, more rounded parts are fruiting bodies. *(Photomicrograph courtesy of Don Gier, Iowa State Univ.)*

example. On the other hand, some fungi are detrimental in that they are the cause of several kinds of root rot in plants.

Fungi generally tolerate acid conditions and low calcium supply better than other microbes and are present in about the same numbers in acid soils as in neutral soils. They dominate the microbial population of most forest soils because other organisms decline in numbers in acid conditions. Fungi produce branched and tangled masses of hyphae like actinomycetes, but cell walls can be detected in the fungal hyphae. Fungi produce various types of fruiting bodies containing spores that help them to reproduce and to survive adverse conditions such as hot, dry soil. Very wet conditions are also unfavorable to fungi.

The population of fungi in the soil is quite sensitive to the amount of decomposable organic matter present. The fungi are prevalent in soils rich in plant residues where competition for food and energy is not too keen but decline rapidly as the readily decomposable material disappears. Bacteria persist longer and consume the fungi.

Algae   Algae are a group of both macro- and microorganisms that contain chlorophyl. Those exposed to light can transform light energy into energy-bearing organic compounds. Algae occurring inside the soil, however, must obtain their energy from oxidation of other materials just like other microbes.

Algae range in size from single-cell forms to the long strands of algae in a pond. The most familiar types are the slimy fibrous masses that grow in stagnant water. Some, such as the blue-green algae, are able to fix nitrogen from the air. Algae are probably not responsible for very much nitrogen fixation in well-drained soils, but they may be important in maintaining the nitrogen levels in soils used for paddy rice production. Algae tolerate a wide range of moisture conditions. They do very well in wet soils and are also important in desert soils.

Algae are most abundant at the soil surface where their ability to utilize light energy and to withstand dry conditions permits them to compete most effectively with other microbes. They produce enough sticky materials in some soils to hold the soil particles together in a crust that helps resist wind erosion.

Algae combine with fungi to produce the *lichen* crust that often is the first encroachment of plant life on bare rock. The relationship is symbiotic with the algae producing carbohydrates and fixing nitrogen from the air. The fungus provides water absorption and storage and sinks hyphae into the rock surface to provide anchorage and absorb mineral nutrients. Lichen growth is slow under the harsh conditions on the rock surface and may persist for hundreds or thousands of years before the rock is weathered enough to produce soil and support other plant life.

Protozoa   Protozoa are single-celled animals. Most soils contain a much smaller number of these microanimals than of the microflora because protozoa need more water in order to thrive. Nevertheless, there are many types of protozoa. Some of them appear to bridge the gap between plants and animals in that they can carry on photosynthesis like plants yet they ingest solid food particles like animals. Protozoa feed on bacteria (among other things), and it has been noted that numbers of protozoa are higher in soils that have large populations of bacteria.

Nematodes   Nematodes are small worms that represent a higher form of animal life than protozoa. Most of them live on decaying organic matter, but some infect plant roots and live as parasites (still others are parasites on animals including human beings). The parasitic types are the best known because of the damage they cause. Some do serious damage to the roots of fruit trees. Species of nematodes that attack the roots of alfalfa, sugar beets, and potatoes create serious problems in the western United States. Still others attack corn, soybeans, and tobacco, especially in the southern United States.

Finely Divided Nonliving Material

The finely divided nonliving material has been the object of much speculation regarding its precise origin. It is relatively resistant to decomposition. The question is, does it come from the macroorganisms or from the microbes? Probably a large part of it is microbial remains and exudates, but there may also be exudates from plant roots, earthworms, etc. Some of this material may be resistant compounds remaining from decomposed plant tissue, but such compounds have surely been altered by microbial action. Probably over 90 percent of the finely divided nonliving material is either derived from microbial tissue or strongly altered by the action of microbes.

Finely divided nonliving materials are major components of soil humus. They tend to coat the mineral particles of the soil and are especially associated with clay. Their small size permits a small amount of organic material to cover the mineral particles as effectively as a coat of paint. About 4 or 5 percent humus is all that is required to impart a black color to soil.

MICROBIAL ACTIVITY

Microbial activity is essential for the release of plant nutrients from dead plant materials. Without such release the available plant nutrient supply would soon be used up and the soil would become sterile. Microbes complete the cycle so that nutrients taken up by the plant return again to the soil. Thus the same nutrient ions can be used over and over. An active, thriving microbial population is usually a good indication of a fertile soil.

One of the most important factors affecting microbial activity is the energy supply. Other factors such as aeration, water supply in the soil, temperature, pH (the degree of acidity or alkalinity), and the supply of each plant nutrient (nitrogen, calcium, phosphorus, potassium, etc.) are important, but the energy supply is very frequently the limiting factor in microbial activity. For most microbes the energy supply is plant and animal residues.

Most microbes wait for the plant or animal to die before they utilize its tissue as an energy source. Unfortunately, some microbes are able to get inside the plant or animal while it is living and cause infection and disease. Benefits accrue even from some infections, as in the case of nitrogen-fixing bacteria living in the roots of legumes.

The most favorable soil water content for microbial activity occurs when about one-half or two-thirds of the pore space is filled with water. Microbes live in a water medium, but most of them require atmospheric oxygen as well.

The effect of temperature on microbial activity was shown in Figure 3-16. The activity increases with rising temperature until there is some interference with life processes of the microbes unless the soil first becomes relatively dry. Microbial activity nearly halts in dry soil.

Reactions near neutral (or slightly alkaline) with a good supply of calcium present are favorable to most microbes. Liming a soil to raise its pH is likely to increase the microbial activity. This can speed up the release of plant nutrients making a larger supply available within a few weeks or months but at the expense of nutrient release during the next few years. Liming can also increase the activity of some disease-producing microbes such as the actinomycete that causes potato scab. The more acid soils generally produce better potatoes in areas where the scab organism is present.

Decomposition of Chemicals

Decomposition of chemicals added to the soil has received much attention in recent years. It is desirable for weed killers, insecticides, and other chemicals to decompose after they have done their work. It was once assumed that any such chemical that reached the soil could be decomposed by some soil microbe. This is true of many chemicals, but some are so different from natural soil materials and are so resistant to decomposition that such an assumption is questionable. It has been found that accumulations of certain chemicals over several years' time can poison the environment for some living things.

The ability of a chemical to perform a specific function and its ability to resist decomposition are often related to different parts of its molecular structure. One of the first applications of this discovery was in making laundry and dishwashing detergents that can be decomposed much more readily than the earlier detergents could. A major source of water contamination was thereby drastically reduced without any reduction in the effectiveness of the detergents for their intended uses. Much the same must be done for many other chemicals. DDT is an example. It is so effective that it rapidly became the widest-used insecticide. Its resistance to decomposition, however, is so great that it threatens unintended victims. Drainage water carries DDT from the soil into streams. Worms living in the soil ingest DDT and then birds eat the worms. Each step concentrates the DDT more, and it appears that some birds have been killed by DDT. Foodstuffs contain varying concentrations of DDT; some fish have been declared unfit for human consumption because of it. A chemical that could be readily decomposed by soil microorganisms would be unlikely to cause such problems.

Humanity needs effective pesticides but some, like DDT, can be overused and misused. A limited amount of DDT residue may be more acceptable than are the insects that it controls. Research is needed to determine how fast it breaks down and whether the decomposition can be accelerated. Most of the decomposition probably takes place in the soil. The rate of decomposition is an important factor in determining how much DDT can reasonably be used. The same applies to many other chemicals.

Decomposition of Refuse

Much of what has been said of chemicals also applies to solid refuse materials. Some materials decompose readily in the soil, whereas others are very resistant. For example, iron cans will rust but aluminum cans remain unchanged for years. Many plastics are very resistant to decomposition. Such materials accumulate and are a significant environmental pollution problem merely because of their large bulk. Efforts are being made to develop plastics and other materials that will decompose after they have served their purpose.

Much refuse is readily decomposable by soil microbes and can therefore be eliminated by being mixed with soil. The results range from a desirable fertilization effect to creation of pollution problems when the material is too concentrated. Many communities bury refuse in landfills where it may decompose. The main products of decomposition of organic wastes are carbon dioxide and water. These two products present no problem other than a possible settling of the fill. Many less abundant products of decomposition, however, will probably enter the groundwater and may become pollutants. Some such materials might be toxic to some forms of life, but the main ones are more likely to be plant nutrients (nitrates, for example). These can create problems by promoting the growth of unwanted algae and other plants in bodies of water.

CHEMICAL COMPOSITION OF HUMUS

The chemical composition of humus can be considered from the point of view of its elemental constituents or of the classes of organic compounds it contains. The principal elemental constituents are carbon, hydrogen, oxygen, and nitrogen with smaller amounts of phosphorus and sulfur. Humus contains every element absorbed by growing plants but not in the same proportions as in plants. Single-charged cations such as potassium and sodium are easily leached from dead organic matter because they do not take part in the covalent bonding of organic compounds. These cations soon return to either the mineral portion of the soil or to a living portion of the organic matter.

Decomposition processes gradually change the elemental composition of the soil organic matter. Microbes readily decompose the simpler compounds, and carbon dioxide and water are produced. The decomposition reduces the amount of organic matter but leaves it enriched in certain constituents, especially nitrogen. How far this process has gone can be roughly gaged by the ratio of carbon to nitrogen in the material. Table 5-2 shows such ratios for several organic materials. Except for legumes, most plant residues have much wider carbon-to-nitrogen ratios than the soil microbes and their residues have.

Approximate soil organic-matter contents can be calculated by multiplying percent organic carbon or percent nitrogen by suitable factors. A few such factors derived from the data in Table 5-2 are:

Grassland A1 horizon: % organic carbon × 1.9 = % organic matter
Most B2 horizons: % organic carbon × 2.5 = % organic matter
Either of the above: % nitrogen × 20 = % organic matter

Such calculations are sometimes useful, but they should be considered as approximations only. The factor 1.724 was long accepted for calculating organic matter from

Table 5–2   Average Carbon and Nitrogen Relations in Some Organic Materials on an Oven-dry Weight Basis

Organic material	% C	% N	C:N
Plant tissue:			
Alfalfa hay	40	3.0	13:1
Cornstalks	40	0.7	60:1
Oat straw	37	0.5	80:1
Microbial tissue:			
Bacteria	50	10.0	5:1
Actinomycetes	50	8.5	6:1
Fungi	50	5.0	10:1
Average microbes	50	6.2	8:1
Soil organic matter:			
Forest A1	50	2.5	20:1
Grassland A1	52	5.0	10.4:1
Average B2	40	5.0	8:1

organic carbon based on the assumption that organic matter is 58 percent carbon. Few samples actually contain that much carbon, and the data are much too variable to justify such a precise factor. The errors are likely to be even larger in calculations based on percent nitrogen because the factor is larger and more variable.

ORGANIC MATTER DECOMPOSITION

A question naturally arises as to how the relatively wide C:N ratios of plant materials are changed to the much narrower ratios of the microbes. A part of the answer is that much of the carbon is utilized as an energy source and converted to carbon dioxide. Most of the nitrogen becomes part of the microbial tissue, and the result is a much narrower C:N ratio. Sometimes the ratio is so wide that some nitrogen is drawn from the soil and utilized along with the plant carbon in forming microbial tissue. Decomposition is slowed if there is not enough nitrogen available in the soil and the plant tissue combined.

An example may help to clarify the preceding. Suppose that 6,000 kg/ha of cornstalks are left standing in a field after harvest. If these cornstalks are incorporated into the soil, the addition to the soil will be 6,000 kg of cornstalks containing:

2,400 kg of carbon (40%)
40 kg of nitrogen (60:1 C:N ratio)

If nitrogen is available in the soil and if three-fourths of the carbon is utilized for energy and one-fourth for microbial tissue, the resulting soil microbes will contain:

600 kg of carbon (¼ of plant carbon)
75 kg of nitrogen (8:1 C:N ratio)

This will require the immobilization of 35 kg (75–40) of nitrogen from the soil. If no nitrogen is available from the soil, the decomposition is limited to forming soil microbes containing:

40 kg of nitrogen (total available)
320 kg of carbon (8:1 C:N ratio)

An additional 960 kg of carbon will have been utilized for energy, making a total consumption of 1,280 kg of carbon. The other 1,120 kg of carbon must wait for a later cycle of decomposition. Later cycles occur as the microbes die and their bodies decompose. Additional carbon from the undecomposed residues is used to meet part of the needs of later generations of microbes until the supply is exhausted.

The preceding is certainly an oversimplification. The decomposition takes place over a period of time rather than suddenly as might be inferred. Also, no allowance was made for the changes that might occur in the microbial population with time. Fungi would probably be more important at first while the carbon supply for energy is plentiful. The C:N ratio resulting from predominantly fungal decomposition would be near 10:1. Many of the fungi would die as the supply of readily decomposable material became scarce. Bacteria would then decompose the fungal tissue along with some of the remaining plant tissue. Millions of bacteria can live on a few fungi because of the vast differences in size. The bacterial decomposition changes the C:N ratio toward (but seldom to) a limiting value of 5:1.

Nitrogen Release

Decomposition releases nitrogen if the organic material has a sufficiently narrow C:N ratio. By the assumptions of the preceding example the break-even point would be at 32:1 (25 percent efficiency in carbon conversion and 100 percent efficiency in nitrogen conversion reduces 32:1 to 8:1). This figure is consistent with the experimental results shown in Table 5-3. Waksman used rye plants of various ages to obtain materials of various C:N ratios (young plants usually contain a higher percentage of nitrogen than older plants). Table 5-3 also shows that the materials with the narrower C:N ratios decomposed about 50 percent faster than those that were deficient in nitrogen. This is an important principle in the use of green-manure crops. Young vegetation worked into the soil will decompose more rapidly and release available nitrogen much sooner than more mature crop residues.

The decomposition of 1 ton of alfalfa residues with a C:N ratio of 13:1 will serve as a field example of nitrogen release. Using the same assumptions as before, we have 1,000 kg of alfalfa containing:

400 kg of carbon (40%)
31 kg of nitrogen (13:1 C:N ratio)

Soil microbes decomposing the alfalfa will contain:

100 kg of carbon (¼ of plant carbon)
13 kg of nitrogen (8:1 C:N ratio)

The decomposition of each ton of alfalfa residues should therefore release about 18 kg of nitrogen to the soil where it is available for the current year's plant growth or, perhaps, to decompose other residues. Over a long period of time the microbial tissue should gradually decompose and release additional nitrogen.

Table 5–3   Influence of C:N Ratio of Rye Plants, as Controlled by Age, upon Their Decomposition during a 27-day Period

C:N ratio of plants	Relative decomposition C as CO_2 liberated, mg	N as NH_3 liberated, mg	N consumed from added nitrogen, mg
20:1	287	22.2	0
28:1	280	3.0	0
50:1	200	0	7.5
200:1	188	0	8.9

Source: Waksman, 1942.

Decomposition of Humus

The microbes become a part of the soil humus along with materials that have partially or entirely resisted the process of decomposition. The higher resistance of these remaining materials slows but does not stop decomposition. The continuing slow release of plant nutrients from decomposing humus is a very important part of the ability of the soil to supply the needs of plants.

The amount of humus decomposing each year is approximately the same as the amount of new humus formed during the year (unless some environmental factor such as climate or vegetation has changed recently). The average annual decomposition and renewal of humus is 2,000 to 4,000 kg/ha in most soils (this is only a portion of the residues added to the soil each year, but much of the residues decompose without ever becoming humus). The amount decomposed might represent anywhere from 1 percent or less of the humus present in the soil of a very cool area to over 25 percent of the humus present in a tropical soil. Temperate regions average about a 3 percent annual turnover in soil humus.

An understanding of humus decomposition helps explain why fertilizers become necessary after a field has been farmed for a few years. A farmer growing corn on a previously uncultivated field in a temperate region will serve as an illustration. Only nitrogen will be considered, for simplicity.

Suppose the soil being considered contained 5 percent organic matter (100,000 kg per hectare-furrow slice) of which 3,000 kg would normally decompose each year. The first year of cultivation, however, accelerates the decomposition so that at least 4,000 kg of humus are decomposed and 200 kg (5 percent of 4,000 kg) of nitrogen are released. This is about 4 percent of the nitrogen present and should be enough nitrogen to produce at least 60 q/ha of corn without fertilizer if other conditions are favorable. Some of the nitrogen will return to the soil in the corn residues and form new humus, but about half of it will be removed when the crop is harvested. The soil is likely to suffer a net loss of about 2,000 kg/ha of humus as a result of the first year of cultivation.

Suppose, further, that the farmer continues to grow corn on this field for 30 or 40 years during which time the soil-organic-matter content drops to about 3 percent (60,000 kg per hectare-furrow slice). At the same time, the rate of organic-matter decomposition and nitrogen release drops from the previous high rate of 4 percent per year to a new rate of only 2 percent per year because the remaining humus is composed almost entirely of materials that decompose very slowly. Of the 3,000 kg (5 percent of 60,000) of nitrogen present in the humus, only about 60 kg are released for plant growth. The corn yield potential *without fertilizer* has therefore declined to about 20 q/ha.

Fresh organic material in the form of plant residues decomposes much faster than soil humus. Jenkinson (1965) studied this with ryegrass residues labeled with carbon 14. His results are shown in Figure 5-3. Residues from young ryegrass were added to field soils, and the losses from decomposition were measured. Jenkinson found that about two-thirds of the carbon from the ryegrass residues was lost during the first 6 months. He further noted that the remainder of the decomposition closely followed a half-life concept wherein the amount of decomposition each year is

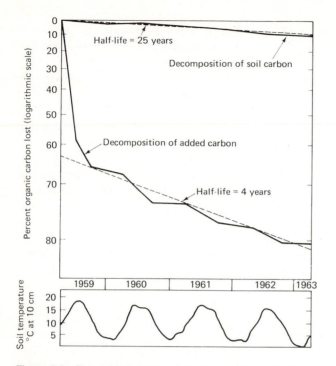

Figure 5-3   The decomposition rates of carbon from added ryegrass residues versus that from soil humus. (*Jenkinson,* 1965.)

proportional to the amount of material remaining. The ryegrass residues had a half-life of about 4 years, and the soil humus had a half-life of about 25 years. These half-lives correspond to annual decomposition rates of 16 percent per year for the ryegrass residues and 2.7 percent per year for the soil humus.

ORGANIC COMPOUNDS IN HUMUS

Humus is complex and variable. Thousands of organic compounds can be identified in soil organic matter, but these identifiable compounds constitute only 10 or 15 percent of the total organic weight. Most of the identifiable compounds are too readily decomposable to be considered humus. The components that last long enough to be true humus are sometimes referred to as *humic* materials. The most discussed, and probably most important, of the humic materials is the *humic acid fraction.* This is a group of substances that have several characteristics in common. They dissolve and make a dark brown or black solution when treated with weak NaOH or NH_4OH. Adding HCl causes the humic acid fraction to precipitate; the precipitate will not dissolve in alcohol.

Within their structures the humic acid materials contain several kinds of reactive groups including:

1 *Carboxyl groups:* – COOH. These are the characteristic groups of organic acids. The hydrogen can be ionized away leaving a negatively charged site that can attract any cation. The one bond of the carbon that is not tied to an oxygen (Figure 5-4) ties the carboxyl group to an organic structure.

2 *Phenolic hydroxyl groups:* – carbon ring with OH. The circle of carbon atoms (Figure 5-4) represents a benzene ring. The fourth bond of each carbon can tie to another part of an organic molecule, to a hydrogen atom, or to some other atom or radical (group of atoms). The H from the phenolic OH has a slight tendency to ionize away and leave a negatively charged site.

3 *Amine groups:* – NH_2 (Figure 5-4). Nitrogen typically enters organic structures with three covalent bonds to carbon and hydrogen atoms. Six electrons are shared in these three bonds but there are still two more electrons in nitrogen's outer shell and a hydrogen ion can attach itself to them. The addition of a hydrogen ion to the group results in a positively charged site. It is also possible for amine groups to react with carboxyl groups and produce a *peptide linkage* (Figure 5-4). Such linkages are the means by which amino acids are linked together to form proteins.

The reactive groups combined with nonreactive chains and rings form large molecules. Something of the complexity of the resulting molecular structure is illustrated in Figure 5-5. Such structures are suggestive only of the nature of the humic acid fraction; these materials are too variable to be accurately depicted in an illustration. The molecules are often many times as large as the one shown.

The presence of various types of reactive groups in soil organic matter is very important. These enable the organic materials to attract and store plant nutrients in forms available to plants. Such groups also tend to link the organic matter to clay particles, thus stabilizing soil structure and helping the organic material resist decomposition.

The 10 to 15 percent of soil organic matter that can be specifically identified includes a wide range of organic compounds. A simple classification is as follows:

Carboxyl group

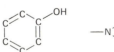

Phenolic hydroxyl group

Amine group

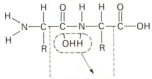

Peptide linkage

Figure 5-4 Important reactive groups and the peptide linkage.

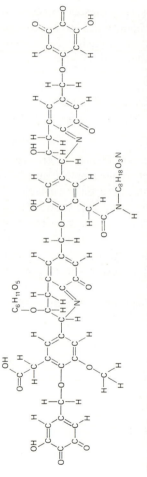

Figure 5-5 Dragunov's suggestion of the structure of a humic acid molecule. (*Redrawn from Kononova, 1961.*)

1 Polysaccharides—including cellulose and its decomposition products
2 Polypeptides—including proteins and their decomposition products
3 Polyphenols—including lignins and tannins
4 Simple organic compounds—organic acids, esters, alcohols, aldehydes, hydrocarbons, etc.

The lignins (polyphenols) are relatively resistant to decomposition and commonly become incorporated in humus. The other types of identifiable materials listed are relatively easily decomposed and would soon disappear from the soil if they were not frequently replenished. They are replenished when fresh residues reach the soil; in addition, much protein is formed as body tissue of microbes. Thus humus becomes enriched in lignin and protein and is sometimes referred to as a *ligno-proteinate complex*.

DISTRIBUTION OF ORGANIC MATTER IN THE SOIL

The distribution of organic matter with depth in a typical grassland soil is shown in the left diagram of Figure 5-6. Most of the roots occur in the upper part; the soil organic matter is naturally most concentrated in this same zone. Probably the relative difference in root concentration from one depth to another is greater than the difference in organic-matter content. More decomposition occurs in the upper layers because more organic matter is added there; furthermore, aeration is more adequate there than it is below. The organic-matter content of any horizon of any soil depends partly on how much organic matter is turned over to the soil every year and partly on what percentage of the organic matter decomposes during the year. The organic-matter content is stable when these two processes are balanced (plus or minus allowance for eluviation of humus in certain highly leached soils).

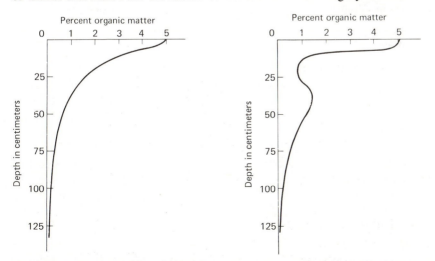

Figure 5-6 The distribution of organic matter in the profile of a grassland soil in a subhumid temperate region (left) and in the profile of a forested soil containing an accumulation of illuvial humus (right).

Effect of Vegetation

Changes in vegetation alter the pattern of organic matter accumulation within the soil. The pattern for a grassland soil is shown on the left side of Figure 5-6; that for a forest soil is shown on the right side. Important organic-matter features of a forest soil are the O horizons on the soil surface, the thin A1 horizons with organic-matter percentages comparable with grassland soils, eluviated A2 horizons of low organic-matter contents, and, in some, illuvial humus in the upper B horizons. The contrast with the simpler distribution pattern in the grassland soils is readily apparent.

Much of the organic matter of forest soils is derived from leaf fall. Tree roots are less important sources of organic matter than grass roots because much of the tree root system lives for many years. The annual turnover of organic matter from dying tree roots is therefore small.

The leaves and twigs in the O horizon (or litter layer) commonly blanket the soil surface so thoroughly as to make it almost immune to soil erosion (in forms other than leaching). The lower part of the litter decomposes gradually and is classified by forestry students as humus. This finely divided, dark-colored organic matter becomes mixed with the top 5 to 15 cm of mineral soil by the movement of the myriad of insects, worms, and other small animals living there. These same few centimeters of soil receive more organic matter from decomposing roots than do other layers because the macro- and microanimals damage these roots and cause sloughing.

The smaller accumulations of organic matter at lower levels in forest soils come mostly from root residues plus some illuvial humus. The illuvial humus is most important in Spodosols (a type of forested soil formed under acid-leaching conditions that cause the humus to move). The slight bulge in the organic-matter content in the middle of the forested soil profile shown in Figure 5-6 represents an illuvial humus B horizon. Younger and less acid soils commonly lack this bulge.

Effect of Climate

The organic-matter contents of grassland soils of drier climates average less than those illustrated in Figure 5-6 because less plant growth results and less organic matter is turned over to the soil. The relative distribution with depth in the soil, however, remains about the same except that all the horizons tend to be thinner in drier climates. Approximate contents of organic matter in the A1 horizons of well-drained grassland soils in the northern United States are shown in Figure 5-7.

Warmer climates have much the same effect on organic-matter contents as drier climates because they accelerate decomposition (Figure 5-8). The freezing weather during the winters of cooler climates tends to preserve the organic matter of soils by preventing microbial activity. This preservative effect more than offsets the decreased annual growth associated with the cooler temperatures. The organic-matter contents continue to increase even where the average annual temperature drops below freezing as in much of Alaska. The organic matter is produced during the warm summer and preserved during the cold winter.

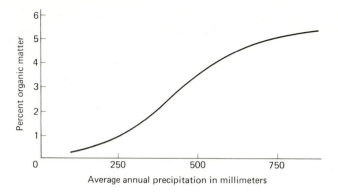

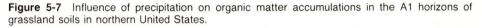

Figure 5-7 Influence of precipitation on organic matter accumulations in the A1 horizons of grassland soils in northern United States.

Forested soils do not show such dramatic increases in organic-matter content with increasing precipitation as do the grassland soils. In fact, the organic-matter content is more likely to decline in the very humid areas. The increased leaching decreases the fertility of the soils and thereby reduces plant growth. The additional water does little good for plant growth because most of it percolates through the solum into the substrata.

Topography

Topography modifies the microclimate and influences the vegetation, thereby producing a strong effect on the amount of organic matter in the soil. Topography also has a marked effect on the movement of water and soil both at and below the surface.

Soils on steep slopes have more runoff and make less water available to plants. Not only is the organic-matter content less because of the reduced plant growth, but also some of the organic matter produced is lost by erosion from the steep slopes, particularly from the upper part of the slopes. The result is a relatively shallow soil with a reduced percentage of organic matter and a much-reduced total content of

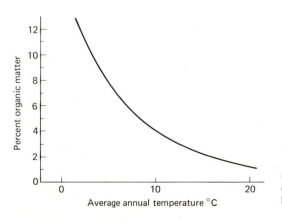

Figure 5-8 Influence of temperature on organic-matter accumulations in the A1 horizons of humid grassland soils.

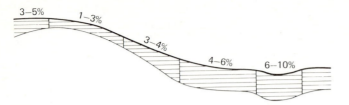

Figure 5-9 Soil-organic-matter contents and thickness of A1 horizons as related to topographic positions in a subhumid temperate climate.

organic matter. The soil, water, and organic matter moved by runoff and erosion accumulate at the foot of the slope, as shown in Figure 5-9.

The shape of the slope is also important in determining most of the above factors. The upper part of a slope commonly blends into the hilltop area with a convex segment. Much of this convex segment usually has even shallower soils and lower organic-matter contents than the steeper straight slope below. The reason is that the soil at any point in the convex area receives less soil and water moving to it from the flatter area above than it loses to the steeper area below. The result is much the same where the convexity is across the slope as on the nose of a hill. The water and soil moving there are spread out as they move downslope. The reverse effect occurs where the surface is concave.

The aspect (direction the slope faces) is important on steeply sloping soils. The average temperature of a soil that faces toward the warm early afternoon sun may be several degrees above that of a soil on the opposite slope. The soils on the sunny side are similar to soils of a warmer and drier climate where less organic matter accumulates. The soils on slopes receiving less sunshine are similar to soils of a cooler, more humid climate where more organic matter accumulates.

Parent Material

The organic-matter content of soil is influenced by parent material thickness, texture, and mineral content. Thin soils resulting from parent materials formed from hard rock produce less plant growth and contain less organic matter than deeper soils. In climates where either grasses or trees may grow, the finer-textured soils are likely to have a higher fertility level that is favorable to grass vegetation. Trees are more likely to be the dominant vegetation on the more leached and less fertile, sandier soils.

Other things being equal, the parent materials that provide a more adequate supply of mineral nutrients generally produce more plant growth and therefore more organic matter. The factor "opposing" organic-matter accumulation is decomposition. A warmer, better-aerated soil has more rapid decomposition than a cooler, less-aerated soil. Sandy soils are usually warmer and better aerated than finer-textured soils and permit more rapid decomposition of organic matter than finer-textured soils if, again, other things are equal. Actually, other things usually reinforce this effect. Sandy soils store less water and are therefore likely to produce less plant growth than soils with more clay. Moreover, unfertilized sandy soils are inherently likely to be less fertile than the soils with more clay, and this, too, leads to less plant growth. All these factors combined result in a good general correlation between the organic-matter contents and the clay contents of the soils of a given climate. There are exceptions to this general rule, of course. For example, some clay soils have characteristics that limit plant growth, and therefore the organic-matter accumulation in such soils may be less than that in adjacent coarser-textured soils. Such cases are exceptions. There are many examples of data showing good correlations between clay and organic-matter contents of soils.

Another important aspect of the relation between clay and organic matter has already been mentioned. The reactive groups of organic compounds possess electric charges that may be attracted to the electrically charged clay particles. This attraction is of great significance because the reactive groups are the parts of the organic compounds that are first attacked by enzymes produced by microorganisms. Where the reactive group is attracted to a clay particle, it is less accessible to enzymatic attack. Clay, therefore, tends to stabilize organic compounds against decomposition.

Effect of Time

To understand the time factor in organic-matter accumulation, one may visualize the result of scraping off the solum down to the parent material. Such exposures are frequently observed along new highways. Assume that no soil is replaced and no amendments are added to encourage plant growth but there is a reasonable thickness of parent material remaining. There is little plant growth on the parent material during the first few years of exposure because of problems of high bulk density, water supply, and available nitrogen deficiency (rock materials usually can provide most of the essential nutrients except nitrogen). The physical conditions gradually improve under the loosening action of wetting and drying, warming and cooling, and freezing and thawing. Water begins to infiltrate more readily and is more available to plants as the pore space increases. There is a gradual addition of nitrogen to the soil by rainfall and through fixation of nitrogen by microorganisms. The growth of plants becomes accelerated by the improvement in physical conditions and the increase in nitrogen and organic matter. After a period of several decades, possibly several centuries, the accumulation of organic matter begins to slow down and finally levels off to an approximately constant amount of organic matter for the particular set of environmental conditions.

The history of organic-matter accumulation can be divided into five phases as shown in Figure 5-10. The first phase is the gradual improvement of physical conditions and accumulation of enough nitrogen to support a cover of vegetation. The second phase is an accelerated rate of organic-matter accumulation because of increased availability of nitrogen and other nutrients. The third phase is the slowing down of the annual rate of organic-matter accumulation. The fourth phase is the equilibrium phase of maturity, where the amount of organic matter remains nearly constant over a very long period of time. Eventually a fifth phase occurs in which declining soil fertility leads to a decreased plant growth and a gradual decrease in organic matter in the soil.

The maturity phase lasts so long (probably tens or hundreds of thousands of years) that some of the other factors influencing soil organic matter may change. The

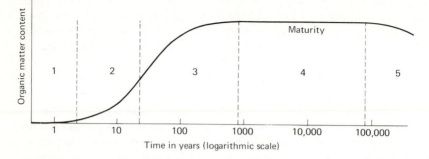

Figure 5-10 A schematic diagram illustrating the five stages in the accumulation and eventual decline of the organic matter in soils. The time scale is indicative only. The actual length of time of each stage varies from one soil to another.

climatic changes associated with the Pleistocene ice ages must surely have caused changes in climatic conditions and in soil-organic-matter contents. These effects would extend hundreds of kilometers from the nearest glacier. Also, many soils have undergone changes in vegetation at one time or another and have certainly undergone changes in their organic-matter contents.

Soils undergo a drastic change when they are first placed under cultivation. Some also undergo important changes later when cropping practices are changed. Each change leads to a gradual change in the soil-organic-matter content toward a new equilibrium level. Two possible shifts are shown in Figure 5-11. Similar changes might result from changes in fertilizer applications or other practices that influence plant growth. In each instance the rate of change is more rapid for the first decade or two than later because the discrepancy between the soil-organic-matter content and the new equilibrium level decreases with time. After several decades this discrepancy becomes so small as to be negligible.

The effects of different crops and cropping systems make it possible to manage the organic-matter content of cropland. High-producing legumes and grasses tend to increase the organic-matter contents and improve the physical properties of soils. Such crops are often needed to offset the detrimental effects of frequent cultivation for corn or other row crops. However, management is equally as important as the crop grown. A well-fertilized crop will produce more growth and return more organic matter to the soil than one that suffers from a deficiency. Use of the growth

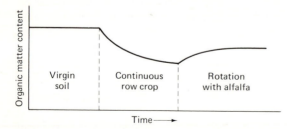

Figure 5-11 Changes in soil-organic-matter contents caused by cropping.

produced is also important. Mixing crop residues into the soil adds to the soil organic matter. Complete removal by burning or by harvesting all the growth as hay or silage without returning manure is soil-depleting no matter what crop is grown.

THE RHIZOSPHERE

Considerable study has been concentrated on the *rhizosphere* in recent years. Important processes take place within this zone in the immediate vicinity of the plant root. The roots absorb water and plant nutrients to meet the needs of the entire plant. With the exception of a few plants (such as rice) with special structures, the roots must also absorb oxygen for respiration. Then, too, some materials move from the root to the soil. Respiration produces carbon dioxide that escapes to the soil atmosphere. Hydrogen and hydroxyl ions produced in the roots may exchange places with some of the plant nutrient ions being absorbed. A variety of organic compounds are exuded from the roots either as waste products or to serve some function outside the plant roots.

An electron micrograph such as that shown in Figure 5-12 shows that a coating called *mucigel* builds up outside the cell walls of roots. Mucigel is probably mostly water, but it contains organic materials that give it body. Bacterial populations also surround the roots and may be seen within the mucigel, and so it is impossible to say how much of it is produced by the plant root and how much by the bacteria. The mucigel frequently bridges the gap between a plant root and a nearby soil particle. Probably most of the nutrients absorbed by plants pass through a layer of mucigel, many of them without passing through the simple water film that was long supposed to be the important carrier in the nutrient absorption process. The bacteria

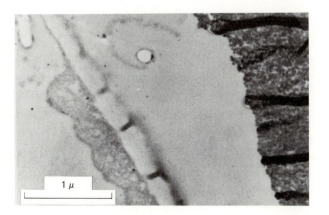

Root cell   Cell wall   Mucigel   Soil particle

Figure 5-12   The contact zone between a plant root and a soil particle. The curving gray line in the upper part of the mucigel is a bacterium. *(Electron micrograph courtesy of Dr. H. Jenny, Berkeley, Calif.)*

living in the mucigel undoubtedly influence its properties and may have much to do with nutrient availability and the movement of ions through the mucigel.

The fungi called *mycorrhizae* are abundant in the rhizosphere. Guttay (1975) suggests that mycorrhizae grow on the roots of nearly all plants and behave almost like a filamentous extension of the roots. Plants grown without mycorrhizae have been severely stunted in comparison to normal plants with mycorrhizae. The mycorrhizae absorb nutrients and pass them on to the plant, secrete hormones that stimulate plant growth, and help protect the plant roots from disease organisms. In return, the mycorrhizae depend on the plant for carbohydrates. Mycorrhizae are favored by moist, well-aerated soil and by the presence of decomposable organic matter. They are harmed by excessively acid or alkaline conditions, by excess salts, and by some soil sterilants and fungicides.

ORGANIC SOILS

All soils contain some organic matter, but relatively few contain enough to be classified as organic soils. The basic criteria for this classification is that enough organic matter be present to dominate the soil properties. The amount required varies according to the clay content of the soil because it takes more organic matter to coat a gram of clay particles than it does to coat coarser particles. The definition is based on the carbon content of the organic matter with 12 percent organic carbon being the limit between organic and mineral soils that contain no clay. The limit is raised as the clay content increases and reaches 18 percent for any soil containing more than 60 percent clay. The limit for soils containing less than 60 percent clay is calculated as:

$$\text{Minimum \% organic C in organic soil} = 12\% + \frac{\%\text{clay}}{10}$$

A soil is considered to be organic if its content of organic carbon exceeds either 18 percent or the variable limit calculated from the equation. It may be inferred from the equation for the variable limit that 1 percent of organic carbon (nearly 2 percent of organic matter) is able to influence the soil properties about as much as 10 percent clay.

An organic soil is dominated by organic layers in its root zone. The organic layers constitute at least two-thirds of the depth of a soil that is shallow to rock or gravel. Deeper soils must have organic layers totaling at least 40 cm thick within the upper 80 cm of soil. The 40-cm minimum is increased to 60 cm if the organic materials have a bulk density less than 0.1 g/cm^3.

Organic soils form where decomposition is slow because of wet conditions or cold temperatures. A combination of coldness and wetness causes organic soils called *Tundra* to blanket large areas in Alaska and Siberia. Tundra soils are shallow over permafrost. The deeper organic soils form in bogs in warmer climates. More plant material is produced in a warmer climate and much of it is preserved in the saturated conditions of a bog. Bogs are called *swamps* if they have trees or *marshes* if the vegetation is grasses, sedges, and other nontree species.

Peat

The soils formed in bogs are called *peats* and are classified according to the nature of the plant parts found in them. Relatively fresh materials in which the original plant fibers are still identifiable have been called *fibrous peat* and are classified as *fibric* soil materials by the Soil Survey Staff (1975). Peat may also be named for the plant material it contains such as *reed and sedge peat* or *sphagnum moss peat*. Sphagnum moss forms a very low-density peat (less than 0.1 g/cm^3) with a low ash content. Sphagnum moss peat is useful for holding water in a flower pot but something else should be added to provide fertility. Sphagnum moss has a very low fertility requirement and the peat it forms provides little fertility for anything else to grow.

Peat in which the plant fibers are no longer identifiable has been called *disintegrated peat*. The Soil Survey Staff distinguishes two stages of disintegration and classifies the in-between material *hemic* and the more disintegrated material *sapric.*

Another contrasting type of peat is called *sedimentary peat* and is classified as a *limnic material*. Sedimentary peat accumulates on the bottom of standing water from the residues of aquatic organisms. Sedimentary peat is undesirable for agricultural purposes because its gelatinous nature results in very low permeability.

Pieces of wood such as branches, logs, and stumps are considered to be coarse fragments comparable to gravel and other stony materials in mineral soils. Peat containing wood pieces more than 2-cm across is called *woody peat*.

Muck

The conditions under which peat forms are too wet for agricultural use (many such areas serve very well for waterfowl and other wildlife). Nevertheless, many peat bogs have been made into excellent farmland through artificial drainage. Drainage causes some important changes in the material. The first is a degree of compaction resulting from removal of the buoyancy and swelling effects of the water. The surface of the organic soil may drop as much as 30 cm during the first year the soil is drained.

The second important change is oxidation resulting from entry of air into the drained peat deposit. Oxidation changes the nature of the organic material so significantly that its name is changed from peat to *muck*. Oxidation decomposes a part of the material and releases plant nutrients. It also increases greatly the number of reactive groups, thereby increasing the capacity of the material to store plant nutrients in available forms. The physical characteristics such as pore space, aeration, and water-holding capacity are usually favorable to plant growth. Mucks are among the most productive soils known.

Many muck soils are used to produce vegetables and other high-value crops. The high values of these crops (and of muck soils) and the good results usually obtained frequently justify the application of more fertilizer and lime on the muck than on adjoining mineral soils.

Unfortunately, organic soils cannot be farmed without gradual subsidence (dropping of the surface level). Drainage is essential to farming operations, and drainage permits the organic matter to oxidize much faster than it is replenished. Also, mucks are very subject to wind erosion because the material is of low density. Subsidence at the rate of about 3 cm per year is fairly common. An organic soil 3 m thick may be nearly gone after 100 years of farming. Subsidence can be minimized by providing the minimum amount of drainage (maximum height of water table) consistent with the use made of the land, by flooding the land when it is not being cropped, and by providing protection against wind erosion.

REFERENCES

Alexander, M., 1965, Persistence and Biological Reactions of Pesticides in Soils, *Soil Sci. Soc. Am. Proc.* **29**:1–7.

Armstrong, D. E., G. Chesters, and R. F. Harris, 1967, Atrazine Hydrolysis in Soil, *Soil Sci. Soc. Am. Proc.* **31**:61–66.

Baxter, F. P., and F. D. Hole, 1967, Ant (*Fomica cinerea*) Pedoturbation in a Prairie Soil, *Soil Sci. Soc. Am. Proc.* **31**:425–428.

Boelter, D. H., 1969, Physical Properties of Peats as Related to Degree of Decomposition, *Soil Sci. Soc. Am. Proc.* **33**:606–609.

Broadbent, F. E., 1953, The Soil Organic Fraction, *Adv. Agron.* **5**:153–183.

Clark, F. E., 1954, A Perspective of Soil Microflora, *Proc. Soil Microbiology Conf.,* Purdue Univ.

Gascho, G. J., and F. J. Stevenson, 1968, An Improved Method for Extracting Organic Matter from Soil, *Soil Sci. Soc. Am. Proc.* **32**:117–119.

Giddens, J., S. Arsjad, and T. H. Rogers, 1965, Effect of Nitrogen and Green Manure on Corn Yield and Properties of a Cecil Soil, *Agron. J.* **57**:466–469.

Guttay, A. J. R., 1975, Fungus Helps Plants Grow, *Crops Soils* **27**(9):14–17.

Harris, R. F., G. Chesters, and O. N. Allen, 1966, Soil Aggregate Stabilization by the Indigenous Microflora as Affected by Temperature, *Soil Sci. Soc. Am. Proc.* **30**:207–210.

Jenkinson, D. S., 1965, Studies on the Decomposition of Plant Material in Soil. I. Losses of Carbon from [14]C Labelled Ryegrass Incubated with Soil in the Field, *J. Soil Sci.* **16**:104–115.

Jenny, H., and K. Grossenbacher, 1963, Root-soil Boundary Zones as Seen in the Electron Microscope, *Soil Sci. Soc. Am. Proc.* **27**:273–277.

Kononova, M. M., 1961, *Soil Organic Matter* (translated from Russian), Pergamon, New York, 450 p.

Leehneer, J. A., and P. G. Moe, 1969, Separation and Functional Group Analysis of Soil Organic Matter, *Soil Sci. Soc. Am. Proc.* **33**:267–269.

Macura, J., 1974, Trends and Advances in Soil Microbiology from 1924 to 1974, *Geoderma* **12**:311–329.

McFee, W. W., and E. L. Stone, Jr., 1965, Quantity, Distribution, and Variability of Organic Matter and Nutrients in a Forest Podzol in New York, *Soil Sci. Soc. Am. Proc.* **29**:432–436.

Malcolm, R. L., and R. J. McCracken, 1968, Canopy Drip: A Source of Mobile Soil Organic Matter for Mobilization of Iron and Aluminum, *Soil Sci. Soc. Am. Proc.* **32**:834–838.

Meredith, H. L., and H. Kohnke, 1965, The Significance of the Rate of Organic Matter Decomposition on the Aggregation of Soil, *Soil Sci. Soc. Am. Proc.* **29**:547–550.

Orlov, D. S., Y. M. Ammosova, and G. I. Glebova, 1975, Molecular Parameters of Humic Acids, *Geoderma* **13**:211–229.

Pal, D., and F. E. Broadbent, 1975, Influence of Moisture on Rice Straw Decomposition in Soils, *Soil Sci. Soc. Am. Proc.* **39**:59–63.

Schnitzer, M., 1970, Characteristics of Organic Matter Extracted from Podzol B Horizons, *Can. J. Soil Sci.* **50**:199–204.

Slavnina, T. P., 1971, Biochemical Processes in the Rhizosphere of Crops, *Sov. Soil Sci.* **3**:50–57.

Soil Survey Staff, 1975, Horizons and Properties Diagnostic for the Higher Categories: Organic Soils, Chap. 4 in *Soil Taxonomy*, Agriculture Handbook 436, Soil Conserv. Ser. USDA.

Tinker, P. B. H., and F. E. Sanders, 1975, Rhizosphere Microorganisms and Plant Nutrition, *Soil Sci.* **119**:363–368.

Unger, P. W., 1968, Soil Organic Matter and Nitrogen Changes during 24 years of Dryland Wheat Tillage and Cropping Practices, *Soil Sci. Soc. Am. Proc.* **32**:427–429.

Verma, L., J. P. Martin, and K. Haider, 1975, Decomposition of Carbon-14-labeled Proteins, Peptides, and Amino Acids; Free and Complexed with Humic Polymers, *Soil Sci. Soc. Am. Proc.* **39**:279–284.

Vimmerstedt, J. P., and J. H. Finney, 1973, Impact of Earthworm Introduction on Litter Burial and Nutrient Distribution in Ohio Strip-mine Spoil Banks, *Soil Sci. Soc. Am. Proc.* **37**:388–391.

Waksman, S. A., 1942, The Microbiologist Looks at Soil Organic Matter, *Soil Sci. Soc. Am. Proc.* **7**:16–21.

Chapter 6

Soil Mineralogy

A *mineral* is an inorganically formed solid composed of specific elements in a structural arrangement characteristic of the mineral. The mineral structure may permit some elements to partly substitute for each other although individual crystals are homogeneous. Several different minerals are usually present in any one rock and are inherited by soils formed from the rock.

Earlier chapters have stressed the importance of soil minerals as the original source of most plant nutrients. Sometimes the input of newly available plant nutrients is at a rate adequate to meet the needs of plants; more often it is inadequate for maximum plant growth. The adequacy of this process depends upon the soil minerals and their environment. The makeup of the minerals and the changes brought about by weathering are treated in this chapter.

Much of the basis for understanding soil mineralogy could be considered as solid-phase chemistry. As will be seen, this type of chemistry is quite different from solution chemistry. The size of ion formed by each element present is very important in the solid phase. The concept of molecules as the smallest units of compounds, so important in most concepts of chemistry, is replaced by a consideration of how the ions of the various elements present fit together in a mineral structure. There are no discrete molecules in minerals.

The elemental composition of soil minerals is initially the same as that of the rocks of the crust of the earth. However, the relative abundance of elements in the earth's crust differs from that of the earth as a whole because of gravitational sorting. Quantities of the initial supply of lightweight elements such as hydrogen and helium have risen into the atmosphere and escaped into space. Similarly, heavy elements like iron and nickel have tended to sink toward the center of the earth. The result is that only the nine elements listed in Table 6-1 are very abundant in the crust of the earth. The atoms and ions of the nine abundant elements represent over 99 percent of the atoms and ions present in the earth's crust (the tenth-most-abundant element, titanium, represents only 0.2 percent of the atoms and ions). These nine elements are necessarily the principal building blocks of minerals.

An important difference between cations and anions causes the disparity among the elements to be even greater on a volume basis than on a numerical basis. A cation is formed when an atom loses one or more of its electrons and thus becomes positively charged. The positive charge pulls the remaining electrons closer to the nucleus so the effective size decreases. A cation is therefore smaller than the atom. Conversely, an expansion occurs when an atom gains one or more electrons and becomes an anion. The size of the anion is larger than the atom.

Oxygen is not only the most abundant element in soil minerals, it is the only one of the nine abundant elements that forms an anion. The large size and number of oxygen anions cause them to occupy over 90 percent of the volume of the earth's crust. Therefore, most rock and soil minerals can be considered as essentially a systematic arrangement of oxygen anions with various cations (mostly silicon and aluminum) in the holes between. Minerals having frameworks of oxygen and silicon are called *silicates;* silicates that include aluminum in their frameworks are called *aluminosilicates.* These minerals warrant attention in the study of soils because of their great abundance.

BUILDING BLOCKS OF MINERALS

Minerals are composed of orderly arrangements of ions. These orderly arrangements can best be understood by considering a few rather simple basic units that occur repeatedly within mineral structures. These basic units are groups of oxygen ions

Table 6–1 The Nine Most Abundant Elements in the Crust of the Earth on the Basis of Numbers of Atoms and Ions Present

Element	Chemical symbol	Ions	Percent of all atoms and ions
Oxygen	O	O^{--}	60
Silicon	Si	Si^{++++}	20
Aluminum	Al	Al^{+++}	6
Hydrogen	H	H^+	3
Sodium	Na	Na^+	3
Calcium	Ca	Ca^{++}	2
Iron	Fe	Fe^{++} or Fe^{+++}	2
Magnesium	Mg	Mg^{++}	2
Potassium	K	K^+	1

packed closely together and enclosing cations in holes of appropriate sizes. For discussion purposes, each ion is assumed to be a sphere of a diameter that depends on the atomic number of the element and the charge of the ion. Larger atomic numbers produce larger-sized atoms; positively charged ions are smaller than the corresponding atom, and negatively charged ions are larger. The ions act as though they have very definite sizes even though they are composed mostly of empty space.

The three basic structural units of greatest interest are the tetrahedron, octahedron, and cube. Some features of these units are represented in Figure 6-1.

A *tetrahedron* is a three-dimensional figure having four triangular sides. It may be visualized as a pyramid with a triangular base. Such a figure has four points; four oxygen ions centered at these four points produce the tetrahedral unit of mineral structures. The normal spacing of the oxygen ions in tetrahedra and other compact structures is 2.8 Å[1] (the radius of an oxygen ion is 1.4 Å). The tetrahedral arrangement can be represented by three balls placed in a triangle in contact with each other and with a fourth ball resting on top above the center of the triangle. The space inside a tetrahedron made of oxygen ions is so small that only a very small cation such as Si^{++++} or Al^{+++} can fit inside. Even Al^{+++} is crowded inside oxygen tetrahedra and is more frequently found in octahedral spaces. Cations occurring inside tetrahe-

[1] Å stands for Angstrom unit and is equal to one ten-millionth of a millimeter.

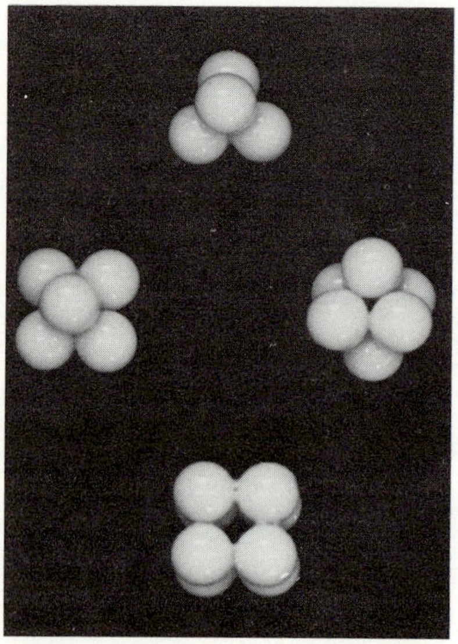

Structural unit	Form	Cations in center	Ionic radius (Å)	Coordination number
Tetrahedron		Si^{++++} or Al^{+++}	0.42 0.51	4 4
Octahedron		Al^{+++} Mg^{++} or Fe^{++}	0.51 0.66 0.74	6 6 6
Cube (or larger)		Na^+ Ca^{++} or K^+	0.97 0.99 1.33	8 8 >8

Figure 6-1 Basic structural units of minerals.

dra are said to have coordination numbers of 4 because there are four anions (oxygen ions) adjacent to them.

An *octahedron* can be visualized as two four-sided pyramids placed base to base. It has eight sides and six points. The arrangement of ions can be simulated by placing four spheres in a square so that each sphere touches two of the others. One additional sphere centered above the square completes one pyramid, and another below completes the other pyramid. Exactly the same figure as seen from a different viewpoint can be made from two triangular arrangements of spheres (Figure 6-1). The space inside an octahedron is too large for Si^{++++} ions because the bonds would be longer and therefore not as strong as in a tetrahedron. Nor does it hold such large cations as Na^+, Ca^{++}, or K^+. But it does contain medium-sized cations such as Al^{+++}, Mg^{++}, or Fe^{++} (as well as several other less common cations of similar size). Octahedral cations have coordination numbers of 6 because there are six anions (oxygens) adjacent to them.

Cubes are the simplest of several possible arrangements of oxygen ions that provide space for large cations such as Na^+, Ca^{++}, and K^+. Many different kinds of large spaces occur because they represent the leftover space after tetrahedra and octahedra have been formed. The smaller structures form first because they have shorter (and therefore stronger) cation-to-anion bonds than the larger structures. It is, of course, the total assemblage that constitutes the structure of the mineral, accounts for all of the bonding, and has a balanced electrical charge. Individual tetrahedra and other units cannot be removed from a mineral without breaking bonds and destroying the charge balance.

Tetrahedra, octahedra, and large structures provide places for all the abundant ions except hydrogen. For practical purposes, hydrogen ions occupy no space at all. Most hydrogen ions consist of a lone proton. This proton is held so close to an oxygen ion that the only significant effect it has is to neutralize one of the negative charges of oxygen. The result is a hydroxyl ion (OH^-) that fits into mineral structures as though it were an oxygen ion with only one negative charge.

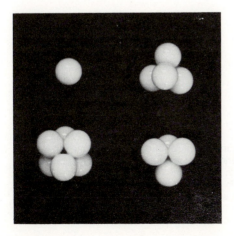

Figure 6-2 The single oxygen ion and the upper half of the octahedron (left) will form a tetrahedron when the gap is closed; three oxygen ions from each of the tetrahedra (right) form an octahedron when the gap is closed.

LINKAGE OF STRUCTURAL UNITS

Oxygen ions are usually part of two structural units rather than merely one. For example, each side of an octahedron is a triangle composed of three oxygen ions. These three oxygen ions plus an additional oxygen ion located outside the octahedron can form a tetrahedron (Figure 6-2). Similarly, an octahedron is formed when the triangular sides of two tetrahedra are placed next to each other. Such occurrences are repeated countless times in mineral structures. Appropriate cations may then occupy each type of space thus formed, for example, Fe^{++} or Mg^{++} in the octahedra and Si^{++++} in the tetrahedra. Ionic bonding between the cations and the oxygen anions holds the structures together.

Tetrahedra can be linked to each other at their corners to produce pairs, rings, chains, sheets, or three-dimensional frameworks. Certain of these are illustrated in Figure 6-3, and some of their significant characteristics are presented in Table 6-2. The manner of linkage of tetrahedra serves as a basis for classifying silicate minerals. Detailed study of most of these minerals belongs to the field of geology, and so only a brief review is included in this book. One group of silicate minerals, however, has particular importance in soils. This group is the *phyllosilicates,* sometimes known as the *layer silicates.* It includes the silicate clay minerals and micas. In addition, soils contain abundant amounts of tektosilicates, inosilicates, and sometimes nesosilicates inherited from their parent materials. All classes of silicates will be discussed in the next section beginning with the simplest and emphasizing the minerals most common in soils.

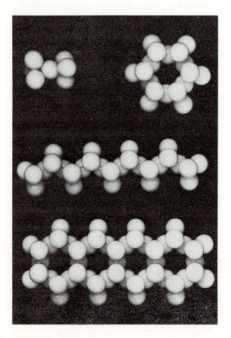

Figure 6-3 Tetrahedra are linked by sharing oxygen ions producing (*a*) pairs as in sorosilicates (upper left); (*b*) rings as in cyclosilicates (upper right); (*c*) single chains as in pyroxene inosilicates (center); and (*d*) double chains as in amphibole inosilicates (bottom).

Table 6–2   Classes of Silicate Minerals

Class	Tetrahedral groupings	Si:O ratio	External charge per Si tetrahedron	Example
Nesosilicates	Independent	1:4	−4	Olivene
Sorosilicates	Pairs	2:7	−3	Hemimorphite
Cyclosilicates	Rings	1:3	−2	Beryl
Inosilicates	Single chains	1:3	−2	Augite
	Double chains	4:11	−1.5	Hornblende
Phyllosilicates	Sheets	2:5	−1	Mica and clay minerals
Tektosilicates	Three-dimensional frameworks	1:2	0	Quartz and feldspars

SILICATE MINERALS

Nesosilicates, the simplest type, are composed of independent tetrahedra. Each tetrahedron has four oxygen ions of its own without any of the sharing that occurs in the other classes of silicates. The most common nesosilicate is olivene, a dark-colored mineral. The chemical formula of olivene, $(Fe,Mg)_2SiO_4$, reveals the 1:4 ratio of silicon to oxygen. The structure can be considered as an organized group of tetrahedra with half pointing up and half pointing down (Figure 6-4), all packed tightly together forming a dense mineral. Octahedra are formed from the adjacent sides of pairs of tetrahedra. Magnesium and ferrous (+2) iron are interchangeable in the octahedral spaces, and most olivene contains some of each. Weathering action in soils gradually releases these cations and makes them available for plant growth as the nesosilicate structure breaks apart. Olivene occurs in basalt and other rocks high in iron and magnesium but relatively low in silicon. Such rocks are termed *basic* or *ultrabasic* because of their low contents of silicon.

The *sorosilicates* are relatively rare minerals and are unimportant in soils. They are formed of pairs of tetrahedra (such as that shown in Figure 6-3) bonded together with appropriate cations in octahedral spaces and containing water molecules in otherwise vacant spaces.

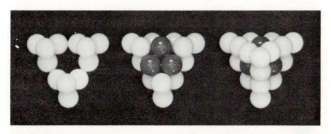

Figure 6-4   Nesosilicates such as olivene are formed by close packing of tetrahedra, some pointing up (light colored in photograph) and some down (dark colored). Octahedra are formed between the tetrahedra and contain either Mg^{++} or Fe^{++}.

The *cyclosilicates* have rings of tetrahedra. Cyclosilicates are not common in soils, but some of them have significance as gems and as ores. Beryl ($Be_3Al_2Si_6O_{18}$) is the commercial ore for the element beryllium; emeralds and aquamarines are gem forms of beryl. The structure of a cyclosilicate may be considered as stacks of Si_6O_{18} rings such as the one shown in Figure 6-3 arranged in parallel tubes with cations such as beryllium and aluminum, and some impurities in the remaining spaces. The impurities have much to do with the color and the quality of the gems.

The *inosilicates* are composed of either single or double chains. They commonly occur as dark-colored minerals in rocks and soils (though there are also other dark-colored minerals and some inosilicates are light colored). Those composed of single chains are termed *pyroxenes* and those of double chains are called *amphiboles,* but the distinction is often difficult to make and is usually not very important in soils. The inosilicate structure is essentially a bundle of parallel chains held together by cations in the octahedral and larger spaces between the chains (Figure 6-5). The chemistry is very complex. Somewhat idealized formulas[1] are $Ca(Mg,Fe,Al)$ $(Al,Si)_2O_6$ for augite (a single-chain mineral) and $NaCa_2(Mg,Fe,Al)_5(Si,Al)_8O_{22}$-$(OH)_2$ for hornblende (a double-chain mineral). These minerals are sources of several essential plant nutrients that are released by weathering.

The simplest *tektosilicate* is quartz (SiO_2). The structure of quartz is a three-dimensional framework of tetrahedra with every tetrahedron linked to four other tetrahedra (one at each corner). The linked tetrahedra cannot pack tightly together. Some fairly large spaces occur in the structure. Water molecules and other impurities are sometimes found in these spaces, but they are not held there by ionic bonds because the SiO_2 formula has no leftover electric charge. Pure quartz does not contain any essential plant nutrients. Quartz occurs in a rock only if the formation of other minerals leaves a surplus of silicon and oxygen. Rocks containing quartz are said to be acidic.[2]

Quartz is too often equated with sand. It is true that certain highly weathered sands and sandstones are nearly pure quartz, but quartz particles can be of any size from fine clay to sizable stones. Some quartz in soils is inherited from the parent rock and some forms in the soil as a residue from the weathering of other silicate minerals. Some quartz crystallizes inside plant cells from dissolved silicon absorbed along with plant nutrients. These small, elongated quartz crystals are called *plant opal* and give clues to the soil's past vegetation.

Tektosilicates other than quartz result when some tetrahedra contain aluminum rather than silicon. Each Al^{+++} provides one less charge than Si^{++++} so that additional positive charges are required to balance the negative charges from the oxygen. These positive charges must come from large cations because the tektosilicate structure does not provide any octahedral sites. The resulting minerals are the feldspars, the most abundant of all minerals. The most important feldspars are

---

[1]In these and other mineral formulas cited the cations are grouped according to those that occupy large spaces, octahedral spaces, and finally tetrahedral spaces; those cations enclosed by parentheses can substitute for one another in the appropriate spaces.

[2]The terms *acidic* and *basic* as applied to rocks have nothing to do with pH. This terminology was developed under early theories of rock crystallization. The theories have been replaced, but the terminology is still used.

Figure 6-5 An inosilicate structure showing the arrangement of three double chains pointing alternately up and down. The chains would be much longer and would merge more closely together in the actual mineral structure.

orthoclase (with an ideal formula of $KAlSi_3O_8$) and plagioclase (any intermediate between $NaAlSi_3O_8$ and $CaAl_2Si_2O_8$).

The *phyllosilicates,* sometimes called *layer silicates,* are of special importance in soils because they include the silicate clay minerals and micas. These minerals have much to do with soil chemistry and soil fertility. They have been left until last because they will be discussed in greater detail than the other classes of minerals.

Phyllosilicates are composed of layers that are strongly bonded internally but weakly bonded between layers. The layers are only three or four oxygen ions thick and give mica its well-known ability to be split into ultrathin layers. The internal layers in phyllosilicates can be described as sandwiches of tetrahedral and octahedral sheets. In some phyllosilicates the tetrahedral and octahedral sheets alternate and occur in equal numbers. These are known as 1:1 layer silicates in contrast with the 2:1 layer silicates that have tetrahedral sheets both above and below the octahedral sheet in each "sandwich" or layer.

Tetrahedra sharing oxygen at three of their four corners form a tetrahedral sheet. The unshared oxygen of the tetrahedra all point in the same direction and form rings as shown in Figure 6-6.

Octahedral sheets lack the holes of the tetrahedral sheets, but they are slightly expanded (to match the oxygen spacing in tetrahedral sheets) as shown in Figure 6-7. The octahedral sheet shown in the figure shows only two-thirds of the octahedral sites filled with cations. This condition corresponds to trivalent cations such as Al^{+++} occupying the sites. Octahedral sheets also occur in which all the octahedral sites are occupied by divalent cations such as Mg^{++} and Fe^{++}. Octahedral sheets formed of hydroxyl ions instead of oxygen ions form stable minerals such as gibbsite, $Al_2(OH)_6$, and brucite, $Mg_3(OH)_6$ [these formulas are often simplified to $Al(OH)_3$ and $Mg(OH)_2$].

Kaolinite

Kaolinite is the simplest layer silicate. Each of its layers contains one tetrahedral and one octahedral sheet united by the apex (unshared) oxygen ions from the tetrahedral

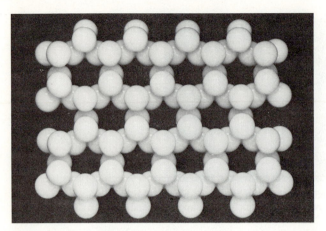

Figure 6-6 The arrangement of oxygen ions in a tetrahedral sheet.

sheet replacing some of the hydroxyl ions of the gibbsite [$Al_2(OH)_6$] form of octahedral sheets. These apex oxygen ions have one charge neutralized by the silicon ions in the tetrahedral sheet and so carry the same effective charge as the OH^- ions they replace. The kaolinite structure is illustrated in Figure 6-8. The ions add up to a theoretical formula of $Al_4Si_4O_{10}(OH)_8$; the composition of kaolinite deviates very little from this formula. Each of the layers is electrically neutral except for broken bonds at crystal edges. Broken bonds give kaolinite a small amount of charge that ultimately serves to store plant nutrients in the soil. All the positive and negative charges within the layer are balanced so that no ionic bonds are left to tie adjacent layers together. Nevertheless, kaolinite crystals or particles are many layers thick

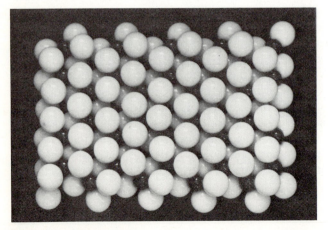

Figure 6-7 The arrangement of oxygen (or hydroxyl) and aluminum ions (the small dark spheres) in an octahedral sheet.

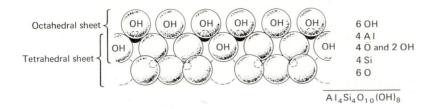

Octahedral sheet {   OH   OH   OH   OH   OH   OH   6 OH
                                                        4 Al
Tetrahedral sheet {   OH   OH   4 O and 2 OH
                                                        4 Si
                                                        6 O
                                                        ─────────────
                                                        $Al_4Si_4O_{10}(OH)_8$

Figure 6-8 A side view of a layer of kaolinite.

because hydrogen bonding holds them together. The hydrogen of the hydroxyl ions in the octahedral sheet of one layer is also weakly attracted to the oxygen in the adjacent tetrahedral sheet of the next layer just as hydrogen bonding links water molecules together. Hydrogen bonding is strong enough to hold the layers together and prevent water and ions from entering between the layers. Kaolinite is therefore a nonexpanding clay mineral and has little or no tendency to swell when wet or shrink when dry.

Kaolinite is a very important clay mineral (or, more correctly, an important group of closely related clay minerals). Kaolinite is described as a 1:1 layer silicate with a nonexpanding lattice and a low cation-exchange capacity. Its particle size is often that of coarse clay or fine silt. Such particles are composed of many layers held rigidly together with a spacing of 7.2 Å from the center of one layer to the center of the next. Electron micrographs show a marked tendency for kaolinite to form six-sided crystals (Figure 6-9). The reason for this can be seen in the arrangement of the oxygen ions illustrated in Figure 6-6.

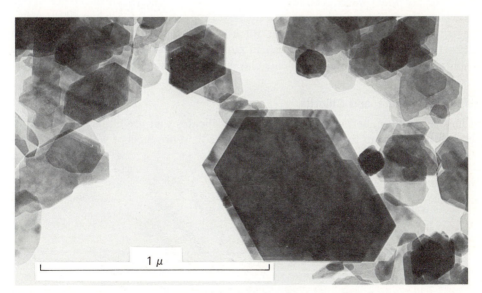

1 μ

Figure 6-9 An electron micrograph of kaolinite. *(Iowa State Univ. Eng. Res. Inst.)*

Montmorillonite

Montmorillonite is another very important clay mineral (or group of clay minerals). Each of its layers contains two tetrahedral sheets and one octahedral sheet. The two tetrahedral sheets are held on opposite sides of the octahedral sheet by apex oxygen ions replacing hydroxyl ions as in the kaolinite structure. The basic arrangement is shown in Figure 6-10 but the ideal formula shown is that of pyrophyllite, a mineral that has an equal number of positive and negative charges in each layer. Montmorillonite differs from the pyrophyllite structure in its content of both cations and water. Some of the aluminum may be replaced by iron or magnesium and some of the silicon may be replaced by aluminum. Variable amounts of water are absorbed between the layers. The actual composition of montmorillonite layers is variable, but something like $(Mg_{0.2}Al_{1.8})(Al_{0.1}Si_{3.9})O_{10}(OH)_2 \cdot xH_2O$. The cation substitutions lead to a deficiency of positive charges in the structure with the result that the layers are negatively charged. The negative charges give montmorillonite a relatively large capacity to attract cations such as Ca^{++} to the layer surfaces, but the attraction is loose enough to permit other cations to replace the ones already there in an exchange process. These loosely held cations, called *exchangeable cations,* are an important reserve of nutrients for plant growth.

Montmorillonite has an expanding lattice that permits both water and cations to enter between its layers. There is no hydrogen bonding to lock the layers together because the hydroxyl ions remaining in the octahedral sheet are completely enclosed by the tetrahedral sheets. The exchangeable cations provide an attractive force between the layers but the attraction is too weak to prevent water from entering. The ease with which water enters and exits from between the layers of montmorillonite leads to large volume changes. Soils containing montmorillonite clay swell or shrink considerably when water is added or removed. The distance from the center of one montmorillonite layer to the center of the next ranges from 9.6 Å when all the water is removed to 21.4 Å or more when much water is present. Still, there is enough attraction to hold the layers together in particles several layers thick. Usually the montmorillonite particles are of fine-clay size, about one-tenth as large as most kaolinite particles.

Illite, Vermiculite, and Chlorite

Illite, vermiculite, and chlorite are 2:1 layer silicates that are either primary minerals inherited from the parent rock or intermediate products of weathering. Wea-

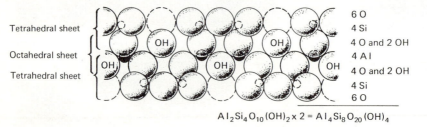

Tetrahedral sheet	6 O
	4 Si
	4 O and 2 OH
Octahedral sheet	4 Al
	4 O and 2 OH
Tetrahedral sheet	4 Si
	6 O

$$Al_2Si_4O_{10}(OH)_2 \times 2 = Al_4Si_8O_{20}(OH)_4$$

Figure 6-10   A side view of the structure of a 2:1 layer silicate.

thering action that will eventually change these minerals to montmorillonite and kaolinte is in progress in many soils. Illite, especially is often mixed with montmorillonite.

Illite is a variable form of muscovite (white mica) occurring as small particles (mostly of clay size). Illite layers have more negative charge than montmorillonite layers because one-fourth of the tetrahedral positions in illite are occupied by Al^{+++} instead of Si^{++++}. The negative charge is offset by K^+ ions held between the layers. The potassium ions fit into the holes in the tetrahedral sheets and hold adjacent layers together too tightly for water to enter. The illite lattice is therefore nonexpanding. The K^+ ions in the lattice are not available to plants but gradually become available as they are released by weathering.

Vermiculite occurs in rocks and soils that have high contents of magnesium. It has Mg^{++} in the octahedral sheet instead of Al^{+++}. Vermiculite is like illite in having Al^{+++} replacing one-fourth of the Si^{++++} in the tetrahedral sheets. The lattice is partially expanded, and two layers of water molecules occupy the space between tetrahedral-octahedral "sandwiches." Enough Mg^{++} ions occur in these water layers to offset the excess negative charges in the tetrahedral sheets. A further very dramatic expansion of vermiculite occurs when it is quickly heated to 250 to 300°C. The vaporizing water makes the vermiculite explode to as much as thirty times its original volume. Expanded vermiculite is widely used as an insulating material.

Vermiculite is not as abundant in soils as kaolinite, montmorillonite, and illite, but it is important in soils where magnesium was abundant at the time of clay formation. Such concentrations of magnesium are most likely to occur under slightly alkaline conditions. The presence of vermiculite in a soil is important because the Mg^{++} ions in the water layers can be exchanged with other cations and give vermiculite an even higher cation exchange capacity than montmorillonite for storing available plant nutrients.

Chlorite is another partially expanded layer silicate occurring in both soils and rocks. It has Mg^{++} and Al^{+++} ions in two layers of OH^- ions between the tetrahedral-octahedral "sandwiches." Enough Mg^{++} and Al^{+++} ions are present to balance the negative charges of the OH^- ions as well as a small negative charge from the tetrahedral-octahedral layers. The charges produce bonding that holds the entire assembly together so that it does not expand further.

Table 6–3   A Summary of Some Important Characteristics of Silicate Clay Minerals

Mineral	Lattice	Bonding between layers	Layer spacing
Kaolinite	1:1 nonexpanding	Hydrogen	7.2 Å
Chlorite	2:1 partially expanded	Magnesium and aluminum ions	14.3 Å
Illite	2.1 nonexpanding	Potassium ions	10.0 Å
Montmorillonite	2:1 expanding	Exchangeable cations	9.6—21.4 Å
Vermiculite	2:1 partially expanded	Magnesium ions	14—15 Å

Table 6-3 summarizes some important characteristics of the silicate clay minerals. The spacings between layers shown in the table are important means of identifying clay minerals. These spacings can be measured by x-ray diffraction techniques.

NONSILICATE MINERALS IN SOILS

Only a few nonsilicate minerals are important in soils. Many others are either of very restricted occurrence or are too soluble to persist in most soils. Some are too inert to contribute to soil fertility. Only calcite and some very stable oxides will be considered here.

Calcite is the most abundant form of calcium carbonate and the principal component of limestone. It is a common constituent of soil parent materials and is also synthesized in many soils. Calcium carbonate is very slightly soluble in water, and the depth to which free $CaCO_3$ has been leached from a soil is often taken as an indicator of the depth of effective leaching and of the thickness of the solum. In arid and semiarid regions, too little water passes through most of the soils to remove $CaCO_3$ from the profile. It then accumulates in a layer, usually in or beneath the lower part of the solum. Such layers are designated as B3ca and Cca horizons. Calcium carbonate is completely removed from the sola in humid regions.

Carbonate ions consist of a triangular arrangement of three oxygen ions with the tiny carbon ion in the center (carbon is even smaller than silicon). Equal numbers of Ca^{++} and CO_3^{--} are packed together to form calcite or other forms of calcium carbonate. Calcite crystals sometimes bridge the space between soil particles and cement them together. Such cementation sometimes hardens a Cca horizon into a hard or semihard mass.

Gibbsite is an aluminum-rich mineral that accumulates in highly weathered soils. It has layers composed of Al^{+++} ions in octahedral coordination between two sheets of OH^- ions. Gibbsite crystals tend to be flaky because adjoining layers are weakly attracted to each other. The flakes are often too small to be noticeably crystalline.

The chemical formula of gibbsite is $Al(OH)_3$. It usually occurs with two related aluminum minerals: boehmite, $AlO(OH)$, and diaspore, $HAlO_2$. These minerals are either called hydrous oxides or grouped with hematite, anatase, etc., as oxide minerals. The earthy-smelling aluminum ore called *bauxite* contains these three aluminum minerals. Bauxite is formed by intense tropical weathering of rocks containing aluminosilicate minerals. At least the upper part of the deposit was once soil. These same three aluminum minerals occur in present-day highly weathered soils along with other minerals that resist weathering.

Hematite (Fe_2O_3) is an oxide of ferric iron. The Fe^{+++} ions occupy octahedral sites between closely packed oxygen ions. The close packing plus the iron content make hematite about twice as dense (5.26 g/cm³) as most soil minerals. Particle densities of soils, however, are never this high, partly because of the presence of silicate minerals and aluminum minerals and also because much of the iron is hydrated with variable amounts of water. One specific iron mineral that forms in the presence of water is goethite, $FeO(OH)$.

Hematite is the most common red coloring matter in soils. Goethite and related compounds are called *limonite* and constitute the usual yellowish-brown coloring matter. Small amounts of these iron minerals and of the aluminum minerals occur in almost all soils. They are formed by weathering processes and are very insoluble. They tend to accumulate in highly weathered soils and are most abundant in tropical soils. Iron ores are presumably end products of weathering of iron-rich rocks in ages past.

Anatase is a form of titanium oxide, TiO_2. Titanium is the tenth most abundant element in the earth's crust, and a small amount of its oxide occurs in almost all soils. Titanium occurs inside oxygen octahedra and makes a very stable structure. The anatase particles are usually very small but are even more resistant to weathering than are the oxides of iron, aluminum, and silicon.

Anatase is sometimes used as a reference mineral to estimate how much material has been lost from a soil by weathering. The anatase percentages in the A and C horizons are compared. A soil that contains twice as much anatase in its A horizon as in its C horizon must have lost at least half of its original A horizon source material by weathering (assuming uniform parent material).

ALLOPHANE

Allophane is the material in the fine-clay fraction that does not have an identifiable structure. The constituents of allophane are too small to be identified but presumably include the building blocks of minerals—tetrahedra containing silicon and octahedra containing aluminum, magnesium, and iron.

Allophane is most prominent in young soils formed in volcanic ash. This is probably true because material spewed into the atmosphere by a volcano cools instantaneously. There is no time for the fluid material to be organized into mineral crystals large enough to be identified. Soils formed from volcanic ash are high in fine clay and usually quite fertile.

Allophane can be produced in any soil as mineral particles weather. Structural bits too small to be identified are broken from the mineral particles and exist as tiny fragments until they are either built into new minerals or broken completely apart and dissolved. These fragments may bond together into poorly organized units of fine-clay size.

CATION-EXCHANGE CAPACITY

Silicate clay minerals, allophane, and humus all have the important characteristic known as *cation-exchange capacity*. Each of these materials has negative charges that attract cations. Such cations are called *exchangeable* if they can be replaced with other cations dissolved in water surrounding the particles. Such replacement is possible if the bonding is not too strong and if the sites are accessible to the soil solution.

The negative charges responsible for cation-exchange capacities arise in three distinct ways: (1) cation substitutions within mineral layers; (2) broken-edge bonds;

and (3) H^+ ionizing from organic materials. Ionization of H^+ from carboxyl groups and phenolic—OH groups was discussed in Chapter 5. The resulting negatively charged sites give humus a high cation-exchange capacity.

Cation substitutions are also called *isomorphic substitutions* because the substitute cations fit into the mineral structure with no change in external form. Cation substitutions occur within the layers of all 2:1 layer silicates (kaolinite has little or no cation substitution). One negative charge from the surrounding O^{--} ions is left over wherever an Mg^{++} or an Fe^{++} ion replaces an Al^{+++} ion in an octahedral sheet or an Al^{+++} replaces an Si^{++++} ion in a tetrahedral sheet. These negative charges are balanced by cations held between or outside the layers. Charges arising from cation substitutions are part of the cation-exchange capacity if the cations are held at the outer surfaces of the particle. Cations held between layers are exchangeable if the lattice expands (as in montmorillonite and vermiculite). Cations inside the nonexpanding lattices are nonexchangeable because of inaccessibility. Weathering processes, however, gradually open lattices and release the previously nonexchangeable cations.

Broken-edge bonds occur in all minerals because mineral lattices are held together by continuous sequences of ionic bonds. There is no place to terminate the structure without leaving some ions with unsatisfied negative or positive charges at the edges. Bonds at broken edges would tie additional ions to the structure if it were larger. Broken-edge bonds are unimportant in sand- and silt-size particles because of their relatively small surface area per gram. The amount of charge attributable to broken-edge bonds increases with decreasing particle size and reaches a maximum in allophane. The disorganized nature of allophane results in large numbers of broken-edge bonds.

Broken-edge bonds can be either positive or negative and therefore give rise to both anion- and cation-exchange capacities. There is, however, a tendency for some anions to be held too tightly to be truly exchangeable. The presence of such strongly adsorbed anions as $H_2PO_4^-$ reduces the effective anion-exchange capacity.

Silicate clay minerals tend to have mostly negative charges in their broken-edge bonds and therefore attract mostly cations. The reverse is true in the oxide clays where positive charges are dominant and mostly anions are attracted to the clay particles.

The amount of cation-exchange capacity possessed by a soil clay is partly pH dependent. The cation-exchange capacity is lower under acid conditions and rises as the pH rises. The abundance of H^+ ions in an acid soil can change some exposed O^{--} ions to OH^- ions and thus reduce the cation-exchange capacity. The constant part of the charge probably comes from cation substitutions, and the variable part from broken-edge bonds.

The cation-exchange capacity of a soil is an important component of soil fertility, or at least of potential soil fertility. The usual procedure for measuring cation-exchange capacity is to leach a soil sample with enough ammonium acetate to cause ammonia to occupy all the exchangeable cation sites. Excess ammonia is then removed by leaching with alcohol. The amount of ammonia held in the sample is a measure of the cation-exchange capacity of the soil sample. The ammonia in the sample can be replaced with another cation (often Na^+ or Mg^{++}) and distilled into

Table 6-4 Cation-exchange Capacities of Soil Clays and Humus

	Cation-exchange capacities, meq/100 g	
	Representative	Usual range
Humus	200	100–300
Vermiculite	150	100–200
Allophane	100	50–200
Montmorillonite	80	60–100
Illite	30	20–40
Chlorite	30	20–40
Peat	20	10–30
Kaolinite	8	3–15

another bottle where it can be titrated. The result is expressed in terms of milliequivalents per 100 grams of soil (usually abbreviated as meq/100 g).[1]

Representative cation-exchange capacities of clay minerals and humus are given in Table 6-4. These values can be used for calculating an approximate cation-exchange capacity of a soil of known or estimated composition. Laboratory methods are available to measure the amount of organic matter and the amount and types of clay present. Reasonable estimates of the amounts of organic matter and clay can be based on soil color and texture. The types of clay are usually consistent across broad climatic areas and can be assumed to be the same as those occurring in related soils.

As an example, consider a dark-colored loamy soil estimated to contain 4 percent organic matter and 20 percent clay. If the clay is half montmorillonite and half illite, the soil has a calculated cation-exchange capacity of 19 meq/100 g:

$$
\begin{aligned}
4\% \text{ organic matter:} \quad & 0.04 \times 200 = 8 \text{ meq} \\
10\% \text{ montmorillonite:} \quad & 0.10 \times 80 = 8 \text{ meq} \\
10\% \text{ illite:} \quad & 0.10 \times 30 = \underline{3 \text{ meq}} \\
& 19 \text{ meq}
\end{aligned}
$$

The above composition is representative of many grassland soils of central United States and of other subhumid temperate zones. It would also represent soils of similar clay content in grassland soils of drier climates if the figure for organic matter were reduced. In contrast, some of the soils of the southeastern United States have cation-exchange capacities of only about 9 meq/100 g of soil resulting from:

$$
\begin{aligned}
2\% \text{ organic matter:} \quad & 0.02 \times 200 = 4 \text{ meq} \\
5\% \text{ montmorillonite:} \quad & 0.05 \times 80 = 4 \text{ meq} \\
12\% \text{ kaolinite:} \quad & 0.12 \times 8 = \underline{1 \text{ meq}} \\
& 9 \text{ meq}
\end{aligned}
$$

[1]A milliequivalent is the amount of material that will combine with or replace one milligram of hydrogen. The number of milligrams in a milliequivalent is calculated by dividing the atomic weight or ionic weight by the valence. Avogadro's number (6.0×10^{23}) of reactive charges corresponds to 1 equivalent or 1,000 milliequivalents.

A wet soil high in both clay and organic matter might have a much higher cation-exchange capacity:

15% organic matter: $0.15 \times 200 = 30$ meq
30% montmorillonite: $0.30 \times \ 80 = 24$ meq
10% illite: $0.10 \times \ 30 = \ \ 3$ meq
                                            57 meq

The preceding examples are illustrative only and are not intended to cover the total range of cation-exchange capacities in soils. Some very sandy soils have almost no cation-exchange capacity, and some muck soils have much higher capacities than any of the examples calculated.

A capacity to hold a certain number of milliequivalents does not ensure the presence of that great a concentration of valuable plant nutrients. Some of the exchange sites may be occupied by H^+, Al^{+++}, or some other nonessential ions. The available concentrations of such nutrient cations as Ca^{++}, Mg^{++}, and K^+ may be determined by leaching the soil with ammonium acetate (as for the cation-exchange determination) and analyzing the leachate (rather than the ammonium-saturated soil).

The *percent base saturation* of a soil is a good measure of how much of the cation-exchange capacity is being utilized to store plant nutrients. It is defined as the percentage of the cation-exchange capacity occupied by basic cations (usually Ca^{++}, Mg^{++}, K^+, and Na^+; it specifically excludes H^+ and Al^{+++} because they produce acid reactions). Base saturation is treated in more detail in Chapter 8.

WEATHERING

A young soil consists of primary minerals inherited from its parent material plus air, water, and organic matter. These constituents react together as time passes. The air, water, and organic matter are continually renewed, but the primary minerals undergo weathering, and secondary minerals are formed. The soil composition shifts toward more of the minerals that are most stable at the surface of the earth. The most stable minerals are usually not the same as the primary minerals contained in the parent rocks because the environment has changed. Most of the primary minerals were formed at high temperatures and pressures and in the virtual absence of air and water.

The chemical reactions occurring during weathering are varied and quite complex. They can be represented in a very general way as follows:

The soluble salts are mostly the larger cations such as Na^+, K^+, and Ca^{++} along with some Mg^{++} and Fe^{++} combined with such anions as CO_3^{--} and HCO_3^- (from carbon dioxide in the air), Cl^-, and SO_4^{--}. Smaller quantities of many other ions are also present, among them such relatively insoluble materials as SiO_2. Leaching removes most of the soluble salts from the soils of humid regions and carries them through the groundwater into the rivers and thence to the ocean.

In arid regions little percolating water reaches the groundwater. Most of the soluble salts remain in the lower part of the solum and in the underlying C and R layers. Usually there is some separation according to solubility, with the more soluble materials being carried to greater depths. Concentrations of very slightly soluble compounds like SiO_2, Fe_2O_3, and $CaCO_3$ sometimes cement B3 or C horizons into rocklike layers known as *duripans*.

The *secondary minerals* formed in soils are mostly of clay size. They include several varieties of phyllosilicates (especially the silicate clay minerals, montmorillonite and kaolinite) and the nearly insoluble oxides of aluminum, iron, and silicon.

Some weathering transformations are simple and direct. A prime example of a relatively simple transformation is the conversion of muscovite mica to montmorillonite. Since both minerals have 2:1 lattice structures, the principal change required to convert muscovite to montmorillonite is the removal of potassium ions from between the layers. Water and dissolved cations can then enter between the layers, and the resulting mineral has the expanding lattice characteristic of montmorillonite. Generally there are also some changes within the layers; the resulting charge on the lattice is only about a third of that which would be produced by simple removal of all the K^+. Examples of such charge-reducing changes would be the oxidation of small amounts of Fe^{++} that might be present in octahedral spaces to Fe^{+++} and the addition of H^+ to oxygen ions to form hydroxyl ions.

Many weathering transformations involve changing primary minerals with one type of structure to secondary minerals with an entirely different structure. An example would be the conversion of feldspars (tektosilicate primary minerals) to montmorillonite or kaolinite (phyllosilicate secondary minerals). Major changes are required whenever kaolinite is formed because none of the important primary minerals has kaolinite's 1:1 layer structure. These changes require that even strong bonds like those between silicon and oxygen must be broken and relocated in the conversion process. Structural fragments are formed that probably exist as allophane for a time after the old mineral structure is broken apart and before the new structure is formed.

Intermediate types of minerals also occur during the weathering process. For example, removal of a portion of the K^+ from mica produces intermediates between mica and montmorillonite.

WEATHERING STAGES

Progressive changes occur in the mineral composition of soil as it ages. These changes can be divided into a series of weathering stages. Such stages can be detected not only in soils of varying age but also in soils of equivalent age but from climates of different intensities of weathering. Different stages of weathering can also be detected in the profile of a single soil where the A horizon is strongly weathered, the B horizon is less weathered, and the C horizon is only beginning to weather.

Table 6–5   The Sequence of Removal of Minerals from the Fine-clay Fraction of Soil by Weathering Processes

1. Gypsum	$CaSO_4 \cdot 2H_2O$ (and other more soluble salts)
2. Calcite	$CaCO_3$
3. Hornblende	$NaCa_2(Mg,Fe,Al)_5(Si, Al)_8O_{22}(OH)_2$
4. Biotite mica	$K(Mg, Fe)_3 (AlSi_3)O_{10}(OH)_2$
5. Feldspars	
Plagioclase	$CaAl_2 Si_2O_8 - NaAlSi_3O_8$
Orthoclase	$KAlSi_3O_8$
6. Quartz	SiO_2
7. Muscovite	$KAl_2 (AlSi_3)O_{10}(OH)_2$
8. Clay-mica intermediates	
(including illite, vermiculite, and chlorite)	
9. Montmorillonite	$(Mg, Al)_2Si_4O_{10}(OH)_2 \cdot xH_2O$
10. Kaolinite	$Al_4 Si_4 O_{10}(OH)_8$
11. Gibbsite	$Al(OH)_3$
12. Hematite	Fe_2O_3
13. Anatase	TiO_2

Source: Adapted from Jackson and Sherman, 1953.

Jackson and Sherman (1953) identified the minerals present in the clay fractions of the A horizons of many different soils. They found it possible to arrange the minerals in a sequence according to the weathering time or intensity required to eliminate each mineral from the soil. This mineral sequence is listed in Table 6-5. It should be emphasized that this sequence of removal applies to clay-size particles; the sequence would be different for coarser sizes. Quartz would appear as a more resistant mineral in a sand-size list than it does in the clay-size list.

Some of the differences in rates of weathering are readily explainable. Gypsum and calcite (items 1 and 2 in Table 6-5) are removed by simple solution with no structural changes necessary. Hornblende (item 3) has silicon-to-oxygen bonds in only one dimension, whereas biotite (item 4) has silicon-to-oxygen bonds in two dimensions, and the feldspars and quartz (items 5 and 6) have them in all three dimensions. The silicon-to-oxygen bonds are the strongest bonds in silicate minerals and therefore tend to make the minerals more resistant to weathering.

The presence of large cations that can be gradually dissolved and removed is a structural weakness that causes feldspars to weather faster than quartz. The presence of Al^{+++} in tetrahedral positions also weakens the feldspar structure. Oxidizable iron (Fe^{++}) is a weak point in hornblende and biotite.

The weathering sequence in the micas and secondary minerals goes from a higher to a lower externally balanced charge on the layers and an increasing inclusion of water and/or hydroxyl ions in the structure. At the bottom of the list are three extremely insoluble oxides.

Almost any soil contains several different minerals even in its clay fraction. Jackson and Sherman found that the dominant minerals in the clay of any particular soil are usually near one another in the list in Table 6-5. The first minerals in the list disappear before the last ones are formed in significant amounts. It is therefore possible to classify soils in terms of their weathering as follows:

1 *Young.* Composed almost entirely of primary minerals. May have lost soluble salts including gypsum and calcite. Still contains all other minerals that were present in the parent material.

2 *Early maturity.* Most of the clay-size minerals containing Ca^{++}, Mg^{++}, and Fe^{++} (hornblende, biotite, and plagioclase feldspar) are lost from the A horizons. Silicate clay minerals are forming and accumulating in B horizons; 2:1 lattice types predominate.

3 *Late maturity.* Most of the primary minerals have been weathered from the clay fraction and from much of the silt fraction; the 2:1 lattice clays are being replaced by 1:1 lattice kaolinite. Aluminum and iron oxides are beginning to increase.

4 *Old.* Nearly all clay and silt have been removed from the A horizon by weathering and eluviation; any primary aluminosilicate minerals remaining are distinctly weathered even in the sand size; A horizons are reduced to little more than quartz sand. B horizons are high in kaolinite and oxide clays.

Important changes occur in the ability of a soil to supply fresh plant nutrients and to store nutrient ions as it passes through the different weathering stages. Most of these changes are in the direction of declining fertility in older soils.

Soils in stage 1 generally provide abundant amounts of the essential plant nutrients. Fresh supplies of nutrient elements are continually supplied because weatherable minerals are still present. Some associated soils in low topographic positions may contain excessive amounts of soluble salts, but most of the soils of stage 1 have good supplies of plant nutrients without excess soluble salts. Stage 1 soils occur mostly in subhumid or drier climates where relatively fresh parent materials exist. These parent materials include glacial till and loess associated with the Wisconsin glaciation (the most recent one; it ended in the United States about 10,000 years ago), alluvium of comparable or more recent age, and some older materials that have been exposed by erosion within the last few thousand years. Many soils of the western half of the United States (with a notable bulge eastward to include southern Minnesota and northern Iowa) are in weathering stage 1. To this should be added a sizable area of bottomlands along the Mississippi and other rivers. Fertilizers are not as essential for stage 1 soils as they are for soils of the other stages, but they may be useful nevertheless. Many of these soils have very high productive potentials; realizing these potentials frequently requires that even fertile soils be fertilized. Much of the area is not fertilized, however, because the climate is too dry and irrigation water is lacking to make use of a high fertility level.

Soils in stage 2 occupy most of the northeastern United States with the border-line area crossing Missouri, Kentucky, Pennsylvania, and New England. These soils are acid in reaction. Most of them have high capacities to store plant nutrients, but these capacities are not fully utilized. From 20 to 50 percent of the capacity to store nutrient cations may be occupied by hydrogen ions. Many of these soils have glacial till parent material of Wisconsin age comparable with that of many soils of stage 1 with the difference being caused by more intense weathering in the more humid climate. Leaching increases the need for lime and fertilizer in stage 2 soils. The potential productivity may still be high, but the cost of attaining it is frequently higher than with the soils of stage 1.

Stage 3 soils occur in much of the southeastern United States. These soils are more weathered than most of the others in this country because of the warm, humid climate, and most are older because glaciation did not reach this area. Red and yellow colors resulting from the presence of iron oxides in these soils are quite noticeable. The plant nutrient storage capacities are low because of the reduced amount of 2:1 lattice clays and because hydrogen ions may occupy more than 50 percent of the storage sites for nutrient cations. The nutrients that were once there have been lost by leaching. The soils of stage 3 therefore have a great need for added soil fertility in the form of fertilizers. Lime is needed but not so much as might be anticipated. Kaolinite and oxide clays give reactions much nearer to neutral than the reactions would be with 2:1 lattice clays in the same conditions.

Soils in stage 4 are so highly weathered that almost all their fertility is in the organic matter. Everything else is weathered and leached out in the tropical climates where almost all these soils are found. Even the organic matter is subject to rapid decomposition, and so most of the nutrient supply is either held in living organic matter or lost. It is therefore important to have something growing on these soils at all times except where a dry season precludes both growth and leaching. Fertilizer must be applied at times and in amounts that will ensure its being promptly utilized, or it will be leached away. Red colors are common in these soils because the ferric iron content is generally high. Some B and C horizons in tropical soils have a combination of oxide and kaolinite clays called *plinthite* that hardens irreversibly when it dries. Great care must be taken to keep plinthite from drying when it is farmed. The dried material is so hard that bricks are made by cutting and drying pieces of it.

It must not be supposed that the preceding generalizations about weathering stages apply to all the soils in any given area. Too many other factors influence soils: one is the variable mineralogy of the parent material; another is wetness caused by topographic position. The type of clay formed in a particular soil depends in part on both of these factors. The formation of vermiculite, for example, requires abundant magnesium. Biotite mica might readily be altered to form vermiculite, especially if other magnesium-containing minerals are present. Muscovite mica would be more likely to form montmorillonite. Kaolinite will form where well-drained soils are highly weathered and leached, whereas montmorillonite will form in low, wet areas nearby where the supply of silica and of basic cations is higher.

Soils of markedly different stages of weathering often occur in close proximity to one another. Parent material differences can be partly responsible for this, but the main factor is usually erosion and deposition. Erosional processes continually uncover fresh material in some places. Deposition may cover older materials with younger materials eroded from other sites. The most weathered soils usually occur on upland flats where neither erosion nor deposition is active.

REFERENCES

Andersson, A., and L. Wiklander, 1975, Release of Crystal Constituents by Chemical Weathering of Some Soil Minerals, *Soil Sci.* **120**:13–19.

Barnhisel, R. I., and C. I. Rich, 1967, Clay Mineral Formation in Different Rock Types of a Weathering Boulder Conglomerate, *Soil Sci. Soc. Am. Proc.* **31**:627–631.

Borchardt, G. A., F. D. Hole, and M. L. Jackson, 1968, Genesis of Layer Silicates in Representative Soils in a Glacial Landscape of Southeastern Wisconsin, *Soil Sci. Soc. Am. Proc.* **32**:399–403.

Brindley, G. W., and D. L. Gibbon, 1968, Kaolinite Layer Structure: Relaxation by Dehydroxylation, *Science* **162**:1390–1391.

DeMumbrum, L. E., and G. Chesters, 1964, Isolation and Characterization of Some Soil Allophane, *Soil Sci. Soc. Am. Proc.* **28**:355–359.

Eswaren, H., and F. de Coninck, 1971, Clay Mineral Formations and Transformations in Basaltic Soils in Tropical Environments, *Pedologie* **21**:181–210.

Eswaren, H., and C. Sys, 1970, An Evaluation of the Free Iron in Tropical Basaltic Soils, *Pedologie* **20**:62–85.

Farmer, V. C., J. D. Russell, and B. Velde, 1970, Replacement of OH by OD in Layer Silicates, and Identification of the Vibrations of These Groups in Infra-red Spectra, *Mineral. Mag.* **37**:869–879.

Fey, M. V., and J. Le Roux, 1976, Electric Charges on Sesquioxidic Soil Clays, *Soil Sci. Soc. Am. Proc.* **40**:359–364.

Jackson, M. L., and G. R. Sherman, 1953, Chemical Weathering of Minerals in Soils, *Adv. Agron.* **5**:221–318.

Johnson, L. J., R. P. Matelski, and C. F. Engle, 1963, Clay Mineral Characterization of Modal Soil Profiles in Several Pennsylvania Counties, *Soil Sci. Soc. Am. Proc.* **27**:568–572.

Jones, R. L., and A. H. Beavers, 1964, Variation of Opal Phytolith Content among Some Great Soil Groups in Illinois, *Soil Sci. Soc. Am. Proc.* **28**:711–712.

Lai, Sung-Ho, and L. D. Swindale, 1969, Chemical Properties of Allophane from Hawaiian and Japanese Soils, *Soil Sci. Soc. Am. Proc.* **33**:804–808.

McNeal, B. L., 1968, Limitations of Quantitative Soil Clay Mineralogy, *Soil Sci. Soc. Am. Proc.* **32**:119–121.

Maeda, T., and B. P. Warkentin, 1975, Void Changes in Allophane Soils Determining Water Retention and Transmission, *Soil Sci. Soc. Am. Proc.* **39**:398–403.

Marshall, C. E., 1949, *The Colloid Chemistry of the Silicate Minerals,* Agronomy Monograph 1, Academic, New York.

Miller, G. A., F. F. Riecken, and N. F. Walter, 1975, Use of an Ammonia Electrode for Determination of Cation Exchange Capacity in Soil Studies, *Soil Sci. Soc. Am. Proc.* **39**:372–373.

Rai, Dhanpat, and W. L. Lindsay, 1975, A Thermodynamic Model for Predicting the Formation, Stability, and Weathering of Common Soil Minerals, *Soil Sci. Soc. Am. Proc.* **39**:991–996.

Tucker, B. M., 1974, Displacement of Ammonium Ions for Cation Exchange Capacity Measurements, *J. Soil Sci.* **25**:333–337.

Twiss, P. C., Erwin Seuss, and R. M. Smith, 1969, Morphological Classification of Grass Phytoliths, *Soil Sci. Soc. Am. Proc.* **33**:109–115.

van der Plas, L., and L. P. van Reeuwijk, 1974, From Mutable Compounds to Soil Minerals, *Geoderma* **12**:385–405.

Chapter 7

Soil Chemistry

Soil chemistry is a vital link between soil fertility considerations to be discussed in the next few chapters and the physical aspects of soil that have been the topics of previous chapters. The reactions that maintain dilute solutions of nutrient elements are indispensable for continued plant growth. The nutrient elements and the means of maintaining them at suitable concentrations in the soil are discussed in this chapter. Certain nonnutrient elements are also included because they are abundant in soils and because they interact with the nutrient elements.

Soil chemistry deals with aspects of solution chemistry and of solid phase chemistry (or mineralogy). The contact zone between the solid and liquid phases is very important in soil chemistry. Many aspects of soil chemistry relate to colloidal chemistry in which surface forces play a significant part. Ions adsorbed[1] on the surfaces of solid particles greatly outnumber the ions in true solution in most soils. Adsorbed ions have a slow-acting equilibrium with absorbed ions inside mineral

---

[1]*Adsorbed* refers to being attracted and held next to a surface as contrasted with an *absorbed* material being taken inside and probably held more tightly than the adsorbed material. The word *sorbed* is sometimes used to include both absorbed and adsorbed.

particles and a rapid equilibrium with dissolved material. Soil chemistry includes all these equilibrium reactions.

CATION EXCHANGE

Cation exchange consists of an interchange between cations adsorbed on charged surfaces and cations in the soil solution. The cation-exchange mechanism (discussed in Chapter 6) is a dominant and pervasive factor in soil chemistry. Over 99 percent of the cations are adsorbed on the colloidal surfaces and less than 1 percent are present in solution (excluding saline soils and a few very highly weathered soils). Equilibria existing between the adsorbed cations and the soil solution allow the concentrations of adsorbed cations to determine what concentrations of ions in solution will be maintained. Even changes in the soil solution brought about by external factors such as plant removal or fertilizer application are usually small because the adsorbed cations act as a buffer to the solution. Larger changes can be made, but they require very large applications of the ions involved.

Cation exchange enters into most of the other topics in this and several other chapters of this book. The terminology used in relation to cation exchange must be understood. The cations held by the system are often called *adsorbed* cations. They are held by small, negatively charged particles of clay and organic matter called *micelles.* Micelles can be considered as polyanions because they carry large numbers of negative charges.

THE LIQUID MEDIUM

The significance of water as the liquid medium in soil chemistry can hardly be overemphasized. Without water none of the living things present in the soil could exist, chemical weathering would virtually cease, and the soil would be static and lifeless.

Water is the most universal solvent known, particularly when acids or bases are present in it. Water dissolves at least a trace of anything it contacts. The soil solution is a source from which plants can obtain some of every element present in the soil. Percolating water likewise transports some of every element present, but not the same amount nor always the same proportions of each. The materials that are more weatherable and soluble tend to be removed more rapidly from the soil, especially from the A horizon, and the more resistant ones are left behind.

Organic and mineral soil components supplement each other in holding some of each element in the soil in spite of the potency of water as a leaching agent. Every essential plant nutrient occurs in either cation or anion form or both. Anions are built into the structure of organic matter and held there until the organic material decomposes. Cations escape from dead organic materials, but the cation-forming elements are held in mineral structures. Cations in minerals cannot be leached until they are released by weathering. Even then they are partially protected from leaching by the cation-exchange mechanism.

IMPORTANT ELEMENTS

The list of elements most important in soil chemistry is not the same as the list for soil mineralogy. The amounts of the various elements involved are very different both relatively and absolutely. Large amounts of oxygen, silicon, and aluminum and smaller amounts of other elements are chemically inert because they are rigidly held in mineral structures. The addition or removal of a few tons per hectare of this kind of soil material makes little difference. Soil chemistry deals with the reactive materials. As little as 50 g/ha of certain of these materials is very significant to plant growth.

The forms of the elements are as important as the elements themselves. All the elements important in soil chemistry form one or more ions. The ions are adsorbed by micelles, occur in solution, enter into chemical reactions, and are absorbed as plant nutrients. The most important elements in soil chemistry are listed in Table 7-1 along with their most common ionic forms. It is noteworthy that only one of the cations listed in Table 7-1 contains more than one element (NH_4^+), whereas all but one (Cl^-) of the anions contain oxygen plus other elements. Thus most of the anions qualify as radicals (ions containing groups of atoms). Polyvalent cations like Al^{+++} and Fe^{+++} may also become radicals by attracting one or two hydroxyls and forming cations of lower valency (lesser charge) such as $Al(OH)_2^+$ and $Fe(OH)_2^+$. The process is analogous to CO_3^{--} absorbing H^+ to form HCO_3^- radicals or to HPO_4^{--} absorbing H^+ to form $H_2PO_4^-$ radicals. These processes are reversible, depending on the relative abundance of H^+ and OH^-.

Table 7-1 Elements Important in Soil Chemistry and Their Chemical Symbols and Principal Forms

Element	Symbol	Principal ions
Aluminum	Al	Al^{+++}
Boron	B	$B_4O_7^{--}$
Calcium	Ca	Ca^{++}
Carbon	C	CO_3^{--}, HCO_3^-
Chlorine	Cl	Cl^-
Cobalt	Co	Co^{++}
Copper	Cu	Cu^{++}
Hydrogen	H	H^+, OH^-
Iron	Fe	Fe^{++}, Fe^{+++}
Magnesium	Mg	Mg^{++}
Manganese	Mn	Mn^{++}, MnO_4^-
Molybdenum	Mo	MoO_4^{--}
Nitrogen	N	NH_4^+, NO_2^-, NO_3^-
Oxygen	O	With other elements
Phosphorus	P	$H_2PO_4^-$, HPO_4^{--}
Potassium	K	K^+
Silicon	Si	$H_3SiO_4^-$
Sodium	Na	Na^+
Sulfur	S	SO_4^{--}
Zinc	Zn	Zn^{++}

Oxygen is clearly the number one element in soils. It is the principal building block of soil minerals, it constitutes almost all the bulk of water molecules, and it complexes with other elements to form important ions and organic compounds.

Hydrogen is important in balancing the electric charge of oxygen, not only in water molecules but also in the OH$^-$ of mineral structures and in complex ions such as $H_2PO_4^-$. It bonds with nitrogen in NH_4^+ and with carbon in organic compounds. Hydrogen ions can balance charges without producing any increase in volume.

Carbon is the core element in the large organic molecules that characterize all living things. The unique ability of each carbon atom to share in four covalent bonds with adjacent carbon, hydrogen, oxygen, nitrogen, phosphorus, or sulfur atoms makes possible the formation of carbohydrates, fats, proteins, and other essentials of life. Much of the carbon supply is tied up in both living and dead organic materials and is made available to new living things when the former ones are decomposed. Sometimes the cycle of a carbon atom going from one living thing through decomposition processes and on to another living thing is brief enough to be measured in minutes; other times it is so long it takes millions of years. An example of a long cycle is when organic materials are buried and transformed into coal or into an organic component of shale. Eventually these rocks may be reexposed at the surface where the carbon can return to circulation.

Carbon also unites with oxygen to form CO_2 and carbonates. Large amounts of carbon occur as $CaCO_3$ in limestone and marble; smaller amounts occur in other rocks and in the CO_2 of the atmosphere. Some CO_2 dissolves in the soil water yielding H^+ and HCO_3^- ions that increase the weathering potential of the water.

Silicon and *aluminum* are not considered essential for plant growth but are very abundant in soil minerals. Silicon in the form of quartz sand and aluminum in the form of gibbsite are very resistant to weathering. Nevertheless, aluminum and silicon ions enter the soil solution as weathering decomposes aluminosilicate minerals. Only small amounts of these ions are in solution at any one time because their solubilities are low. Loss from solution occurs by the formation of secondary minerals such as quartz, gibbsite, or kaolinite. Some of these dissolved materials are eluviated to a lower horizon where the secondary minerals crystallize and act as cementing agents in certain soils. Considerable amounts of silicon and aluminum are absorbed by plants even though the plants do not need them. Some silicon crystallizes as plant opal (a form of quartz) inside plant cells. Aluminum sometimes becomes soluble enough under strongly acid conditions to become toxic to plant growth. The Al^{+++} ion is so related to acid conditions that it is considered acidic along with the H^+ ion.

Magnesium, iron, and *manganese* all occur as divalent cations in octahedral sites of mineral structures. All three of these elements are essential for plant growth but in different amounts. Magnesium is a macronutrient; iron and manganese are micronutrients. Iron is required in larger amounts than any other micronutrient, and some of its forms are so insoluble that it sometimes becomes deficient even when large amounts are present. Both iron and manganese may be oxidized to form other ions, the iron to the trivalent ferric ion and the manganese to complex anions. Most of the ferric iron precipitates as an oxide or hydroxide of even lower solubility and greater resistance to weathering than the corresponding aluminum compounds. The magnesium ions do not change valence. They are either adsorbed by cation exchange, absorbed by growing plants, or leached from the solum.

Calcium, sodium, and *potassium* are large cations that are abundant in feldspars and other minerals. Calcium and potassium are both macronutrients essential for plant growth. Sodium is considered a nonessential element (nonessential for plants; essential for animals), but it has been shown to be beneficial in small amounts (Chapter 12). Sodium deficiencies never occur in nature because the amount of sodium in soils is comparable with some of the macronutrients. In fact, sodium can partially substitute for potassium in meeting plant needs. Potassium deficiencies are sufficiently common to make potassium one of the three principal fertilizer elements. Calcium is seldom deficient as a plant nutrient but is often applied to soil in soil amendments (Chapter 8). All these large cations are removed faster from minerals by weathering than are the smaller cations. These three large cations plus magnesium are the principal cations in drainage water and are the dominant basic cations adsorbed by cation exchange.

Cobalt, copper, and *zinc* are micronutrients. Cobalt has not yet been shown to be essential to higher plants, but it is essential in the symbiotic relation between legumes and Rhizobia. These three elements occur as impurities in crystal structures and in minerals of their own. They are concentrated in some places as ore deposits, but, more important to soils, small crystals are widely dispersed in rocks. These elements are released by weathering and form cations that can be adsorbed by cation exchange. They are also subject to gradual leaching and to the possibility of becoming deficient in soils that are either highly weathered or formed in parent materials that happen to be deficient in one or more of these elements.

Chlorine, like sodium, was long cited as an element essential to animals but not to plants. Thanks to the work of Broyer and others (1954) we now know that minute amounts of chlorine are essential to some plants. The amounts needed by plants are so small and chlorine is so abundant that deficiencies never occur in nature. Chlorides are highly soluble; they are abundant in the soil solution and in the leachate from soils.

Nitrogen is the nutrient most often supplied as a fertilizer. Plants need large amounts of nitrogen but only a trace of it is present in igneous rocks. About 78 percent of the air is nitrogen, but nitrogen gas is unavailable to higher plants. To become available, it must be combined with hydrogen or oxygen. The combining can be accomplished by certain bacteria, by lightning, or synthetically. About 99 percent of the nitrogen in the soil is in the structure of proteins and other organic compounds. When NH_4^+ ions are present, they can be held by cation exchange, but usually they are soon converted to NO_3^- ions by nitrifying bacteria.

Phosphorus is one of the three principal fertilizer elements. Most soils contain phosphorus in both organic and inorganic forms, but the availability is often low. The organic phosphorus is firmly tied into large organic molecules. Inorganic phosphorus comes mostly from the mineral apatite, which occurs as microscopic crystals dispersed in rocks. Fortunately, apatite also occurs in concentrated form in certain places. Apatite from these deposits is mined, ground, treated with acid to make the phosphorus more soluble, and applied to soil as fertilizer. Untreated apatite is very insoluble, as are many other phosphorus compounds.

There is usually less than a kilogram of dissolved phosphorus per hectare-furrow slice. Loss by leaching is usually considerably slower than that of Ca^{++}, Mg^{++}, Na^+, or K^+ but a little faster than that of Si and some of the other elements that form resistant oxides. The form and solubility of phosphorus vary with the acidity or alkalinity of the soil. Under acid conditions the $H_2PO_4^-$ ion is dominant. This form is available to plants when it is in solution, but it frequently reacts with or is adsorbed by iron and aluminum compounds. In slightly alkaline conditions the HPO_4^{--} ion becomes the dominant form, but its solubility is low. In highly alkaline conditions the PO_4^{---} ion begins to appear, but it combines with calcium and precipitates as calcium phosphates.

Sulfur is a macronutrient constituent of organic compounds and, like nitrogen and phosphorus, is an integral part of their structures. Inorganic sulfur compounds are not as resistant to weathering as phosphorus minerals. The sulfides are of low solubility but are oxidized to sulfates, most of which are quite soluble except for calcium sulfate (gypsum). Significant amounts of sulfur are leached as the SO_4^{--} ion, but most of this loss is replaced by sulfur compounds in the air that dissolve in rainwater, thus changing from air pollutants into fertilizer. Sulfur is more often applied as a soil amendment (Chapter 8) or as an incidental component of fertilizer than as a remedy for a nutrient deficiency for sulfur.

Boron and *molybdenum* are micronutrients that combine with oxygen to form anions. A kilogram per hectare of each of these elements in available form is adequate for most plants, but sometimes not even this much is available. Too much is as bad as not enough. Excesses of boron can become toxic to plant growth. Such excesses are most likely to occur in evaporite deposits formed by groundwater evaporating in very arid conditions. Such was the mode of formation of borax deposits in Death Valley, California.

ACIDS, BASES AND SALTS

Acids are water solutions that contain more H^+ cations than OH^- anions (other anions are also present so that the total numbers of + and − charges are equal). The concentration of active H^+ ions in solution determines the acidity.

Bases are water solutions containing more OH^- anions than H^+ cations. The greater the concentration of OH^- ions in solution, the more alkaline (basic) it is.

Salts are compounds formed of cations other than H^+ and anions other than OH^-. When an acid and a base react they form water and a salt. The resulting solution may be acidic, alkaline, or neutral, depending on the relative amounts and strengths of the acid and base.

The term *micelle* is used to signify a charged solid particle, whether clay or organic in nature. Micelles must be considered a part of the acid-base-salt system in the soil. Micelles contain most of the negative charges in soils but only a small portion of the positive charges. The soil water surrounding the solid particles usually contains a preponderance of cations—enough to offset the excess negative charges of the micelles and make the soil electrically neutral (in some highly weathered tropical soils the situation is reversed, with the micelles having a small excess positive charge and the soil water containing the corresponding excess of anions).

THE PRINCIPAL CATIONS

The acidity or alkalinity of a soil depends largely on the balance between the negatively charged micelles and the positively charged basic cations (mostly Ca^{++}, Mg^{++}, K^+, and Na^+). These cations are called *bases* because enough OH^- ions are generally present to make the system alkaline when the positive charges from these cations outnumber the negative charges of the micelles. The soil is acid when the reverse situation occurs and many of the negative charges of the micelles are offset by H^+ and Al^{+++} ions.

The amount of each basic cation present in the soil may be divided into three parts: (1) Structural parts of minerals (unavailable for plant growth), (2) exchangeable cations (adsorbed on micelles and made available by the cation-exchange process), and (3) cations in solution (the smallest amount but the most readily available).

Cations contained in mineral structures are gradually made available by weathering processes. The rates of release, however, are different for different cations and different minerals. Minerals containing calcium generally weather faster than average; those containing potassium are about average in rate of weathering. The cations released by weathering either can be held by cation exchange or move into solution (unless they precipitate as a secondary mineral). They may be utilized by growing plants or microbes. Any excess cations may be leached to the bottom of the solum in an arid region soil or leached into the drainage water from a humid region soil. Most of the dissolved materials in river water have been leached from soil; the amount is so large that in humid regions it exceeds the weight of silt and other solids carried to the ocean by the rivers. The dominant cations in river water are Ca^{++}, and the dominant anions are CO_3^{--} (Table 7-2).

Table 7-2   Average Composition of River Water Expressed as Percent of the Ions in the Water

	North America	South America	Europe	Asia	Africa	World average
CO_3	33.40	32.48	39.98	36.61	32.75	35.15
SO_4	15.31	8.04	11.97	13.03	8.67	12.14
Cl	7.44	5.75	3.44	5.30	5.66	5.68
NO_3	1.15	0.62	0.90	0.98	0.58	0.90
Ca	19.36	18.92	23.19	21.23	19.00	20.39
Mg	4.87	2.59	2.35	3.42	2.68	3.41
Na	7.46	5.03	4.32	5.98	4.90	5.79
K	1.77	1.95	2.75	1.98	2.35	2.12
$(Fe, Al)_2 O_3$	0.64	5.74	2.40	1.96	5.52	2.75
SiO_2	8.60	18.88	8.70	9.51	17.89	11.67

Source: Clarke, 1924.

The percentages of the ions shown in Table 7-2 must be closely proportioned to their net rates of release by weathering. Otherwise the amounts present in the soil would not be stable. Soil composition does change but over time periods comparable with the stages of weathering discussed in Chapter 6. Some idea of the resulting differences may be had by comparing the data for Africa and South America (both of which include large tropical regions) with the data for the other continents. The losses of SiO_2 and $(Fe,Al)_2O_3$, which weather with difficulty, are much larger percentages in these two continents than in the others.

Weathering in arid regions is slower, and less material enters the soil solution than in humid regions. Nevertheless, the solution is more concentrated in the arid region soils because the water content is reduced more than the salt content of the soil. Also, dissolved materials are carried downward in the soil, but they are not leached from the soil in an arid region. Instead, they collect at the bottom of the B horizon as salts of Ca^{++}, Mg^{++}, Na^+, K^+, and other cations combined with CO_3^{--}, SO_4^{--}, and other anions. Some of these salts move back upward in the profile through capillary action when the upper horizons are drier than the lower ones. Highly soluble salts such as those of Na^+ are the ones moved upward in the largest amounts.

Preferential Adsorption

The assemblage of cations adsorbed by cation exchange on the soil colloids depends on (1) the available supply of each cation, (2) the intensity of leaching action and other removal processes, and (3) the strength with which each cation is adsorbed.

The available supply of each cation consists of (1) ions that have been held for some time, plus (2) ions that have cycled through plants, animals, and microbes but returned to the soil, plus (3) fresh supplies coming from either mineral weathering or outside sources such as fertilizer. The fresh supplies may not be large, but they are very important for offsetting the slow removal of plant nutrients by leaching losses and harvesting of crops. Without fresh supplies of plant nutrients the soil would become impoverished and sterile.

Leaching action readily removes excess soluble salts (excess soluble salts are those not attracted to the soil colloids). Leaching occurs at a much slower rate when nearly all the ions present are attracted to the soil colloids. An ion exchange must be made to remove adsorbed cations by leaching. Leaching makes soil more acidic because H^+ ions are exchanged for basic cations on the micelles. Supplies of H^+ ions come from CO_2 dissolving in water, from life processes, and from decomposition products of organic materials.

Other removal processes include crop removal and volatilization. Crop removal can be a very large factor and frequently leads to a need to fertilize soils that once produced good crops without fertilizer. Plant growth that is not harvested does not deplete the soil because the plant nutrients are returned to the soil in the residues. Plant growth even counteracts leaching by absorbing ions from the subsoil and depositing them in residues on the surface.

Volatilization occurs in largest amounts in wet soils. Certain anaerobic microorganisms produce volatile materials (mostly nitrogen gas, partially reduced oxides of nitrogen, hydrogen sulfide, and methane) under wet conditions. Nutrient deficiencies can result from these gases escaping to the atmosphere.

The strength of the attractive force between a cation and a micelle depends on the charge of the cation, the charge density on the micelle, and how close the cation can get to the micelle. The charges are single for H^+, Na^+, K^+, and NH_4^+ ions; double for Mg^{++} and Ca^{++} ions; and triple for Al^{+++} ions. The closeness of approach is limited by the amount of water of hydration carried with the ion. Higher charge and smaller nonhydrated size both lead to larger hydrated sizes. Exact hydrated sizes are difficult to determine, but the size sequence from largest to smallest is $Al^{+++} > Mg^{++} > Ca^{++} > Na^+ > K^+ = NH_4^+$. The sequence of relative attractive forces, reflecting both the number of charges and the hydrated sizes, is $Al^{+++} > Ca^{++} > Mg^{++} > K^+ = NH_4^+ > Na^+$. This list, sometimes called the *lyotropic series,* may be considered as a preferential adsorption sequence. Cation-exchange sites adsorb more of an ion early in the sequence than of one later in the sequence if the two are present in equal amounts. Preferential adsorption, however, can be overcome by mass action when an ion is much more concentrated than the others in the soil solution.

Tendencies of ions to flocculate or disperse clay particles are related to the lyotropic series. The Al^{+++} and Ca^{++} ions at the beginning of the series have strong tendencies to flocculate clay, whereas the Na^+ ions at the other end are noted for causing dispersion.

The position of H^+ ions relative to the lyotropic series is a matter of dispute. Adding H^+ (acid) to a suspension of clay particles normally produces flocculation. This tendency to flocculate suggests that H^+ belongs near Ca^{++} in the lyotropic series. But a highly hydrated monovalent cation such as H^+ should appear at the Na^+ end of the series. The discrepancy may arise because Al^{+++} is released from minerals under acid conditions, or it may arise from the ability of H^+ ions to interact with water molecules and move in close to the micelle in spite of hydration.

Usual Proportions of Cations

The pH of a soil is closely related to the relative amounts of acidic cations (H^+ and Al^{+++}) and bases on its cation-exchange sites. The pH rises when the concentrations of bases increase and drops when the concentrations of acidic cations increase. Variations also occur within these two groups. The Al^{+++} ion is much less common than H^+ at pH values above 5 but becomes the dominant ion in extremely acid soils. Aluminum apparently is released from the crystal lattice of clay minerals and moves into exchange positions when the pH drops below 5. Aluminum ions also go into solution and sometimes injure plant growth. Concentrations of 1 ppm or more of Al^{+++} in culture solutions are unfavorable to barley and corn (Pierre et al., 1932). Concentrations of 4 ppm of Al^{+++} in nutrient solutions produce obvious damage to sugar beet roots as well as reducing growth (Keser, Neubauer, and Hutchinson, 1975). Aluminum toxicity can be avoided by maintaining a soil pH above 5.

The dominant bases in cation exchange are calcium and magnesium. Calcium usually constitutes 75 to 85 percent of the exchangeable bases. There are two main reasons for this predominance:

Table 7–3   Usual Proportions of Bases in Soils Dominated by 2:1 Lattice Clays

Base	Ca^{++}	Mg^{++}	K^+	Na^+	Other bases
Percentage of total exchangeable bases	75—85	12—18	1—5	1	1

1 Weatherable calcium is abundant in feldspars and other minerals, so it is released at a relatively fast rate.
2 Calcium is adsorbed by the clay and humus micelles more strongly than any of the other basic cations.

Table 7-3 shows the usual proportions of bases present in soils. These proportions are surprisingly constant in soils of many different origins having different types of clay, cation-exchange capacities, and base saturations.

The four cations shown in Table 7-3 constitute about 99 percent of the exchangeable bases in soil (excluding the acidic cations). The other 1 percent is also important because it includes micronutrient cations such as Fe^{++}, Co^{++}, Cu^{++}, Mn^{++}, and Zn^{++}. Trace amounts of these ions are adequate to meet the needs of plants.

The most important exceptions to the proportions of bases shown in Table 7-3 occur in the sodic soils (soils high in exchangeable Na^+). These soils occur as problem spots in arid regions. Dispersion of the colloids is likely to occur and cause low permeability to water, air, and roots where the exchangeable Na^+ exceeds 15 percent of the total exchangeable cations. Fortunately, the usual dominant cations (Ca^{++} in neutral soils; H^+ and Al^{+++} in acid soils) tend to cause colloids to flocculate rather than disperse.

pH

The pH scale serves as a measure of acidity and alkalinity. It uses as a reference or "neutral" point the concentration of H^+ in pure water at 24°C. This concentration is not zero because water has a slight tendency to ionize:

$$H_2O \rightleftharpoons H^+ + OH^-$$

When the temperature is 24°C and no other materials are dissolved in the water, there are 1.0×10^{-7} g of H^+ ions per liter and an equal number of OH^- ions weighing 17.0×10^{-7} g per liter of water. Only 1.8 molecules per billion are ionized in pure water, but the H^+ and OH^- ions produced are chemically active and very important.

The numbers of H^+ and OH^- ions usually do not remain equal when other ions are also present. What does remain constant is the product of the concentrations[1]

---

[1]The more precise term is *activities,* which may be considered as what the concentrations appear to be rather than what they really are. Two other factors that must be considered for precise work are temperature and salt concentration in solution. More water molecules ionize at higher temperatures and fewer ionize when large salt concentrations are present.

of the H^+ and OH^- ions. These concentrations are usually expressed in normalities (symbolized as N). Normalities are equal to the number of gram-equivalent weights present in a liter of solution. A liter of 1 N acid contains 1 g of H^+, and a liter of 1 N base contains 17 g of OH^-. A liter of pure water is 1.0×10^{-7} N (0.000,000,1 N) in both H^+ and OH^-. The product of these two normalities is 1.0×10^{-14} and remains constant (subject to the limitations mentioned in the footnote on p. 167). The concentration of either H^+ or OH^- in a water solution can be readily calculated if the concentration of the other is known.

The pH Scale

Numbers such as 0.000,000,1 and 1.0×10^{-7} are inconvenient to use. The pH scale simplifies these by means of a negative logarithm; the sign of the -7 exponent is changed to give pH 7 at neutrality. The logarithm of the multiplier number is subtracted from the exponent to give fractional pH values. The logarithm of 1.0 is zero, and so pH 7.0 corresponds to 0.000,000,1 and 1.0×10^{-7} N H^+. The logarithm of 2 is 0.3, and so pH 6.7 corresponds to 0.000,000,2 and 2.0×10^{-7} N H^+. Doubling the H^+ concentration lowers the pH by 0.3 units at any point along the pH scale; multiplying the H^+ concentration by 100 lowers pH by 2.0 units, etc.

The OH^- concentration can be measured on a pOH scale just as H^+ is measured by pH. The sum of pH plus pOH is always 14 (subject to salt and temperature limitations). When pH goes up, pOH goes down by a corresponding amount, and vice versa. This relation is shown in Table 7-4.

Soil pH

Most soils have pH values between 4 and 8, as shown in Figure 7-1. Nearly all soils with pH values above 8 have a high percentage of Na^+ ions on their cation-exchange sites. Soils with pH values below 4 generally contain sulfuric acid.

Table 7-4   Relation of pH and pOH to Normalities of Acidic and Alkaline Solutions

pH	Acidity (normality of H^+)	Alkalinity (normality of OH^-)	pOH
0	1.0	0.000,000,000,000,01	14
1	0.1	0.000,000,000,000,1	13
2	0.01	0.000,000,000,001	12
3	0.001	0.000,000,000,01	11
4	0.0001	0.000,000,000,1	10
5	0.00001	0.000,000,001	9
6	0.000001	0.000,000,01	8
7	0.000,000,1	0.000,000,1	7
8	0.000,000,01	0.000001	6
9	0.000,000,001	0.00001	5
10	0.000,000,000,1	0.0001	4
11	0.000,000,000,01	0.001	3
12	0.000,000,000,001	0.01	2
13	0.000,000,000,000,1	0.1	1
14	0.000,000,000,000,01	1.0	0

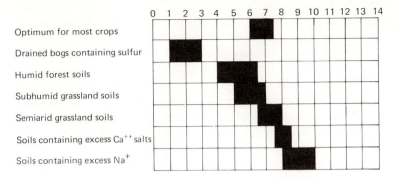

Figure 7-1 Typical pH ranges for various types of soils.

An evaluation of pH in descriptive terms is given in Figure 7-2. Any such terminology is necessarily arbitrary, but the terms are useful. By strict definition, any pH below 7.0 is acid and any pH above 7.0 is alkaline. More practically, a small zone near 7.0 may be considered to be neutral.

Soil pH depends on a variety of factors including all five soil-forming factors plus the season of the year, cropping practices, the soil horizon sampled, the water content at sampling time, and the way the pH is determined.

Leaching processes remove bases from the soil and therefore tend to lower the pH with time. This gradual lowering of pH is most significant in young soils and loses effect by the time weathering processes remove most of the 2:1 lattice silicate-clay minerals. Fertilizers containing sulfur or nitrogen tend to lower the soil pH at a rate that often produces a noticeable effect within a few years. The pH can be raised by an application of lime if it is too low. Extremely high pH values caused by excess Na^+ can be lowered by applying gypsum; almost any soil pH can be lowered by applying sulfur because it will be oxidized by soil microorganisms to form sulfuric acid. The practice of altering pH with soil amendments is discussed in Chapter 8.

Vegetation influences soil pH in complex ways because it produces organic matter and influences leaching. The addition of decomposable organic matter to a soil results in the formation of organic acids. These acids add to the cation-exchange capacity, but the percent base saturation and pH are lowered. Bases released from the organic matter and from accelerated weathering of the soil minerals may or may not be enough to prevent the soil from becoming acid.

Leaching processes are accelerated under acid conditions because more cations are released by acid weathering and fewer are held by cation exchange. Furthermore, acid conditions usually occur in humid climates where leaching is intense. However, leaching is partially offset by plant growth. Plant nutrients are absorbed from all

Figure 7-2 Descriptive terms for soil pH ranges.

parts of the root zone, but nutrients in the plant tops are deposited on the soil surface when the plant dies. In temperate regions, soils formed under grass are usually less acid than soils formed under trees. Climate causes part of the difference—grasses predominate in drier climates where leaching is less intense. But grassland soils are usually less acid than forested soils even where they are intermixed. The reason appears to be that grasses producing new growth each year utilize more bases and therefore deposit more bases on the soil surface than do trees. Thus the grasses help to keep the soil from becoming strongly acid. There are, however, species differences. Many coniferous trees are particularly light feeders on bases and tend to produce the most acid soils. Forest vegetation is usually more effective in offsetting the intense leaching action in the humid tropics than is grass vegetation; the deeper penetration of tree roots is an important factor there.

Relation of Base Saturation to Soil pH

Adding an acid or an acid-forming material lowers the soil pH; adding bases raises the soil pH. But the amount of pH change is relatively small compared with the amount of acid or base added because most soils are well buffered.

The cation-exchange capacity is the principal buffer mechanism in soil. Usually the exchangeable cations are partly basic (Ca^{++}, Mg^{++}, etc.) and partly acidic (H^+, Al^{+++}) but their acidic or basic properties are expressed only when they are in solution. Most soils have over 100 times as many cations adsorbed on the micelles (inactive) as in true solution (active) at any one time. Only the active portion of the H^+ ions is measured in the pH, but there is a continual exchange or equilibrium between exchangeable adsorbed ions and ions in solution.

Only a fraction of the added H^+ ions remain in solution after acid is added to soil. The others exchange places with adsorbed bases on the micelles. The reverse reaction takes place when a base is added to soil; H^+ ions from the micelles are replaced by cations from the base and move into the solution where they react with OH^- ions and form water. Thus the pH usually changes only slightly when an acid or a base is added to soil, but there is also a change in the percentage of cation-exchange sites occupied by H^+ versus those occupied by bases. The percentage of cation-exchange sites occupied by bases is referred to as the *percent base saturation.* Figure 7-3 shows curves relating pH to percent base saturation in several different types of clay and organic materials. The types of bases present also influence the locations of the pH-base saturation curves, but this complicating factor is usually ignored. It can be seen from Figure 7-3 that the materials with the highest cation-exchange capacities have the lowest pH values at any specified degree of base saturation. For example, kaolinite has a pH of 7 at 60 percent base saturation, but the pH of montmorillonite at the same saturation is below 5. It takes 95 percent base saturation to give montmorillonite a pH of 7. Part of the reason for this difference is the large number of all kinds of ions present in the systems when the materials with high cation-exchange capacities are present. There is also a difference in the ease with which H^+ and other cations escape or ionize from the various colloids. A higher percentage of the H^+ ionizes from montmorillonite than from kaolinite.

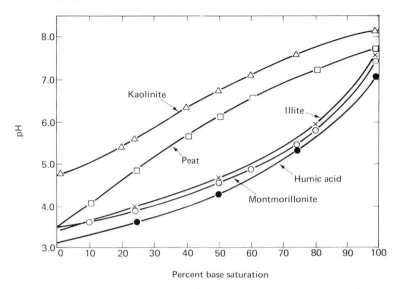

Figure 7-3   pH—percent base saturation curves for soil colloids. (*After Mehlich,* 1941 *and* 1942.)

Measuring pH

Two kinds of methods are used to measure pH. The electrometric method is the most accurate and the most used in the laboratory. The colorimetric method is the most portable, the cheapest, and the most used in the field.

The *electrometric* method uses a sensitive meter to measure the difference in the electric potential between a special glass electrode and a reference calomel electrode. The electric potential of the glass electrode varies with the H^+ ion concentration, but the calomel electrode has a constant potential. The two electrodes are immersed in a mixture of soil and water (or other material whose pH is to be determined) and an electric current is turned on. The difference between the potentials of the two electrodes shows on a dial marked in pH units.

Most pH meters can be read to 0.01 pH unit. Determinations approaching this accuracy require control of the soil-to-water ratio in addition to good technique in handling the equipment. Many workers prefer a minimum of water, usually a saturated paste (all pore space filled with water but no excess), or a 1:1 weight ratio of soil to water. Ratios of 1:2, 1:5, and 1:10 are sometimes used. The soil-to-water ratio must be specified along with the result because the pH reading is higher when more water is added.

A pH that is nearly independent of dilution and closer to that of the soil at its normal water content can be obtained by mixing the soil with 0.01 molar $CaCl_2$ solution rather than water.

The *colorimetric* method uses indicators that change color gradually over a known pH range. These indicators are weak organic acids (or bases, but the bases are less used because the soil is more likely to absorb their color) that serve as indicators because their ions have a different color than their molecules. The organic-acid indicators ionize much more if the H^+ ion concentration of the soil tested is low than if it is high. For example, bromthymol blue is very little ionized at pH 6 or below and therefore shows its molecular color (yellow); above pH 7.5 it is almost completely ionized, and it shows its ionic color (blue); shades of green indicate degrees of partial ionization that occur between pH 6.0 and 7.5.

Colorimetric pH readings are usually made in small depressions in white porecelain "spot plates" (Figure 7-4). A small amount of soil is placed in a depression, and enough indicator solution is added to saturate the soil plus about two drops excess. The mixture is shaken or stirred to obtain a uniform color. The color of the excess solution against the white porcelain background is then compared with a color chart for estimating pH. The useful range of any one indicator is limited to a little over 1 pH unit because its color change is too gradual outside this range. This problem is overcome by the use of other indicators, each with its own color change and pH range. Mixed indicators are sometimes used to cover a wide pH range if accuracy is not critical. The mixed indicators can be read to about 0.5 pH unit and the single indicators to about 0.1 pH unit.

IMPORTANCE OF pH TO PLANT GROWTH

There is evidence that pH has little or no *direct* effect on plant growth. Varying the concentrations of H^+ and OH^- ions over a wide range seems to make no difference

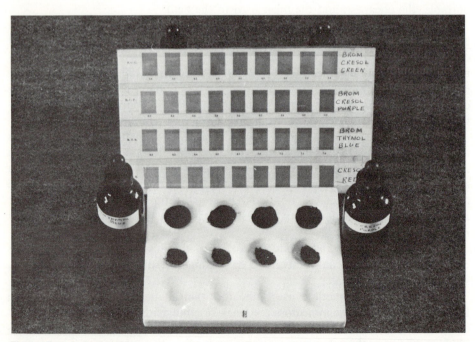

Figure 7-4 Indicator solutions are mixed with soil in a spot plate for testing pH in the field. The color of the solution around the soil is compared to a chart to read the pH.

to plants so long as other factors remain favorable. Indirect effects, however, are numerous and potent. Aluminum toxicity has already been mentioned. A soil at pH 4 probably has enough soluble Al^{+++} to be very detrimental to most plants, whereas nutrient solutions at pH 4 show no such problem. Similar effects are experienced with manganese at low pH.

Effect of pH on the Availability of Plant Nutrients

The most universal effect of pH on plant growth is nutritional. The soil pH influences the rate of plant nutrient release by weathering, the solubility of all materials in the soil, and the amounts of nutrient ions stored on the cation-exchange sites. The pH is therefore a good guide for predicting which plant nutrients are likely to be deficient. Figure 7-5 shows the relative probabilities of plant nutrient deficiencies at various soil pH values. Figure 7-5 is, of course, merely a guide to probable deficiencies related to pH. Some parent materials are so deficient in certain elements that nutrient deficiencies occur in the soil no matter what the pH may be. The deficient elements can be supplied only by fertilizer applications in such places, whereas in pH-related deficiencies there is a choice. The soil can be amended to a more favorable

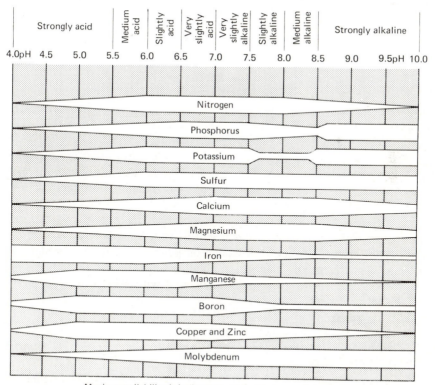

Maximum availability is indicated by the widest part of the bar

Figure 7-5 A schematic illustration of the relation between plant nutrient availability and soil reaction. (*After Truog,* 1946.)

pH (Chapter 8) or enough fertilizer can be supplied to overcome the deficiency in spite of the pH. The optimum pH varies with lime and fertilizer prices and with the crop being grown. Nutrient requirements vary with plant species and so does the best pH. Usually the optimum pH is somewhere between 6.0 and 7.5 because all plant nutrients are reasonably available in that range. As an exception, "iron-loving" plants such as azaleas, rhododendra, blueberries, and strawberries thrive in soils with pH between 5 and 6.

Most *nitrogen* and *sulfur* mineral compounds (except calcium sulfate) are soluble at any pH and subject to leaching. The supply of these nutrients therefore comes mostly from organic sources and is limited by slow decomposition of organic matter. The rate of mineralization of these elements from organic matter is fastest between pH 6 and 8. Much of the available sulfur in soils more alkaline than pH 8 comes from mineral sources rather than organic.

Phosphorus compounds in soils have low solubilities. Even when soluble phosphates are added as fertilizers, they tend to react in the soil and form low solubility materials. A pH between 6.5 and 7.5 is usually best for phosphorus availability. Iron and aluminum phosphates precipitate at low pH, and calcium phosphates precipitate at high pH. Phosphates become available again above pH 8.5 because such high pH values indicate a high percentage of sodium salts; sodium phosphates are soluble in water.

Potassium is soluble at any pH but is removed from solution by sorption. Some of it is exchangeable, but some is held more tightly. As a soil is leached and becomes acid, the total potassium declines, and the amount available also declines. Raising the pH by liming may drive the potassium into nonexchangeable positions and further suppress the potassium availability.

Calcium and *magnesium* are the most abundant bases. They are frequently added in soil amendments but are rarely required as fertilizers.

Iron, manganese, copper, and *zinc* all form metallic cations that precipitate in low-solubility compounds at high pH. Deficiencies of these elements often limit plant growth in high-lime or alkaline soils. Iron deficiency is sometimes brought about by liming an acid soil. Iron availability increases as the pH is lowered because soils contain too much iron for it to be depleted by leaching. The other elements in this group may be leached out and become deficient at very low pH.

Boron may be leached at very low pH, and its solubility is low at very high pH. However, since boron is usually plentiful in alkaline soils, the low solubility is not a problem. Boron deficiencies are usually associated with soils formed in parent materials very low in boron.

Molybdenum solubility is increased by liming. Molybdenum is precipitated by iron and aluminum at low pH values.

Other Effects Related to pH

The physical condition of the soil is related to pH and has an important effect on plant growth. Soil pH values above 8.5 indicate the presence of considerable Na^+ and the likelihood of dispersed soil colloids. Plant growth is markedly reduced where colloids are dispersed and is completely eliminated where the condition is severe. Improving the fertility of a dispersed soil does no good because the problem is primarily physical. The dispersed clay particles plug the soil pores, and water and air movements become so slow that plants cannot grow. The remedy is to amend the soil as discussed in Chapter 8.

Soil pH values near 7.5 or 8.0 are usually associated with saline soils (those containing excess soluble salts). The soil colloids are usually flocculated when excess salts are present but may become dispersed if the salts are removed and a high percentage of Na^+ remains. The presence of excess soluble salts also implies that most plant nutrients are present in abundance. The principal problem with most saline soils is a shortage of water available to the plants. Not only are most saline soils located in arid regions but also the osmotic effect of the salts makes the water less available. Usually the salts are leachable, but this requires drainage and a water supply.

Few saline soils occur in humid regions, and the salts in these soils are usually less soluble, often dominated by calcium carbonate. Drainage and fertilizer application offset most of the detrimental effects from calcium carbonate. These "high-lime" soils need more phosphorus and potassium fertilization than neighboring soils. Iron and zinc deficiencies are more common on high-lime soils than on other soils.

Saline soils are often produced by people growing house plants. Watering the plants adds dissolved salts to the soil. The salts accumulate and can be identified by crystals forming on high points at the soil surface. Salt accumulation can be avoided by applying excess water and letting it drain from the soil.

SOLUBILITIES

The solubility of a material determines how fast and how far it will move in a particular environment. Materials are often classified as either soluble or insoluble in water, but these should be considered as relative terms rather than absolutes. Even the "insoluble" materials can usually be shown to have slight solubilities.

Soluble salts include chlorides, sulfates, nitrates, and some carbonates and bicarbonates of sodium, potassium, magnesium and calcium (and many other elements in lesser amounts). Excess amounts of these salts are soon leached from the upper horizons of most soils. Soluble salts are of concern only in the saline soils where they are detrimental to plant growth because of their osmotic effect. A total concentration of dissolved molecules and ions equivalent to 1 mole per liter of soil solution produces an osmotic pressure equivalent to about 24 atm at 20°C. The osmotic pressure adds to the soil moisture tension and decreases the availability of water to plants. High osmotic pressure can cause permanent wilting to occur at field capacity. Such high concentrations of soluble salts are unusual even in saline soils and must be avoided if plants are to be grown.

Solubilities vary with temperature and pH. The pH variation is much more important in soil than is the temperature variation, especially for relatively insoluble compounds containing more than one OH^- ion. For example, the solubility product constant $(K\text{sp})^1$ of ferric hydroxide at 18°C is 1.1×10^{-36}. Using parentheses to represent concentrations (actually activities) gives:

$$(Fe^{+++})(OH^-)^3 = 1.1 \times 10^{-36} \quad \text{or} \quad (Fe^{+++}) = \frac{1.1 \times 10^{-36}}{(OH^-)^3}$$

At pH 7 the concentration of OH^- is 1×10^{-7} moles/liter; the maximum possible concentration of Fe^{+++} in solution at pH 7 is therefore:

$$(Fe^{+++})_{max} = \frac{1.1 \times 10^{-36}}{(1 \times 10^{-7})^3} = 1.1 \times 10^{-15} \text{moles/liter}$$

Lowering the pH to 6 gives a concentration of OH^- of 1×10^{-8} moles/liter and (by calculating as above) increases the maximum possible concentration of Fe^{+++} in solution by 1,000 times to 1.1×10^{-12} moles/liter. Lowering the pH to 5 increases the potential Fe^{+++} concentration by another factor of 1,000 to 1.1×10^{-9} moles/liter.

Solubilities of $Al(OH)_3$ respond to pH in the same way as those of $Fe(OH)_3$ but at a higher level because $Al(OH)_3$ has a higher $K\text{sp}$ (1.6×10^{-34}). However, the Al^{+++} ion concentration is limited in other ways at low pH values. Values of $K\text{sp}$ for some important low-solubility compounds in soils are given in Table 7-5. $K\text{sp}$ values permit one to calculate maximum concentrations for various ions under specified circumstances. The actual concentrations are lower if there is not enough of the ion available to saturate the solution or if some other low-solubility compound containing the ion limits its concentration. Precipitation will occur if the $K\text{sp}$ of any material is exceeded (but the precipitation process is slow for many materials, and so $K\text{sp}$ may be exceeded for a time). The amount of precipitate (solid) that can be present is limited only by volume considerations; there must be a place to put it.

[1]The $K\text{sp}$ of a compound is the product of the concentrations of all the ions of the compound in a saturated solution. It is a more precise means of expressing solubility than methods based on the weight or the molarity of the compound as a whole.

Table 7–5   Solubility Product Constants for Some Relatively Insoluble Soil Constituents

Constituent	Ions	Ksp
Calcite	$(Ca^{++}) (CO_3^{--})$	0.87×10^{-8} at 25°C
Calcium sulfate	$(Ca^{++}) (SO_4^{--})$	1.95×10^{-4} at 10°C
Hydroxy-apatite	$(Ca^{++})^5 (PO_4^{---})^3 (OH^-)$	3.7×10^{-58}
Aluminum hydroxide	$(Al^{+++}) (OH^-)^3$	1.6×10^{-34}
Ferric hydroxide	$(Fe^{+++}) (OH^-)^3$	1.1×10^{-36} at 18°C
Variscite	$(Al^{+++}) (OH^-)^2 (H_2PO_4^-)$	3×10^{-31}

Sources: Compiled from *Handbook of Chemistry and Physics* and from private communications from M. Peech, Cornell University, and C. A. Black, Iowa State University.

Some of the precipitate will dissolve again when the product of the ion concentrations drops below the Ksp of the material. But many dissolving processes are too slow to keep solutions saturated.

DISTRIBUTION OF IONS

The actual number of ions present in even a very dilute solution is very large. For example, ferric hydroxide is usually considered insoluble in water, but the concentration of Fe^{+++} at pH 7 can be over 600,000,000 ions per liter [1.1 X 10^{-15} (from the previous section) X 6.0 X 10^{23} (Avogadro's number)]. The number of ions involved is so large that their distribution in solution can be predicted reasonably well by mathematical equations. These equations are beyond the scope of this text, but some of the fundamentals involved are less complex.

The distribution of ions in the soil solution is greatly influenced by the concentration of negative charges on the micelles. The situation is one of solid particles, each containing a large number of negative charges, surrounded by a swarm of cations. These cations are attracted to the micelles but not too tightly to be exchanged. Most of them are adsorbed at the micelle surface, but some of them are various distances away in the soil solution. Their concentration gradually diminishes with distance, as illustrated in Figure 7-6.

Anions are much less concentrated than cations in the system and follow an inverse pattern to that of the cations (assuming negatively charged micelles). The negative charge of the anions is repelled by the negative charge of the micelles; hence anions are least concentrated near the micelle surfaces. Their maximum concentration is in the area farthest from the negatively charged surfaces and is nearly large enough to balance the positive charges of the cations in that area.

One inference drawn from Figure 7-6 is that the soil solution can never be removed from the soil and studied as a separate entity. The electric-charge balance cannot be maintained without the micelles. The nature of the solution and the concentrations of all the ions in it would depend on whether all or part of the water were removed. The study of the colloidal system and the resultant emphasis on cation exchange have proven to be more meaningful ways of learning about soil chemistry.

The electrical charges of ions attract water molecules that orient themselves in shells around the ions. This *water of hydration* combined with the concentration of cations near a silicate clay particle produces a pressure from crowding in that vicinity. Clay particles that are not tightly bonded together are forced apart into a dispersed condition when the pressure is too high.

The difference between monovalent cations such as Na^+ and divalent cations such as Ca^{++} can also be related to Figure 7-6. The higher-charged cations are held more strongly to the micelle and therefore constitute a much larger percentage of the adsorbed cations than of the cations some distance out in the solution. A large amount of Na^+ must be present in solution before it will replace many of the adsorbed cations. But most sodium salts are highly soluble, and so it is possible to increase the amount of adsorbed Na^+. High Na^+ concentrations cause the clay

Figure 7-6 An illustration of how ions are distributed in the soil solution near a negatively charged clay particle.

particles to disperse because a pressure buildup results from replacing Ca^{++} ions with twice as many Na^+ ions.

A plant can absorb a cation from solution easier than from the micelle surface even when the root makes contact with the ion. Work must be expended to pull the cation away from the electrical attraction of the micelle. The amount of such work required decreases with an increase in distance between the cation and the micelle. The plant root normally accomplishes this work by using energy from respiration. Adequate soil aeration for respiration is therefore very important in the absorption of cations by plant roots.

The nature of the soil solution is quite dynamic. Rapid plant growth reduces the concentrations of ions in solution because the equilibrium processes are generally slow acting. The soil pH changes with the seasons as the growing plants absorb cations and release H^+ during periods of growth; processes of solution and of ion exchange restore the equilibrium during other periods. Additions of water by rain or irrigation tend to dilute the soil solution, whereas removal of water by evaporation and transpiration tend to concentrate the ions in it. These changes in concentration cause exchanges to occur between adsorbed cations and cations in solution. When the solution is more concentrated, there is less preference for higher-charged cations to be adsorbed by the micelles instead of those carrying single charges. The availabilities of nutrients for plant growth vary as the ion concentrations and distributions vary.

ANION EXCHANGE

Anion exchange is usually masked by the much larger cation-exchange capacity present in most soils. Where both types of exchange sites are present, there is a tendency for them to bond the colloidal particles together producing a strong structure and using up the exchange sites. Toth (1937) showed that such bonding existed by finding an increased cation-exchange capacity after he removed free ion and aluminum from clays. The ultimate in this direction is plinthite where the bonding hardens the material irreversibly into a rocklike mass when it dries.

Anion-exchange sites hold phosphate ions more rigidly than cation-exchange sites hold cations. One way of determining anion-exchange capacity is by measuring the amount of $H_2PO_4^-$ that can be adsorbed by the soil colloids. The problem is that only a small fraction of the $H_2PO_4^-$ ions thus adsorbed is readily exchangeable for other ions. They may be exchanged later, but the process is very slow.

Anion-exchange sites can result from (1) amine groups in humus, (2) bonding terminating with a cation at the edge of a silicate-clay material, and (3) OH^- ionizing from such materials as $Al(OH)_3$ or $Fe(OH)_3$. The likelihood of ionizing significant amounts of OH^- depends on the abundance of minerals containing it and on pH. It increases in old, highly weathered, and leached soils because the Al and Fe compounds accumulate in such soils and because most of the ionized OH^- combines with H^+ to form water under acid conditions.

Some exchange capacity for both cations and anions occurs on the edges of silicate-clay crystals. This type of exchange capacity probably accounts for most or

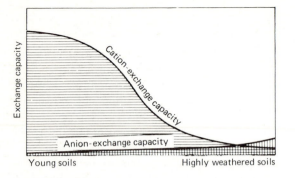
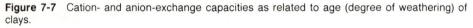

Figure 7-7 Cation- and anion-exchange capacities as related to age (degree of weathering) of clays.

all of the exchange capacity of kaolinite and of the anion-exchange but not of the cation-exchange capacities of montmorillonite, vermiculite, and illite. The likelihood of a structure terminating with an exposed positive charge or an exposed negative charge should be about equal. Kaolinite should therefore have nearly as much anion-exchange capacity as cation-exchange capacity. Dean and Rubins (1947) found this to be true. Variations in pH cause deviations from equality in the numbers of positive and negative edge charges. Anion-exchange capacity increases at low pH, and cation-exchange capacity increases at high pH.

Figure 7-7 schematically relates the cation- and anion-exchange capacities to the degree of weathering of the clay. Relatively unweathered clays have a high proportion of 2:1 layer silicate minerals such as illite, vermiculite, and montmorillonite. The cation-exchange capacities of these clays are relatively high. Kaolinite increases as weathering proceeds, and eventually cation- and anion-exchange capacities are equal. Finally the free iron and aluminum that accumulate in the clay fraction may cause the anion-exchange capacity to exceed the dwindling cation-exchange capacity. This condition exists in some tropical soils.

REFERENCES

Barrow, N. J., 1972, Influence of Solution Concentration of Calcium on the Adsorption of Phosphate, Sulfate, and Molybdate by Soils, *Soil Sci.* **113**:175–180.

Broyer, T. C., A. B. Carlton, C. M. Johnson, and P. R. Stout, 1954, Chlorine—A Micronutrient Element for Higher Plants, *Plant Physiol.* **29**:526–532.

Cate, R. B., Jr., and A. P. Sukhai, 1964, A Study of Aluminum in Rice Soils, *Soil Sci.* **98**:85–93.

Chao, T. T., M. E. Harward, and S. C. Fang, 1962, Soil Constituents and Properties in the Adsorption of Sulfate Ions, *Soil Sci.* **94**:276–283.

Clarke, F. W., 1924, The Data of Geochemistry, *U.S. Geol. Survey Bull.* 770.

Dean, L. A., and E. J. Rubins, 1947, Anion Exchange in Soils, *Soil Sci.* **63**:377–406.

DeVilliers, J. M., and M. L. Jackson, 1967, Cation Exchange Capacity Variations with pH in Soil Clays, *Soil Sci. Soc. Am. Proc.* **31**:473–476.

Doss, B. D., and Z. F. Lund, 1975, Subsoil pH Effects on Growth and Yield of Cotton, *Agron. J.* **67**:193–196.

Foy, C. D., W. H. Arminger, L. W. Briggle, and D. A. Reid, 1965, Differential Aluminum Tolerance of Wheat and Barley Varieties in Acid Soils, *Agron. J.* **57**:413–417.

Greig, J. K., and F. W. Smith, 1962, Salinity Effects on Sweetpotato Growth, *Agron. J.* **54**:309–313.

Iwata, S., 1974, Thermodynamics of Soil Water. III. The Distribution of Cations in a Solution in Contact with a Charged Surface of Clay, *Soil Sci.* **117**:87–93.

Keser, M., B. F. Neubauer, and F. E. Hutchinson, 1975, Influence of Aluminum Ions on Developmental Morphology of Sugarbeet Roots, *Agron. J.* **67**:84–88.

Khasawneh, F. E., and Fred Adams, 1967, Effect of Dilution on Calcium and Potassium Contents of Soil Solutions, *Soil Sci. Soc. Am. Proc.* **31**:172–176.

Marshall, C. E., 1949, *The Colloid Chemistry of the Silicate Minerals,* Agronomy Monograph 1, Academic, New York.

Mehlich, A., 1941, Base Unsaturation and pH in Relation to Soil Type, *Soil Sci. Soc. Am. Proc.* **6**:150–156.

Mehlich, A., 1942, The Significance of Percentage Base Saturation and pH in Relation to Soil Differences, *Soil Sci. Soc. Am. Proc.* **7**:167–174.

Miller, R. W., 1967, Soluble Silica in Soil, *Soil Sci. Soc. Am. Proc.* **31**:46–50.

Pierre, W. H., G. G. Pohlman, and T. C. McIlvaine, 1932, Soluble Aluminum Studies. I. The Concentration of Aluminum in the Displaced Soil Solution of Naturally Acid Soils, *Soil Sci.* **34**:145–160.

Sawhney, B. L., C. R. Frink, and D. E. Hill, 1970, Components of pH Dependent Cation Exchange Capacity, *Soil Sci.* **109**:272–278.

Schalscha, E. B., P. F. Pratt, and L. de Andrade, 1975, Potassium-Calcium Exchange Equilibria in Volcanic-Ash Soils, *Soil Sci. Soc. Am. Proc.* **39**:1069–1072.

Schuffelen, A. C., 1974, A Few Aspects of 50 Years of Soil Chemistry, *Geoderma* **12**:281–297.

Shainberg, I., and W. D. Kemper, 1966, Hydration Status of Adsorbed Cations, *Soil Sci. Soc. Am. Proc.* **30**:707–713.

Toth, S. J., 1937, Anion Adsorption by Soil Colloids in Relation to Change in Free Iron Oxides, *Soil Sci.* **44**:299–314.

Truog, E., 1946, Soil Reaction Influence on Availability of Plant Nutrients, *Soil Sci. Soc. Am. Proc.* **11**:305–308.

Chapter 8

Amending the Soil

Any user of soil who finds his soil to be exactly what is best for his needs is indeed fortunate. Usually he finds that some of its characteristics are not ideal. What, then, should he do? One alternative is to use the soil in its nonideal state and accept the lowered productivity or other limitation that it imposes. Another alternative is to modify the soil in ways that make it more suitable for the intended use.

Soil modification may be aimed at some specific item of fertility such as supplementing the supply of one or more plant nutrients (Chapters 9 through 14) or it may be aimed at a more general change in the chemical and physical properties of the soil. The latter, termed *amending the soil,* is the subject of this chapter.

Some extreme cases where the soil is essentially remade are discussed in the latter part of this chapter. More generally, amending the soil involves the application of a number of tons per hectare of some soil amendment to change the soil properties. Usually the direct effect is to change the soil pH. The indirect effect is to change a multitude of properties that vary with pH. Several indirect effects of pH on plant growth were discussed in Chapter 7.

RAISING SOIL pH

Most soils occurring in climates humid enough to grow large crops without irrigation have acid reactions (Figure 8-1). The acidity is caused by the leaching of bases by percolating water. The result is a lowered soil fertility and a less-than-ideal medium for the growth of most plants. Raising the soil pH to a reaction near neutral may be highly profitable in spite of the cost involved. This procedure is known as *liming.*

Liming will help maintain soil fertility if used in a well-managed cropping system; but in a poorly managed cropping system, liming acts as a stimulant producing good crops immediately, followed by gradual impoverishment of fertility. An old, time-tested German expression illustrates the effect of lime used on soils without supplemental fertility practices: "Kalk macht die Väter reich aber die Söhne arm" (lime makes the fathers rich but the sons poor). Somewhat faster action but the same result are implied in another old saying: "Lime, and lime without manure, will make both farm and farmer poor." The effect of lime alone on yields is illustrated in Figure 8-2.

Liming and fertilization usually go together as complementary practices, at least in humid regions. Liming increases the usage of nutrients because more crop is harvested. Microbial activity is also favored by liming, and organic-matter decomposition is accelerated. The accelerated decomposition causes the organic-matter content of the soil to decline. Liming therefore increases the available plant nutrient

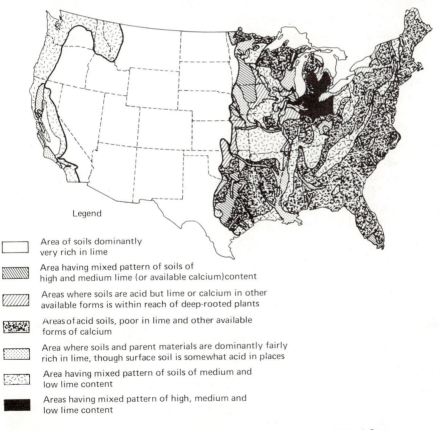

Legend

☐ Area of soils dominantly very rich in lime

▨ Area having mixed pattern of soils of high and medium lime (or available calcium)content

▧ Areas where soils are acid but lime or calcium in other available forms is within reach of deep-rooted plants

▨ Areas of acid soils, poor in lime and other available forms of calcium

⋯ Area where soils and parent materials are dominantly fairly rich in lime, though surface soil is somewhat acid in places

⋙ Area having mixed pattern of soils of medium and low lime content

■ Areas having mixed pattern of high, medium and low lime content

Figure 8-1   The relative lime content and acidity of the soils of the United States.

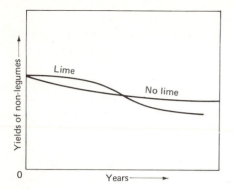

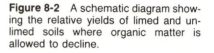

Figure 8-2 A schematic diagram showing the relative yields of limed and unlimed soils where organic matter is allowed to decline.

supply for a time but decreases the future supply. That is why yields with liming are first higher but later lower than without liming. The lowering of yields in later years can usually be avoided by using fertilizer as well as lime. The decline in soil organic matter may be averted, or at least limited, if enough crop residues are returned to the soil from the increased production.

Fertilization often increases the need for lime. The ammonium form of nitrogen fertilizer is oxidized in the soil, and the nitrate form is produced. Ammonia is a base, but the nitrate form has an acidic effect on the soil. An application of lime is generally required after an ammonium form of fertilizer has been used for a few years. Ammonia is generally the cheapest nitrogen fertilizer, and so it is usually more economical to use ammonia and correct the acidity by liming rather than to avoid its use. Also, organic nitrogen forms such as urea acidify the soil by the same process as ammonia.

Liming Materials

A satisfactory material for raising soil pH meets several requirements:

1 It should have a mild alkalizing (pH increasing) effect. The intent is to raise the pH to near neutral. The ideal material should have an action mild enough to cause no harm where an overdose is applied.
2 It should result in a desirable proportion of cations adsorbed on the cation-exchange sites. The added cations should be mostly Ca^{++}, although some Mg^{++} is good. Little or no Na^+ should be included.
3 It should have a favorable effect on soil structure. The most favorable base for good soil structure is Ca^{++}.
4 It should not be too expensive.

Almost all liming material used is ground limestone rock (Figure 8-3) because it best meets the requirements. Limestone is mainly $CaCO_3$; dolomitic varieties contain some $MgCO_3$. The desired cations are supplied, and the carbonate anions cause no ill effects. The alkalizing effect is mild but effective. Few materials are cheaper or more abundant.

Some highly effective materials such as Na_2CO_3 have too strong an alkalizing effect, supply the wrong cations, and are too expensive to be used as liming materials.

Figure 8-3 Limestone occurs in abundance under soils in many parts of the country. The ground limestone provides a cheap and abundant source of lime.

Some materials such as $CaCl_2$ and $CaSO_4 \cdot 2H_2O$ (gypsum) contain Ca^{++} but are neutral salts and have no alkalizing effect. Neutral calcium salts may be useful as calcium fertilizers on oxide clays where the pH is already satisfactory, but they are not considered to be liming materials.

The alkalizing effects of $CaCO_3$ and $MgCO_3$ are mild because they are slightly soluble salts of moderately strong bases and a weak acid. A much stronger alkalizing effect results from Na_2CO_3 because it is a highly soluble salt of a strong base and a weak acid. The effectiveness of $CaCO_3$ and $MgCO_3$ in spite of their mildness results from the escape of CO_2 gas permitting the reaction to go to completion:

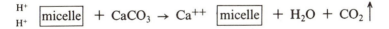

$$\begin{matrix} H^+ \\ H^+ \end{matrix} \boxed{\text{micelle}} + CaCO_3 \rightarrow Ca^{++} \boxed{\text{micelle}} + H_2O + CO_2 \uparrow$$

Calcium Carbonate Equivalent

No two limestone deposits are alike nor are they equally effective in neutralizing soil acidity. Each material is therefore compared with pure $CaCO_3$ as a standard with an assigned value of 100 percent. Some impurities such as quartz are always present in limestone and reduce its equivalency. The presence of $MgCO_3$ in dolomitic limestone increases its equivalency because $MgCO_3$ has a lower molecular weight than $CaCO_3$; a gram of $MgCO_3$ therefore contains more molecules than a gram of $CaCO_3$. Pure $CaCO_3$ has a molecular weight of 100. Pure $MgCO_3$ has a molecular weight of 84. Therefore, a sample of pure $MgCO_3$ has a calcium carbonate equivalent of $100 \div 84$, or 119 percent.

A ratio of one part $MgCO_3$ to about six parts $CaCO_3$ would maintain the same ratio of Ca^{++} to Mg^{++} ions as normally exists in unlimed soils. The use of dolomitic lime (that containing some $MgCO_3$) is particularly beneficial where magnesium deficiencies occur.

Burned lime can be used instead of limestone where it is desirable to minimize weight. The heat drives off CO_2 and reduces the weight of 100 g of $CaCO_3$ to 56 g of CaO:

$$CaCO_3 \xrightarrow{\text{heat}} CaO + CO_2 \uparrow$$

The CaO reacts with water to produce $Ca(OH)_2$ (sometimes known as slaked lime):

$$CaO + H_2O \rightarrow Ca(OH)_2$$

The calcium carbonate equivalent of CaO is $100 \div 56$, or 179 percent. The calcium carbonate equivalents of other possible liming constituents such as MgO, $Ca(OH)_2$, $Mg(OH)_2$, and CaC_2 can be calculated from their molecular weights in similar fashion. Occasionally a source containing one or more of these compounds may be cheaper or easier to obtain locally than ground limestone. Examples of such sources are refuse lime, sugar-factory lime, water-softening-process lime, building (slaked) lime, blast furnace slag, marl, and oyster shells. However, the use of ground limestone as a liming material far exceeds all the others combined.

Effective Calcium Carbonate Equivalent

Lime must be finely ground to be effective. The finer it is, the more quickly it reacts with the soil. Limestone pebbles over 5 mm in diameter will persist for so many years in the soil that they are ineffective for practical liming purposes; lime ground fine enough to pass through a 60-mesh sieve[1] will completely react with the soil within 3 years. Most ground limestones contain a range of particle sizes. A means of estimating the percent effectiveness of liming materials is illustrated below:[2]

Percent of material coarser than 4 mesh   X 0   =   0% effectiveness
Percent of material between 4 and 8 mesh   X 0.1 =   % effectiveness
Percent of material between 8 and 60 mesh X 0.4 =   % effectiveness
Percent of material finer than 60 mesh   X 1.0 =   % effectiveness

Total percent effective during first 3 years   =   % effectiveness

[1]A 60-mesh sieve has 60 wires and 60 openings per inch. Each opening is about 0.21 mm across.
[2]This method is based on the requirements set forth in the Iowa Agricultural Limestone Act of 1967. It requires the determination of the amounts of the material that will pass through 4-mesh, 8-mesh, and 60-mesh sieves. Many states have similar laws.

A sample of lime, all of which passes the 4-mesh sieve but 10 percent of which is retained on the 8-mesh sieve, 35 percent retained on the 60-mesh sieve, and 55 percent passes through the 60-mesh sieve, gives 70 percent effectiveness during the first 3 years:

Coarser than 4 mesh	0% × 0	=	0%
Between 4 and 8 mesh	10% × 0.1	=	1%
Between 8 and 60 mesh	35% × 0.4	=	14%
Finer than 60 mesh	55% × 1.0	=	55%
Total percent effective during first 3 years		=	70%

The *effective calcium carbonate equivalent* (ECCE) or *effective neutralizing value* (ENV) depends on both the fineness and the chemical nature of the lime. If the above lime sample has a calcium carbonate equivalent of 90 percent, its ECCE is 63 percent (70 percent of 90 percent). The best way to determine the relative worth of liming materials for raising soil pH is to compare their ECCE values.

Farmers nearly always have to apply more lime than they would need of pure, finely ground $CaCO_3$. The ECCE value of the liming material permits them to calculate how much to apply. A farmer will need nearly 1,600 kg of the lime with an ECCE of 63 percent to equal 1,000 kg of pure, finely ground $CaCO_3$ (1,000 kg/0.63 = 1,587 kg).

Lime Requirement

The amount of lime needed depends on the crop, the soil, and the ECCE of the liming material. Crops differ in their needs, and there is not much reason to apply lime unless the crops to be grown in the next few years will respond. A compilation of relative yields obtained from various crops grown in Ohio at five different pH levels is shown in Table 8-1. All these crops did best at pH values near neutral (6.8 or 7.5 in these data), but there was a large difference in the sensitivity of the crops to lower pH values. Oats still produced 93 percent of their maximum yield at pH 5.0, whereas sweet clover would hardly grow at that pH.

Most legumes utilize large amounts of calcium and respond well to liming. Grasses usually obtain enough calcium for their needs even from acid soils. The main benefits of lime for grasses are through increased availability of nutrients other than calcium. A farmer growing continuous corn or wheat or some other cropping system without legumes may therefore have an alternative to liming. The farmer can supply additional nutrients as fertilizer, or can apply lime to make the nutrients that are there more available. The farmer can choose between lime and fertilizer or some combination of the two on the basis of costs and probable returns. The advantage usually swings toward more lime when there are legumes in the rotation.

Certain plants are injured by the addition of lime. Such plants have frequently been called "acid-loving," but they are probably favored by the increased availability of certain nutrients such as iron and manganese rather than by the high concentra-

Table 8–1.   Relative Yields of Crops at Different Soil Reactions

	Average yields* at pH of :				
Crop	4.7	5.0	5.7	6.8	7.5
Corn	34	73	83	100	85
Wheat	68	76	89	100	99
Oats	77	93	99	98	100
Barley	0	23	80	95	100
Alfalfa	2	9	42	100	100
Sweet clover	0	2	49	89	100
Red clover	12	21	53	98	100
Alsike clover	13	27	72	100	95
Mammoth clover	16	29	69	100	99
Soybeans	65	79	80	100	93
Timothy	31	47	66	100	95

* Highest yield shown as 100. All other yields are shown as percent of the highest yield.

tion of H^+ ions at low pH. Table 8-2 shows a classification of plants according to probable response to the addition of lime on an acid soil. Certain berries and flowers may be injured by liming; strawberries and watermelons usually show no effect. By far the majority of plants show some degree of positive response to liming.

White potatoes thrive on soils high in lime, but so do the actinomycetes that cause potato scab. Potatoes tolerate acid soils, but the scab organism does not; an effort is therefore sometimes made to keep the soil acid for scab-free potatoes (the alternative is to grow a variety that resists scab). Ammonium fertilizers are particularly advantageous for potatoes susceptible to scab because of their acidifying effect.

With a few exceptions such as those noted, farmers in humid regions need to apply lime from time to time. The desired pH level must be determined before liming recommendations can be made. A knowledge of the crops to be grown plus yield data as influenced by pH are vital information. The returns from liming are spread over a number of years, and so the time element must be considered. For example, improved crop yields in Iowa will usually give maximum long-time returns if the soil is limed to pH 6.9. But, if maximum profit over a 3-year period is sought, it is probably better to reduce the liming cost by raising the pH only to 6.5. If only a single year is considered, it is more economical to use fertilizer alone without any lime. A graphic evaluation of the returns from liming a corn-oats-meadow rotation under typical conditions in Iowa is shown in Figure 8-4. The greatest profit is where the solid line representing returns is farthest above the dashed line representing cost.

The type of soil also influences the optimum soil pH. The pH optima already given apply to temperate region soils but are too high for tropical soils. For example, soybeans in temperate regions grow best in soils with reactions near neutral but their optimum soil pH in the tropics is near 5.5 (Martini et al., 1974). Similar results have been reported by others working with several different crops. The results indicate that soils dominated by oxide clays should be left at a lower pH than soils dominated by silicate minerals.

Table 8–2 Grouping of Plants According to Response to Lime on Acid Soils

High	Medium	Low	None	May be injured
Alfalfa	Red clover	Lespedeza	Strawberry	Azalea
Sweet clover	White clover	Soybeans	Watermelon	Rhododendron
Sugar beets	Corn	Alsike clover		Cranberry
	Barley	Oats		Blueberry
	Wheat	Rye		
	Timothy	Flax		
	Orchard grass	Cotton		
	Bluegrass	Tobacco		

Testing Soils for Lime Requirement

A soil test should be made after the most appropriate pH for the plants to be grown has been decided. The pH tests described in Chapter 7 will tell whether the pH is already adequate or not. Actual pH values more than 0.5 unit below the desired pH indicate that lime probably should be applied at the next opportunity.

The amount of lime required can be determined in several different ways with varying degrees of accuracy. Two of the better methods will be described here. One is the pH-base saturation method and the other is the buffer-solution method.

pH-Base Saturation Method The pH-base saturation curve, the cation-exchange capacity, and the soil pH must all be known for this method. Several pH-base saturation curves were given in Figure 7-3. The curve for any particular soil is an average of the curves for its constituent colloids. Where the colloid proportions are reasonably consistent, a single curve will often serve for all the soils in a state or a major section of a state. The montmorillonite curve will usually serve for soils in which nearly all the cation-exchange capacity comes from montmorillonite, illite, and humus. This curve is reproduced in Figure 8-5 for use in the example that follows.

Methods for measuring or estimating the cation-exchange capacity are discussed in Chapter 6. Methods for measuring pH are discussed in Chapter 7. A cation-exchange capacity of 24 meq/100 g of soil, a pH of 5.2, and a pH-base saturation relation represented by the curve in Figure 8-5 will be assumed for purposes of an example. The desired pH is assumed to be 6.9. Figure 8-5 indicates that the desired base saturation is 95 percent (at pH 6.9) and the initial base saturation is 70 percent (at pH 5.2). The number of milliequivalents of H^+ to be replaced by Ca^{++} and Mg^{++} is therefore:

$$(0.95 - 0.70) \ (24 \text{ meq}/100 \text{ g}) = 6 \text{ meq}/100 \text{ g of soil}$$

The milliequivalent weight of $CaCO_3$ is its molecular weight in milligrams divided by the valence of calcium or:

$$100 \text{ mg}/2 = 50 \text{ mg } CaCO_3/\text{meq}$$

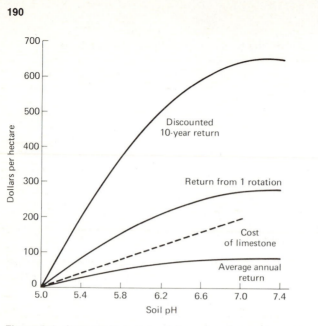

Figure 8-4 Increase in average dollar value of crops in a corn-oats-meadow rotation for 1 year, one rotation, and discounted at 7 percent for 10 years at soil pH above 5.0. Assumed prices: corn, $10 per quintal; oats, $10 per quintal; alfalfa, $50 per ton; lime, $10 per ton, with 1 ton/ha required for each 0.1 increase in pH. (*After Voss, Hanway, Pesek, and Dumenil,* 1965.)

The 6 meq/100 g of soil can therefore be converted to milligrams of ECCE by multiplying by 50. The result is converted to kilograms per hectare-furrow slice as follows:

$$
\begin{aligned}
6 \text{ meq/100 g} \times 50 \text{ mg } CaCO_3/\text{meq} =\ & 300 \text{ mg ECCE/100 g of soil} \\
=\ & 300 \text{ mg ECCE/100,000 mg of soil} \\
\times\ 20/20 =\ & 6{,}000 \text{ mg ECCE/2 million mg of soil} \\
=\ & 6{,}000 \text{ kg ECCE/2 million kg of soil} \\
=\ & 6{,}000 \text{ kg ECCE/hectare-furrow slice}
\end{aligned}
$$

As noted in Chapter 3, the units pounds per acre (lb/ac) are roughly comparable to kilograms per hectare (kg/ha) and can be inserted in the last two steps of the example to give a lime recommendation of 6,000 lb ECCE/acre-furrow slice.

The preceding stepwise procedure can be greatly abbreviated by noting that the two conversion factors used are 50 and 20 and these multiplied together equal 1,000; therefore, when the amount of cation exchange to be neutralized is 6 meq/100 g of soil:

$$6 \text{ meq/100 g} \times 1{,}000 = 6{,}000 \text{ kg } CaCO_3/\text{hectare-furrow slice}$$

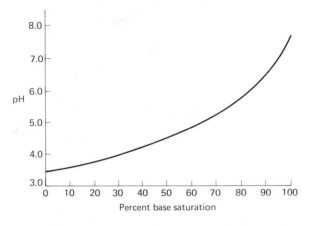

Figure 8-5 pH-base saturation curve for soils containing a mixture of illite, montmorillonite, and humus. (From Figure 7-3.)

This abbreviated procedure is very convenient, but its assumptions must be remembered. The factor 1,000 applies to a starting point of the number of milliequivalents of cation-exchange capacity to be neutralized. The result obtained is the number of kilograms or pounds of 100 percent effective calcium carbonate or its equivalent (ECCE) required per 2 million kg or lb of soil (a typical hectare- or acre-furrow slice).

Buffer-solution Method   The buffer-solution method of determining the lime requirement of soils is based on the ability of a soil to lower the pH of a carefully prepared buffer solution. Other characteristics of the soil need not be known; the buffer solution characteristics are known instead. The soil and the buffer solution are mixed so that any acidity present in the soil will lower the pH of the buffer. The pH of the mixture (called buffer pH) is then determined, and the ECCE requirement is read from a table prepared to match the characteristics of the buffer solution. One reason the buffer-solution method works well is its similarity to the buffering action when lime is mixed with soil. The soil acidifies the lime while the lime raises the pH of the soil.

When to Apply Lime

The reaction of applied lime with soil is distributed over a period of many years. The rate of reaction is most rapid during the first year or two, then gradually declines. Usually it takes about 2 years to reach the maximum pH that will result from liming. After that, the reaction rate becomes slower than the leaching rate and the pH gradually lowers until the next lime application.

Lime could be applied every year to some soils, but this is rarely done because the cost of application might exceed the cost of the lime. The rate of pH change is slow enough that once in 4 to 8 years is usually adequate.

The best time to apply lime should be considered in two parts—the year and the season. The year of application is determined in some systems by the time when soil pH drops below a certain chosen value. This is the only criterion needed where a monoculture (the same crop grown every year) is practiced. But under a rotation system there is some advantage to applying lime about a year ahead of the crop that needs it most (often a legume seeding). This timing should be planned for use on a regular basis as a means of obtaining maximum profit from the lime.

Applying lime during the optimum year in a rotation is advantageous if the pH is never low enough to reduce the value of other crops in the rotation. Needs can be great enough, however, that lime should be applied as soon as possible. Such conditions are most likely to occur when lime needs are ignored for several years. There can be a good response to lime even if it is applied on the same day as the crop needing it is planted. Lime applications on acid soils have a valuable starter effect for many crops.

The season when lime is applied is often more critical than the year in the rotation, not because of the crop but because of the soil. The solubility of $CaCO_3$ is so low that it cannot act like soluble salts to damage the crop, but heavy trucks carrying lime can damage the soil. The application should be at a time when the soil is either relatively dry or frozen. The frozen-soil period is ideal where there is no danger of erosion in spring runoff but should be avoided on hilly topography. Applications on dry soil are often made shortly after harvest. Lime is often disked into the soil along with cornstalks in the corn belt.

Applied lime will not move significantly in the soil, and so it must be mixed with the soil. A disc is probably the best implement for this purpose because it tends to mix the soil. A plow does not mix much but rather tends to place the lime at the bottom of the plow layer. Some cultivators have a mixing action but not as much as a disc has. ECCE applications of more than 15 tons per hectare need to be mixed into a greater depth of soil than most discs penetrate and should be split into two parts. After the first half is applied, the soil is disked and then plowed to turn over the plow layer. The second half is then applied and disked into the soil.

Too many people assume that liming the soil once corrects the soil acidity permanently. The same conditions that made the soil acid once will continue to act and make it acid again. Crop removal, leaching action, and the effect of nitrogen fertilizers are likely to add up to the equivalent of 250 to 500 kg of $CaCO_3$ per hectare per year. A 4-year rotation on a well-drained soil will therefore require 1 to 2 tons of ECCE once during each cycle of the rotation. Poorly drained soils generally suffer less leaching and are less likely to require frequent liming than are well-drained soils.

Lime for Permanent Vegetation

Areas with permanent vegetation may also need lime. Lawns, golf courses, and pastures are subject to as much leaching action as the same soils would be if they were cropped. In fact, the top few centimeters of soil in such areas may become acid more rapidly than in plowed fields because plowing brings some lime back to the surface from the bottom of the plow layer.

Acidity in the top few centimeters of soil can be critical for establishing seedlings. A problem that occurred in an Iowa pasture will illustrate this situation. The soil in the pasture was tested and found to need no lime. It was then disked and reseeded but a very poor stand was obtained. Additional soil tests showed that the top 5 cm of soil had become quite acid and needed lime. Lower layers still contained adequate lime but the disc had not penetrated that deeply. The first soil test did not reveal the problem because the samples were taken at too great a depth.

Lime can be applied and left at the soil surface when necessary for permanent vegetation. The amounts should be limited to 2 to 4 tons per hectare and applied often enough so that these rates are adequate. The reaction rate can be expected to be slower than where the lime is mixed with the soil.

Overliming

Too much liming can be worse than not enough. Excess lime reduces the availability of iron, phosphorus, manganese, boron, and zinc. Too much lime also suppresses the availability of potassium. Crops often are stunted and turn yellow on high-lime spots. In general, the pH of acid soils should not be raised above 7.0.

Overliming is most likely to occur where there are contrasting soil conditions. Spots of very sandy soils with low cation-exchange capacities will be overlimed if treated like neighboring heavier soils. The hazard is serious on the sandy soils because they are also more likely to be deficient in micronutrients whose availability is reduced by overliming.

Another example of overliming occurs where certain low-lying wet soils are already rich in lime. Some of these, such as the Harps soils of Minnesota and Iowa (Figure 8-6) contain enough free lime to lighten their color from black to gray and to result in strong effervescence when acid is applied. Nevertheless, some of the nearby soils need lime. Sometimes farmers will lime an entire field including an area of Harps soils. The lime applied to the Harps area costs as much as that on the rest of the field but yields no return. Harps soils should have special fertilizer applications instead of lime. Areas of distinctly different soils should be identified on a soil map so they can be sampled, tested, and treated separately.

Careful determination of the lime requirements of different soils will do little good unless the lime is properly applied. A good spreader is essential to avoid overliming some spots and underliming others. Most dealers have trucks equipped with suitable spreaders and can spread the lime with little more time and effort than is required to deliver it. Usually the farmers will also be present to offer guidance in how much lime is to be applied in different places, and sometimes the farmers will elect to own the necessary equipment and spread the lime themselves. One advantage of the farmers doing the spreading is that they may be able to choose a time when there is less wind for spreading the lime. Too much wind makes it impossible to achieve a uniform application of finely ground lime.

ACIDIFYING SOILS

It is sometimes desirable to acidify a soil in order to grow a particular plant or crop. This is particularly true in gardening because such flowers as azalea and rhododendron and berries such as blueberries and cranberries thrive on acid soils. Some potato growers have also chosen to lower the soil pH to avoid potato scab.

Figure 8-6 Two strips of Harps soil flanking the black soil in this photograph are identifiable because their high $CaCO_3$ content makes them lighter colored than other low-lying soils.

Applications of sulfur or sulfur compounds are the usual means of lowering soil pH. Sulfur must be applied long enough ahead of time to be oxidized to sulfuric acid, but it is the most effective material available per unit weight. Agricultural sulfur is the cheapest acidifier, unless an industrial by-product is available. Sulfuric acid from copper smelters, for example, is used to acidify soils in the western United States. The acid acts immediately and produces good results. Concentrated H_2SO_4 applied to bare soil or a 3 percent H_2SO_4 solution applied on Bermuda grass produced no ill effects in trials reported by Ryan, Stroehlein, and Miyamoto (1975). Ferrous sulfate and aluminum sulfate (alum) are also used at times. Ferrous sulfate is advantageous when the soil is being acidified to grow plants that have a high iron requirement.

Nitrogen fertilizers containing ammonia considerably reduce the pH of the soil over a period of years whether acidification is desired or not. Urea will acidify neutral or acid soils but has no effect on soils containing enough free carbonates to keep the urea from hydrolyzing (Chandler and Abrol, 1973).

The amount of sulfur or other acidifying material to be applied to lower soil pH can be calculated in the same manner as the lime requirement to raise pH. Assume that the soil to be acidified has a cation-exchange capacity of 24 meq/100g, that the present pH is 6.5, and that the desired pH is 5.0. Assuming also that the curve in Figure 8-5 represents the soil allows conversion of the pH values to a present base saturation of 90 percent and a desired base saturation of 65 percent. The calculations are:

$$(0.90 - 0.65) \times 24 \text{ meq}/100 \text{ g} = 6 \text{ meq to be acidified}/100 \text{ g of soil}$$

The equivalent weight of sulfur is 16 and the conversion factor to convert meq/100g to kg/ha is 20:

$$6 \text{ meq}/100 \text{ g} \times 16 \times 20 = 1{,}920 \text{ kg of S/hectare-furrow slice}$$

Similar calculations can be made on a square meter basis. Assuming a bulk density of 1.33 g/cm^3, a square meter of soil to a depth of 15 cm weighs 200 kg:

$$100 \text{ cm} \times 100 \text{ cm} \times 15 \text{ cm} \times 1.33 \text{ g/cm}^3 = 200{,}000 \text{ g} = 200 \text{ kg}$$

Expanding 6 meq/100 g to 200,000 g and allowing for the milliequivalent weight of sulfur gives:

$$6 \text{ meq}/100 \text{ g} \times 200{,}000 \text{ g/m}^2 \times 0.016 \text{ g S/meq} = 192 \text{ g of S/m}^2$$

The amount of H_2SO_4 that would be needed (5,880 kg/ha or 588 g/m^2) can be calculated by inserting its equivalent weight (49) in place of the equivalent weight of sulfur (16) in the computations. A similar procedure can be used for other acidifying agents.

An application of 1,920 kg of sulfur or 5,880 kg of H_2SO_4 per hectare should acidify this particular soil to a depth of about 15 cm. Acidification, like liming, usually applies only to the plow layer; the upper part of the soil has the most plant roots and therefore is most important. Deeper treatments would increase the amount of material required and increase the cost and difficulty of mixing the amendment into the soil.

Soil acidification by ammonium nitrogen in fertilizers, whether desired or unintentional, can also be calculated. Suppose that 200 kg of nitrogen per hectare are applied as ammonia each year. Nitrification converts the ammonia to nitric acid:

$$NH_3 + 2O_2 \xrightarrow[\text{bacteria}]{\text{nitrifying}} HNO_3 + H_2O$$

The nitric acid ionizes releasing one H^+ ion for each atom of N applied as NH_3. The equivalent weight of N is therefore 14, the same as its atomic weight. The number of milliequivalents of cation exchange acidified is calculated by dividing by 14 and by the conversion factor, 20:

$$\frac{200 \text{ kg of N/hectare}}{14 \times 20} = 0.7 \text{ meq acidified}/100 \text{ g of soil}$$

That rate of acidification would lower the pH of the soil with 24 meq of CEC (cation-exchange capacity)/100 g from 6.5 to 5.0 in 8 or 9 years. Liming at the rate of 700 kg of ECCE each year or 2,000 to 3,000 kg every 3 or 4 years would be required to compensate for annual acidification of 0.7 meq/100 g. Acidification by leaching should be added, but the total may still be less than 0.7 meq/100 g because there are processes that reduce acidification. Denitrification removes nitrate ions from the soil and eliminates their acidifying effect along with their fertility value.

Crop removal has a variable effect depending on the balance between basic cations and anion-forming elements (nitrogen, phosphorus, sulfur, and chlorine) removed. The usual excess of basic cations over anion-forming elements influences the chemistry of food products and has been called *excess base* by food chemists and animal nutritionists. Pierre and Banwart (1973) adopted the term *excess base* to also mean a measure of the effect of crop removal on soil acidification. Pierre, Webb, and Shrader (1971) found that nitrogen-fertilized corn actually produced about 35 percent of the theoretical acidification of the nitrogen fertilizer as a long-term effect on the soil. Most grain crops give results similar to corn; other crops result in anything from 20 to over 100 percent of the theoretical acidification from nitrogen fertilizer. W. H. Pierre (private communication) found buckwheat producing 180 percent of the theoretical acidification.

Many soils, especially those with low cation-exchange capacities, are fertilized with enough nitrogen fertilizer during a 10- to 15-year period to make them quite acid unless lime is added. Changing the form of nitrogen to a solid such as NH_4NO_3 or to urea or some other organic form produces as much acidifying effect as NH_3 per unit of nitrogen. Applications of $(NH_4)_2SO_4$ give twice as much acidification per unit of nitrogen. Even using the nitrogen-fixing ability of legumes instead of fertilizer nitrogen results in acidification. The fixed N eventually becomes NO_3^- and produces as much acid effect as if it had been ammonia. One of the early nitrogen fertilizers, $NaNO_3$, raises soil pH, but the Na^+ ion can be detrimental to soil structure. The use of KNO_3 is a possibility when potassium is needed as well as nitrogen, but it costs too much to use for nitrogen alone. Calcium nitrate also has a neutral or alkaline effect, but it, too, is more expensive than the fertilizers containing ammonia.

AMENDING ALKALINE SOILS

Alkaline soils form because of one or more of the following conditions:

1 In arid areas there is not enough precipitation to leach the soil.
2 Groundwater carries bases into soils in low topographic positions and leaves the bases there when it evaporates.
3 Certain parent materials provide exceptionally large amounts of bases that maintain a high base status in young soils.

A combination of the first two conditions is particularly effective in forming alkaline soils. The most likely place to encounter such soils is in the valleys of arid regions.

Alkaline soils are of great concern in irrigation agriculture because the valleys of arid regions are likely to be irrigated.

Problems with Alkaline Soils

The problems resulting from soil acidity are strictly chemical in nature, but the problems resulting from soil alkalinity can be physical as well as chemical. The physical problems are not always present, but when they occur they are very detrimental to the growth of anything on the soil. There are always some chemical problems with alkaline soils, but these may or may not be serious. Generally the severity of the problems depends on how much the pH is above 7.0.

Chemical problems result from the reduced availability of phosphorus, potassium, and most of the micronutrients. Iron deficiencies are a common problem in alkaline soils. In addition, some alkaline soils contain so many soluble salts that plants have difficulty absorbing water; the osmotic concentration can become higher in the soil solution than in normal plant cells. Bernstein, Francois, and Clark (1974) found that increasing salt concentrations reduced the response of grain and vegetable crops to fertilizer applications.

Physical problems result from the dispersion of the soil colloids. The dispersed particles block pores and form a crust. The soil permeability drops so low in some dispersed soils that it becomes less than the rate of evaporation. Such low permeability makes plant growth impossible and also makes it very difficult to get a chemical soil amendment into the soil to correct the condition.

The effects of an increasing percentage of sodium on the growth of wheat seedlings are shown in Table 8-3. Adding $NaHCO_3$ to the soil increased the concentrations of both Mg and Na and raised the soil pH while lowering the Ca concentration, the percent germination, and the amount of dry matter produced. The most severe treatment reduced germination to 80 percent and dry matter production to little more than half of the control.

Types of Alkaline Soils

There are four distinct types of alkaline soils. Each one has a different set of actual or potential problems. The treatments required to make them productive differ

Table 8–3   The Effect of Increasing Na^+ on Soil pH and on the Growth of Wheat in Pots Containing a Sandy Loam Soil

Treatment	meq/100 g			Soil pH	Germination % of control	Dry matter after 50 days g/pot
	Ca	Mg	Na			
1	36.3	25.6	3.45	8.1	100	32.5
2	31.5	26.0	7.15	8.9	97	27.2
3	25.6	30.2	8.79	9.3	85	19.7
4	27.4	30.8	14.33	9.5	80	17.1

Source: Data from Poonia, Virmani, and Bhumbla, 1972.

according to the types of problems. No single soil has all the problems described in the preceding section.

High-lime soils occur mostly as spots in humid regions, whereas the other three types of alkaline soils occur mostly in arid or semiarid regions. Most soils high in lime are young soils forming in parent material high in calcium and occurring in low topographic sites with shallow water tables. Some of the high-lime soils are as much as 10 percent $CaCO_3$, but they are not saline because the solubility of $CaCO_3$ is low. The more soluble salts rarely accumulate in soils of humid regions. The pH is almost always between 7.5 and 8.0 because of the buffering action of $CaCO_3$. The problems with high-lime soils result from wetness and plant nutrient deficiencies and are much less serious than the problems associated with alkaline soils of arid regions.

Saline soils have a high content of soluble salts. The U.S. Salinity Laboratory uses the electrical conductivity of a saturation extract as a measure of soil salinity. The extract is prepared by first adding just enough water to the soil to fill all its pore space, allowing the resulting "saturated paste" to equilibrate, and then removing the saturation extract by vacuum. Soils have traditionally been classed as saline if the conductivity of the saturation extract exceeds 4 millimhos/cm.[1] Recently the limit has been lowered to 2 millimhos/cm for some horticultural crops (McNeal, 1976). Higher salinity classes are defined for soils with saturation extracts having conductivities of more than 8 and 12 millimhos/cm.

Saline soils usually contain more than 0.2 percent soluble salts. The presence of the salts produces an osmotic pressure that makes it more difficult, and in extreme cases impossible, for plants to absorb water from saline soils. Most of the salts are neutral or nearly neutral in reaction, but some are alkaline, and the soil pH values are generally between 7.3 and 8.5. The salts move up and down in the soil along with the soil water. White salt crusts form on the soil surface during dry seasons but are temporarily washed down into the soil by rain. These soils are often called *white alkali* because of the white salt crusts.

Sodic soils have more than 15 percent of their cation-exchange sites occupied by Na^+ ions. They are low in soluble salts. This combination results in dispersed colloids and a pH above 8.5. The high pH results from the strongly basic influence of Na^+ ions in solution. Much more than 15 percent of the cations in solution are Na^+ because almost any other cation is adsorbed by the micelles in preference to Na^+. Certain sodium salts such as Na_2CO_3 absorb H^+ from water and leave Na^+ and OH^- ions producing the strongly alkaline reaction of NaOH in solution. Sodic soils are the most alkaline of all soils and the hardest to reclaim of the alkaline soils. They have dispersed colloids, very low permeability, and support little or no plant growth. Some of the soil organic matter dissolves at the high pH of sodic soils and accumulates on the soil surface as thin black deposits that give sodic soils the common name *black alkali.*

Saline-sodic soils combine the high salt content (greater than 4 millimhos/cm in the saturation extract) of the saline soils with the high percent Na^+ (more than 15 percent on the cation-exchange sites) of the sodic soils. Usually they have pH

---

[1]A millimho is equal to 0.001 mho, the basic unit of electrical conductivity. When measurements are made with plates 1 cm square spaced 1 cm apart, the resulting units are millimhos per centimeter.

values between 8.0 and 8.5. Their other properties match those of the saline soils except that leaching will turn them into sodic soils. The saline and the saline-sodic soils look alike, but it is important to identify which is which. Saline soils can be reclaimed simply by leaching, whereas saline-sodic soils deteriorate if they are leached without a soil amendment.

Treatment of Alkaline Soils

Alkaline soils need good drainage. It is futile to correct alkalinity if excess water will still be there restricting root development and causing the alkaline condition to recur. A good drainage system removes enough water to permit the soil to be used and prevents accumulation of soluble salts. Drainage systems are discussed in Chapter 18.

High-lime Soils   High-lime soils with adequate drainage and proper fertilization usually will produce nearly as much plant growth as nonalkaline soils. The appropriate fertilizer to be added depends in part on the crop. Corn tends to suffer from potassium deficiency on high-lime soils because K^+ has been driven into nonexchangeable positions and the excess Ca^{++} tends to cause more Ca^{++} and less K^+ to be absorbed by the plants. Aeration is usually restricted enough to decrease the ability of the plant to absorb K^+ selectively. Additional potassium fertilizer should be applied to high-lime soils when corn is to be grown on them. Soybeans, on the other hand, are likely to be chlorotic when grown on high-lime soils. The chlorotic condition results from either iron or manganese deficiencies caused by low solubilities at high pH. The chlorotic condition can usually be remedied by spraying the plants with ferrous sulfate. Manganese deficiency is less common than iron deficiency but equally serious. The remedy is similar—spray the plants with manganese sulfate.

Nitrate release by decomposition of organic matter is likely to be more rapid in high-lime soils than in others because microbial activity is favored in the slightly alkaline pH range. It is therefore possible to reduce nitrogen fertilization on high-lime soils that have high organic-matter contents. Sometimes the NO_3^- level is high enough that it, combined with moist conditions and low K^+, results in lodging of small grain crops.

Legume forage crops such as sweet clover usually do well on high-lime soils. They are favored by the high Ca^{++} supply. The conditions may be too wet, however, for alfalfa.

Manures are beneficial when applied to high-lime soils because they are sources of available K^+ and they produce organic acids that have some neutralizing effect on soil alkalinity. Gypsum and sulfur should not be used to lower the pH of these soils because this would tend to increase the salt content of the soil solution. Raising the salt content tends to decrease the amount of K^+ in solution and increase the likelihood of a potassium deficiency.

Saline Soils   The treatment of either saline or sodic soils to make them more productive is known as *reclamation*. Saline soils that are not sodic can be reclaimed very quickly if they can be drained and if adequate water is available for leaching. Too often an irrigator will try to leach saline soils without adequate drainage. This is likely to raise the water table and increase the rate of salt accumulation. Good drainage is the key point. Excess soluble salts can be leached from a soil with a few heavy irrigations. The greater difficulty is to keep the problem from recurring.

The salt content of some irrigation water is so high that it tends to create saline soils. When such water is used, the amount applied must be considerably in excess of the amount used by evaporation and transpiration. The amount of excess water must be enough to dissolve the salts from all the applied water and carry them into the drainage system.

Some crops can be grown on saline soils. Table 8-4 shows the degree of sensitivity of various crops to salinity. When growing plants, a saline soil should be kept at a higher moisture content than is necessary with other soils. The presence of salts causes an osmotic pressure that adds to the soil moisture tension and increases the water content at the wilting point. Unfortunately, field capacity is not increased, and so the available water-holding capacity is decreased in soils containing salts.

A saline soil can sometimes be leached by increasing the effectiveness of the natural precipitation without applying any irrigation water (Sandoval and Benz, 1973). The soil is leached by keeping it free of vegetation for about a year to accumulate enough water to penetrate beyond the depth of the solum. The effectiveness of this procedure can be improved by the use of a mulch on the soil surface to reduce evaporation losses. The small amount of precipitation received annually can remove a large amount of salts if the movement is downward at all times.

Saline-sodic Soils   The most important difference between saline soils and saline-sodic soils is that a soil amendment must be applied to the latter before they are leached. Leaching without soil amendment will convert saline-sodic into sodic soils, a much more difficult condition to reclaim.

The most commonly used amendments for alkaline soils are gypsum and sulfur. Gypsum ($CaSO_4 \cdot 2H_2O$) is usually favored. Most of the Na^+ ions present on the cation exchange should be replaced by Ca^{++} so that dispersion will not occur:

$$\begin{matrix} Na^+ \\ Na^+ \end{matrix} \boxed{\text{micelle}} \; + CaSO_4 \rightleftharpoons Ca^{++} \boxed{\text{micelle}} \; + 2Na^+ + SO_4^{--}$$

The sodium sulfate formed is soluble and can be leached from the soil. Enough excess Ca^{++} should be present in solution to cause the above reaction to replace nearly all of the Na^+ on the micelles. Also, enough gypsum should be provided to remove most of the carbonate ion from solution:

$$2Na^+ + CO_3^{--} + CaSO_4 \rightarrow CaCO_3 \downarrow + 2Na^+ + SO_4^{--}$$

This reaction is important because Na_2CO_3 in solution produces a very high pH and all the undesirable characteristics that accompany it, whereas Na_2SO_4 is a neutral

Table 8–4 A Grouping of Plants According to Salt Tolerance

High salt tolerance	Medium salt tolerance	Low salt tolerance
Date palm	Olive	Pear
Garden beets	Grape	Apple
Kale	Cantaloupe	Orange
Asparagus	Tomato	Grapefruit
Spinach	Cabbage	Prune
Saltgrass	Cauliflower	Plum
Bermuda grass	Lettuce	Almond
Rhodes grass	Potato	Apricot
Canada wild rye	Carrot	Peach
Western wheatgrass	Peas	Strawberry
Barley	Squash	Lemon
Bird's-foot trefoil	Sweet clover	Avocado
Sugar beet	Mountain brome	Radish
Rape	Strawberry clover	Celery
Cotton	Dallis grass	Green beans
	Sudan grass	White Dutch clover
	Alfalfa	Meadow foxtail
	Tall fescue	Alsike clover
	Rye	Red clover
	Wheat	Ladino clover
	Oats	Field beans
	Orchard grass	
	Meadow fescue	
	Reed canary	
	Smooth brome	
	Rice	
	Sorghum	
	Corn	
	Sunflower	

Source: U.S. Salinity Laboratory Staff, 1954.

salt. The reaction goes nearly to completion because calcium carbonate precipitates. Leaching will remove the excess Na^+ and SO_4^{--} ions.

Sulfur is likely to be used as a soil amendment to reduce alkalinity when the soil contains free lime as well as too much sodium. The advantage of sulfur is the much lighter weight of material that must be transported and applied. Less than one-fifth as much sulfur is required because the equivalent weight of sulfur is only 16 compared with 86 for gypsum. One disadvantage is the time required for the sulfur to be oxidized in the soil:

$$2S + 3O_2 + 2H_2O \xrightarrow[\text{action}]{\text{microbial}} 2H_2SO_4$$

The time lag can be eliminated by applying H_2SO_4 as the amendment. Of course, any sulfuric acid released in an alkaline soil or applied as an amendment will immediately react with some of the salts present:

$$H_2SO_4 + 2Na^+ + CO_3^{--} \rightarrow 2Na^+ + SO_4^{--} + H_2O + CO_2\uparrow$$

$$H_2SO_4 + CaCO_3 \rightarrow CaSO_4 + H_2O + CO_2\uparrow$$

Bubbles of carbon dioxide give evidence of such reactions when an acid is applied to soils containing carbonates. Any $CaSO_4$ formed in the soil can react in the same ways as applied gypsum.

A saline-sodic soil can be made suitable for plant growth by leaching after it has been amended with gypsum or sulfuric acid. Sulfur serves the same purpose, but sufficient time for microbial action must be allowed before leaching when sulfur is used. The amount required is the sum of the amount needed to displace Na^+ from the micelles plus that required to react with Na^+ in the soil solution.

The amount of soil amendment required for the soil solution is variable, but some estimates can be made if chemical data are available and some assumptions are made. If, for example, a 2-million-kg hectare-furrow slice contains 0.25 percent Na_2CO_3 that is to be neutralized with sulfur, there are

2 million kg $\times$ 0.0025 = 5,000 kg of Na_2CO_3

The sulfur is oxidized to H_2SO_4 and then reacts with the Na_2CO_3 as already indicated. The equivalent weight of S is 16 and that of Na_2CO_3 is 53, and so the sulfur requirement is:

5,000 kg $\times$ 16/53 = 1,510 kg of S/hectare-furrow slice

The gypsum requirement would be much larger because the equivalent weight of gypsum is 86:

5,000 kg $\times$ 86/53 = 8,100 kg of gypsum/hectare-furrow slice

The additional amount of soil amendment required to replace the exchangeable Na^+ will be calculated in the next section. Reclamation of saline-sodic soils is likely to require several tons of gypsum, a few tons of sulfur, or some combination of the two. It should never be attempted without the addition of the necessary soil amendment.

Sodic Soils   The sodic soils are the most difficult to reclaim and the least likely to be worth the cost in time, money, and effort. The problem is that the high sodium percentage causes an unfavorable physical condition that makes it difficult to distribute a soil amendment through the soil. Leaching is an extremely slow process because of the low permeability. Nevertheless, the reclamation effort may be justified when there is enough difference in purchase price between the sodic soils and soils already suitable for use. Also, sodic soils usually occur as spots intermixed with saline soils and other soils (Figure 8-7). Reclaiming the sodic spots may be worthwhile because it makes the entire area more usable.

Drainage and soil amendments are essential for the reclamation of sodic soils. The amount of soil amendment required can be calculated in much the same manner

Figure 8-7 Corn growth on alkaline soils in Idaho. The pH is below 8.5 where the corn is growing well, about 9.0 where the corn is stunted, and above 9.5 where the corn would not grow.

as the lime requirement of acid soils. Only the exchangeable sodium is considered because most of the soluble salts have already been leached from the soil. A soil with a cation-exchange capacity of 20 meq/100 g will serve as an example. If this soil has an initial Na^+ saturation of 50 percent that is to be reduced to 5 percent, it will have:

$$20 \text{ meq/100 g} \times (0.50 - 0.05) = 9 \text{ meq to be replaced/100 g}$$

The milliequivalent weight of gypsum is 86, and so it takes:

$$9 \times 86 = 774 \text{ mg of gypsum/100 g of soil}$$

If a hectare-furrow slice weighing 2 million kg is to be reclaimed, then

$$774 \times 20 = 15,480 \text{ kg of gypsum/hectare-furrow slice}$$

Over 15 tons of gypsum per hectare would be required to reclaim this sodic soil to a depth of about 15 cm. Unfortunately, 15 cm may not be adequate. The subsoil is likely to be as alkaline as the plow layer. Leaching the soil is a slow and difficult process if the subsoil is left in a dispersed condition.

Reclaiming sodic subsoils not only requires additional tons of soil amendment, but also presents the difficulty of getting the material into contact with the bulk of the soil. Deep plowing is expensive, often impractical, and does not mix the soil and soil amendment very well. Small spots are sometimes treated by excavating the soil with a backhoe, mixing, and replacing. This procedure frequently yields a bonus in the form of free gypsum and lime at the bottom of the solum that can be mixed into the soil. Such excavations are useful on small spots, but they are very expensive on a field basis.

Almost any kind of plant growth should be encouraged during the process of reclaiming a sodic soil. The plant roots open channels that improve the soil permeability and make it easier to treat. Added organic matter is also beneficial because it helps to improve the soil structure and to provide a nitrogen reserve for future plant growth.

ARTIFICIAL SOILS

Changing pH and removing soluble salts are significant ways to amend soils. Many properties important to plant growth are changed through such amendments. Still, the original soil remains in place, amended but basically the same.

When a particular soil is used intensively, its nature may be changed drastically to match the needs of the user. Many such changes involve the shape of the soil surface. A farmer may want to shape the surface of irrigated fields so that water will flow across them more uniformly. Usually several centimeters of soil are moved from places that are too high into areas that are too low. Occasionally a cut or a fill may be a few meters deep over a small area. Large earth-moving equipment is used for land shaping. The whole operation must be carefully planned so that the amount of soil taken from the high points will match that needed to fill the low points.

Other examples of land shaping are terrace construction, road building, and landscaping around homes and other buildings. The nature of the soil is very important in land shaping. All the soil horizons should be considered. It is important that the surface soil be suitable for establishing plant growth. Some B and C horizons are suitable for use as topsoil, and some are not. Some soils have a heavy B2 or a hard B3ca that will not support plant growth, whereas the horizons above and below are usable. Calcareous horizons may require special fertilization as discussed for high-lime soils. Whatever the case may be, material in which it is difficult to grow plants should be removed from areas where it would be exposed. It can often be buried in the bottom of a deep fill. Then suitable topsoil material is brought in to cover both the cut and the fill.

Sometimes it is advisable to stockpile all the A horizon, perform the land shaping with the B and C horizons, and then spread the A horizon on top. The thickness of A horizon placed as a topsoil is unimportant in some soils but may be quite critical in others. A homeowner may find that the lawn grows very well where there are 10 to 15 cm of A horizon but is sparse and weedy where there are only 5 cm of A horizon. Highway departments often encounter the same results in right-of-way strips. Some gardeners find that their gardens grow better if they increase the topsoil thickness in their garden plots.

Land shaping is costly. A typical cost for smoothing a farmer's fields is about $500 per hectare. Landscaping and highway construction usually involve moving more soil and much higher costs. Stockpiling the A horizon and placing it on top again also adds to the cost. Adding special fertilizers or mulches to control erosion until vegetation is established is also costly but often a necessary supplement to land shaping.

Golf-course Greens

Golf-course greens are some of the most expensive land on earth. Specially constructed soils are required to meet the special requirements of greens. Basic requirements include a soil that resists compaction under heavy traffic and has a resilient surface. It must be highly fertile and well drained so that it can support a strong stand of close-growing grass in spite of trampling and short mowing. It should be so permeable that neither rain nor irrigation makes it unplayable.

Natural soils are not likely to meet the rigid requirements of a heavily used golf green. Many different types of artificial soils have been tried with varying degrees of success. A cross section of a standard type of green recommended by the U.S. Golf Association Green Section Staff (1960) is shown in Figure 8-8. Construction of such a green begins with a base formed to the same shape that the final surface will be but about 35 cm lower at all points. Trenches are dug and tiles are laid in them in a gravel bed so that all parts of the green are well drained. Then successive layers of gravel, sand, and soil of the indicated thicknesses are placed for the green (the thicknesses indicated in Figure 8-8 allow for about 10 cm of settling).

All the materials used in making the green are carefully chosen for specific purposes. The gravel must provide free passage for water to reach the nearest tile, yet its pores must not be so large that sand particles will move through it. Pea gravel about 5 mm in diameter is preferred. The sand acts as a filter and a support for the soil. It must be too large to pass through the pores in the gravel, yet its pores must be too small for the soil to pass through. Sand particles about 1 mm in diameter are usually about the right size.

The top 30 cm of the green is a mixture composed of about 85 percent sand, 5 percent clay, and 10 percent peat or some similar organic material. Brown and Duble (1975) recommend a similar mixture but with a clay content of only 3 percent to ensure a playable surface after heavy water applications. Most of the sand should

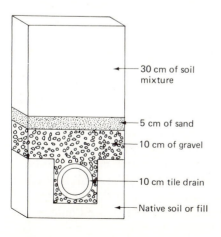

30 cm of soil mixture

5 cm of sand

10 cm of gravel

10 cm tile drain

Native soil or fill

Figure 8-8 Sectional diagram of a golf green. (*After U.S. Golf Assn.*, 1960.)

be coarse to provide aeration, water permeability, and resistance to compaction. Some clay is considered essential to give the mixture some cohesion and to provide water-holding capacity and cation-exchange capacity for soil fertility considerations. Silt is not desired because it fills up pore space and does not add to the fertility. The peat helps make the mixture friable and adds both fertility and water-holding capacity. Fertilizer and soil amendments are added to the mixture and, later, to the green as needed.

Indoor Soils

Plants need water, air, nutrients, and mechanical support whether grown indoors or outdoors. These needs may be met through natural soil, a nutrient solution plus some kind of support mechanism, or some kind of "root media." Nutrient solutions are outside the scope of this text, but root media will be considered here as a type of artificial soil.

Indoor soils represent the most intensive use people make of soil or soil-like materials. Yields produced on greenhouse benches or in clay pots are usually several times as great as can be grown in an open field on an equal area. However, environmental control is the main reason for growing plants in a greenhouse. Indoor costs are too high for economic competition when the same plants can be grown outdoors. Moreover, the high yields obtained in greenhouses should be discounted somewhat because the space occupied by walkways, walls, etc., has usually been omitted from the calculations.

Many different reasons justify the expense of growing plants indoors. Often the reason is aesthetic in that the plants are used as ornaments for a home. Sometimes there is economic justification, as there is for "hothouse tomatoes." Sometimes the plants are grown for research purposes. All these uses justify a much greater investment in the root media than would be profitable for outdoor use.

The value of an ideal root medium is great, and natural soil is seldom the best choice for indoor use. Most soils are inadequately aerated and too low in fertility to be ideal. Other materials are therefore mixed with the soil to improve it.

Root media commonly contain an organic material to provide aeration, water-holding capacity, and perhaps some plant nutrients. Coarse sand or fine gravel is usually added to resist compaction and promote good aeration. Some typical mixtures are one-third soil, one-third coarse sand, and one-third organic material or one-half soil and one-fourth each of the others. Peat moss, muck, and sphagnum moss are commonly used organic materials. Leaf mold and rotted manure are also used. Sometimes the soil is omitted and the other materials take over completely. Expanded vermiculite or perlite may be added or substituted for some of the other constituents.

Two common problems with indoor soils are drainage and salt accumulation. The root medium becomes saturated whenever too much water is added to a container that has no bottom outlet. A compact root medium often becomes saturated even with a bottom outlet because there is no downward capillary pull beyond the bottom of the medium. The capillary action of small pores in the medium easily offsets the gravitational force. Large pores produced by coarse sand and organic materials help to avoid saturated conditions.

Salt accumulations are common in indoor soils because of failure to leach adequately. The water supplied for plant growth normally adds salt to the root media. Fertilizer applications also add to the salt concentration. Evaporation at the surface of a root medium frequently leaves a white or brown salt accumulation revealing that the medium has become a saline soil. It can usually be reclaimed by leaching. Of course, the indoor soil often presents an alternative that is seldom feasible with outdoor soils—it can be discarded and replaced with a new medium.

Ideal conditions for plant growth are also ideal for microbes, including harmful types. Various wilts and nematodes thrive in indoor soils. Most root media are therefore sterilized before plantings are made. Steaming a medium long enough to cook a potato buried in it is a popular means of sterilization. Certain chemical sterilants are also used. Pots and other containers should be sterilized, as should shovels, screens, and whatever else will contact the media.

Fertilizer use on fields has increased in recent decades but still does not compare with the concentrations of fertilizer used in root media. The rule indoors is to use the maximum safe amount rather than the minimum. Enough fertilizer may be used to furnish the entire needs of the plant with no consideration of the ability of the media to release new nutrients by decomposition processes. Such release is small compared with the amount being added as fertilizer.

The principal fertilizer elements are the same indoors as outdoors: nitrogen, phosphorus, and potassium. These are commonly supplied in a ratio of $3:1:2$[1] when expressed as $N:P_2O_5:K_2O$ (the method of expression is explained in Chapter 9). The easiest way to make periodic fertilizer applications is usually in the water. Highly soluble forms of fertilizer are therefore favored.

An endless variety of things can be done with artificial soils. These artificial soils can be specially made to fit the needs of particular plants. The media used and the methods of treating them are highly variable because many kinds of plants are grown indoors and much is yet to be learned in this field. Understanding gained by working with artificial soils can help us to better understand natural soils. Some of the ideas developed indoors may well serve as guides for improved use of outdoor soils, particularly for soils that are intensively used.

REFERENCES

American Society of Agronomy, 1967, *Soil Acidity and Liming,* Agronomy Monograph 12, Madison, Wis., 288 p.

Bernstein, Leon, L. E. Francois, and R. A. Clark, 1974, Interactive Effects of Salinity and Fertility on Yields of Grains and Vegetables, *Agron. J.* **66**:412–421.

Brown, K. W., and R. L. Duble, 1975, Physical Characteristics of Soil Mixtures Used for Golf Green Construction, *Agron. J.* **67**:647–652.

Chandler, Harish, and I. P. Abrol, 1973, Effect of Three Nitrogenous Fertilizers on the Solution Composition of a Saline Sodic Soil, *Commun. Soil Sci. Plant Anal.* **3**:51–56.

Dahiya, I. S., and I. P. Abrol, 1974, The Redistribution of Surface Salts by Transient and Steady Infiltration of Water into Dry Soils, *J. Indian Soc. Soil Sci.* **22**:209–216.

[1]Private communication from Dr. Charles Sherwood, Horticulture Department, Iowa State University.

Lunin, J., M. H. Gallatin, and A. R. Batchelder, 1964, Effects of Supplemental Irrigation with Saline Water on Soil Composition and on Yields and Cation Content of Forage Crops, *Soil Sci. Soc. Am. Proc.* **28**:551–554.

McNeal, B. L., 1976, Managing Salt-affected Soils: Recent "Dissolution" of Some Myths, *Crops Soils* **28**(4):22–23.

Martini, J. A., R. A. Kochhann, O. J. Siqueira, and C. M. Borkert, 1974, Response of Soybeans to Liming as Related to Soil Acidity, Aluminum and Manganese Toxicities, and Phosphorus in Some Oxisols of Brazil, *Soil Sci. Soc. Am. Proc.* **38**:616–620.

Nossaman, N. L., and D. O. Travis, 1966, Grain Sorghum Production in a Calcareous Cut Site as Influenced by Phosphorus, Zinc, and Iron Fertilization, *Agron. J.* **58**:479–480.

Pierre, W. H., and W. L. Banwart, 1973, Excess-base and Excess-base/Nitrogen Ratio of Various Crop Species and Parts of Plants, *Agron J.* **65**:91–96.

Pierre, W. H., J. R. Webb, and W. D. Shrader, 1971, Quantitative Effects of Nitrogen Fertilizer on the Development and Downward Movement of Soil Acidity in Relation to Level of Fertilization and Crop Removal in a Continuous Corn Cropping System, *Agron. J.* **63**:291–297.

Pionke, H. B., 1970, Effect of Climate, Impoundments, and Land Use on Stream Salinity, *J. Soil Water Conserv.* **25**:62–64.

Pionke, H. B., R. B. Corey, and E. E. Schulte, 1968, Contributions of Soil Factors to Lime Requirement and Lime Requirement Tests, *Soil Sci. Soc. Am. Proc.* **32**:113–117.

Poonia, S. R., S. M. Virmani, and D. R. Bhumbla, 1972, Effect of ESP of the Soil on the Yield, Chemical Composition and Uptake of Applied Calcium by Wheat, *J. Indian Soc. Soil Sci.* **20**:183–185.

Reeve, R. C., and E. J. Doering, 1966, The High Salt-water Dilution Method for Reclaiming Sodic Soils, *Soil Sci. Soc. Am. Proc.* **30**:498–504.

Ryan, J., J. L. Stroehlein, and S. Miyamoto, 1975, Sulfuric Acid Applications to Calcareous Soils: Effects on Growth and Chlorophyll Content of Common Bermuda Grass in the Greenhouse, *Agron. J.* **67**:633–637.

Sandoval, F. M., and L. C. Benz, 1973, Soil Salinity Reduced by Summer Fallow and Crop Residues, *Soil Sci.* **116**:100–105.

Sinha, B. K., and N. T. Singh, 1974, Effect of Transpiration Rate on Salt Accumulation around Corn Roots in a Saline Soil, *Agron. J.* **66**:557–560.

Swartz, W. E., and L. T. Kardos, 1963, Effects of Compaction on Physical Properties of Sand-soil-peat Mixtures at Various Moisture Contents, *Agron. J.* **55**:7–10.

U.S. Golf Association Green Section Staff, 1960, Specifications for a Method of Putting Green Construction, *U.S. Golf Assoc. J.* **13**:1–5.

U.S. Salinity Laboratory Staff, 1954, Diagnosis and Improvement of Saline and Alkali Soils, *USDA Handbook* 60, 158 p.

Voss, R. D., J. J. Hanway, J. T. Pesek, and L. C. Dumenil, 1965, A New Approach to Liming Acid Soils, *Iowa State Univ. Coop. Ext. Ser. Pam.* 315, 12 p.

Woodruff, C. M., 1947, Determination of Exchangeable Hydrogen and Lime Requirement of the Soil by Means of the Glass Electrode and a Buffered Solution, *Soil Sci. Soc. Am. Proc.* **12**:141–142.

Chapter 9

Fertilizers

Fertilizers are sources of plant nutrients that can be added to soil to supplement its natural fertility. They are intended to supply plant needs directly rather than indirectly through modification of such properties as soil pH and structure. There is usually a very dramatic improvement in both quantity and quality of plant growth when appropriate fertilizers are added. The change from an agrarian society and a subsistence type of agriculture to a modern mechanized society is accompanied by a marked increase in the use of fertilizers.

Proper use of fertilizer leads to the production of more nutritious food. Generally the kind of fertilizer is not of much importance as long as it supplies the needed nutrients. Organic and mineral fertilizers are equally good for plants and for animals or people eating the plants for food. Health problems enter when needed elements are deficient in the soil and are not provided in fertilizer applications. The increasing use of fertilizer in recent decades can be credited with improving food quality as well as quantity.

The use of fertilizer is a very important way to reduce the unit cost of producing food and fiber. A farmer who uses fertilizer effectively has a great competitive advantage over one who does not. Much of this advantage is ultimately passed on to the consumer in the form of lower prices. Food prices in the United States and many other countries would be much higher than they are if fertilizer were not used.

Crop removal is an important reason for the use of fertilizers. Removing a crop breaks into the natural cycle and prevents the nutrients contained in the crop from returning to the soil which provided them (Figure 9-1). The supply of available nutrients in the soil is thereby depleted and future plant growth is impoverished. Weathering of soil minerals can usually supply enough nutrients to compensate for small losses such as leaching but not for the larger losses resulting from harvesting crops. Adding appropriate fertilizers compensates for the nutrients removed by the crops.

HOME USE OF FERTILIZERS

Homeowners use fertilizers for lawns, gardens, and house plants. The fertilizer elements they need are the same as those used by farmers for field crops. These elements are contained in common household items, as the following recipe that has been suggested for a household fertilizer illustrates:

5 cc baking powder ($NaHCO_3$, starch, $Ca(H_2PO_4)_2$, etc.)
5 cc epsom salts ($MgSO_4 \cdot 7H_2O$)
5 cc saltpeter (KNO_3)
2.5 cc household ammonia (NH_3 in H_2O)
4 liters tepid water

This solution contains all of the macronutrients needed by plants and probably contains traces of micronutrients as well. Results from its use, however, are not guaranteed because of variable plant needs and variable composition of some of the ingredients.

Small containers usually increase the cost of any product, and fertilizers are no exception. Homeowners may wonder if they could save money by purchasing fertil-

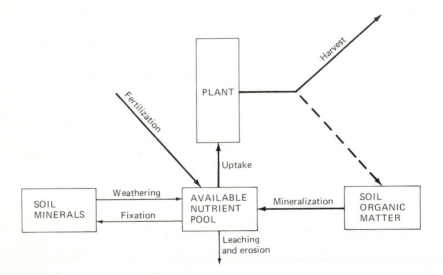

Figure 9-1 The available nutrient cycle is broken when crops are harvested. Fertilization replenishes the soil's available nutrient pool.

izer from the same source as farmers. Often they can if they will buy it in a large bag and if they know how much of each fertilizer nutrient they want to apply. Fertilizer bagged for farm use has a guaranteed analysis but does not tell how many square meters of lawn or garden it should cover. Conditions vary, and only generalities can be given here, but a 10-10-10[1] fertilizer might be used on a lawn or garden at rates between 20 and 100 g/m².

The fertilizer needs of lawns, gardens, and house plants are as varied as those of agricultural crops. A 10-10-10 fertilizer might or might not be suitable for a specific situation. The choice of a particular fertilizer should be based on a knowledge of the plant to be fertilized plus either a soil test or experience with the soil being used.

Lawn grasses have more predictable fertilizer needs than many other plants that homeowners may grow. Grasses respond to nitrogen fertilization. They may or may not need phosphorus and potassium, but a good guide is to supply these elements if they are needed by farm crops in the area. In fact, lawns can usually be fertilized according to the same guidelines as pastures grown on similar soils.

Fertilizers containing weed-killer additives can be used on lawns but must be kept away from most garden and house plants. Farmers purchase and apply weed killers separately from their fertilizer. Homeowners can do this, too. Even so, it is entirely possible for fertilizer (especially fertilizer intended for farm use) to be contaminated with small amounts of weed killer because both are handled by the same processors and dealers. Such contamination is usually of no consequence for farm use or lawn use, but it can be devastating to sensitive ornamental house plants. Horticulturalists will often pay a premium for fertilizer that assuredly contains no weed killer.

[1] This is a fertilizer analysis. The numbers will be explained later in the chapter.

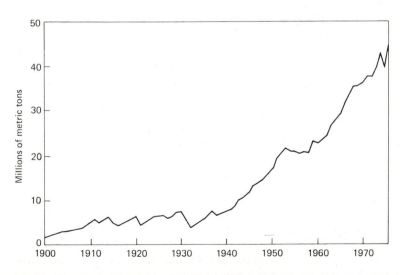

Figure 9-2   Gross fertilizer usage in the United States from 1900 to 1976. (Based on data from the National Fertilizer Association and the USDA Crop Reporting Board.)

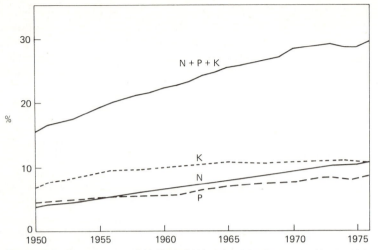

Figure 9-3 Percentages of N, P, and K in mixed fertilizers used in the United States from 1950 to 1976. *(Based on data from USDA Crop Reporting Board.)*

FERTILIZER CONSUMPTION IN THE UNITED STATES

The use of fertilizers in the United States increased from 2.0 million tons in 1900 to 16.6 million tons in 1950 and to 44.3 million tons in 1976. This growth in fertilizer use is plotted in Figure 9-2. Dramatic increases in fertilizer consumption were registered during the 1940s and again during the 1960s. The decline in fertilizer use in 1975 was in response to prices having approximately doubled.

A steady increase in the plant nutrient content of the fertilizers in percent by weight has accompanied the increase in fertilizer consumption. Figure 9-3 shows the percentages of the three principal fertilizer elements contained in fertilizers used in the United States from 1950 to 1976. Increases in these percentages have been accomplished by using compounds containing more N, P, and K relative to other elements [for example, NH_4NO_3 or urea instead of $(NH_4)_2SO_4$] and by decreasing the amounts of inert or "filler" materials in fertilizers. The results of increasing both tonnage and nutrient percentages have been very rapid increases in the amounts of these three nutrient elements supplied in fertilizers. Data for the United States from 1950 to 1976 are shown in Figure 9-4.

There are, of course, limits to the amount of fertilizer that can and should be used. Some persons are already using too much fertilizer. But excessive use of fertilizer is still rare even in the United States. There is much room for increased fertilizer usage among those who are not yet using any and those who are using less than optimum amounts.

FERTILIZER CONSUMPTION IN THE WORLD

About 80 percent of the fertilizer used in the world is used in "developed" countries. The United States alone has been consuming between 25 and 30 percent of the

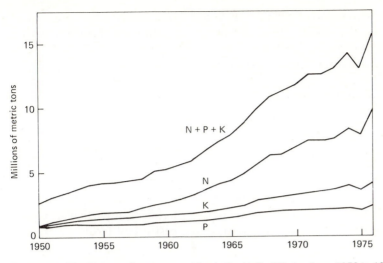

Figure 9-4 Fertilizer nutrient consumption in the United States from 1950 to 1976. *(Based on data from USDA Crop Reporting Board.)*

fertilizer used in the world during recent years. However, many of the "developing" nations are beginning to use fertilizer in significant amounts. These nations are likely to show much larger percentage increases in fertilizer use in future years than are the "developed" nations. World fertilizer consumption from 1950 to 1973 is shown in Figure 9-5.

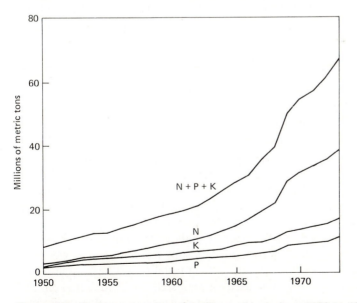

Figure 9-5 World consumption of fertilizer nutrients from 1950 to 1973. *(Based on data from Harre, 1969, and from the Production Yearbook, 1974, Food and Agricultural Organization of the United Nations.)*

The use of nitrogen as a fertilizer has been increasing faster than the use of either phosphorus or potassium. A major part of the reason for this is the decline in the price of nitrogen during the 1960s. Cheaper nitrogen resulted in much substitution of nitrogen fertilizer for nitrogen-fixing legumes in crop rotations. Using nitrogen fertilizer permits farmers to grow corn and small grain crops either continuously or more frequently than they previously did.

FACTORS AFFECTING FERTILIZER USE

The most profitable type and amount of fertilizer to use depends on crop, soil, climate, economic, and management factors. The actual amount used depends on these factors plus the personal characteristics of the users. Some are prejudiced either for or against fertilizers in general or some types of fertilizer in particular. Others do not have the necessary information to make the best use of fertilizer.

Crop Factor

Fertilizer can hardly be profitable unless the crop will respond to it. Some crops need relatively large amounts of certain nutrients. For example, a properly inoculated crop of alfalfa obtains nitrogen from the air but needs large amounts of phosphorus and potassium. The same is true of most legumes. Grass-family crops, on the other hand, require large amounts of nitrogen.

The crop variety also makes a difference. Much work in plant breeding has been aimed toward producing varieties that respond well to fertilizer. These varieties will produce much higher yields than the older varieties if adequate plant nutrients are available. The newer varieties, however, are likely to do poorly when they are not adequately fertilized.

The growth of intensively farmed crops such as truck crops, citrus, potatoes, and tobacco requires large amounts of fertilizer. The highest fertilizer use is likely to occur where some of these high-value crops are grown with irrigation. Sometimes the fertilizer applications are over 2 tons per hectare per year.

Soil Factor

The ability of soils to supply plant nutrients varies from soil to soil and from time to time. Weathering and leaching reduce soil fertility as a soil becomes old. The fertility of most cropped soils declines considerably during the first several years of cultivation because harvesting crops removes nutrients. It becomes necessary, then, to supply more and more of the plant needs by fertilizer.

Many soils have physical properties such as shallow depth or low permeability that limit their productivity. Crops grown on such soils will not yield as well as on better soils and are likely to be less responsive than usual to fertilizer applications. The optimum amount of fertilizer to be used on such soils is less than it is on more productive soils. A farmer should apply the most fertilizer to the soils that are most responsive and should not attempt to obtain the same crop yield on soils that have different potentials. Large applications of fertilizer can be profitably used on soils that have high potential but low fertility.

Climatic Factor

Relatively little fertilizer is used in regions receiving less than 400 mm of precipitation per year except on irrigated land. Soils in these regions have lost little by leaching, and their inherent fertility level is relatively high. The limited amount of water available for plant growth in such regions does not justify using fertilizers to raise their fertility level.

Crops grown in humid regions (or with irrigation) usually need fertilizer to produce their best yields. Most soils of humid regions have lost significant amounts of plant nutrients by weathering and leaching. The water supply is adequate for high production, but productivity is usually limited by plant nutrient supplies unless fertilizer is used.

The soils of the eastern United States are more leached and less fertile than those of the western United States. Figure 9-6 shows at a glance that much more fertilizer is used in the eastern half of the United States than in the western half. Much of the fertilizer that is used in the western United States is applied on irrigated land.

Economic Factor

Fertilizer use is increased by low prices and decreased by high prices like any other commodity. Crop prices have the opposite effect because a high price for the crop will give a profitable return from larger fertilizer applications. Most countries that use little fertilizer have high fertilizer prices relative to the prices of their crops.

Yield increases from fertilizer follow a curve of diminishing returns. Small applications of needed fertilizers result in the greatest return per kilogram of nutrient applied. Additional amounts of fertilizer give progressively smaller increases in yield. Eventually a point is reached where the last increment of fertilizer added

Figure 9-6 Distribution of fertilizer use by states in 1968. Each dot represents 50,000 tons.

barely increases yield enough to repay the cost of adding it. Wise managers try not to add more fertilizer than the break-even point. Those with adequate capital to purchase fertilizer should add as nearly as possible the amount of fertilizer indicated by the break-even point. A graphical approach for estimating this point is explained later in this chapter.

Management Factor

The personal preferences of different managers for doing things in particular ways have a large influence on the use of fertilizers. Furthermore, there are many biased persons endeavoring to influence people either favorably or unfavorably toward fertilizers. Fertilizer companies obviously encourage the use of fertilizers and sometimes oversell their product. On the other hand, some persons are strongly opposed to the use of any fertilizers or of some specific fertilizer. For example, one widely circulated magazine discourages the use of any fertilizer which has been treated with sulfuric acid. The editors of this magazine ignore the fact that sulfur is an essential plant nutrient needed in large amounts by plants. Such writers generally provide no scientific evidence to prove their claims against fertilizers but nevertheless convince thousands of people that commercial fertilizers are detrimental to soils and crops.

Managers choose the input-output level at which they will operate. Increased crop outputs usually require increased fertilizer inputs. More fertilizer is needed whether the increased production is sought by increasing the amount of cropland or by increasing the yield of the crop.

Top yields depend on many factors. The best managers learn how to either control or adjust to as many of these factors as possible. Among these factors are soil type, climate, present and past crop type and variety, present and past applications of fertilizers and soil amendments, tillage practices, weed and insect control, and timing of operations. Seeding rate, depth of seeding, and row spacings can have much to do with the resulting crop production. Irrigation, drainage, and erosion control are important where needed.

Anyone using fertilizers needs to be aware that fertilizers are good only for curing fertility problems. Adding more and more fertilizer will bring about little or no increase in production when other factors are limiting. Too-large applications of some fertilizers can reduce yields by unbalancing the nutrient supply. Excessive fertilizer rates can lead to large nutrient losses by leaching and erosion. Leached and eroded nutrients may enter streams and cause pollution.

Managers who most effectively control all limiting factors obtain high yields and make the most profit. These managers can make profitable use of larger fertilizer applications than can poorer managers.

WHAT IS IN A FERTILIZER?

A "complete" fertilizer includes nitrogen, phosphorus, and potassium in appropriate forms for increasing soil fertility. State fertilizer laws require that every bag of fertilizer bear the name and address of the manufacturer, the weight, and the guaranteed analysis of the fertilizer. The guaranteed analysis, usually printed in large type, is a combination of numbers such as 8-32-16. This particular analysis would guarantee that the fertilizer contained the following:

8 percent nitrogen (total N)
32 percent phosphate (citrate-soluble P as P_2O_5)
16 percent potash (water-soluble K as K_2O)

Phosphorus that is soluble in a dilute ammonium citrate solution is considered to be available to plants. Citrate-soluble phosphorus is equal to or greater than the amount of water-soluble phosphorus but may be less than the total phosphorus.

Fertilizer phosphorus and potassium have traditionally been expressed in the oxide forms. The terms *phosphate* and *potash* indicate that this tradition is being followed. This means of expression is strictly a mathematical manipulation of the percentages. Fertilizers do not contain either P_2O_5 or K_2O. Some manufacturers are now printing fertilizer analyses both in oxide forms and as percentages of N, P, and K. Conversion factors based on the atomic weights of P, K, and O are:

$$\%P \times 2.29 = \%P_2O_5 \qquad \%K \times 1.20 = \%K_2O$$
$$\%P_2O_5 \times 0.44 = \%P \qquad \%K_2O \times 0.83 = \%K$$

Bags of "complete" fertilizers usually do not indicate what chemical compounds are in the fertilizer. This information is usually not needed because the guaranteed analysis tells the purchaser how much of each plant nutrient is being bought. Various chemicals used as fertilizers are discussed in Chapters 10, 11, 12, and 14.

Some fertilizers contain only one or two fertilizer elements. The guaranteed analyses of these fertilizers contain zeros for elements omitted. For example, 0-20-0 means that the fertilizer contains phosphorus but nitrogen and potassium are not included. Most single-nutrient and some double-nutrient fertilizers are single compounds. These compounds are often named on the bag. For example, 33.5-0-0 is ammonium nitrate and 13.5-0-38 is potassium nitrate.

Comparing Fertilizer Prices

Fertilizer costs should be compared on the basis of needed nutrients actually supplied to the crop. Both purchase costs and application costs should be considered. The least expensive fertilizer combination to purchase sometimes involves additional application cost that keeps it from being the most economical choice.

The lowest purchase price per unit of nutrient usually comes from high-analysis single-nutrient fertilizers. For example, the cheapest nitrogen fertilizer per unit of nutrient is anhydrous ammonia. But anhydrous ammonia must be applied with special equipment. It cannot be mixed with other fertilizer elements that may be required. Choosing anhydrous ammonia as a fertilizer is therefore likely to result in two application costs instead of one. A large application will save more than enough to pay for the application cost, but a small application may not.

The price per unit is easy to calculate for single-nutrient fertilizers. All that is required is to divide the costs by the number of units of nutrient. Such calculations make it easy to decide which single-nutrient fertilizer is the cheapest source of the desired element.

Comparing the prices of mixed fertilizers is more complicated. One approach is to first calculate the cost per unit of N, P, and K obtained from single-nutrient fertilizers. These values can then be multiplied by the analysis figures for the mixed fertilizer. Care should be used to include application costs in the calculations. Some allowances should be made for the convenience of applying all the needed nutrients at once, but convenience is difficult to evaluate.

The above procedure is adequate if the mixed fertilizer contains the desired ratio of nutrient elements. It can be modified to fit other situations. Sometimes the mixed fertilizer is supplemented with another to obtain the right ratio. The costs can then be summed for comparing alternatives. Sometimes a moderate excess of one nutrient is applied in order to obtain enough of another nutrient from a mixed fertilizer. The fertilizer should then be valued on the basis of the amount of nutrient needed or beneficial rather than on the excess amount applied.

Fertilizer Ratios

Fertilizer analyses are sometimes reduced to fertilizer ratios for the purpose of comparing relative amounts of each nutrient supplied by the fertilizer. The ratio is obtained by dividing the analysis by a factor that produces the smallest possible whole numbers. A 6-24-24 fertilizer thereby becomes a 1-4-4 ratio. A 5-20-20 fertilizer also has a 1-4-4 ratio. These two fertilizers could therefore be substituted for one another simply by adjusting the rate of application.

TYPES OF FERTILIZERS

Fertilizers can be classified in several different ways. A classification by fertilizer analysis is one way; a division into mineral or organic classes is another. The method of preparation is the basis for the classification to be considered here. On this basis, fertilizers may be single compounds (or perhaps natural mixtures of compounds), manufactured fertilizers, bulk blends, or liquid mixes. Most ready-made fertilizer sold in bags is either the single-compound or the manufactured type. Bulk blends and liquid mixes are usually specially prepared for each user.

Single-compound Fertilizers

A fertilizer consisting of only one compound has the advantage of having a known constant analysis. Every grain of fertilizer has the same composition. Uniform composition is an advantage where it is necessary or desirable to attain a uniform application of fertilizer.

Most single-compound fertilizers supply only one fertilizer element. Examples are anhydrous ammonia, ammonium nitrate, monocalcium phosphate, and potassium chloride. A few fertilizers combine two fertilizer elements in one compound such as diammonium phosphate or potassium nitrate. These latter compounds help to make high-analysis fertilizers.

The use of single-compound fertilizers has been increasing in recent years. The amounts of plant nutrients now being supplied by single compounds is approximately equal to those being supplied in mixed fertilizers (Figure 9-7). The single compounds containing two elements are a small portion of the total but are increasing rapidly in importance. The leading combination is nitrogen and phosphorus in the ammonium phosphates.

Manufactured Fertilizers

Most manufactured fertilizers are marketed in a *granulated* (sometimes called *pelleted*) form. The ingredients are thoroughly mixed in a moist condition and then passed through a rotating drum that granulates the material as it dries. Each fertilizer granule has the same analysis as each of the others.

The ingredients of manufactured fertilizers are chosen not only to contain the desired concentration of nutrients but also to granulate well. The granules should not stick together even after storage. Some manufacturers apply a protective coating to their granulated fertilizer to maintain a desirable physical condition.

Manufactured fertilizers are the traditional means of marketing mixed fertilizers. The manufacturer chooses a fertilizer ratio that fits the needs of a market area, selects appropriate ingredients, and mixes them in the right proportions. Micronutrients can be included if desired. Manufactured fertilizers are mass produced to minimize production costs. Many manufactured fertilizers are sold under copyrighted trade names.

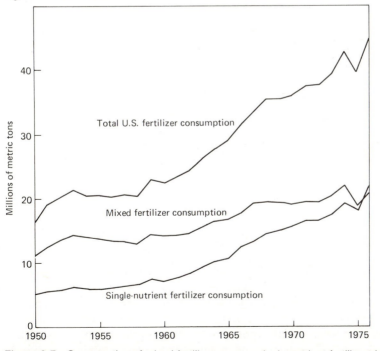

Figure 9-7 Consumption of mixed fertilizers versus single-nutrient fertilizers in the United States from 1950 to 1976. *(Based on data from USDA Crop Reporting Board.)*

Bulk-blend Fertilizers

Bulk blending is one of the most flexible ways of producing fertilizer of any specified ratio. A bulk-blending plant stores single-compound fertilizers and mixes them just before they are loaded into a truck to be hauled to the field. Micronutrients may be added in a liquid form sprayed on the dry fertilizer as it flows into the truck.

Bulk-blend fertilizers are less homogeneous than the other major types because the ingredients are mixed in the solid state. Some mixtures tend to segregate again after they are mixed because the various ingredients may have different particle sizes and densities. Segregation may occur during the loading, hauling, and spreading operations even after thorough mixing at the blending plant. This disadvantage can be minimized by using a uniform particle size and is offset by the flexibility of the bulk-blending system. The mixture can be varied at will so that one truck can be loaded with one fertilizer analysis and the next truck with another.

Bulk-blending plants are usually much smaller than fertilizer manufacturing plants. They are geared to serve a more local area than the manufacturing plants. Bulk-blending plants began to appear in the United States in the late 1950s and have increased rapidly since then. They marketed about a third of the mixed fertilizer sold in the United States in 1967. There is a marked concentration of bulk-blending plants in and near the corn belt.

Liquid Mixed Fertilizers

All nutrient elements are available in forms that are soluble in water and therefore can be applied in liquid form. Liquids can be sprayed on to give a uniform distribution. The liquid form can be added to irrigation water and distributed without any equipment crossing the field. The greatest handicap to liquid fertilizers is the need for corrosion-resistant equipment.

The first liquid fertilizer plant in the United States was built in 1923. The use of this type of fertilizer increased very slowly at first, but its popularity has increased somewhat recently. Nearly 9 percent of the mixed fertilizer (about 5 percent of all fertilizer) used in the United States during 1967 was applied in liquid form.

Anhydrous ammonia and phosphoric acid are both liquids that mix readily with water. Mixing the two produces ammonium phosphate, which is also soluble unless too much ammonia is used. Urea or ammonium nitrate can be added to the solution if more nitrogen is desired than the solubility of ammonium phosphate permits. Potassium can be added, usually in the form of potassium chloride, to make a complete fertilizer.

APPLYING FERTILIZER

Fertilizer may be applied before, during, or after the planting of a crop. The decision in this matter depends partly on the amount of fertilizer to be applied. A small amount can be applied conveniently at the time of seeding; large applications usually are made earlier or later, or perhaps divided into two or more parts.

Nitrogen fertilizers are usually applied a short time before they are needed to avoid loss by leaching. Phosphorus and potassium are less mobile and may be applied before planting. They can then be mixed with the soil or placed at the bottom of the tilled layer. This possibility is a good one to remember for permanent grass plantings such as pasture or lawn. It is a good idea to plow under a fairly heavy application of P and K where these elements are needed before planting a permanent grass cover. A large early application of P and K will benefit the grass as it is becoming established, and will carry over for future years as well. Subsequent fertilizer applications can therefore contain proportionally more N and less P and K than would otherwise be used.

Placement of Fertilizer

Fertilizer applied before planting is usually "plowed down." This practice places most of the fertilizer in a zone several centimeters below the surface and is superior to broadcasting and disking in the fertilizer. Plant roots grow downward and soon encounter the fertilizer that has been plowed down. Large applications of P and K can be made by the plow-down method.

Small applications known as "starter" fertilizer are made at planting time. Starter fertilizer is placed about 5 cm from the seed, usually slightly below and to one side (Figure 9-8). The plant roots should reach it very soon to give the crop an early start. The amount of starter fertilizer that can be used, however, is limited by salt hazard. Too much soluble salt near a seed or a plant will dehydrate the seed.

Later applications of fertilizer may be either top-dressed or side-dressed, according to the crop. Close-growing crops such as pasture or hay are top-dressed by simply broadcasting fertilizer on the surface. Some of this fertilizer will land on the plants rather than on the soil, but it does no harm.

Row crops are side-dressed when it is desired to add fertilizer after the crop is established. The fertilizer is placed in a band a short distance to the side of the crop row. The band may be placed on the soil surface, or a cut may be made to place the fertilizer at a shallow depth. Some fertilizers such as anhydrous ammonia must be placed beneath the surface to avoid loss. Care must be taken not to cut off many roots when placing fertilizer in the soil near the crop row.

IDENTIFYING FERTILIZER NEEDS

Fertilizers should be carefully chosen to supply the needed amounts of the nutrients that are deficient. Plant nutrients that are already available to the plants in adequate

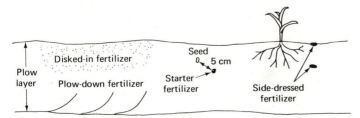

Figure 9-8 A schematic illustration of various types of fertilizer placement. Top-dressed fertilizer would be in a thin layer on the soil surface.

amounts should not be added. An oversupply of a nutrient can prove detrimental to plant growth, especially if another nutrient is deficient and in some instances can lead to leaching losses that pollute underground water.

The surest way to demonstrate a need for a nutrient suspected of being deficient is to apply it as a fertilizer and see what happens. Several different kinds, amounts, and combinations of fertilizer elements can be applied for a more complete test. One or more areas should be left as untreated checks for reference to show what happens without fertilizer.

Factorial designs help to determine the number of test plots needed in an experiment. For example, if nitrogen, phosphorus, and potassium are each to be tested at 0-, 50-, and 100- kg/ha levels, the factorial design will require $3 \times 3 \times 3 = 27$ separate plots to give all possible combinations. A simple 2×2 factorial design with the actual results from one experiment is shown in Figure 9-9. Factorial experiments have greatly increased our information about fertilizer needs on different soils. The data shown in Figure 9-9 indicate that both P and K were needed for the best growth of alfalfa on the soil tested. Either P or K alone gave some increase, but the combination of the two gave more.

Interaction between fertilizer elements is often shown by factorial experiments. Interaction may be either positive or negative. Data from a $2 \times 2 \times 2$ factorial experiment by Dumenil (1951) show both types. Dumenil's data appear in Table 9-1. There was a negative interaction between phosphorus and potassium in the no-nitrogen block. Either of these elements alone gave a 1-q (quintal) increase in corn yield (essentially no change), but used together and without nitrogen they caused a 4-q *decrease* in yield.

Nitrogen alone increased corn yields by 12 q in Dumenil's data. In this block there was a positive interaction between phosphorus and potassium. Potassium alone produced no increase, and phosphorus alone increased yields by only 7 q, but the combination increased yields by 14 q.

The most limiting nutrient for corn in Dumenil's experiment was nitrogen. Phosphorus was needed after the nitrogen level was increased. A significant need for potassium was shown only after both of the other nutrients had been supplied. These data show that all three of these nutrients were needed on this soil to produce the

Check plot	K treatment
No P No K	No P 67 kg K
5.48 tons alfalfa hay/ha	7.57 tons alfalfa hay/ha (2.09 ton increase)
P treatment	P and K treatment
67 kg P No K	67 kg P 67 kg K
8.04 tons alfalfa hay (2.56 ton increase)	9.79 tons alfalfa hay (4.31 ton increase)

Figure 9-9 A 2×2 factorial design showing the influence of P and K fertilization of alfalfa hay production on Iowa State University's Clarion-Webster experimental farm. *(Data from Schaller and Webb in Annual Progress Rept. OEF 69–14, 22, Iowa State Univ. Agr. and Home Econ. Exp. Sta.)*

Table 9–1 Effect of N, P, and K (Alone and in All Combinations) on the Yield of Corn

Fertilizer treatment	No nitrogen, q/ha		67 kg of N, q/ha	
	No phosphorus	67 kg of P_2O_5 (30 kg of P)	No phosphorus	67 kg of P_2O_5 (30 kg of P)
No potassium	29	30	41	48
67 kg of K_2O	30	25	41	55

Source: Dumenil, 1951.

top yield of corn. Similar results have been determined on many but not all other soils.

The data in Table 9-1 do not show how much N, P, and K were needed nor what the maximum possible yield might be under these particular soil, crop, and climatic conditions. Further testing with various rates of each element would help to supply this information.

Using fertilizer trials is a laborious way to identify fertilizer needs. The information comes too late to benefit the crop tested, but it is useful for the future and for interpreting the results of other types of tests. Soil tests and tissue tests are cheaper and more practical to apply on an extensive basis than are fertilizer trials.

Deficiency Symptoms

Another way to identify fertilizer needs is to watch for deficiency symptoms. Some of the best-known symptoms are a pale yellowish-green color for nitrogen deficiency, purplish spots or streaks for phosphorus deficiency, and brown dead spots for potassium deficiency. Deficiency symptoms are discussed in more detail in Chapters 10 to 14.

Deficiency symptoms are seldom conclusive unless the deficiency is severe. Plants may respond favorably to the application of one or more fertilizer elements even when they show no obvious deficiency symptoms. Growth retardation occurs before deficiency symptoms show. Fertilizer applied to correct the deficiency may benefit the affected plants but not as much as it would if it had been applied earlier.

Tissue Tests

Two different types of tissue tests are used as indicators of the nutrient status of plants. The first, tissue analysis, is a laboratory test and can be made quite precise. The second is a quick test that can be done in the field and is sometimes called a "sap test." The plant material selected for either test must be identified in terms of the kind of plant, the portion of the plant sampled, and the stage of maturity.

Tissue tests can be made only after the plant has attained a good share of its growth. They are sometimes useful for determining needs for supplemental applications of nitrogen fertilizer because plants respond quickly to nitrogen. The main application of nitrogen and all of the phosphorus and potassium should have been applied earlier for maximum effectiveness. Tissue tests on annual crops often are of more value for deciding what to do the next year than for adjusting the current year's program. Perhaps their greatest value is for perennial crops, because additions of the fertilizer nutrients these crops need at the moment are certain to benefit their future growth. Use of tissue tests requires supplementary information to tell how much plant nutrient is needed even after a needed nutrient is identified.

Tissue Analysis   The plant material selected for tissue analysis is finely ground and tested chemically to determine its total content of the nutrients in question. These contents are then compared with the nutrient levels in the same parts of normal plants at the same stage of maturity. The comparison is used to determine whether there is a nutrient deficiency, or if the supply is adequate or excessive for each of the nutrients tested. The test is meaningless without a standard for comparison. For example, a test showing 2 percent N might indicate a deficiency of nitrogen during the early stages of growth of corn. But, 2 percent N in the stalk of a mature corn plant is a more-than-adequate supply of nitrogen.

Sap Tests   A sap test quickly determines whether a measurable amount of a particular ion is dissolved in the plant sap. One such test is illustrated in Figure 9-10. Tests like this may be used as an indication of the amount of each plant nutrient the plant has available for immediate use. It should be recognized that the supplies of anion-forming elements such as nitrogen, phosphorus, and sulfur are mostly combined in organic forms that are undetected by a sap test.

Sap test results are more variable over time than are results from tissue analyses. One reason is that the amounts of soluble nutrients accumulated in a plant are related to the moisture status of the soil. For example, a sap test made 2 or 3 days after a rain is likely to show a relatively high nitrogen level. Nitrification is usually rapid after a rain, and the plants promptly absorb the freshly released nitrates. Day-night cycles are another cause of variations. Nutrients absorbed at night build up to maximum sap-test levels in early morning. The levels decline during the day as the nutrients are built into organic compounds.

Soil Tests

Several quick-test kits are available for testing soils, but the results from these are of limited usefulness. A soil test must be interpreted meaningfully after it is made. The interpretation requires data and experience that only a soil-testing laboratory is likely to have.

Most states have soil-testing laboratories under the direction of the state agricultural college. Much research goes into determining the best type of soil test to use. The test results are correlated to field trials so the fertilizer recommendations are based on experimental results. Usually several items of information such as soil type and cropping history are considered along with the soil-test results in making the recommendations. Many laboratories use computer processing to make their recommendations as speedily and reliably as possible.

Many private soil-testing laboratories are now providing services similar to those of the state laboratories. Often the same testing procedures and interpretation

Figure 9-10 Nitrate tests can be made at the base of the leaf midrib with diphenylamine without destroying the entire plant. This is an important consideration in making numerous tests on small experimental plots. The height at which nitrates are present in the plant as well as the intensity of the blue color gives an indication of the nitrate status of the plant. *(Courtesy of W. L. Nelson, B. A. Krantz, and L. F. Buckhart.)*

criteria are used in both places. Most private laboratories are financed by fertilizer companies.

Soil-Test Samples

The most important limitation in testing soils is the soil sample. Test data are practically worthless if the sample is not representative of the area to be fertilized. Separate samples should be taken for each soil type and slope condition. Fields with different cropping or fertilizing histories must be tested separately. Large areas should be subdivided into smaller parts even if they appear to be uniform. Each soil sample should represent no more than 4 ha (less for areas of intensive land use). The cost of having soils tested is small compared with the loss resulting from the application of either too little or too much fertilizer.

Soil samples can be collected with any tool that samples a cross section of the furrow slice (about 15 cm of soil). Soil augers and coring tools are very satisfactory. A shovel can be used but is not as fast. Several small subsamples should be collected from points distributed across the area of uniform soil to be represented by the sample. The subsamples should be thoroughly mixed.

Unusual areas such as low wet spots, rock outcrops, and severely eroded spots should be omitted from soil samples unless they are large enough to sample and treat separately. Strips within 30 m of a gravel road should be avoided, especially if limestone gravel was used on the road.

A sketch map and notes should be kept to identify each area sampled. The notes should tell the soil type (or describe the soil), slope percentage, soil drainage conditions, liming and fertilizer history, cropping history, and the date of sampling.

Soil-Test Classes

Chemical tests used for assessing the availability of nitrogen, phosphorus, and potassium are discussed in Chapters 10 to 12. The tests are intended to place numerical values on the amount of each nutrient that will be available to growing plants. Total amounts present in the soil are meaningless for this purpose unless the fraction available can be predicted.

Soil-test results are classified on a scale ranging from very low to very high for each nutrient. The probability of making a profit from applying fertilizer is about 90 percent on soils that have very low soil tests, 50 percent for medium soil tests, and 10 percent for very high soil tests.

Interpretations of soil-test data are made for each soil-test class rather than for each numerical value. Using the soil-test classes permits the pooling of crop-response data to make sound fertilizer recommendations.

Choosing the Rate of Fertilizer Application

A need for fertilizer can be identified by deficiency symptoms or by any of the tests that have been discussed. There is still a question of the amount to be applied. Determining the optimum amount of fertilizer requires experience that can serve as a basis for predicting how the crop will respond to fertilizer applied to the soil on which it is grown.

The available data from similar soils with the same soil-test class are plotted on a graph to determine a fertilizer response curve. For example, the yield increases obtained from nitrogen fertilizer on two sets of corn test plots in Iowa are shown in Figure 9-11.

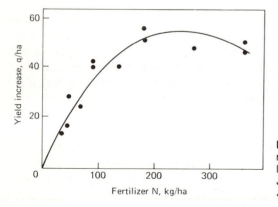

Figure 9-11 Corn yield responses from nitrogen fertilizer applications on certain Iowa soils. (*Data from annual reports of J. R. Webb and A. J. Englehorn, Iowa State Univ.*)

The curve in Figure 9-11 is drawn to represent the data as closely as possible by a single curve. It shows that the yield increase per unit of additional fertilizer declines at the higher rates of fertilization. A point is reached where the additional fertilizer barely increases production enough to be worth its cost. This point is the maximum fertilizer rate that should be considered. Figure 9-12 shows how to determine this point by graphic analysis of cost and returns data. The curve is the same as the one shown in Figure 9-11, but the yield scale has been changed to dollars.

A farmer may rationally choose either the minimum or the maximum fertilizer rate shown in Figure 9-12, or possibly some rate between these limits. The maximum rate gives the largest profit per hectare of land and is the best rate if enough capital is available. The minimum rate gives the highest return per dollar of fertilizer investment and is the best choice if money for fertilizer is limited. It is more profitable to fertilize part of the crop area at the minimum rate than it is to fertilize the entire area at a lower rate.

The curve shown in Figures 9-11 and 9-12 represents average corn yield response to nitrogen fertilizer on a certain group of Iowa soils. Comparable soils with higher soil tests are likely to give lower response curves and a lower probability of a profitable return from fertilizer.

Another factor that influences fertilizer rates is the cost of the fertilizer. A higher price gives a steeper cost line in Figure 9-12 and reduces the size of the profit section on the graph. The tangent showing the maximum recommended fertilizer rate then shifts to the left giving a lower recommended rate for the higher-priced fertilizer.

The approach illustrated in Figure 9-12 enables a farmer to make the best choice of fertilizer rate if such a curve, or the information to make it, is available for a particular crop and soil combination. Often, however, farmers must estimate their fertilizer needs on the basis of the yield level they think will be optimum. Fortunately, there is a range of nearly optimum fertilizer rates that will give good results. Figure 9-12 shows that any of a wide range of nitrogen fertilizer rates would give a good profit on these particular soils.

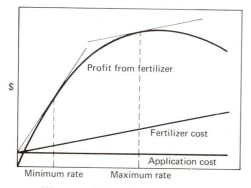

Figure 9-12 Graphic analysis of nitrogen fertilizer costs and returns based on the response curve in Figure 9-10. The minimum fertilizer rate recommended is determined by the tangency point of a line originating at the fixed cost of fertilizer application. The maximum recommended rate is determined by the tangency point of a line parallel to the fertilizer cost line.

Procedures similar to those illustrated for nitrogen can be used to determine appropriate amounts of phosphorus or potassium to be applied. The approach for micronutrients is also similar except that the cost of the small amounts that may be required can often be ignored. The amount of the micronutrient to be applied then becomes the amount required to produce maximum growth.

Combinations of fertilizers require more data and larger experiments. Several different rates of each fertilizer nutrient must be tried in combination with various rates of each of the other nutrients applied. Each nutrient element has its own effect, but these effects usually show some degree of interaction. The true optimum amount of each nutrient can be determined only when the other nutrients are also near optimum.

Some farmers will, of course, obtain higher fertilizer response than others because of differences in management methods. Some soil-testing laboratories make three levels of fertilizer recommendations: low for farmers with limited financial resources for purchasing fertilizer; medium for average farmers; and high for farmers aiming for top yields through superior management practices combined with high fertility.

Adjustments in rates of fertilizer application may also be made for the amount of water that will probably be available to the growing plants. Such adjustments may be made on the basis of the amount of available water in the soil at planting time or on the availability of irrigation water.

MANURE

The fertilizer value of manure depends on its source, the way it is managed, and the pricing of plant nutrients. The pricing factor has varied considerably in the last few decades. The production of synthetic ammonia reduced the cost of nitrogen to a United States farmer from about 50 cents per kilogram in 1940 to about 12 cents per kilogram in 1970, then rose to about 50 cents per kilogram again in 1975 as a result of rising petroleum prices. The value of 5 kg of nitrogen fell and rose by $1.90 whether it was in a bag of fertilizer or a ton of manure.

A ton of an average manure contains about as much nitrogen, phosphorus, and potassium as a 50-kg bag of 10-5-10 fertilizer. The nutrients in the 10-5-10 form cost about $1.75 in 1970 but their price had about doubled by 1975. Handling costs of over $1 per ton to load, haul, and spread the manure left only a small margin of real saving at 1970 prices as compared with commercial fertilizers. This comparison is not completely valid, however, because the manure also contributes other nutrients such as calcium, sulfur, and the various micronutrients and provides a bonus in the form of organic matter added to the soil.

A part of the cost of handling manure can rightly be charged to the livestock operation as waste disposal. Not only is a large manure pile an unsightly obstruction on the premises, but it also is a common source of water pollution. Runoff from the area must go someplace. Such runoff may reach the farmer's well and contaminate it. Very likely the runoff will reach a nearby stream and contaminate it. Eutrophication (increased growth of undesired vegetation such as algae resulting from an enriched supply of plant nutrients) will occur in backwater areas. The oxygen supply of the flowing water will be depleted as the organic matter decomposes.

Cheap mineral fertilizers are indirectly responsible for some of the pollution described above. Operators who would rather buy fertilizer than spread manure allow their manure to accumulate and cause pollution. Such indirect effects are much larger and more widespread than pollution resulting directly from fertilizers applied to soil.

Composition of Manure

There is a great deal of variation in the composition of manure. The variation depends on such diverse factors as the kind and age of livestock, the use of bedding in the manure, whether the liquid excrement has been retained, and how much decomposition and leaching have taken place during storage. Table 9-2 shows the average composition of solid and liquid excrement from different kinds of livestock.

The advisability of adding straw or other bedding to manure is sometimes questioned. The bedding obviously has a diluting effect and increases the tonnage of manure, but the bedding helps to retain the liquid excrement. About half of the nitrogen and over half of the potassium is in the liquid excrement. The use of bedding is therefore a worthwhile practice. The average manure is assumed to include bedding and to have a composition of about 0.5 percent N (5 kg per ton), 0.125 percent P (1.25 kg per ton), and 0.4 percent K (4 kg per ton). All other essential plant nutrients are also contained in manure in varying amounts, mostly in rough proportions to the needs of plants.

The phosphorus content of manure is relatively low compared with plant needs. Furthermore, some of the phosphorus (and some nitrogen and other nutrients) in manure is not immediately available for plant growth. It is often advisable to supplement manure applications with a phosphorus fertilizer, which is sometimes mixed with the manure producing what has been called *reinforced manure.*

Some persons have suggested that phosphorus fertilizer mixed with manure might be partially transformed to less soluble forms. Research done by Midgley and Dunklee (1945) indicates that such transformations are not a problem. They grew tomatoes in several different soils in a greenhouse. Pots receiving reinforced manure generally produced more tomato growth than pots receiving equal but separate applications of phosphorus fertilizer and manure.

Table 9–2   Manure Production and Composition

Kind of livestock	kg per day per 1,000 kg live weight	% N		% P		% K	
		Solid	Liquid	Solid	Liquid	Solid	Liquid
Cattle	70—100	0.5	0.25	0.11	0.06	0.41	0.21
Hogs	70	0.5	0.1	0.13	0.42	0.37	0.09
Poultry	60	1.5		0.43		0.41	

Source: Calculated from data cited by Hinish, 1974.

Conservation of Manure

Manure contains valuable nutrients that become available for plant growth if the manure is applied to soil. McIntosh and Varney (1973) found that the phosphorus from manure tested more available than that from mineral fertilizer. But much of the value of manure is readily lost by decomposition and leaching. Decomposition produces many gases including CO_2, NH_3, and H_2S. It also releases soluble mineral compounds that may be lost by leaching. Manure left in the open may lose most of its potassium, some of its phosphorus, much of its nitrogen, and varying amounts of other nutrients.

The best method of storing manure is to pack it tightly into a lined pit with no drainage. Conditions should be anaerobic to minimize volatilization losses. Lacking such a pit, the manure may be allowed to accumulate in a barn or shed where it is protected from leaching. Packing by trampling or other means will help to keep it anaerobic.

Manure should be hauled to the field as soon as it is removed from a building or a storage pit. In the field it should be spread and plowed under as soon as possible. Leaching is no longer a problem after the manure is spread, but erosion losses and volatilization may still occur. It has long been known that large nitrogen losses occur if the manure is left on the soil surface. Heck (1931) exposed cow manure for different periods of time at 20°C. He found that 7.7 percent of the total nitrogen volatilized in 12 h, 23.4 percent in 36 h, and 36.2 percent in 7 days.

A covered feedlot is considered the most practical way of storing manure accumulated from steers. The trampling by the steers compacts the manure and causes anaerobic conditions. The use of bedding helps to prevent loss of liquid excrement.

If manure must be stored outside, it should at least be piled high and rounded over to shed water and thus reduce leaching losses. The pile should be kept as compact as possible.

Applying Manure as a Fertilizer

Manure is primarily a nitrogen-potassium fertilizer. It will release these nutrients fastest when the soil provides warm, moist conditions favorable for microbial decomposition. Manure applications are therefore most effective on a warm-weather crop needing nitrogen and potassium.

Manure applications are normally made at rates of several tons per hectare. The largest yield increases per ton of manure are at relatively low rates of application. But, the largest profit per hectare resulting from manure applications comes from higher rates. Rates of about 8 to 10 tons of manure per hectare have been recommended for corn. About 5 tons per hectare are more appropriate for cooler-weather crops such as wheat or oats. Maximum profit usually comes from covering a maximum area at about these rates rather than using higher rates on less land. Additional nutrients needed may be applied as inorganic fertilizer.

The best time to apply manure depends on what factor is more important. Loss of nutrients from the manure can be minimized by rapid handling. Daily applications of manure would be best for minimizing nutrient losses. Such a practice is usually not an efficient use of time and often is not practical in terms of field use of the manure. It is easier to find time to haul and spread manure during some seasons than others. It is often expedient to time manure applications accordingly.

Loaded manure spreaders are heavy, and care should be taken not to cause excessive soil compaction. A field is therefore unavailable for manure application when it is too wet as well as when crop use prevents entry. The ideal time for a field to receive manure is just before plowing for a crop that will benefit from the nitrogen and potassium in the manure.

The timing of manure applications is necessarily some kind of compromise between saving the nutrients in the manure and the availability of personnel and fields. Some have added an unnecessary restriction by applying manure only to fields near the barn or feed lot. The additional cost of hauling to another field is usually small compared with loading and spreading costs. Applying the manure to a greater variety of fields usually means that it can be applied at more varied times during the year. The benefit received from the manure may thereby be increased.

Structural Benefits from Manure

Manure provides organic materials that have a favorable influence on soil structure. Hafez (1974) showed that manures from various domestic animals increased the aggregation of soil particles and reduced the bulk density. Working with Dinuba fine sandy loam, he reduced the bulk density progressively from 1.43 to 1.10 g/cm^3 as he added 2.5, 5.0, and 10.0 percent manure to the soil. Hafez concluded that fibrous materials in the manure were important in altering physical properties of soils.

Tiarks, Mazurak, and Chesnin (1974) added feedlot manure to a silty clay loam and increased its permeability as well as decreased its bulk density. They also measured soil aggregation and found that it increased when manure was added.

Lagoon Systems

The manure from some livestock operations is caught in trenches from which it is washed out to a pond called a *lagoon*. Bacteria in the lagoon decompose much of the material and release plant nutrients. Several researchers have shown that lagoon water makes an effective fertilizer. Turner (1971) was able to pump material containing up to 9 percent solids from a lagoon containing dairy cattle wastes. The material was sprayed on cropland in amounts up to 2.5 cm on untilled land or 10 cm on plowed corn ground. Larger amounts would coat the soil with fiber creating a thatch and increasing runoff. The material could also be sprayed on vegetated land but excessive amounts can coat the plants and smother them. Turner recommends that at least 0.2 ha of disposal land be used per dairy cow.

COMPOST

Mixtures of soil and decomposing organic matter called *compost* are often used by gardeners as a fertilizer. Usually a pile is made by alternating layers of soil or sod with organic materials such as manure, garbage, grass clippings, etc. Some add chemicals such as lime, superphosphate, or other fertilizer to the pile to adjust the pH or to increase the concentration of desired nutrients. Nitrogen fertilizer may be added to narrow the C:N ratio and accelerate decomposition (Chapter 10).

Compost piles are kept moist while the organic matter is decomposing. Sometimes they are turned and mixed once or twice. After several weeks or a few months they are spread on the garden or other plot to be fertilized.

Compost provides gardeners with a fertilizer well adapted to their needs. Much of the carbohydrate has been decomposed. Some nitrogen is lost, but the carbohydrate loss is much larger, and the percent nitrogen in the compost is increased. The compost will gradually release plant nutrients by decomposition after it is spread on the soil. The supplies of various nutrients can be varied by changing the ingredients and their proportions in the compost.

The nutrient release from a well-rotted compost is probably better balanced and regulated than that from fresh manure. Gardeners can therefore apply larger amounts of compost than of fresh manure without danger of injuring plants. The use of compost also results in humus formation and promotes good soil structure.

The advantages of compost as a safe, effective fertilizer are real, but some of the claims made for it are overly enthusiastic. The nutrients released from compost are no better for plants than those released from other sources. Some nutrients have been lost by volatilization. The composting process has a high labor requirement relative to the amount of plant nutrients involved. Composts are practical for gardening but not for extensive use on field crops.

OTHER ORGANIC WASTES

Manure and compost are the principal organic fertilizers but not the only ones. Other waste materials can be used as fertilizer. Some of these are major stream pollutants. Diverting these to fertilizer use provides a twofold benefit.

Cities produce large amounts of sewage and must dispose of it in some manner. Usually it goes through some degree of processing that produces a thick sludge and a watery effluent. The effluent is usually dumped into a nearby river. Both the sludge and the effluent contain plant nutrients. The nutrients contained in the effluent are pollutants to the river but would be fertilizer to the land. The Agricultural Research Service (1969) has shown that plant-soil filters can purify the sewage effluent at about one-tenth the cost of equivalent chemical treatment. Braids, Hinesly, and Molina presented a paper on the use of sludge as a fertilizer to the 1969 meetings of the American Society of Agronomy in Detroit. They found that plants responded well to sludge applied to soil. Disagreeable odors were avoided by storing the sludge for 2 weeks of anaerobic digestion at 35°C. The sludge was applied to the land by including it in irrigation water.

Dried sludge is also used as a fertilizer, most often for small areas such as gardens or lawns. Larson (1974) indicates good crop response from sludge. He found that dried sludge averages 5 percent N, 3 percent P, and 0.5 percent K.

The presence of heavy metals in sewage and some other wastes causes a limitation in the amount that should be applied to any one soil. Repeated heavy applications can cause lead, mercury, or other heavy metals to accumulate in the soil to levels toxic to plant growth. The hazard is illustrated in some old orchard sites where sprays containing heavy metals were once used as fungicides. Weekly sprayings during critical periods over a number of years poisoned the soil so growth is still spotty even several decades later.

Some industrial by-products have fertilizer value. Sharratt, Peterson, and Calbert (1962) estimated that about 2 billion kg of whey per year are dumped in rivers, sewers, and ditches in the United States. They found that whey applications increased corn yields and improved the physical condition of the soil. One centimeter of whey per hectare contains 150 kg of N, 50 kg of P, and 180 kg of K.

It is sometimes tempting to apply very large amounts of some materials to the soil when waste disposal is the primary object. Heavy applications minimize the area that must be covered. However, heavy applications also minimize any benefits that may be derived from the material and increase the likelihood of soil damage and groundwater pollution. For example, Sharratt, Peterson, and Calbert found that applications of more than 10 cm of whey per year could cause salt accumulations that would inhibit plant growth.

Many other types of organic wastes are available in various places. The soil is probably the best place we have to dispose of many of them. Usually they will not damage either soil or crop if reasonable rates are applied. Returning these materials to the soil facilitates the recycling of the elements contained in them and thereby obtains a fertilizer value from them.

REFERENCES

Agricultural Research Service, 1969, Clear Water from Wastes, *Agric. Res.* **18**(6):10–13.

Aldrich, S. A., 1972, Fertilizing for Optimum Yields Will Give You Minimum Pollution, *Crops Soils* **24**(5):17–18.

Bower, C. A., and L. V. Wilcox, 1969, Nitrate Content of the Upper Rio Grande as Influenced by Nitrogen Fertilization of Adjacent Irrigated Lands, *Soil Sci. Soc. Am. Proc.* **33**: 971–973.

Childs, F. D., and E. M. Jencks, 1967, Effect of Time and Depth of Sampling upon Soil Test Results, *Agron. J.* **59**:537–540.

Consumption of Commercial Fertilizers in the United States, 1969, *USDA Stat. Reporting Ser. SpCr* 7(10–69).

Duell, R. W., 1964, Fertilizer-seed Placement with Birdsfoot Trefoil and Alfalfa, *Agron. J.* **56**:503–505.

Dumenil, Lloyd, 1951, Don't Starve Your Corn, *Iowa Farm Sci.* **6**:150–152.

Evans, C. E., 1973, Quick Tissue Testing, *Crops Soils* **26**(3):9–11.

Foote, L. E., 1969, Fertilizer Placement with Sodding, *Agron. J.* **61**:965–966.

Garcia, R. L., and J. J. Hanway, 1976, Foliar Fertilization of Soybeans during the Seed-filling Period, *Agron. J.* **68**:653–657.

Hafez, A. A. R., 1974, Comparative Changes in Soil-Physical Properties Induced by Admixtures of Manures from Various Domestic Animals, *Soil Sci.* **118**:53–59.

Harre, E. A., 1969, *Fertilizer Trends—1969,* National Fertilizer Development Center, Tennessee Valley Authority, Muscle Shoals, Ala., 103 p.

Heck, A. F., 1931, Conservation and Availability of the Nitrogen in Farm Manure, *Soil Sci.* **31**:335–359.

Hermanson, H. P., 1965, Maximization of Potato Yield under Constraint, *Agron. J.* **57**: 210–213.

Herron, G. M., and A. B. Erhart, 1965, Value of Manure on an Irrigated Calcareous Soil, *Soil Sci. Soc. Am. Proc.* **29**:278–281.

Hinish, W. W., 1974, Manure Doesn't Smell So Bad Anymore, *Crops Soils* **27**(3):12–15.

Jones, J. B., Jr., 1973, Soil Testing in the United States, *Commun. Soil Sci. Plant Anal.* **4**: 307–322.

Jones, J. B., Jr., R. L. Large, D. B. Pfleiderer, and H. S. Klosky, 1971, The Proper Way to Take a Plant Sample for Tissue Analysis, *Crops Soils* **23**(8):15–18.

Larson, W. E., 1974, Cities' Waste May be Soils' Treasure, *Crops Soils* **27**(3):9–11.

Lodge, F. S., and Sidney Wald, 1950, U.S. Fertilizer Consumption Registers 11th Consecutive Yearly Increase, *Fert. Rev.* **25**(2):8–11.

McIntosh, J. L., and K. E. Varney, 1973, Accumulative Effects of Manure and N on Continuous Corn and Clay Soil. II. Chemical Changes in Soil, *Agron. J.* **65**:629–633.

Martin, W. P., 1970, Soil as an Animal Waste Disposal Medium, *J. Soil Water Conserv.* **25**:43–45.

Midgley, A. R., and D. E. Dunklee, 1945, The Availability to Plants of Phosphates Applied with Cattle Manure, *Vermont Agric. Exp. Sta. Bull.* 525.

National Fertilizer Association, 1949, Fertilizer Consumption in 1948 at All Time High, *Fert. Rev.* **24**(3):7–9, 14.

National Fertilizer Association, 1951, Fertilizer Consumption for 1950 Soars to a New Record, *Fert. Rev.* **26**(2):8–11, 14.

National Plant Food Institute, 1955, *Plant Food Rev.* **1**(1), Washington, D.C.

Pesek, John, 1968, *Fertilizing: Fertilizer Levels,* McGraw-Hill Yearbook of Science and Technology, pp. 174–175.

Reuszer, H. W., 1957, Composts, Peat, and Sewage Sludge, *Soil,* 1957 Yearbook of Agriculture, U.S. Government Printing Office, pp. 237–245.

Rothwell, D. F., and C. C. Hortenstine, 1969, Composted Municipal Refuse: Its Effects on Carbon Dioxide, Nitrate, Fungi, and Bacteria in Arredondo Fine Sand, *Agron. J.* **61**: 837–840.

Sharratt, W. J., A. E. Peterson, and H. E. Calbert, 1962, Effect of Whey on Soil and Plant Growth, *Agron. J.* **54**:359–361.

Thompson, S. O., W. F. Wedin, and O. J. Attoe, 1968, Effect of Oil Treatment of Fertilizers on Crop Yields and Uptake of Fertilizer Constituents, *Agron. J.* **60**:241–242.

Tiarks, A. E., A. P. Mazurak, and Leon Chesnin, 1974, Physical and Chemical Properties of Soil Associated with Heavy Applications of Manure from Cattle Feedlots, *Soil Sci. Soc. Am. Proc.* **38**:826–830.

Turner, D. O., 1971, Disposing of Animal Wastes, *Crops Soils* **23**(5):10–11.

Voss, Regis, and John Pesek, 1967, Yield of Corn Grain as Affected by Fertilizer Rates and Environmental Factors, *Agron. J.* **59**:567–572.

Chapter 10

Nitrogen

Soils rarely contain enough nitrogen for maximum plant growth. The pale green color of nitrogen-deficient plants is the most common deficiency symptom exhibited by growing plants. Such plants respond dramatically to applications of nitrogen fertilizer. It is not surprising that nitrogen is the most widely applied fertilizer element.

The concentration of nitrogen in igneous rocks is so low that it is negligible for meeting plant needs. The atmosphere is 78 percent nitrogen, but this nitrogen cannot be used by higher plants until it is chemically combined with hydrogen, oxygen, or carbon. The process of combining nitrogen with another element is known as *nitrogen fixation*. Nitrogen fixation is accomplished in nature by certain microorganisms and by lightning, but the amount of nitrogen fixed is usually small and seldom as much as plants could use.

About 99 percent of the combined nitrogen in the soil is contained in the organic matter. The organic matter occurs mostly in either partly humified but still recognizable particles or in humus components associated with clay minerals (Swift and Posner, 1972). Both the humus content and the percent nitrogen in the humus are higher in the smaller size fractions. The close association between clay and humus slows the rate of decomposition of the humus.

The organic nitrogen in large, complex molecules is unavailable to higher plants and would remain so if it were not released by microorganisms. Microbial activity gradually breaks down the complex organic materials into simple inorganic ions that can be utilized by growing plants.

The potential rate of nitrogen use by growing plants generally exceeds the rate at which nitrogen becomes available. Consequently, the amount of available nitrogen in the soil is usually very small. Occasionally a soil may contain 100 kg/ha or more of available nitrogen, but the average amount is probably less than 30 kg/ha. The amount of organic nitrogen is much larger, somewhere near 3,000 kg/ha in an average furrow slice. The organic nitrogen can be considered as a reservoir of which between 1 and 5 percent is likely to become available each year in temperate climates and up to 50 percent under tropical conditions.

So much of the soil nitrogen is in organic forms that the nitrogen distribution in the soil profile is approximately the same as that of the organic matter. Distributions of organic matter in the profiles of two different soils are shown in Figure 5–6.

NITROGEN AND PLANT GROWTH

Most organic compounds in plants contain nitrogen. Among the nitrogen compounds are the amino acids, the nucleic acids, many enzymes, and energy transfer materials such as chlorophyl, ADP (adenosine diphosphate), and ATP (adenosine triphosphate). A plant cannot carry on its life processes if it lacks nitrogen to form these vital constituents.

Growing plants must have nitrogen to form new cells. Photosynthesis can produce carbohydrates from CO_2 and H_2O, but the process cannot go on to the production of proteins, nucleic acids, etc., unless nitrogen is available. Thus a severe shortage of nitrogen will halt the processes of growth and reproduction. Nitrogen deficiencies are among the many causes of stunted plant growth.

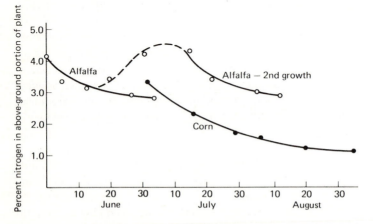

Figure 10-1 Nitrogen content of alfalfa and corn at different dates under Iowa conditions. *(Data from Drs. W. F. Wedin and J. J. Hanway, Iowa State Univ.)*

Rate of Nitrogen Uptake by Plants

Plants absorb nitrogen whenever they are actively growing but not always at the same rate. The amount of nitrogen absorbed per day per unit of plant weight is at a maximum when the plants are young and gradually declines with age. Nitrogen, therefore, constitutes a larger percentage of the dry weight of a young plant than of an older plant, as shown in Figure 10-1.

The decline in percent nitrogen shown in Figure 10-1 is really a dilution factor caused by the continued growth of the plant. The amount of nitrogen in the plant is still increasing as shown in Figure 10-2. This figure shows that nitrogen absorption runs relatively ahead of dry-matter accumulation during most of the life of the plant. Growth cannot get ahead of nitrogen uptake because the plant must have nitrogen on hand before it can make new cells. Also, plants can absorb extra nitrogen when it is available and store it to be used later if needed. The maximum rate of nitrogen absorption on an absolute basis usually comes during the early stages of vigorous growth.

Nitrogen Deficiency Symptoms

Plants grow slowly when nitrogen is deficient; they appear spindly, stunted, and pale when compared with healthy plants. The nitrogen deficiency limits the production of protein and other materials essential for the production of new cells. The rate of growth then becomes very nearly proportional to the rate at which nitrogen is supplied.

The pale green color of nitrogen-deficient plants results from a shortage of chlorophyl. Slow growth follows because chlorophyl is needed for carbohydrate production by photosynthesis.

Paleness caused by nitrogen deficiency is usually most pronounced in the older leaves, especially along the veins. Chlorophyl breaks down and disappears from these areas because it is not replaced. A yellowish-brown color begins along the veins at the tips of the older leaves and progresses inward. Part of the nitrogen from these areas is translocated and used in the parts of the plant that are still growing.

Excess Nitrogen

Plants can have too much nitrogen, especially if some other factor such as phosphorus, potassium, or water supply is inadequate. The rapid growth resulting from

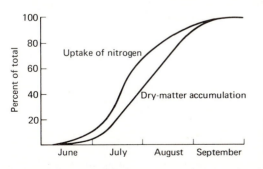

Figure 10-2 Nitrogen and dry-matter accumulation in corn. (*Redrawn from Hanway,* 1966.)

a high nitrogen level requires a good supply of all other requirements for plant growth.

An oversupply of nitrogen generally produces a dark green, succulent, vegetative growth. Vegetative growth may be at the expense of seed production in grain crops, fruit production in tomatoes and some tree crops, and sugar content in sugar beets. Too much nitrogen can cause potatoes to be watery. Perennial crops such as fruit trees may enter the cold season with such succulent growth that they are more susceptible than usual to frost damage if nitrogen is too abundant.

The negative effects of too much nitrogen on growing plants can be lessened if the phosphorus and potassium supplies are adequate for the rate of growth produced. High levels of phosphorus and potassium help to avoid the succulent growth and delayed maturity that can be caused by excess nitrogen.

Excess nitrogen may be the result of large fertilizer applications, or it may result from conditions under which the rate of release of soluble nitrogen by microbial activity exceeds the rate of use by growing plants. Excess nitrogen in the soil usually accumulates in the nitrate form. The nitrate levels reached in most soils are not nearly high enough to cause plants, animals, or people any problem where they are formed. Nitrates, however, are soluble; leaching can carry them to the groundwater table. Nitrogen released from soil organic matter, manure piles, etc., may also enter the groundwater. The nitrates are then transported to a low spot in the landscape where the water reappears at the surface in a spring, a seepy spot, a swamp, or a marsh. Sometimes the dissolved nitrogen reaches a shallow well or a stream that serves as a water supply for animals or people.

Water containing more than 45 ppm of nitrate is regarded unsafe for infants, though usually harmless to adults (Hanway et al., 1963). The nitrite form of nitrogen is much more toxic but is less commonly found in water supplies. Proximity to a silo, a place of disposal for organic wastes, or a reducing agent such as the zinc coating on galvanized iron increases the probability of nitrite contamination.

One particular type of reappearance of groundwater containing nitrates has proved to be a serious problem in a few places. The problem occurs where a pond is formed that has a fluctuating water level but lacks green plants. Vegetation may grow next to such a pond, then be killed by flooding when the water level rises. Later the water level drops, and the dead plants serve as nitrate accumulators. Water from the pond, which may already have an increased nitrate concentration because of evaporation, seeps through the soil into the roots and up the stems of the dead plants. The water evaporates and leaves nitrates in the dead stems and leaves. The nitrate concentration in such plants can be toxic to livestock if eaten.

In extremely arid regions it is possible for nitrates to accumulate on the soil surface as well as on dead plants. A large-scale example of this occurs in Chile. Rainfall in the Andes Mountains dissolves soil nitrates and seeps down into an extremely arid valley where the water evaporates. This evaporation has formed large beds of sodium nitrate that have been mined for use as nitrogen fertilizer.

FORMS OF NITROGEN

Organic nitrogen is combined in protein and other complex molecules. Much of it is present as amine groups ($-NH_2$), and the remainder is tied into ring or chain

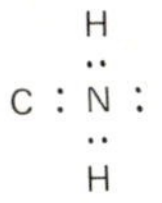

$$C : \overset{\displaystyle H}{\underset{\displaystyle H}{\overset{\cdot\cdot}{\underset{\cdot\cdot}{N}}}} :$$

Figure 10-3 The eight outer electrons of the nitrogen atom in an amine group provide a place to absorb an H^+ ion.

structures with carbon. Either way, the nitrogen is held by covalent bonding and cannot ionize. The organic matter must be at least partially decomposed before the nitrogen becomes available again for plant growth.

The nitrogen in an amine group has covalent bonds to one carbon and two hydrogen atoms (Figure 10-3). Six of the eight electrons in the outer shell are involved in these three bonds. Two electrons remain, and these can accept another hydrogen ion and bond to it. A positively charged site is thus formed. These positive sites can bond to negatively charged clay surfaces and thus help to stabilize soil structure. Such bonding also stabilizes the organic compound and makes it more resistant to decomposition. The positively charged sites that are not bonded to clay particles can serve as sites for anion exchange.

Microbial action gradually decomposes the organic matter. An amine group generally absorbs a fourth hydrogen when the carbon-to-nitrogen bond is broken. An amine group that already had a positive charge from absorbing a third hydrogen becomes an ammonium ion (NH_4^+) that enters the soil solution. A molecule of ammonia (NH_3) is released if the amine group was not positively charged, but the molecule soon absorbs a hydrogen ion and becomes an ammonium ion. This release of ammonium ions from decomposing organic materials is known as *ammonification*.

Ammonification is dependent on microbes and the enzymes they produce. Most microbes can decompose organic matter; ammonification therefore takes place rapidly when conditions favor microbial activity in general. Favorable conditions include a moist, warm soil well supplied with nutrients and organic matter. The process slows down when the soil becomes dry or the temperature drops toward freezing.

Ammonium ions are available to microbes and to higher plants. Some ammonium ions remain in the soil solution for a time, but most of them are adsorbed on cation-exchange sites. Adsorbed cations are difficult to leach, and most of them are held until they are either utilized by plants and microbes or oxidized to other forms. Some ammonium ions become trapped in crystal lattice positions such as the interlayer area of illite normally occupied by potassium ions. Ammonium ions and potassium ions are interchangeable because they are equal in size and both are monovalent.

Optimum conditions for microbial activity are favorable to *nitrification* as well as to ammonification. Nitrification is a two-stage oxidation process in which ammonia is oxidized to nitrite (NO_2^-) and the nitrite to nitrate (NO_3^-):

$$2NH_4^+ + 3O_2 \rightarrow 2NO_2^- + 2H_2O + 4H^+ + \text{energy}$$
$$2NO_2^- + O_2 \rightarrow 2NO_3^- + \text{energy}$$

Almost any microbe can carry out ammonification, but only a few specific ones carry out nitrification. *Nitrosomonas* sp. (and a few others) oxidize ammonia to nitrite and *Nitrobacter* sp. oxidize nitrite to nitrate. These bacteria are aerobic autotrophs; they obtain energy by oxidizing inorganic nitrogen compounds. They are widely distributed in soils, and so nitrification is seldom prevented by lack of bacteria. Nitrification slows or stops when the soil is too cold, too dry, or deficient in oxygen.

Soils rarely contain significant amounts of nitrite because it is immediately oxidized to nitrate. This is fortunate because concentrations of nitrite are toxic to plants. Nitrate is the end product of the reactions and is the principal form of nitrogen utilized by plants. For practical purposes the nitrite step may be omitted and nitrification considered to be the oxidation of ammonia to nitrate:

$$NH_4^+ + 2O_2 \rightarrow NO_3^- + H_2O + 2H^+ + energy$$

The H^+ ions released during nitrification have an acidic effect on the soil reaction (Chapter 8).

Temperature has a marked effect on all microbial processes including ammonification and nitrification. Figure 10-4 shows the relative rates at which nitrification takes place at different temperatures. It is very slow at temperatures near freezing and stops if the soil becomes frozen. The rate increases rapidly between 10 and 30°C. Nitrification declines rapidly above 35°C because nitrifying bacteria cannot tolerate such high temperatures.

Ammonification shows basically the same temperature response as nitrification except at the extremes. Ammonification is likely to be slowed even more than nitrification below 10°C. Thompson (1947) found that the rate of ammonification continues to increase above 35°C by the action of heat-tolerant bacteria. In fact, ammonia continues to be released at temperatures too high for any bacteria because organic matter decomposes chemically at high temperatures.

THE NITROGEN CYCLE

Nitrogen passes repeatedly through its various forms as it moves from the soil into the bodies of living organisms and back again. Figure 10-5 illustrates the complex open cycle in which nitrogen is involved. It is a cycle because the nitrogen can go around and around. It is open because nitrogen can enter or leave the cycle in various ways.

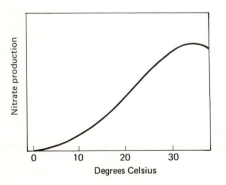

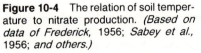

Figure 10-4 The relation of soil temperature to nitrate production. *(Based on data of Frederick, 1956; Sabey et al., 1956; and others.)*

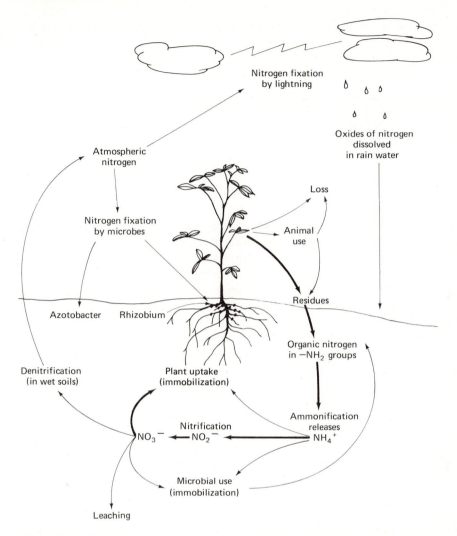

Figure 10-5 The nitrogen cycle. The darker lines indicate the main cycle of mineralization and immobilization.

The portion of the cycle in which organic matter is decomposed and inorganic ions are released is called *mineralization.* The reverse process, in which inorganic ions are converted to organic forms, is called *immobilization.* In the long run these two processes must balance, but in the short run one or the other may dominate for a time. The balance is best measured in terms of the amount of $NH_4^+ + NO_3^-$ in the soil. The production of NH_4^+ constitutes mineralization, but the NH_4^+ is very likely to be converted to NO_3^-, and so both must be measured.

Both higher plants and microorganisms assimilate (absorb) and immobilize NO_3^- and NH_4^+ ions. Immobilization by microorganisms occurs because the microorganisms must have nitrogen to build protein for their own bodies. They will use NO_3^- or NH_4^+ from the soil if there is insufficient nitrogen in the material they are decomposing.

The Significance of the Carbon-to-Nitrogen Ratio

Microbial action can either mineralize or immobilize nitrogen. The principal factor determining which occurs is the carbon-to-nitrogen (C:N) ratio. Microbes use carbon both for building body tissue and as an energy source. Nitrogen is required in a rather fixed ratio to the amount of carbon going into body tissue. It is shown in Chapter 5 that a C:N ratio of about 32:1 is the break-even point for decomposing organic materials in a few weeks time. Wider ratios cause some soil nitrogen to be immobilized, and narrower ratios permit mineralization to occur as the organic matter decomposes.

Nitrogen will eventually be mineralized even though the organic material added has a wide C:N ratio, but a lengthy waiting period is required. The wider the C:N ratio, the longer the period of net immobilization. The narrower the C:N ratio of freshly added decomposable materials, the sooner nitrogen will be mineralized. Decomposition with neither net mineralization nor immobilization indicates that ammonification from decomposing organic matter is equal to nitrogen use by microorganisms.

The relation between time and the C:N ratio is illustrated in Figure 10-6. This figure can be used as a guide for determining a suitable time interval between the time residues are mixed into a soil and the planting of the next crop. Competition for nitrogen between the crop and the soil microbes should be avoided because the crop will suffer until the microbial needs are satisfied. Nitrogen fertilizer should be added whenever such competition is likely to occur. The fertilizer nitrogen reduces the C:N ratio, and the immobilization period is shortened. The time period needed between residue incorporation and planting is shortened accordingly.

A 75-q corn crop will probably leave about 7,500 kg of cornstalks and other residues in the field. These residues should contain about 3,000 kg of carbon and 50 kg of nitrogen. Figure 10-6 indicates that such residues (60:1 C:N ratio) should be incorporated into the soil at 8 weeks before planting to escape the immobilization period. As an alternative, about 50 kg of fertilizer nitrogen may be incorporated along with the residues. This will reduce their C:N ratio to 30:1 and avoid the necessity of a time interval before planting.

The time periods given in Figure 10-6 are necessarily generalized. The rates of the biological processes involved depend on several factors such as water content of the soil, aeration, soil pH, and soil temperature. The rate of breakdown of the residues compared with the amount of residues added to the soil determines how long the time periods actually are in any given instance.

Rate of Mineralization

A C:N ratio narrower than 32:1 in decomposing residues indicates that net mineralization is probably taking place. It does not tell how fast. The rate of mineralization is very important when this source of nitrogen is relied upon to support plant growth.

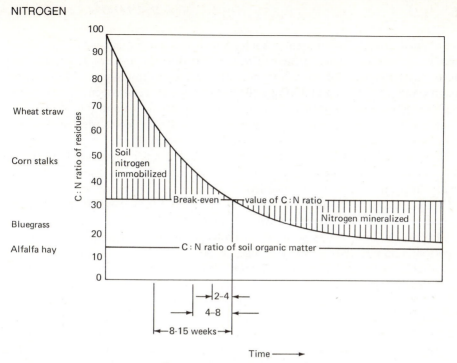

Figure 10-6 A schematic diagram showing the effect of C:N ratio on immobilization or mineralization of nitrogen. The time scale at the bottom indicates how much time should be allowed for residues to decompose before the next crop is planted if nitrogen fertilizer is not used.

Relation of Soil Moisture to Mineralization Microbial activity ceases when the soil water content is near the wilting point. It proceeds again almost immediately when the soil water is renewed. The rate of mineralization is most rapid shortly after the soil is remoistened and then slows down. Nitrate production following the moistening of a dry soil is illustrated in Figure 10-7. This curve results if the soil

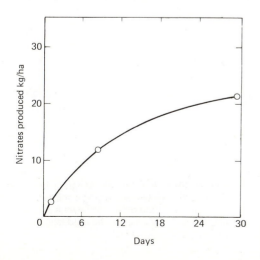

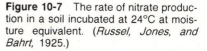

Figure 10-7 The rate of nitrate production in a soil incubated at 24°C at moisture equivalent. (*Russel, Jones, and Bahrt*, 1925.)

is maintained at an optimum water content after it is once moistened. The curve would be lowered somewhat if the soil were allowed to become dry again, but the pattern may be repeated if the soil is moistened again. Eventually, a soil that goes through cycles of wetting and drying is likely to release more nitrogen than one that is continuously moist. Wetting and drying cycles pump air in and out of the soil. This frequent renewal of the oxygen supply aids in the process of decomposition. Figure 10-8 illustrates the effect of a series of rains at intervals of 10 days to 2 weeks.

The significance of the rate of mineralization depends on the supply of plant nutrients from other sources. The importance of mineralization diminishes when nitrogen and other elements are added as fertilizer. Crop yields with adequate fertilizer are likely to be limited by the water supply. The highest yields will then be obtained by maintaining soil moisture near field capacity.

Mineralization is very important when nitrogen is in short supply. An extended dry period is doubly serious when nitrogen is deficient because the water shortage slows mineralization and aggravates the nitrogen shortage. Crops often show nitrogen deficiency symptoms during a prolonged dry period. Corn, for example, will "fire" (lower leaves turn yellow down the midrib and eventually die) under such conditions. Nitrogen starvation is the direct cause of the firing, but the moisture shortage is the basic cause. Firing may develop even when there is still available water in the B horizon. Mineralization nearly stops when the A horizon is dry because most of the organic matter is in the A horizon.

Nitrogen deficiency aggravates the water shortage caused by drought. Plants need nitrogen to build protein for roots as well as tops. Limited nitrogen means limited root development even when moisture is adequate. When dry weather comes, the plants can draw water only from their root zones. Those plants with shallow root systems because of nitrogen deficiency have little soil water within their reach.

During the first few years of cultivation of a soil the rate of mineralization may be so high that no fertilizer nitrogen is needed. When Carroll Brown of Rose Hill, Iowa, plowed under a 50-year-old stand of bluegrass and planted corn in 1948 he applied phosphorus fertilizer but no nitrogen. His corn yield was 141 q/ha. The crop probably used over 450 kg/ha of nitrogen. At least 5 or 6 percent of the organic nitrogen in the soil must have been mineralized during that 1 year. The usual percentage mineralized in climates such as Iowa's is only 2 or 3 percent per year. The bluegrass had built up a supply of readily decomposable organic matter. Aera-

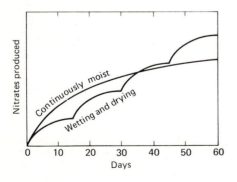

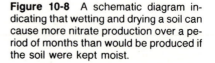

Figure 10-8 A schematic diagram indicating that wetting and drying a soil can cause more nitrate production over a period of months than would be produced if the soil were kept moist.

tion from cultivation and from the wetting and drying cycles of frequent rains during the 1948 season caused an unusually large percentage of that organic matter to decompose. Such effects nearly always occur during the early years of cultivation of a soil but not always to as great an extent as in Mr. Brown's field. Such rapid mineralization causes the soil organic-matter content to decline toward a new equilibrium level (as shown in Figure 5-11). What this new level will be depends on the cropping and management practices as well as the natural soil-forming factors.

It has been estimated that the soils of the United States have lost a third or more of their organic matter and of the plant nutrients stored in the organic matter in the last 100 years. The estimate may be correct, but it does not mean that another third will be lost in the next 100 years. The rate of mineralization declines as the organic matter declines. Many of the soils have reached a new equilibrium level. Mineralization taking place in these soils is completely offset by fresh additions of organic matter. However, the rate of mineralization is usually not adequate to provide enough plant nutrients for large crops. The data in Table 10-1 illustrate the decline in mineralization in six soils. Samples were taken from paired sites where virgin land lay next to old, established fields. Much less nitrogen was mineralized from the cultivated soils than from the virgin soils. The need for fertilizer nitrogen and several other nutrients increases as the rate of mineralization decreases. More fertilizer nitrogen is needed now than in decades past even to maintain the same yields.

Other Factors Affecting the Rate of Mineralization   Mineralized plant nutrients are by-products of microbial metabolism. Anything that affects the microbial action will also affect mineralization. Among the important factors are water, oxygen, pH, and temperature. These factors are discussed in Chapter 5.

Immobilization

Mineralization constitutes one-half of the closed portion of the nitrogen cycle, and immobilization is the other half. Plants and microbes absorb ammonium and nitrate ions and utilize them to build protein. The absorption part of the process is called *assimilation*. Immobilization includes assimilation and protein production so that the inorganic ions are made into building blocks of large organic molecules.

Table 10–1   Nitrogen Mineralized in 30 days at 35°C

Soil	Virgin, kg/ha[*]	Soil	Cultivated, kg/ha[*]
1	295	2	91
3	354	4	86
5	280	6	80
7	239	8	44
9	399	10	145
11	449	12	156

[*]Soils 1 and 2, 3, and 4, were adjacent soils.
Source: Computed from Thompson, 1947.

The relative availability of ammonium and nitrate ions to plants has been the subject of much research and debate. One form or the other may be advantageous for a specific plant species, but the usual conclusion is that plants can utilize whichever form is present. The nitrate form is generally utilized in largest amounts for two reasons. Nitrification acts rapidly when the weather is warm enough for rapid plant growth, and so the amount of ammonia present then is small. Also, most of the ammonium ions are held on cation-exchange sites. The nitrate ions are in the soil solution where they are more likely to be absorbed.

Losses from the Nitrogen Cycle

Nitrogen supply would not be a problem if there were no losses from the mineralization-immobilization cycle. However, there are so many losses that the cycle might be described as a leaky system. One of the biggest losses is from crop harvest. Large amounts of nitrogen are hauled from the field in grain crops, hay crops, etc. A quintal of corn removes about 1.7 kg of nitrogen if only the grain is harvested and about 3 kg if the whole plant is removed for silage. A quintal of wheat removes about 2 kg of nitrogen. Table 10-2 indicates the approximate amounts of nitrogen removed by different crops. These are average values. The nitrogen content of a crop varies considerably with the fertility status of the soil and other factors discussed in Chapter 15. Only a small portion of this nitrogen is returned to the fields. Even a pasture with the manure returned suffers a net removal of about half of the nitrogen in the forage.

Erosion Losses   Every ton of soil lost from a field carries nitrogen with it. For example, 4 percent organic matter in the soil amounts to 40 kg per ton. About 2 kg (one-twentieth) of this is nitrogen. The nitrogen loss can be estimated as half of the product obtained by multiplying the soil loss in tons by the organic-matter content in percent. A 10-ton loss of soil containing 4 percent organic matter would amount to 20 kg of nitrogen. Soil losses of 10 tons per hectare are more than should be tolerated but are not nearly as large as the actual losses on some fields. Methods of conserving soil are discussed in Chapter 19.

Leaching Losses   Nitrate nitrogen is easily leached from permeable soil. Professor A. J. Sterges[1] found that a single rain at College Station, Tex., moved nitrates from the soil surface to a depth of 60 cm during a 24-h period. Additional rain could move it beyond the reach of plant roots. Among the several factors that affect leaching losses are (1) the rate of nitrification (and fertilization), (2) the amount of rain, (3) the permeability and water-holding capacity of the soil, and (4) the crop growing on the soil.

Leaching periodically removes most of the nitrate nitrogen from the profiles of permeable soils in humid regions. Soils of drier regions do not suffer leaching losses where water does not penetrate beyond the depth of the solum. But the nitrate nitrogen can be removed to a lower part of the solum (Troeh, 1952). Older plants

---

[1]Personal communication.

Table 10-2 Average Amounts of Nitrogen Removed by Crops

Crop	Part of crop	Yield, tons/ha	% N	N, kg/ha
Alfalfa	Hay	9	2.4	216
Barley	Grain	3	1.9	57
Corn	Grain	7	1.4	98
Cotton	Lint	0.5		
	Seed	1.0	3.7	37
	Stalks, leaves	1.5	1.8	27
Cowpeas	Hay	5	3.0	150
Lespedeza	Hay	8	2.1	168
Oats	Grain	3	1.8	54
	Straw	3	0.6	18
Peanuts	Nuts	2.5	3.0	75
	Vines	8	1.6	128
Potatoes	Tubers	20	0.4	80
	Tops			70
Red clover	Hay	5	1.9	95
Soybeans	Grain	2.5	6.0	150
	Straw	3	0.6	18
Sugar beets	Roots	40	1.8	72
	Tops			80
Sweet clover	Hay	12	2.6	312
Sweet potatoes	Roots	20	0.3	60
	Tops		2.0	35
Timothy	Hay	4	1.1	44
Tobacco	Leaves	1.5	3.6	54
	Stalks			25
Tomatoes	Fruit	25	0.3	75
	Vines			50
Wheat	Grain	3	2.1	63
	Straw	3	0.6	18

Source: Calculated from various sources.

can use nitrogen from the B horizon, but young plants with small root systems need to have available nitrogen in the upper part of the soil.

Growing plants often keep the nitrate-nitrogen supply depleted to the point that little is lost by leaching. Table 10-3 illustrates the effectiveness of vegetation in conserving nitrogen. Even so, it is possible for mineralization to exceed plant needs at certain seasons of the year. Soils that are subject to cool, moist spring seasons are likely to be leached at that time. The nitrogen supply becomes an important factor limiting plant growth in the summer on some of these same soils.

The practicality of applying fertilizer at a time when there is not too much other work to do can be important. Some farmers apply nitrogen fertilizer in the fall for a crop to be planted the following spring. Leaching losses from such applications can be considerable if the nitrate form of nitrogen is present. The leaching hazard can be markedly reduced by applying the ammonium form of nitrogen after the topsoil temperature is below 5°C. Nitrification is slow at low temperatures and

Table 10-3   Losses of Nitrogen under Different Systems of Management. Lysimeter Studies on Dunkirk Silty Clay Loam.

Management practice	Annual nitrogen losses, kg/ha
Bare	69.6
Rotation	7.0
Grass	2.2

Source: Bizzell and Lyon, 1927.

ceases when the ground is frozen (Figure 10-4). Most of the nitrogen will still be in the ammonia form through the winter if freezing weather comes soon after ammonia is applied.

Volatilization   There are several volatile forms of nitrogen including molecular nitrogen (N_2), ammonia (NH_3), and the various oxides of nitrogen (NO, N_2O, etc.).

Ammonia applied to the soil as a fertilizer absorbs H^+ ions and forms NH_4^+ ions that can be adsorbed by cation exchange. Ammonium ions from NH_4NO_3 or $(NH_4)_2SO_4$ fertilizers are likewise adsorbed and thus held against loss. Fenn and Kissel (1976) found that ammonia volatilization is minimal from ammonium fertilizer incorporated into soils having high cation-exchange capacity, high soil moisture content, and low pH. Mills, Barker, and Maynard (1974) showed that growing plants also help reduce ammonia volatilization. Their results show that large losses can occur when high rates of nitrogen are applied to alkaline sandy soil without any plant growth (Figure 10-9). This specific combination of factors is rare enough that ammonia volatilization from soil is seldom much of a problem. Significant losses of ammonia are more likely to occur from a manure pile, or from manure left on the soil surface.

Denitrification is a volatilization process causing nitrogen to be lost from soils that are saturated with water part of the time or in part of their volume. The aerated soil produces nitrates by normal nitrification processes. The nitrates are later reduced to volatile products by anaerobic bacteria in the water-logged soil. Van Cleemput (1971) found that wet soil lost nitrates rapidly producing gaseous nitrogen in alkaline soils and nitrogen oxides in acid soils. Garcia (1975) found that most of the denitrification occurring in a rice paddy takes place in the rhizosphere. He suggested that the rhizosphere contributed to denitrification by developing anaerobic zones around the roots, producing root exudates, and increasing the population of denitrifiers. Other kinds of plant roots probably have similar effects.

Losses by denitrification are not thought to be very large in well-drained soils. They may be fully offset by natural nitrogen fixation by lightning and by certain microbes. In wet areas, however, denitrification losses may be quite high. The addition of 50 kg or more of nitrogen fertilizer per hectare on spots where the soil has been saturated for a considerable time can make the difference between crop failure and success. It is estimated from crop response to nitrogen fertilizer on such

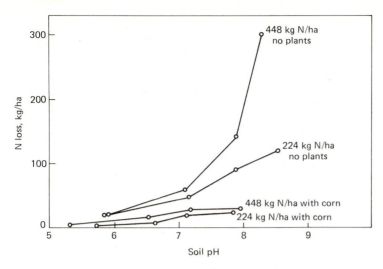

Figure 10-9   The amount of nitrogen volatilized as ammonia at various pH values with and without corn plants at two different rates of fertilization with NH_4Cl. The soil was a fine sandy loam containing 4 percent clay and 4.7 percent organic matter. (*Based on data of Mills, Barker, and Maynard,* 1974.)

spots that denitrification may account for nitrogen losses up to 100 kg/ha in a single season. Denitrification losses can often be drastically reduced by improved soil drainage.

Nitrogen Sources

The soil inherits from its parent rock its initial supply of almost all the nutrient elements it provides for plant growth. Rock weathering replenishes the nutrient-ion supply as time passes. But rock weathering, especially igneous rock weathering, does little or nothing for the supply of available nitrogen. Igneous rocks contain very low concentrations of this element. Sedimentary rocks may contain more nitrogen, but they inherit it from an earlier soil or from some living thing.

Almost all soil nitrogen comes from living things, but they are a secondary source. The primary nitrogen source is the atmosphere. The original soil nitrogen was probably fixed by lightning and carried to the soil in rainwater. This process continues to act but probably accounts for only 5 to 10 kg of nitrogen per hectare per year, an amount that will produce only a small amount of plant growth. Local areas, however, may receive larger amounts of soluble airborne nitrogen. The Agricultural Research Service (1970) reported that a Colorado lake about 2 km downwind from a large feedlot received annually about 33 kg of ammonium nitrogen per hectare. This report suggested that "airborne ammonia from cattle feedlots near lakes and rivers may contribute more nitrogen enrichment to those bodies than runoff and deep percolation from the same sources."

Fixation of Nitrogen by Living Organisms   A few living organisms are able to utilize molecular nitrogen (N_2) from the atmosphere. The best known of these are the Rhizobia ("legume bacteria"), the free-living bacteria such as Azotobacter and

Clostridium, and the blue-green algae. Nitrogen fixation by Rhizobia is a symbiotic process and will be discussed in the next section.

Azotobacter bacteria are aerobic heterotrophs. They obtain energy by the oxidation of dead organic matter. They have the ability to utilize N_2 from the air if the organic matter does not supply enough nitrogen to meet their needs. These bacteria are a part of the soil organic matter, and the nitrogen in their bodies can be mineralized after they die. The Clostridium bacteria function in much the same way but in an anaerobic environment. However, the amount of nitrogen fixed by these and other nonsymbiotic bacteria is not very large, and the conversion efficiency is low. About 100 kg of carbon is used in respiration for every kilogram of nitrogen fixed (Waksman and Starkey, 1931). Free-living bacteria probably fix about 10 kg of nitrogen per hectare per year in soils low in nitrogen, providing there is a supply of decomposable organic matter, adequate moisture, and a pH of 6.5 or above. Fixation drops to zero if the nitrogen supply in the soil is already adequate.

Bacteria of the genus Beijerinckia fix nitrogen, especially where sugar cane is grown (Becking, 1974). Beijerinckia do best in acid soils with pH between 5.0 and 6.4 and seem to have chemical needs related to tropical soils. Their occurrence in temperate regions is negligible.

Nitrogen fixation is also accomplished by several genera of facultative anaerobes. Koch and Oya (1974) indicate that Klebsiella, Enterobacter, and Achromobacter fixed up to 32 kg of nitrogen per hectare annually in some Hawaiian pasture soils. These bacteria can live aerobically but fix nitrogen only under anaerobic conditions.

Koch and Oya indicate that various nitrogen-fixing bacteria are most numerous on or near root surfaces. It is likely that the bacteria utilize root exudates from the plants and are therefore stimulated by being near growing roots. Several researchers are trying to find some kind of bacteria that can fix significant amounts of nitrogen in association with grass-type crops such as wheat and corn. Such bacteria could have great impact in reducing the nitrogen-fertilizer requirements of grass-type crops.

Blue-green algae live in wet environments such as ponds or wet soils. Mitchell (1970) regards them as the most autotrophic of all living things because they can both carry on photosynthesis and utilize atmospheric nitrogen. Their contribution to the nitrogen cycle may be significant in rice culture with a limited nitrogen supply. It is probably insignificant to most other crops.

Blue-green algae can also fix nitrogen in drier environments when living in association with fungi as lichens. Rychert and Skujins (1974) reported that lichen crusts become active in desert grasslands after a rain and provide a major input of available nitrogen. They estimated annual nitrogen fixation under some desert conditions at 10 to 100 kg/ha. Optimum conditions were soil moisture near field capacity, temperature near 20°C, good light, and a ready source of carbon. The ability of lichens to remain dormant for long periods of time and yet grow during favorable periods is important in the harsh conditions of the desert. Nitrogen deficiency would limit or prevent lichen growth if they did not have the facility to fix nitrogen. The same abilities that adapt lichens to desert growth permit them to invade the harsh environment of barren rock surfaces, thus providing a toehold for plant life and soil formation.

Symbiotic Nitrogen Fixation   Several different pairs of organisms are able to utilize atmospheric nitrogen (N_2) through symbiotic (mutually beneficial) relations. Generally one organism contains chlorophyl and uses light energy to produce carbohydrates. The other organism receives some of these carbohydrates and uses them as an energy source in the nitrogen fixation process. Some of the nitrogen compounds seep back to the first organism, and so there is mutual benefit.

The best known examples of symbiotic nitrogen fixation are the relations between legumes (alfalfa, clovers, peas, beans, locust trees, etc.) and various members of the Rhizobium genus of bacteria. These bacteria form nodules (Figure 10-10) on the legume roots and carry on nitrogen fixation there.

The amount of nitrogen fixed by Rhizobium bacteria varies with the carbohydrate supply in the plant and the available nitrogen supply in the soil. The bacteria need the carbohydrates for energy to fix nitrogen. However, they will not fix much nitrogen when it is readily available in the soil even if the carbohydrate supply is high. In addition, the right species of bacteria for the particular legume must be present. A species that nodulates alfalfa roots and fixes nitrogen will not serve for a soybean crop, nor vice versa. The simplest way to overcome this problem is to add

Figure 10-10   Nitrogen is taken from the air by bacteria which live in the nodules on the roots of alfalfa, clover, and other legume plants. *(Soil Conservation Service, USDA.)*

a suitable strain of Rhizobium when a new legume seeding is made. The bacteria can be applied to the seed in a simple process known as *inoculation*.

Legume-Rhizobium combinations can fix sizable amounts of nitrogen. The exact amounts contributed to the soil are variable and difficult to determine precisely. Some reasonable estimates based on what determinations have been made are given in Table 10-4. The potential is at least twice as high as the values given in Table 10-4, but there is usually some nitrogen available in the soil or some other condition unfavorable to maximum fixation. Agboola and Fayemi (1972) reported fixation rates up to 370 kg/ha by calopo (a tropical legume) growing in unfertilized soil and up to 450 kg/ha in a nutrient culture.

The amount of nitrogen fixed by Rhizobia decreases when nitrogen is available in the soil but does not go to zero even when fertilizers are used. The nitrogen obtained by growing legumes is very important when fertilizers are unavailable or prohibitively expensive. It has often been shown that corn yields of 60 q/ha can be obtained without fertilizer by preceding the corn with 2 years or more of alfalfa. Comparable results can be obtained with other grain crops.

Growing a legume does not automatically ensure a net addition of nitrogen to the soil. Many things can go wrong. Growth conditions may be unfavorable, or the right strain of Rhizobium may not be present. The harvested crop may remove more nitrogen than is fixed. A general rule of thumb is that the amount of nitrogen in the aboveground portion of the legume approximately equals the amount of nitrogen fixed. An amount equal to the nitrogen contained in the roots was probably withdrawn from the soil. If all the tops are removed for hay and nothing is returned, there will be no net addition.

Usually some nitrogen is added to the soil by legumes because some residues are left in the field and, of course, manure may be returned. Even so, it is advantageous to incorporate considerable plant material into the soil when it is desired to increase the nitrogen supply of the soil by growing legumes. Green material disked in or plowed under is known as *green manure*. The benefit the next crop receives from green manure will probably exceed the value of the lost hay crop when nitrogen is in short supply. The nitrogen percentage is highest if the green manure crop is still young and succulent when it is incorporated into the soil.

As indicated in Table 10-4, fairly large amounts of nitrogen can be fixed by Rhizobia on legume roots. Nevertheless, the importance of this source of nitrogen

Table 10-4   Estimated Amounts of Nitrogen Fixed during 1 year by Well-nodulated Legumes

Crop	Nitrogen fixed, kg/ha
Alfalfa	150
Sweet clover	120
Red clover	90
Soybeans	60
Field beans	25
Field peas	25

has declined. Methods of producing nitrogen fertilizer have been revolutionized during the last few decades. The price of fertilizer nitrogen dropped drastically in the United States and in many other countries during the 1950s and 1960s. Farmers therefore began buying more and more nitrogen fertilizer and using less and less green manure. Some have even applied nitrogen fertilizer to benefit legume crops such as soybeans. This reduces the amount of nitrogen fixed by the legume bacteria but may increase yields more than enough to pay for the fertilizer.

Fertilizer nitrogen is a very important addition to the nitrogen cycle where it is used. Not only can it be applied in large amounts, it can also be controlled in terms of both amount and time of application.

NITROGEN FERTILIZERS

Nitrogen fertilizers may be either mineral or organic. The organic forms have by far the longest history of use. They include the application of animal manure, a practice so old no one can be sure when it was first used. Another such practice was the American Indian method of placing a fish in each hill of corn. Nitrogen and other nutrients were released as the fish decayed. The compost piles utilized by some gardeners are another source of organic nitrogen fertilizer. The value of the compost is, of course, dependent on the materials placed in it and is therefore extremely variable.

A gradual release of nitrogen is a claimed advantage for the organic nitrogen fertilizers. The release process is the same as for plant residues grown on the soil. The merits of slow release depend largely on when the nitrogen is wanted. Gradual release is advantageous when one large application must last for a long time, possibly for turfgrass. It is a disadvantage when the nitrogen is needed fast, perhaps for a grain crop.

Inorganic Nitrogen Fertilizers

Mineral forms are frequently assumed when the word *fertilizer* is used. They are usually cheaper, more concentrated, easier to haul and spread, and more rapidly available to the plant than are the organic forms.

Most of the early inorganic nitrogen fertilizer in the United States was sodium nitrate ($NaNO_3$) imported from Chile. This is a naturally occurring material containing about 16 percent N. Crop response is rapid and dramatic owing to the immediate availability of the nitrate form of nitrogen. But sodium nitrate is little used now because of several difficulties. The Chilean nitrate beds are large but they could not meet the world's needs for nitrogen for long. Also, 16 percent is a relatively low analysis for a fertilizer that must be shipped a long distance. The shipping cost detracts from its value. Furthermore, the repeated use of large applications of $NaNO_3$ could result in the formation of sodic soils.

The limitations of a fixed amount of natural nitrate deposits were overcome by the development of commercial methods of fixing nitrogen from the air. The first such method to become important in the United States was the production of calcium cyanamide ($CaCN_2$) by reacting nitrogen gas with calcium carbide at a high temperature. Pure calcium cyanamide contains 35 percent N, so it is a much more concentrated fertilizer material than sodium nitrate. It was used as a fertilizer for a number of years but has since been entirely replaced by synthetic ammonia and products made from ammonia.

Ammonia is now the source of most nitrogen fertilizer. It is the cheapest and the most concentrated form of combined nitrogen available. It can be produced in any needed quantity. Probably it would dominate the market even more completely than it does were it not for the special equipment and handling techniques required for its use.

Ammonia is produced in the Haber-Bosch process by the direct reaction of nitrogen and hydrogen. A mixture of the two gases is subjected to a pressure between 150 and 1,000 atm and a temperature of about 500°C in the presence of an iron catalyst. This procedure converts about one-fourth of the nitrogen and hydrogen to ammonia, which is then liquefied and removed. The remaining hydrogen and nitrogen are recycled and eventually converted to ammonia. Most of the hydrogen for this process comes from natural gas. Partly for this reason, many petroleum companies are involved in the fertilizer business.

Ammonia is 82 percent N. It is an odoriferous gas, but it can be liquefied under pressure or dissolved in water. The pure form is known as *anhydrous ammonia* and must be stored and transported under pressure. It is injected into the soil 10 to 20 cm beneath the surface so that it can be dissolved in the soil water and adsorbed on cation-exchange sites (soils with unusually low cation-exchange capacities may not be able to hold much ammonia). The best soil moisture content is near field capacity. The soil should be friable and not cloddy. Large openings in the soil will permit some of the ammonia to escape to the atmosphere. Such escape is easily detected by the ammonia odor.

Liquefied ammonia will mix with water. Adding water reduces the amount of pressure required to keep the ammonia from vaporizing. Solutions containing less than 20 percent N can be handled without pressure and may be sprayed on the soil; more concentrated solutions should be injected. Solutions of ammonia in water are commonly known as *aqua ammonia*.

Many nitrogen compounds can be made from ammonia. Ammonia reacts with acids to form salts. For example, the reaction with sulfuric acid forms ammonium sulfate:

$$2NH_3 + H_2SO_4 \rightarrow (NH_4)_2SO_4$$

Ammonium sulfate contains 21 percent N. It is especially valuable where sulfur is also likely to be deficient.

Ammonia can be oxidized to form nitric acid:

$$NH_3 + 2O_2 \rightarrow HNO_3 + H_2O$$

Ammonium nitrate is formed by reacting ammonia with nitric acid:

$$NH_3 + HNO_3 \rightarrow NH_4NO_3$$

Ammonium nitrate in pure form contains 35 percent N (impurities reduce the analysis to 33 or 34 percent in fertilizer grade) and is one of the most popular forms of solid nitrogen fertilizer.

The most concentrated form of solid nitrogen fertilizer available is urea, a simple organic compound containing 46 percent N. Synthetic urea can be formed by reacting ammonia with carbon dioxide:

$$2NH_3 + CO_2 \rightarrow (NH_2)_2CO + H_2O$$

Joint applications of nitrogen and phosphorus make ammonium phosphates logical materials to apply. Two forms are available:

$$NH_3 + H_3PO_4 \rightarrow NH_4H_2PO_4 \qquad \text{Monoammonium phosphate}$$
$$2NH_3 + H_3PO_4 \rightarrow (NH_4)_2HPO_4 \qquad \text{Diammonium phosphate}$$

Fertilizer grades of monoammonium phosphate contain 11 percent N and 21 percent P; diammonium phosphate contains 18 percent N and 20 percent P.

Another possibility is the combination of potassium and nitrogen in potassium nitrate (KNO_3) containing 13 percent N and 38 percent K. Potassium nitrate has been little used as a fertilizer in the past, but some companies are now producing it.

Calcium nitrate [$Ca(NO_3)_2$] is sometimes used as a fertilizer, particularly in Europe. It contains about 17 percent N, and the calcium content is favorable for maintaining a desirable soil pH.

All the nitrogen fertilizer materials discussed herein are soluble in water. Ammonium nitrate and urea are sometimes applied to the soil as a water solution.

Which Nitrogen Fertilizer Is Best?

All the inorganic forms of nitrogen in fertilizer are readily available for plant growth; all the organic forms become available as they decompose. Each form is equally beneficial to plants if it is supplied at the right time. Even so, some advantages and disadvantages pertain to the various forms.

Anhydrous ammonia is both the highest analysis and the cheapest form of nitrogen fertilizer and is therefore likely to be chosen unless its application is too difficult or too costly in a particular circumstance. It must be injected into the soil because of its gaseous nature. There is a possibility of some of the ammonia being lost by volatilization if the soil fails to seal properly behind the applicator or if the soil is too alkaline to convert the NH_3 to NH_4^+ ions.

Nitrogen fertilizers containing nitrates are the most rapidly available to plants because the nitrate ion readily enters the soil solution and is not absorbed on the cation-exchange complex. Nitrates are subject to loss when leaching occurs and are subject to loss by denitrification under anaerobic conditions. These losses cannot always be avoided by the use of other forms of nitrogen fertilizer because nitrification usually changes them to nitrate.

Urea has the advantage of being the highest analysis solid nitrogen fertilizer generally available. It is highly soluble so it can be applied in a water solution as well as in the solid form. Urea should be incorporated into the soil promptly because it hydrolyzes and releases ammonia:

$$(NH_2)_2CO + H_2O \rightarrow 2NH_3 + CO_2$$

Considerable research has been done with coatings to slow the rate of release of nitrogen fertilizers. The favorite slow-release form has been sulfur-coated urea. Volk and Horn (1975) found that sulfur-coated urea was still releasing available nitrogen 100 days after its application. Slow release is often considered advantageous for keeping turfgrass green and healthy over an extended period of time. Slow release usually has little or no advantage on most field crops because they need nitrogen faster and respond well to rapid-release fertilizers.

A significant side effect of nitrogen fertilizers containing either the ammonium or the amine form was mentioned in Chapter 8. Nitrification changes the nitrogen from a basic (ammonium) form or a neutral (amine) form to an acidic (nitrate) form. This acidifying effect generally calls for an addition of lime to offset it. Applying lime once every few years is usually more economical than avoiding these fertilizers.

Fall Applications of Nitrogen

Fall is often the most convenient time to apply fertilizer even for spring-planted crops such as corn. Dry soil in the fall can support a spreader truck better than wet soil in the spring. Also, time and labor may be more available in the fall and fertilizer may be on sale at a discount. There is some question as to how much nitrogen might be lost by leaching. The loss will be minimal if the nitrogen can be stored as ammonium ions held by cation exchange. One way to keep ammonia in the ammonium form is to apply it when the soil temperature is below 5°C in soils that will soon be frozen. Even then, most of the ammonia is nitrified by planting time (68 to 86 percent in an experiment by Chalk, Keeney, and Walsh, 1975).

Nitrification can be delayed for 30 to 90 days by adding an inhibitor named N-serve to the ammonia tank (Anonymous, 1976). Combined with cold weather, N-serve can make the fall application of ammonia equivalent to a spring application (Boswell and Anderson, 1974).

How Much Nitrogen to Apply

Rates of nitrogen fertilization commonly range from a low of 20 or 30 kg/ha to a high of a few hundred kilograms per hectare. Soil tests for evaluating the available nitrogen in the soil have proven to be less satisfactory than the tests for other nutrients. Available nitrate nitrogen moves about readily as the soil wets and dries, so its position is unreliable. Also, the rate of supply of available nitrogen is more important than the amount in the soil at any one time. The best nitrogen tests require that both surface soil and subsoil samples be tested. The tests are usually designed to measure the nitrogen-supplying capacity rather than the available nitrogen content. Therefore, they begin by leaching all available nitrogen from a soil sample. The sample is then incubated for several days to allow mineralization of nitrogen to occur. The soil is leached again and the amount of either NO_3^- or NH_4^+ released by mineralization is determined by chemical analysis. The results are rated as low, medium, or high and used as a basis for fertilizer recommendations as discussed in Chapter 9.

Most soil-testing laboratories base nitrogen recommendations on a balance sheet approach instead of soil tests. They take into consideration such factors as cropping history, yield level, previous fertilizer applications, and type of soil. This approach saves considerable time because it avoids the necessity of incubating the soil samples.

REFERENCES

Agboola, A. A., and A. A. A. Fayemi, 1972, Fixation and Excretion of Nitrogen by Tropical Legumes, *Agron. J.* **64**:409–412.

Agricultural Research Service, 1970, Airborne Ammonia Eutrophies Lakes, *Agric. Res.* **19**(2):8–9.

Anonymous, 1976, N-Serve Infrared Photos and Purdue Studies, *Fert. Prog.* **7**(1):18–19.

Ashley, D. A., O. L. Bennett, B. D. Doss, and C. E. Scarsbrook, 1965, Effect of Nitrogen Rate and Irrigation on Yield and Residual Nitrogen Recovery by Warm-Season Grasses, *Agron. J.* **57**:370–372.

Becking, J. H., 1974, Nitrogen-fixing Bacteria of the Genus *Beijerinckia, Soil Sci.* **118**:196–212.

Bizzell, J. A., and T. L. Lyon, 1927, Composition of Drainage Waters from Lysimeters at Cornell University, *Proc. Int. Congr. Soil Sci.* **2**:342–349.

Boswell, F. C., and O. E. Anderson, 1974, Nitrification Inhibitor Studies of Soil in Field-buried Polyethylene Bags, *Soil Sci. Soc. Am. Proc.* **38**:851–852.

Broadbent, F. E., and T. Nakashima, 1968, Plant Uptake and Residual Value of Six Tagged N Fertilizers, *Soil Sci. Soc. Am. Proc.* **32**:388–392.

Brown, J. R., and G. E. Smith, 1966, Soil Fertilization and Nitrate Accumulation in Vegetables, *Agron J.* **58**:209–212.

Chalk, P. M., D. R. Keeney, and L. M. Walsh, 1975, Crop Recovery and Nitrification of Fall and Spring Applied Anhydrous Ammonia, *Agron. J.* **67**:33–37.

Colman, R. L., and Alec Laxenby, 1975, Effect of Moisture on Growth and Nitrogen Response by *Lolium Perenne, Plant Soil* **42**:1–13.

Dancer, W. S., and L. A. Peterson, 1969, Recovery of Differentially Placed NO_3-N in a Silt Loam Soil by Five Crops, *Argon. J.* **61**:893–895.

Ensminger, L. E., and R. W. Pearson, 1950, Soil Nitrogen, *Adv. Agron.* **2**:81–111.

Fenn, L. B., and D. E. Kissel, 1976, The Influence of Cation Exchange Capacity and Depth of Incorporation on Ammonia Volatilization from Ammonium Compounds Applied to Calcareous Soils, *Soil Sci. Soc. Am. J.* **40**:394–398.

Frederick, L. R., 1956, The Formation of Nitrates from Ammonia in Soils. I. Effect of Temperature, *Soil Sci. Soc. Am. Proc.* **20**:496–500.

Garcia, J. L., 1975, Effet Rhizosphere du Riz sur la Denitrification, *Soil Biol. Biochem.* **7**: 139–141.

Gersberg, Richard, Kenneth Krohn, Neal Peek, and C. R. Goldman, 1976, Denitrification Studies with ^{13}N-Labeled Nitrate, *Science* **192**:1229–1231.

Hanway, J. J., 1966, How a Corn Plant Develops, *Iowa State Univ. Coop. Ext. Ser. Spec. Rept.* 48.

Hanway, J. J., J. B. Herrick, T. L. Willrich, P. C. Bennett, and J. T. McCall, 1963, The Nitrate Problem, *Iowa State Univ. Coop. Ext. Ser. Spec. Rept.* 34, 20 p.

Heilman, M. D., J. R. Thomas, and L. N. Namken, 1966, Reduction of Nitrogen Losses under Irrigation by Coating Fertilizer Granules, *Agron. J.* **58**:77–80.

Keeney, D. R., and J. M. Bremner, 1966, Comparison and Evaluation of Laboratory Methods of Obtaining an Index of Soil Nitrogen Availability, *Agron. J.* **58**:498–503.

Koch, B. L., and Jean Oya, 1974, Non-symbiotic Nitrogen Fixation in Some Hawaiian Pasture Soils, *Soil Biol. Biochem.* **6**:363–367.

Mills, H. A., A. V. Barker, and D. N. Maynard, 1974, Ammonia Volatilization from Soils, *Agron. J.* **66**:355–358.

Mitchell, R. L., 1970, *Crop Growth and Culture,* Iowa State Univ. Press, Ames.

Morrill, L. G., and J. E. Dawson, 1967, Patterns Observed for the Oxidation of Ammonium to Nitrate by Soil Organisms, *Soil Sci. Soc. Am. Proc.* **31**:757–760.

Perkins, H. F., and A. G. Douglas, 1965, Effects of Nitrogen on the Yield and Certain Properties of Cotton, *Agron. J.* **57**:383–384.

Russel, J. C., E. G. Jones, and G. M. Bahrt, 1925, The Temperature and Moisture Factors in Nitrate Production, *Soil Sci.* **19**:381–398.

Rychert, R. C., and J. Skujins, 1974, Nitrogen Fixation by Blue-Green Algae-Lichen Crusts in the Great Basin Desert, *Soil Sci. Soc. Am. Proc.* **38**:768–771.

Sabey, B. R., W. V. Bartholomew, Robert Shaw, and John Pesek, 1956, Influence of Temperature on Nitrification in Soils, *Soil Sci. Soc. Am. Proc.* **20**:357–360.

Stewart, B. A., L. K. Porter, and D. D. Johnson, 1963, Immobilization and Mineralization of Nitrogen in Several Organic Fractions of Soil, *Soil Sci. Soc. Am. Proc.* **27**:302–304.

Swift, R. S., and A. M. Posner, 1972, The Distribution and Extraction of Soil Nitrogen as a Function of Soil Particle Size, *Soil Biol. Biochem.* **4**:181–186.

Thompson, L. M., 1947, *Mineralization of Organic Phosphorus in Clarion and Webster Soils,* master's thesis, Iowa State University, Ames.

Thompson, L. M., 1950, *The Mineralization of Organic Phosphorus, Nitrogen and Carbon in Virgin and Cultivated Soils,* doctoral dissertation, Iowa State University, Ames.

Troeh, F. R., 1952, *Nitrogen Mobility in Three North Idaho Soils as Influenced by Management Practices,* master's thesis, University of Idaho, Moscow, 47 p.

Van Cleemput, O., 1971, Etude de la Denitrification Dans le Sol, *Pedologie* **21**:367–376.

Volk, G. M., and G. C. Horn, 1975, Response Curves of Various Turfgrasses to Application of Several Controlled-release Nitrogen Sources, *Agron. J.* **67**:201–204.

Waksman, S. A., and R. L. Starkey, 1931, *The Soil and the Microbe,* Wiley, New York.

Chapter 11

Phosphorus

Phosphorus has been called "the key to life" because it is directly involved in most life processes. It is a component of every living cell and tends to be concentrated in seeds and in the growing points of plants.

Phosphorus is second only to nitrogen in frequency of use as a fertilizer element. One or both of these elements are nearly always included when a fertilizer is applied. The phosphorus supply can be even more critical than the nitrogen supply in some natural environments. Certain microbes can make atmospheric nitrogen available to plants, but the initial phosphorus supply must come from rocks.

Phosphorus, like nitrogen and sulfur, forms complex anions with oxygen, but the phosphates are of low solubility. The low availability of nearly insoluble phosphates is a disadvantage. An entire hectare-furrow slice generally contains less than 1 kg of phosphorus in solution (out of a total phosphorus content of about 1,000 kg). In one respect, though, the low solubility is an advantage. It helps keep the leaching losses small.

Phosphorus occurs in the soil in both inorganic and organic forms. The phosphorus of organic matter is tied into the structure of the compounds. Such phosphorus may participate in chemical reactions, but it is nevertheless held firmly in place and unavailable to plants until the organic material decomposes.

Inorganic phosphorus comes from the mineral apatite, $Ca_5(PO_4)_3F$.[1] Apatite occurs as tiny crystals well dispersed through igneous rocks. It can be an infinitesimal speck included inside a crystal of another mineral (Syers et al., 1967). The soil solution comes in contact with the phosphorus as the rocks and minerals decompose. Phosphorus then dissolves if the solution contacting it is sufficiently depleted of phosphorus.

Apatite also occurs in a form known as *phosphate rock* in a few areas of the world. Phosphate deposits are thought to be formed by precipitation in shallow seas or on continental shelves, then lifted above sea level. Such rocks are mined and serve as raw materials for the production of phosphorus compounds.

THE PHOSPHORUS CYCLE

The phosphorus cycle (Figure 11-1) appears simpler than the nitrogen cycle because it does not involve exchanges with the atmosphere. There is only one item in the diagram that does not occur in the nitrogen cycle. Inorganic phosphate ions are adsorbed on positively charged sites of soil clays and of soil organic matter. The adsorbed P and the other solid forms are all in equilibrium with the dissolved P and thereby with each other.

The amount of dissolved phosphorus (mostly $H_2PO_4^-$ and HPO_4^{--}) in the soil at any one time is extremely small; there is less than 1 kg per hectare-furrow slice in most soils and less than 0.1 kg in some soils. Even so, it is believed that plants obtain all or most of their phosphorus from solution. Most crops will utilize between 10 and 30 kg of phosphorus per hectare each year. It is evident that the phosphorus content of the soil solution must be frequently replaced. The replacement, as indicated in Figure 11-1, can come from either mineralization of organic matter or equilibrium reactions with adsorbed phosphate ions and solid phosphorus compounds.

Plants can make satisfactory growth with a very small concentration of dissolved phosphorus if that concentration is maintained. The rates of the equilibria and mineralization reactions making additional phosphorus available seem generally to be fast enough to meet plant needs (Olsen and Watanabe, 1966), but there can be a problem of transport. The $H_2PO_4^-$ and HPO_4^{--} ions must move from the point where they enter the solution to a plant root. Although some phosphorus is intercepted by growing roots, and a small amount is carried to the roots by mass flow as the plant absorbs water, most of the movement is by ion diffusion. All means of transport to roots become very slow if the transport distance is more than 5 or 10 mm or if the soil is dry. The effective transport rate is slowest and distance is shortest for the least soluble phosphorus compounds. Phosphorus compounds with low solubilities must therefore be well distributed through the soil or they will not serve as effective sources of phosphorus.

---

[1]The form indicated here is more specifically known as *fluorapatite*. Other forms such as chlorapatite, $Ca_5(PO_4)_3Cl$, and hydroxyapatite, $Ca_5(PO_4)_3OH$, also exist. The most common form is a fluorapatite containing some carbonate. The simple fluorapatite form will generally be used when a formula is needed to represent apatite in this chapter.

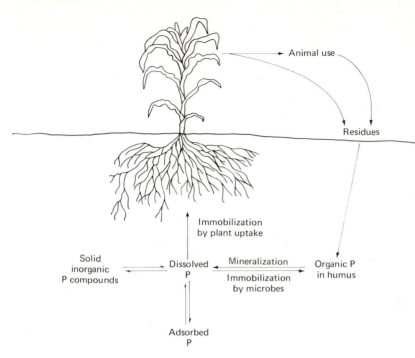

Figure 11-1 The phosphorus cycle.

Phosphorus fertilizers added to the soil are generally provided in relatively soluble forms. It is often advantageous to place soluble fertilizers in concentrated bands rather than to mix them through the soil because high solubility makes them vulnerable to being dissolved and then precipitated in less soluble forms. Mixing accelerates this "reversion" to low solubility.

The rate of plant uptake of phosphorus is influenced by the availability of nitrogen. It has been found that a small amount of nitrogen included in a phosphorus fertilizer makes it more effective (Miller, 1965). The mechanism is not understood but is probably a physiological process in the plant. Stimulated phosphorus absorption can be accomplished with as little as one part of nitrogen added per six parts of phosphorus.

Phosphorus occurring in reactive positions in organic materials may be split from the molecules by microbial enzymes. Much of the phosphorus, however, is built into the structure of the organic compounds and cannot ionize away in the early stages of decomposition. The mineralization of structural phosphorus requires that the organic matter be decomposed. This process is so nonspecific that almost any microbe can perform it, but the rate is often slow.

Organic matter has two types of indirect influence on phosphorus availability. Iron and aluminum ions can be complexed and tied up by organic matter. Complexing leaves less iron and aluminum in solution to precipitate insoluble phosphorus compounds. Decomposing organic matter releases acids that increase the solubility of calcium phosphates. The presence of large amounts of organic matter thus increases the amount of available phosphorus in soils. Sometimes phosphorus fertilizer and organic materials are added to the soil together in order to make the phosphorus more available.

PHOSPHORUS RESERVES IN SOILS

The apatite inherited from rocks is a nearly insoluble compound of calcium and phosphorus. Apatite endures as the principal phosphorus mineral in the soils of arid regions where the calcium content remains high and the pH is alkaline. Limited amounts of phosphorus are converted to organic forms and to ions adsorbed on the surfaces of other minerals. The organic forms and adsorbed ions are more available for plant use than is the phosphorus in apatite crystals.

Weathering and leaching gradually remove calcium from the soils of humid regions and cause changes in the kinds and concentrations of phosphorus compounds. The early changes improve the phosphorus availability because organic forms and adsorbed ions are produced as in arid regions. Later, as illustrated in Figure 11-2, much of the phosphorus becomes less available by combining with iron and aluminum. As shown by its solubility product constant in Table 7-5, the aluminum phosphate compound called *variscite* has very low solubility. The corresponding iron phosphate is called *strengite*. Variscite and strengite are too insoluble to contribute much to plant nutrition. Smeck (1973) suggests that the accumulation of aluminum and iron phosphates is a good indication of the stage of soil development.

ADSORBED PHOSPHATE IONS

The adsorption of phosphate ions by soil minerals has been the subject of much research. Adsorbed phosphate ions are held more strongly by positively charged sites

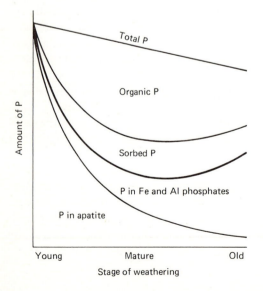

Figure 11-2 Phosphorus reserves in the furrow slice of humid-region soils as influenced by stage of weathering. The forms above the heavy line release available phosphorus more rapidly than those below.

than adsorbed cations are held by cation exchange. Still, the adsorbed phosphate ions are more available than the phosphorus in apatite, variscite, and strengite.

Phosphate adsorption takes place by replacing hydroxyl ions exposed at the edges or other surfaces of silicate clay minerals or of oxide minerals. Parfitt, Atkinson, and Smart (1975) showed that a phosphate ion adsorbed by any of several iron minerals replaces two adjacent hydroxyl ions. This double replacement helps explain why iron minerals hold phosphate strongly but in smaller amounts than some other minerals can adsorb. Galindo, Olguin, and Schalscha (1971) found aluminum and silicon compounds adsorbing relatively more phosphate than iron oxides. Rotini and El-Nennah (1971–1972) found that iron phosphates predominated over aluminum and calcium phosphates in the eight different soils they studied. The phosphorus held by aluminum was 1 to 6 times as available as that held by iron; the iron phosphate was 2 to 18 times as available as that held by calcium. Most of the available mineral phosphorus came from aluminum sites.

ORGANIC PHOSPHORUS IN SOILS

Phosphorus in organic compounds is held by covalent bonds so it cannot ionize. The phosphorus is usually surrounded by oxygen and attached to the rest of the molecule by a carbon-oxygen-phosphorus bond sequence. Decomposition breaks the carbon-oxygen bond and mineralizes the phosphorus into forms that are available for plant growth. For some reason, organic phosphorus is held more tightly in acid conditions than in alkaline conditions. An alkaline-extracting solution can remove much of the phosphorus from soil organic matter.

Identifiable organic phosphorus compounds in soils include inositol phosphates, nucleic acids, and phospholipids. The inositol phosphates are basically sugar molecules with one or more phosphate groups replacing hydrogen. Such sugar compounds contain about one-third of the organic phosphorus in soils.

Nucleic acids from plant, animal, and microbial sources and their decomposition products contain up to 10 percent of the soil's organic phosphorus. Phospholipids (phosphorus combined with fatty compounds) account for about 1 percent of the organic phosphorus. All of the identified forms combined add up to one-half or less of the organic phosphorus in soil; the nature of the remainder is unknown.

SOLUBLE PHOSPHORUS IN SOILS

Three different phosphorus ions deserve consideration here. These three ions can be formed by the ionization of one, two, or all three hydrogens from phosphoric acid (H_3PO_4) to form $H_2PO_4^-$, HPO_4^{--}, or PO_4^{---}. The predominance of one or another of these forms in solution depends largely on pH, as illustrated in Figure 11-3.

Most of the phosphorus absorbed by plants is in the monovalent orthophosphate form, $H_2PO_4^-$.[1] This form is referred to loosely as the *phosphate ion;* a more specific name for it is *dihydrogen phosphate.* Some HPO_4^{--} may be used, or even

---

[1]Unpublished research by E. J. Thompson and C. A. Black of Iowa State University shows that organic phosphorus levels in the soil are depleted by growing plants, possibly by direct absorption of some of the organic phosphorus compounds.

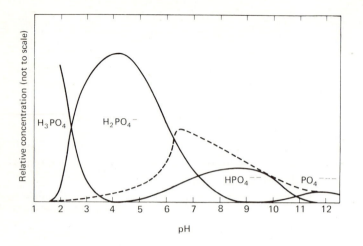

Figure 11-3 A schematic illustration of relative proportions of phosphate ions in solution at different pH levels in a Ca–H_3PO_4 system. The dashed line shows the limits on available P in solution imposed by the solubility of calcium phosphates above pH 6.5 or of iron and aluminum phosphates below pH 6.5.

required, by some plants, but the $H_2PO_4^-$ is the one most readily absorbed by plants. The importance of the HPO_4^{--} ion increases at high pH values because it becomes the dominant ion in solution above pH 7.2. The PO_4^{---} ion occurs at pH values too high for it to be significant in plant nutrition; the H_3PO_4 molecule is not significant, either, because it occurs only at very low pH.

Solubility of the inorganic forms is a more serious problem for phosphorus nutrition than for any other macronutrient. Calcium phosphates become insoluble under alkaline conditions, and iron and aluminum phosphates become insoluble under acid conditions. The most favorable pH for phosphorus availability is near neutral to slightly acid.

Under mildly alkaline conditions there is usually an abundance of calcium in the soil. Soluble phosphorus reverts to hydroxyapatite, $Ca_5(PO_4)_3OH$, or other low-solubility calcium phosphates under such conditions. Calcium phosphates have such low solubilities under alkaline conditions that phosphorus deficiencies commonly result at pH values near 8. There may be an abundance of solid phosphorus compounds in the soil but not enough soluble phosphate to meet plant needs.

Strongly alkaline conditions generally indicate that large amounts of sodium are present instead of calcium. Sodium phosphate is soluble. Phosphorus availability is therefore not a major problem at pH values of 9 or above. Other adverse effects such as poor soil structure, low permeability, and dissolving organic materials are the serious problems where the pH is above 9.

Experiments with nutrient solutions indicate that plants absorb $H_2PO_4^-$ ions most readily at pH values of 4 or below. This leads to the suggestion that OH^- ions may compete with $H_2PO_4^-$ ions for entrance into plant roots. The same may be true of other anions such as NO_3^- and SO_4^{--}.

Plants show no harmful effects when grown in a balanced nutrient solution at a pH of 4.0, but they do not do well in soil with a pH of 4.0. Several problems occur in such acid soils; one such problem is severe phosphorus deficiency. The very low solubility of iron and aluminum phosphates under acid conditions was discussed in Chapter 7.

Phosphorus is also removed from solution by adsorption onto positively charged sites on clay minerals. Such adsorption of anions differs from the adsorption of cations on negatively charged sites in that adsorbed phosphorus ions tend to be held too tightly to be readily exchangeable. Adsorption is most likely to be a problem in soils high in iron and aluminum oxides and hydroxides.

Availability of Inorganic Phosphorus

Phosphorus availability varies with solubility, amount of solution present, and the distance the phosphate ion must move to reach the plant root or microbe that will absorb it. These three variables are in turn dependent on several other factors.

The solubility of phosphorus ions depends greatly on pH because pH influences both the kind of phosphorus ion present and the concentrations of precipitating ions. The dominant form of phosphorus ion can be predicted from Figure 11-3. The solid phase present may differ from the dissolved form either because of equilibrium differences or because another form has been added as a fertilizer and has not had time to fully react with the soil.

The influence of soil pH on phosphorus solubility is one of the good reasons for liming acid soils. The solubility is highest near pH 7. To some extent, then, liming is an alternative to larger and more frequent applications of phosphorus fertilizer. A soil pH of 6 or above precipitates nearly all the iron and aluminum present as hydroxides or other low-solubility compounds. Low concentrations of these ions in solution means that they will not precipitate phosphorus and make it unavailable to plants.

The concentration of phosphorus in solution is about the same whether the soil is moist or dry. The amount of phosphorus in solution is therefore approximately proportional to the water content of the soil. Fine-textured soils hold more water than do coarse-textured soils and are therefore likely to contain more phosphorus in solution. Even in clay soils the phosphorus in solution seldom exceeds 1 kg per hectare-furrow slice.

Neither mass flow nor diffusion can transport phosphorus farther than 5 or 10 mm at an adequate rate to meet growth needs unless some highly soluble fertilizer phosphorus is present. The effective transport distance and rate decrease as the soil dries out. Marais and Wiersma (1975) found that the uptake of phosphorus by soybeans dropped to zero when the soil was near the wilting point. It is fortunate that phosphorus occurs in small particles well distributed in mineral matter. The distance from a plant root to the nearest particle is seldom very far. Furthermore, the large amount of surface area associated with small particle size increases the rate at which the phosphorus can dissolve into an unsaturated solution.

The inorganic forms from which phosphorus may dissolve include minerals such as apatite, any mineral fertilizer that may have been added, and secondary forms in which the phosphate ion is held immobile by attraction to positively charged sites that may occur on clay or organic matter. Adsorbed phosphate anions are less available than adsorbed cations. Their exchangeability is so slow that the term "anion-exchange capacity," while sometimes used, is open to question.

Phosphorus fertilizer applications increase phosphorus availability in several ways. The mere increase in the total amount of phosphorus present helps by increasing the amount of surface area where the soil solution can dissolve phosphorus compounds. Usually the phosphorus fertilizer is placed in a part of the soil where there is a high concentration of roots so that transport distance will be short. Also, fertilizer phosphorus is more soluble than the phosphorus already in the soil. Even when fertilizer phosphorus is adsorbed on micelles, or when the phosphorus is precipitated in a less soluble form, it is more soluble and more accessible to soil water than is the native soil phosphorus. Transport distance, however, can be a factor in the utilization of phosphorus fertilizer. The fertilizer is mixed with only part of the soil, especially where it is applied in a band. Barber's work (1974) indicates that phosphorus uptake from fertilizer is nearly proportional to the length of roots in contact with fertilized soil.

PHOSPHORUS IN PLANTS

Phosphorus is absorbed by plant roots and then distributed to every living cell in the plant. It becomes most concentrated in the plant's reproductive parts. A seed must contain enough phosphorus and other vital nutrients to suffice until roots are formed to obtain a supply from the soil.

Phosphorus in the cell becomes united with carbon, hydrogen, oxygen, nitrogen, and other elements to form complex organic molecules. Phosphorus is an essential component of the genetic material of the cell nucleus. Cells cannot divide unless there is adequate phosphorus (as well as other vital constituents) to form the extra nucleus. Phosphorus deficiency therefore causes stunting, delayed maturity, and shriveled seed.

Phosphorus has the very important capacity of forming bonds of more than one energy level. This permits the storage, transfer, and release of energy within the plant through such materials as adenosine diphosphate and adenosine triphosphate. Lack of phosphorus therefore hampers metabolic processes such as the conversion of sugar into starch and cellulose. The resulting buildup of sugar often leads to the formation of anthocyanins that show as purple spots or streaks in leaves and stems.

Phosphorus is not a component of chlorophyl. High chlorophyl contents often result when phosphorus is deficient and nitrogen is abundant. The resulting plants are dark green in color.

The most common phosphorus deficiency symptoms may be summarized from the preceding discussion. These include stunting, delayed maturity, dark-green coloration, and purple spots or streaks.

Harvested crops contain considerable amounts of phosphorus (Table 11-1). In general, seed crops contain the largest percentages of phosphorus, and forage crops

Table 11-1 Average Phosphorus Contents of Selected Crops

Crop	Part of crop	Yield, tons/ha	% P	P, kg/ha
Alfalfa	Hay	9	0.23	21
Barley	Grain	3	0.38	11
	Straw	2.5	0.10	2
Corn	Grain	7	0.27	19
	Stover	8	0.10	8
Cotton	Lint, seed	1.5	0.53	8
	Stalks, leaves	1.5	0.20	3
Cowpeas	Hay	5	0.27	13
Lespedeza	Hay	8	0.22	17
Oats	Grain	3	0.44	13
	Straw	3	0.16	5
Peanuts	Nuts	2.5	0.30	7
	Vines	8	0.03	2
Potatoes	Tubers	20	0.06	12
Red clover	Hay	5	0.15	7
Soybeans	Grain	2.5	0.70	17
	Straw	3	0.08	2
Sugar beets	Roots	40	0.05	20
Sweet clover	Hay	12	0.10	12
Sweet potatoes	Roots	20	0.04	8
Timothy	Hay	4	0.16	6
Tobacco	Leaves	1.5	0.27	4
Tomatoes	Fruit	25	0.04	10
Wheat	Grain	3	0.38	11
	Straw	3	0.08	2

Source: Calculated from various sources.

contain moderate amounts. Long-term cropping may lead to a need for phosphorus fertilizer in some form even if the soil supply was originally adequate. On fertilized soils, however, the trend is likely to be in the opposite direction. Some soils that have been cultivated and fertilized for several decades or centuries show a considerable increase in phosphorus. Part of the fertilizer phosphorus was used by crops, but part of it accumulated in the soil (Table 11-2).

FACTORS AFFECTING THE ORGANIC-PHOSPHORUS CONTENT OF SOILS

The organic-phosphorus content of soils is related to the organic-matter content of the soil, pH, climate, and cultivation. An increase in the organic-matter content of a soil is an obvious reason to anticipate an increase in the organic-phosphorus content, but the relation of the other factors is not so obvious.

Organic phosphorus tends to be more stable at low pH than at high pH (Figure 11-4). One of the effects of liming acid soils is therefore to "unlock the storehouse" of phosphorus tied up in soil organic matter.

Table 11–2   Organic Carbon, Nitrogen, and Phosphorus Contents of Plots at Rothamsted, England, after more than 100 years of the Specified Treatments

Treatment	Total P added, kg/ha	% C	% N	P, ppm
		Continuous root crop plots		
N	0	0.81	0.092	690
NPKNaMg	3,300	0.90	0.099	1,235
NMan	4,000	2.48	0.246	1,275
NPKMan	7,300	2.58	0.242	1,750
		Permanent grass plots		
Check	0	4.90	0.377	920
Lime	0	5.10	0.398	1,040
NPKNaMg	3,370	4.28	0.362	3,130
NPKNaMgLime	3,370	4.64	0.404	3,110

N = NaNO$_3$, P = superphosphate, K = K$_2$SO$_4$, Na = Na$_2$SO$_4$ or NaCl, Mg = MgSO$_4$, Man = 35 tons manure per hectare annually.
Source: Onioni, Chater, and Mattingly, 1973.

The changes in the organic-phosphorus content of soil organic matter between cool and warm climates resemble the changes between acid and neutral soil reactions. This may be shown by comparing phosphorus to nitrogen (organic-matter contents can be estimated by multiplying nitrogen contents by 20). Data from

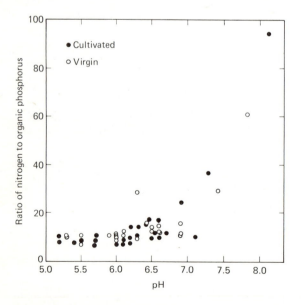

Figure 11-4   Ratio of total nitrogen to organic phosphorus as a function of soil pH. (*Thompson, Black, and Zoellner,* 1954.)

Pearson and Simonson (1939) show that the total nitrogen-to-organic-phosphorus ratio is about 10:1 in Iowa soils. Garman (1948) found this ratio to be about 15:1 in Oklahoma soils. The difference is probably a result of more rapid decomposition occurring in Oklahoma because of higher soil temperatures.

The effect of cultivation on the ratio of nitrogen to organic phosphorus shows in the data of Thompson (1950) plotted in Figure 11-5. Both nitrogen and organic phosphorus declined under cultivation, with phosphorus declining the most. The stability of the remaining phosphorus had evidently increased more than the stability of the nitrogen. The average ratio of nitrogen mineralized to phosphorus mineralized was 7.6:1 for virgin soils and increased to 11.5:1 for cultivated soils.

Thompson's study indicates that there is a fraction of easily mineralized organic phosphorus in virgin soils that soon disappears under cultivation. The remaining organic phosphorus is mineralized more slowly after this fraction is gone. Many soils supply adequate amounts of phosphorus for crops during the first few years of cultivation but develop a need for phosphorus fertilizer when the easily mineralized fraction is depleted.

DISTRIBUTION OF PHOSPHORUS

It has already been mentioned that phosphorus is widely distributed in rocks as tiny crystals of apatite and other minerals. Apatite dissolves very gradually and therefore provides a slow, long-lasting input of available phosphorus. Even so, other minerals such as quartz, the silicate clay minerals, and the insoluble oxides resist weathering

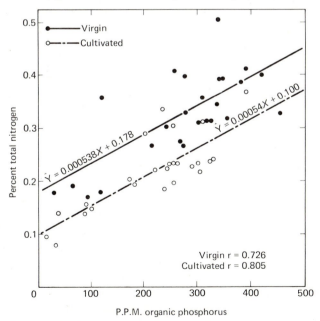

Figure 11-5 The relation of total nitrogen to organic phosphorus in 25 pairs of virgin and cultivated soils of Iowa. (*Thompson,* 1950.)

longer than apatite does. Consequently, the inorganic phosphorus content of the solum gradually declines as leaching occurs. Some of the leached phosphorus often accumulates at a depth of a meter or two in an illuvial layer beneath the solum.

The organic phosphorus content is distributed approximately the same as the soil organic matter. The A1 horizon of a dark-colored soil is likely to contain approximately equal amounts of organic and inorganic phosphorus, as indicated in Figure 11-6. The organic phosphorus often is about equal to the amount of inorganic phosphorus lost from the A1 horizon so that the total phosphorus content there may be about the same as in the unleached material. Some of the organic phosphorus reverts to inorganic forms with the result that even the inorganic phosphorus may be more plentiful at the surface than it is lower in the solum.

The B horizon of a mature soil is likely to contain less phosphorus than any other part of the soil profile. Leaching removes more phosphorus from B horizons than the organic matter replaces. In fact, the presence of plants tends to decrease the phosphorus content of B horizons. The plant roots absorb phosphorus from the B horizon as well as from the A. After being translocated to the top part of the plant, the phosphorus in the plant and animal residues is more likely to return to the A horizon than to the B horizon.

Total Phosphorus Content of Soils

The amount of phosphorus in a soil depends considerably on both climate and parent material but rarely exceeds 0.3 percent. Phosphorus contents of soils from various parts of the United States are shown in Table 11-3. These data show that the phosphorus content of soils varies considerably even among soils of the same soil order.

Regionally in the United States there is a tendency for the phosphorus contents of soils to be highest in the northwestern states where weathering intensity is relatively low and lowest in the southeastern states where the weathering intensity is relatively high (Figure 11-7).

Soils developed from marl, chalk, or other materials containing calcareous skeletal remains generally contain more total phosphorus than other soils. Skeletons

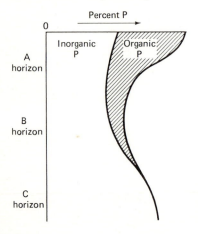

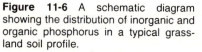

Figure 11-6 A schematic diagram showing the distribution of inorganic and organic phosphorus in a typical grass-land soil profile.

Table 11-3   Total Phosphorus Contents of the A, B, and C Horizons of Several Kinds of Soil

Soil type	Soil order (see Chapter 16)	% P A horizon	% P B horizon	% P C horizon
Miami silt loam	Alfisol	0.035	0.031	0.035
Barnes silt loam	Mollisol	0.100	0.065	0.065
Holdrege silt loam	Mollisol	0.087	0.092	0.096
Macksburg silty clay loam	Mollisol	0.061	0.072	0.086
Marshall silt loam	Mollisol	0.052	0.044	0.070
Otley silty clay loam	Mollisol	0.055	0.055	0.070
Richfield clay loam	Mollisol	0.044	0.017	0.039
Sharpsburg silty clay loam	Mollisol	0.059	0.073	0.081
Nipe clay	Oxisol	0.25	0.14	0.14
Becket fine sandy loam	Spodosol	0.057	0.035	0.031
Davidson clay loam	Ultisol	0.044	0.087	0.105
Maury silt loam	Ultisol	0.136	0.158	0.240
Sassafras sandy loam	Ultisol	0.048	0.035	0.031
Houston black clay	Vertisol	0.065	0.074	0.039

Source: Calculated from various sources.

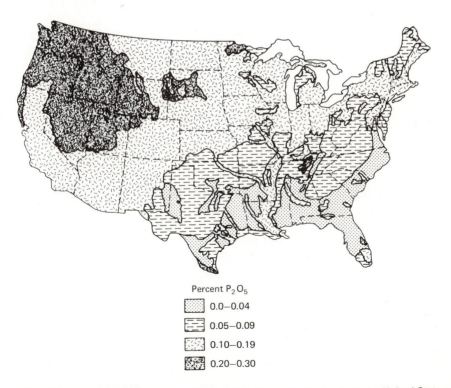

Percent P_2O_5

- 0.0–0.04
- 0.05–0.09
- 0.10–0.19
- 0.20–0.30

Figure 11-7   Percent phosphorus as P_2O_5 in the upper 30 cm of soils in the United States. (*From Parker et al.*, 1946.)

and shells high in both calcium and phosphorus contribute to the fertility of soils formed from fossiliferous rocks.

Phosphorus in Water

Clear water contains very little phosphorus even if it has passed through a fertile soil. The solubility of phosphorus compounds is too low for leaching to be significant. There are, however, other ways by which phosphorus enters streams and lakes in significant amounts.

The phosphorus content of streams has increased as much as 300 percent in the last few decades according to Thomas.[1] Most of the increase appears to have come from detergents containing phosphorus. The phosphorus content of a stream increases markedly near sewage outlets.

Soil erosion, especially from heavily fertilized fields, contributes phosphorus to bodies of water. Sediment is probably most significant in terms of eutrophication—a nutrient-enriching effect causing undesirable growth of algae on lakes and ponds. Other nutrients are also involved in eutrophication, but phosphorus is generally the most critical one. Many undisturbed bodies of water are so pure that little vegetation grows in them. The clear, open water often becomes full of undesired vegetation, especially in the shallow parts, when its nutrient content increases. It should be realized, however, that some eutrophication is natural. Most peat bogs represent bodies of water that were eutrophied by natural processes.

Runoff can cause eutrophication even when it does not erode soils. Runoff from frozen soil carries nutrients, especially where fertilizer was applied and left on the surface. Runoff from feedlots, manure piles, etc., can contribute large quantities of nutrients. Sewage effluent and industrial wastes are dumped into streams and lakes. The total amount of organic material and plant nutrients too often becomes more than the natural purification processes of the water can handle.

IDENTIFYING PHOSPHORUS NEEDS

Persons dealing with plants usually learn to identify deficiencies of certain elements by their effect on plant growth. A tentative identification of a phosphorus deficiency made on the basis of such symptoms as stunting, delayed maturity, dark-green coloration, and purple spots or streaks may be supported by one of the tissue tests or soil tests discussed in Chapter 9. The best possible verification is obtained by applying a phosphorus fertilizer to part of the affected area and observing the growth response.

Soil Analyses

Many different soil tests for available phosphorus have been tried. None has proven universally applicable. The problem is how to extract from the soil the portion of

[1]G. W. Thomas, "Soil and Climatic Factors Which Affect Nutrient Mobility," a paper presented Nov. 12, 1969, at the annual meeting of the American Society of Agronomy in Detroit, Mich.

phosphorus that will become available during a growing season. Leaching with water does not remove enough phosphorus; a dilute-acid extracting solution is therefore generally used for acid soils and an alkaline solution for alkaline soils.

The Bray no. 1 extracting solution is the most used. It contains 0.025 N HCl and 0.03 N NH_4F. The HCl brings some phosphorus into solution. The NH_4F complexes soluble aluminum and iron so that the phosphorus will not be re-precipitated. Phosphorus extracted by the Bray no. 1 solution gives a good correlation to crop response on most young soils whether acid or alkaline. A stronger extracting solution such as Bray no. 2 (0.1 N HCl with NH_4F) works well on older acid soils but gives high results on calcareous soils or soils where rock phosphate has been applied. An alkaline solution containing either Na_2CO_3 or K_2CO_3 works best for extracting available phosphorus from alkaline soils.

Soil-test results for phosphorus are classified and fertilizer recommendations made in the manner described in Chapter 9. Interpretation of the tests requires some knowledge of the soil tested, partly because soil tests are usually run only on the plow layer. The B horizons of some soils are able to supply much more available phosphorus than those of other soils. The need for a high level of phosphorus in the plow layer is less if the B horizon has a high content of available phosphorus.

PHOSPHORUS FERTILIZERS

The solubility problem is as important in evaluating the phosphorus content of fertilizers as it is in the availability of phosphorus in the soil. Some phosphorus fertilizer materials are soluble in water, and others are not. Some of the compounds with low water solubility may nevertheless be available to plants because their roots have a greater extracting power than water has.

A 1N ammonium citrate solution is used to simulate the extracting power of plant roots for phosphorus. Phosphorus fertilizer analyses show the percent citrate-soluble phosphate, and/or the percent total phosphate. The relative importance of these percentages depends partly on how rapidly the growing plants need the phosphorus and partly on soil characteristics such as acidity and ability to adsorb or "fix" phosphorus. The term *available phosphorus* (or *available phosphate*) is applied to the citrate-soluble portion.

Rock Phosphate

Phosphate rock is sometimes ground to a powder and used without any further treatment as a fertilizer called *rock phosphate*. The crystallinity of the material can be a very important factor in the value of rock phosphate as a fertilizer. Howeler and Woodruff (1968) found that certain rock phosphate deposits in Missouri contained such well-crystallized apatite phosphorus that the phosphorus was extremely insoluble. Deposits in Florida and Arkansas containing carbonates in the rock are not so well crystallized and are therefore more soluble.

The suitability of rock phosphate as a fertilizer has been argued for several decades. Its low solubility in both water and ammonium citrate is well known, but some farmers obtain good crop response by means of large applications of rock phosphate. Others obtain little or no response from rock phosphate. Both soil and crop seem to be factors as well as the nature of rock phosphate applied.

Rock phosphate should not be considered as a fertilizer unless the soil is acid. The solubility of apatite is too low in neutral or alkaline soils to make its application worthwhile (Figure 11-8). It slowly becomes available in acid soils by reacting with hydrogen ions and dissolving. The more acid the soil, the more likely it is that this reaction will proceed fast enough to meet plant needs.

Decomposing organic matter produces acids that help apatite dissolve. Manure and rock phosphate can therefore be used to supplement one another. The manure may not contain enough phosphorus to meet plant needs, but acids from the manure help to dissolve the rock phosphate.

Robertson, Thompson, and Hutton (1966) found rock phosphate to be a good fertilizer on certain tropical soils. The beneficial effects of rock phosphate were longer lasting than those of a more soluble phosphorus fertilizer on Oxisols. The soluble fertilizer was largely converted to aluminum and iron phosphates within 2 years.

Liming a soil will normally decrease the effectiveness of rock phosphate by making the soil less acid. Favorable results from rock phosphate on unlimed soils should not, however, be considered a good reason for not liming an acid soil. Still better results can generally be obtained by liming the soil and fertilizing with a more soluble form of phosphorus.

Figure 11-8 This corn was grown on a calcareous soil which was deficient in available phosphorus. The plot on the left received superphosphate; the plot on the right received rock phosphate. *(Iowa State Univ.)*

The results of long-time experiments comparing rock phosphate and super-phosphate in Iowa soils are shown in Table 11-4. Both forms increased yields, but there was a difference in favor of superphosphate for corn and a larger difference in favor of superphosphate for oats. The average annual rates of application were 39 kg/ha of phosphorus in the rock phosphate and 9 kg of phosphorus in the super-phosphate.

Legumes respond more to rock phosphate than do members of the grass family. Most legumes have a high calcium requirement. Some of the calcium may come from the apatite thus weakening the crystals so they will decompose and produce soluble phosphate ions. Table 11-5 contains a list of crops classified according to their relative efficiency in utilizing phosphorus from powdered rock phosphate.

Young seedlings are not as capable as older plants in utilizing phosphorus from rock phosphate. The older plants with their larger root systems are able to draw phosphorus from a larger volume of soil and thus obtain an adequate amount of phosphorus even from a soil containing a relatively low concentration of phosphorus in solution. It is therefore wise to supply some soluble phosphorus fertilizer for starting a young crop even when rock phosphate is chosen as the principal phosphorus fertilizer.

Rock phosphate contains so little soluble material that its analyses are expressed in terms of total phosphorus present. All other phosphate fertilizers are evaluated in terms of citrate-soluble phosphorus. Rates of application on rock phos-

Table 11–4 Response of Corn and Oats to Different Phosphate Fertilizers on Several Iowa Soils Receiving Manure

Soil type	No. of crops	Rock phosphate	Superphosphate
		Increased yield of corn, q/ha	
Carrington loam	15	2.0	3.5
Carrington silt loam	9	3.8	3.4
Clarion loam	19	2.5	2.8
Grundy silt loam	6	3.3	2.3
Grundy silt loam	9	4.6	7.1
Marshall silt loam	6	1.7	2.3
		Increased yield of oats, q/ha	
Carrington loam	9	0.9	1.9
Carrington silt loam	6	1.1	1.1
Clarion loam	8	0.9	1.7
Clarion loam	22	0.9	2.0
Grundy silt loam	4	2.5	3.6
Grundy silt loam	9	4.3	6.8
Grundy silt loam	77	1.1	4.2
Marshall silt loam	3	1.7	2.7

Source: Calculated from an Agronomy Department report, Iowa State University.

Table 11–5   A Classification of Plants in Accordance with Their Efficiency in the
Utilization of Raw-rock Phosphate

Most efficient	Moderately efficient	Least efficient
Lupines	Alfalfa	Cotton
Buckwheat	Peas	Cowpeas
Sweet clover	Rape	Bur clover
Mustard	Cabbage	Rice
Swiss chard		Oats
Vetch		Barley
		Millet
		Rye
		Wheat

Source: Dean and Fried, 1953.

phate range from a few hundred to a thousand kilograms (generally between 25 and 150 kg of phosphorus) per hectare.

Processed Phosphorus Fertilizers

Most commercial phosphorus fertilizers are made by treating rock phosphate with acid. The oldest and still a widely used acid for this purpose is sulfuric acid:

$$2Ca_5(PO_4)_3F + 7H_2SO_4 + 14H_2O \rightarrow$$
$$3Ca(H_2PO_4)_2 + 7CaSO_4 \cdot 2H_2O + 2HF$$

The resulting product is referred to as *superphosphate.* The hydrofluoric acid escapes from the mixture as a gas and is caught as a significant by-product. The monocalcium phosphate is not readily separated from the gypsum ($CaSO_4 \cdot 2H_2O$), and so the two are sold together as superphosphate. The usual superphosphate fertilizer contains about 9 percent P (20 percent P_2O_5 equivalent) by weight.

A more concentrated phosphorus fertilizer can be made by treating the rock phosphate with phosphoric acid:

$$Ca_5(PO_4)_3F + 7H_3PO_4 \rightarrow 5Ca(H_2PO_4)_2 + HF$$

The hydrofluoric acid is again a by-product. The main product, known as *triple superphosphate* (or treble superphosphate or concentrated superphosphate), contains about 20 percent P (46 percent P_2O_5 equivalent). This higher analysis gives triple superphosphate a considerable advantage over "single super" if the fertilizer has to be shipped very far.

The principal limitation to the use of triple superphosphate as a fertilizer is the cost of phosphoric acid. The reaction already cited shows that 70 percent of the phosphorus actually comes from the acid and only 30 percent from the rock phosphate. One way to reduce the expense is to use a smaller proportion of H_3PO_4 and make partially acidulated rock phosphate. Lutz (1973) reported that 20 percent acidulation resulted in fertilizer that was equally as effective for phosphorus uptake by alfalfa as 100 percent acidulated fertilizer. However, partial acidulation has not become a common practice.

High-grade phosphoric acid is produced by reducing rock phosphate to elemental phosphorus in a furnace. The phosphorus is then oxidized and reacted with water to produce H_3PO_4. Such acid is purer than needed for fertilizer purposes and is relatively expensive. A less expensive way of producing phosphoric acid is to treat rock phosphate with enough sulfuric acid to convert all the calcium to gypsum:

$$Ca_5(PO_4)_3F + 5H_2SO_4 + 10H_2O \rightarrow 3H_3PO_4 + 5CaSO_4 \cdot 2H_2O + HF$$

The phosphoric acid is a liquid, the gypsum ($CaSO_4 \cdot 2H_2O$) is a nearly insoluble solid, and the hydrofluoric acid is a gas; separation of the three products is therefore not difficult.

The phosphoric acid is itself a potential fertilizer. Even with some water and impurities present it contains about 25 percent P. However, it has an acidifying effect on the soil and requires very special handling techniques and equipment.

Phosphoric acid applied to alkaline soils can provide several benefits. The acid not only reduces the alkalinity and supplies fertilizer phosphorus but it also helps prevent soil crusting. The Agricultural Research Service (1972) reported that a spray of 12 percent H_3PO_4 solution applied on sugar beet rows decreased crusting and increased sugar beet stands. The suggested mechanism involved a reaction between H_3PO_4 and Ca and Mg ions from the soil to produce phosphates that stabilized soil aggregates.

Phosphoric acid can also be reacted with ammonia to produce fertilizers containing both nitrogen and phosphorus. The fertilizers usually produced are monoammonium phosphate (about 11 percent N and 21 percent P) and diammonium phosphate (about 18 percent N and 20 percent P). The ammonium phosphates are soluble in water. They can be applied as fertilizers either in solid form or in water solution. Intermediate proportions of ammonia and phosphoric acid will produce mixtures of these materials. Like other fertilizers containing ammonia, these have an acidifying effect on the soil.

The most commonly used phosphorus fertilizers are monocalcium phosphate (in superphosphate and triple superphosphate) and diammonium phosphate. However, there are still several more that are either occasionally used or could be used. For example, rock phosphate can be treated with nitric acid:

$$2Ca_5(PO_4)_3F + 14HNO_3 \rightarrow 3Ca(H_2PO_4)_2 + 7Ca(NO_3)_2 + 2HF$$

One problem in all the usual phosphorus fertilizers is their relatively low content of phosphorus. The most concentrated of them contain only about 25 percent P. About 3 kg of other material, mostly calcium and oxygen, must be shipped along with every kilogram of phosphorus.

The ultimate in a high-analysis phosphorus fertilizer would be elemental phosphorus. It has been tested as a fertilizer and found to be effective (Cooke, 1969) but is hazardous to handle. Several compromise materials can be prepared by making salts of special phosphoric acids containing less water than the orthophosphoric acid (H_3PO_4) usually used. Metaphosphoric acid (HPO_3) and pyrophosphoric acid ($H_4P_2O_7$) can be used. Calcium metaphosphate [$Ca(PO_3)_2$] has had some use. Other salts can also be formed and may someday become significant as fertilizers. More exotic possibilities being investigated are materials like phosphonitritic hexamide [$P_3N_3(NH_2)_6$] (Cooke, 1969). Materials such as these would react in the soil with water and oxygen to produce some of the same products as are formed from orthophosphoric acid.

What to Apply for a Phosphorus Deficiency

Some phosphorus deficiencies can be corrected in several different ways and the best choice is not always obvious. Liming an acid soil to pH 6.5 or 7.0 may help to alleviate a phosphorus deficiency (Table 11-6). Liming helps in other ways, too (Chapter 8), and so this might well be one of the first choices to make. Liming alone, however, is likely not to be enough to make an inadequate phosphorus supply entirely adequate. It is more likely to influence the amount of phosphorus that should be added to the soil.

Manuring adds phosphorus to the soil but usually not in large enough amounts. An average ton of wet manure contains only about 1 kg of P (considerably less than its content of N and K). The application of manure may also help to make more soil phosphorus available through the dissolving effect of organic acids. The increase in available soil phosphorus and the readily available phosphorus in the manure are small but possibly significant contributions toward plant needs. They may suffice if the phosphorus needs are small. Larger deficiencies usually call for the addition of a processed phosphorus fertilizer.

The nature of the soil should be considered when a phosphorus fertilizer is chosen. The cheapest source per unit of phosphorus is usually rock phosphate, but rock phosphate is unsuitable for alkaline soils and not very good for neutral soils. Even on acid soils it must be applied in large amounts on the assumption that only

Table 11-6 Effect of Top-dressed Lime and Phosphorus Fertilizer Applications on Available P at Various Depths 2 years after Application

Depth, cm	Available P in kg/ha from applications of:							
	20 kg/ha		40 kg/ha		80 kg/ha		160 kg/ha	
	Unlimed	Limed	Unlimed	Limed	Unlimed	Limed	Unlimed	Limed
0- 2.5	3.4	7.0	10.3	14.7	27.8	25.6	55.1	84.8
2.5- 5	0.2	0.8	0.3	3.8	8.6	6.2	24.4	20.3
5- 7.5	0.0	0.0	0.2	0.2	6.3	5.4	19.8	22.2
7.5-10	0.0	0.0	0.0	0.6	1.3	2.8	6.7	14.4
10-12.5	0.0	0.0	0.0	0.0	1.0	0.3	4.7	6.5

Source: Calculated from Sell and Olson, 1946.

a small percentage of it will become available each year. However, this slow availability can be an advantage in acid soils able to fix large amounts of phosphorus.

The next alternative, and the one usually sought, is to apply a processed phosphorus fertilizer. The most used are superphosphate, triple superphosphate, and diammonium phosphate. Obviously, the need for nitrogen would be an important factor in deciding to use diammonium phosphate. All these fertilizers are soluble in water, and so they are rapidly available for plant growth. Some other phosphorus fertilizers, notably dicalcium phosphate ($CaHPO_4$) and calcium metaphosphate ($Ca(PO_3)_2$), are soluble in ammonium citrate but not in water.

Percent available phosphorus usually means citrate soluble. But there are some circumstances in which water solubility is important. Fast-growing crops such as corn, and short-season crops such as oats and peas, must have a more rapidly available phosphorus supply than is required for longer-season crops, especially if the phosphorus fertilizer has been placed in narrow bands rather than plowed down. High water solubility is helpful in such circumstances. It may also be important in cool, wet soils where all biological processes, including absorption of nutrients, tend to be slow.

APPLYING PHOSPHORUS FERTILIZER

How and when are always important questions in relation to fertilizer applications. Phosphorus does not leach from the soil to any significant extent, but it may become fixed in an unavailable form. A supply adequate for 2, 3, or even more years can be applied all at once if fixation is not a serious problem.

Large phosphorus applications could be useful even on soils that fix phosphorus. A large enough application will saturate (or "quench") the fixation capacity (Younge and Plucknett, 1966), but the amount required is usually too large to be economical. Phosphorus fertilizer needs on soils with quenched fixation capacities are thereafter comparable with those on soils that do not fix much phosphorus.

Techniques of applying rock phosphate are the same for all crops. The rock phosphate is already insoluble, and so fixation is not a major concern. Reaction with the soil generally increases rather than decreases the availability of the phosphorus in rock phosphate. It is therefore advisable to mix rock phosphate into the soil as thoroughly as possible, usually by disking. Processed phosphorus fertilizers may also be applied this way if fixation is not a problem.

Even the highly soluble forms of phosphorus fertilizer do not move much after being applied to the soil. Table 11-6 shows most of the fertilizer phosphorus applied remaining in the top few centimeters of soil 2 years after heavy applications were made. Techniques of placing this fertilizer where the crop can get at it are discussed in the following sections.

Phosphorus Fertilizer for Row Crops

Phosphorus fertilizer should not be placed in the top 5 or 10 cm of soil between the rows of cultivated crops. This soil contains very few roots because it is disturbed at every cultivation and it dries more rapidly than the soil below. The available phosphorus should therefore be placed at a lower depth or in a band near the seed.

Phosphorus fertilizer is often applied and plowed under before the crop is planted. Plowing is preferred to disking to limit the amount of soil contact and fixation that can occur. Downward-growing plant roots will encounter the fertilizer.

The question is often raised as to when phosphorus fertilizer should be plowed under. It is known that soluble phosphate fertilizer is more available when first applied than later. Other factors, however, are usually more important than phosphate application in determining when the soil should be plowed. Sometimes it becomes expedient to apply phosphorus fertilizer and plow it under in the fall even for a spring-seeded crop.

Band application to give even less soil contact than plowing under the fertilizer is recommended by Richards (1975) to give maximum efficiency in utilization of phosphorus. Barber (1974) recommends that banded fertilizer be mixed into a strip of soil 5 cm wide as a compromise between soil fixation and the need to have enough plant roots able to contact the fertilizer.

A band of fertilizer placed about 5 cm from the seed is used as a starter fertilizer to speed up seedling growth. When needed, such fertilizer should be applied at the time of seeding in amounts at least large enough to meet the needs of the young plants. As has already been mentioned, a small amount of nitrogen included in starter fertilizer increases the rate of absorption of the phosphorus.

Modest phosphorus needs can be met entirely by banded fertilizer, if desired. The low solubility of phosphorus makes it unlikely to cause any toxicity problem even when concentrated. However, the volume of fertilizer to be applied may become excessive for a planter attachment to handle. Large applications are therefore usually plowed under.

Phosphorus Fertilizer for Close-growing Crops

It would be desirable to have some phosphorus fertilizer plowed under for close-growing crops wherever it will be needed. But many close-growing crops are seeded on a disked surface rather than a plowed surface. If phosphorus is needed for these crops, part or all of it should be applied at seeding time. This phosphorus can be drilled into the upper part of the soil or even broadcast on the surface. The close-growing crop soon protects the soil from rapid drying, and the surface will not be disturbed because there will be no further tillage operations. Therefore roots can develop in the top part of the soil and absorb the surface-applied phosphorus.

Phosphorus fertilizer can be applied as a top-dressing for established hay or pasture crops. The phosphorus does not penetrate very deeply, but there are roots feeding near the soil surface. Hanway, Stanford, and Meldrum (1953) reported a very efficient recovery by plants of phosphorus applied as a top-dressing on meadow. Such an application would not be recommended for a cultivated row crop.

Home Use of Phosphorus Fertilizer

Lawns that need phosphorus fertilizer should be treated the same as a close-growing crop. Preferably a reasonably large application (possibly 1 or 2 kg of actual P per 100 m^2) should be worked into the soil before the lawn is planted or the sod is placed. Subsequent applications will have to be applied as top-dressing.

Gardens can be fertilized with phosphorus by the same techniques described for row crops. Phosphorus for trees and shrubs can be applied in a circle near the base of a shrub and as much as 1 m out from the trunk of a large tree.

REFERENCES

Anderson, G., 1967, Investigations on the Analysis of Inositol Hexaphosphate in Soils, *Trans. Int. Congr. Soil Sci., 8th,* Bucharest, 1964, **4**:563–572.

Agricultural Research Service, 1972, Combatting Soil Crusts, *Agric. Res.* **20**(8):13.

Barber, S. A., 1974, A Program for Increasing the Efficiency of Fertilizers, *Solutions,* Mar-Apr 1974, 24–25.

Bennett, W. F., John Pesek, and J. J. Hanway, 1962, Effect of Nitrogen on Phosphorus Absorption by Corn, *Agron. J.* **54**:437–442.

Bingham, F. T., 1962, Chemical Soil Tests for Available Phosphorus, *Soil Sci.* **94**:87–95.

Black, C. A., 1969, Phosphorus Nutrition of Plants in Soils, *HortScience.* **4**:314–320.

Cooke, G. W., 1969, Fertilizers in 2000 A.D., *Phosphorus in Agriculture Bulletin De Documentation,* Int. Superphosphate and Compound Mfg. Ass. Ltd., London, **53**:1–13.

Dean, L. A., and Maurice Fried, 1953, Soil Plant Relationships in the Phosphorus Nutrition of Plants, Chap. 2 in W. H. Pierre and A. G. Norman (eds.), *Soil and Fertilizer Phosphorus in Crop Nutrition,* Academic, New York.

Enwezor, W. O., 1967, Significance of the C:Organic P Ratio in the Mineralization of Soil Organic Phosphorus, *Soil Sci.* **103**:62–66.

Galindo, G. G., C. Olguin, and E. B. Schalscha, 1971, Phosphate-sorption Capacity of Clay Fractions of Soils Derived from Volcanic Ash, *Geoderma* 7:225–232.

Garman, W. L., 1948, Organic Phosphorus in Oklahoma Soils, *Oklahoma Acad. Sci. Proc.* **28**:89–100.

Hanway, John, George Stanford, and H. R. Meldrum, 1953, Effectiveness and Recovery of Phosphorus and Potassium Fertilizers Topdressed on Meadows, *Soil Sci. Soc. Am. Proc.* **17**:378–382.

Howeler, R. H., and C. M. Woodruff, 1968, Dissolution and Availability to Plants of Rock Phosphates of Igneous and Sedimentary Origins, *Soil Sci. Soc. Am. Proc.* **32**:79–82.

Huffman, E. O., 1970, Fertilizer-Soil Reactions and the Phosphate Status of Soils, *Phosphorus Agric.* **55**:13–23.

Lutz, J. A., Jr., 1973, Effect of Partially Acidulated Rock Phosphate and Concentrated Superphosphate on Yield and Chemical Composition of Alfalfa and Orchardgrass, *Agron J.* **65**:286–289.

Marais, J. N., and D. Wiersma, 1975, Phosphorus Uptake by Soybeans as Influenced by Moisture Stress in the Fertilized Zone, *Agron. J.* **67**:777–781.

Marbut, C. F., 1935, Soils of the United States, Pt. III, *Atlas of American Agriculture,* USDA, Washington, D.C.

Martin, J. K., 1970, Organic Phosphate Compounds in Water Extracts of Soils, *Soil Sci.* **109**:362–375.

Miller, M. H., 1965, Influence of $(NH_4)_2SO_4$ on Root Growth and P Absorption by Corn from a Fertilizer Band, *Agron. J.* **57**:393–396.

Olsen, S. R., and F. S. Watanabe, 1966, Effective Volume of Soil around Plant Roots Determined from Phosphorus Diffusion, *Soil Sci. Soc. Am. Proc.* **30**:598–602.

Oniani, O. G., Margaret Chater, and G. E. G. Mattingly, 1973, Some Effects of Fertilizers and Farmyard Manure on the Organic Phosphorus in Soils, *J. Soil Sci.* **24**:1–9.

Parfitt, R. L., R. J. Atkinson, and R. St. C. Smart, 1975, The Mechanism of Phosphate Fixation by Iron Oxides, *Soil Sci. Soc. Am. Proc.* **39**:837–841.

Parker, F. W., J. R. Adams, K. G. Clark, K. D. Jacob, and A. L. Mehring, 1946, Fertilizers and Lime in the United States, *USDA Misc. Pub.* 586.

Pearson, R. W., and R. W. Simonson, 1939, Organic Phosphorus in Seven Iowa Soil Profiles, Distributions and Amounts Compared to Organic Carbon and Nitrogen, *Soil Sci. Soc. Am. Proc.* **4**:162–167.

Richards, G. E., 1975, Phosphorus Fertilization: Can We Do a Better Job? *Crops Soils* **27** (4):12–15.

Robertson, W. E., L. G. Thompson, Jr., and C. E. Hutton, 1966, Availability and Fractionation of Residual Phosphorus in Soils High in Aluminum and Iron, *Soil Sci. Soc. Am. Proc.* **30**:446–450.

Rotini, O. T., and M. El-Nennah, 1971–1972, Evaluation of the Relative Availability of Inorganic Phosphorus Fractions in Some Italian Soils Using Isotopic Exchange and Solubility Criteria, *Agrochimica* **16**:23–32.

Sanchez, P. A., and A. M. Briones, 1973, Phosphorus Availability of Some Philippine Rice Soils as Affected by Soil and Water Management Practices, *Agron. J.* **65**:226–228.

Sell, O. E., and L. C. Olson, 1946, The Effect of Surface-applied Phosphate and Limestone on Soil Nutrients and pH of Permanent Pasture, *Soil Sci. Soc. Am. Proc.* **11**:238–245.

Smeck, N. E., 1973, Phosphorus: An Indicator of Pedogenetic Weathering Processes, *Soil Sci.* **115**:199–206.

Syers, J. K., J. D. H. Williams, A. S. Campbell, and T. W. Walker, 1967, The Significance of Apatite Inclusions in Soil Phosphorus Studies, *Soil Sci. Soc. Am. Proc.* **31**:752–756.

Thompson, L. M., 1950, *The Mineralization of Organic Phosphorus, Nitrogen and Carbon in Virgin and Cultivated Soils,* doctoral dissertation, Iowa State University, Ames.

Thompson, L. M., C. A. Black, and J. A. Zoellner, 1954, Occurrence and Mineralization of Organic Phosphorus in Soils, with Particular Reference to Associations with Nitrogen, Carbon, and pH, *Soil Sci.* **77**:185–196.

Vijayachandran, P. K., and R. D. Harter, 1975, Evaluation of Phosphorus Adsorption by a Cross Section of Soil Types, *Soil Sci.* **119**:119–126.

Vyas, M. K., and D. P. Motiramani, 1971, Effect of Organic Matter, Silicates and Moisture Levels on Availability of Phosphate, *J. Indian Soc. Soil Sci.* **19**:39–43.

Younge, O. R., and E. L. Plucknett, 1966, Quenching the High Phosphorus Fixation of Hawaiian Latosols, *Soil Sci. Soc. Am. Proc.* **30**:653–655.

Chapter 12

Potassium

Plants absorb large amounts of potassium, all of it in the form of the K^+ ion. The positive charges of the potassium cations help to maintain electrical neutrality in both soil and plants by balancing the negative charges of nitrate, phosphate, and other anions. Plants require relatively large amounts of potassium and often can use more than the soil can supply. Potassium is the third most likely nutrient element to limit plant growth and is therefore a very common constituent of fertilizers. All "complete" fertilizers contain nitrogen, phosphorus, and potassium.

Potassium in the soil occurs as potassium ions in mineral structures and as hydrated potassium ions either in solution or adsorbed on cation-exchange sites.

The nonhydrated potassium ion is an important constituent of micas and of some of the feldspars. It is nearly as large as an oxygen ion. It fits so perfectly into some of the holes in clay structures that it can become "fixed" there and become almost nonexchangeable. Such immobilized potassium occupies the same kind of spaces as potassium ions that hold mica sheets together, as described in Chapter 6.

Exchangeable potassium occurs as hydrated potassium ions attracted to negatively charged sites of clay and organic matter. These hydrated potassium ions are the same size and held with the same strength as ammonium ions but much less tightly than the other macronutrient cations, calcium and magnesium. Hydrated potassium ions attracted to micelles are readily exchangeable because of the relatively low energy of attraction. They may be able to shift positions along micelle surfaces without too much difficulty.

The soil solution is probably never saturated with potassium ions. Most potassium compounds are readily soluble in water, and potassium ions are withdrawn from solution by adsorption to cation-exchange sites long before saturation of the soil solution occurs. Nevertheless, the potassium in solution is the part that is more readily available to plants. Exchangeable potassium is available if a plant root reaches it but will not move to the root unless an exchange occurs. Even some of the nonexchangeable potassium is either slightly available or slowly becomes available during the growing season. Equilibrium relations among the different forms of potassium are illustrated in Figure 12-1.

DISTRIBUTION OF POTASSIUM IN SOILS

Potassium does not enter into the covalent bonds of organic compounds as do nitrogen, phosphorus, and sulfur. It remains as active ions that are held inside living plant cells but are readily leached out of dead organic matter. Therefore only inorganic potassium need be considered in the distribution of potassium in the soil.

The average rate of weathering of potassium minerals is about the same as that of all minerals in the soil. The percentage of potassium in a soil does not change much even over long periods of time. The A, B, and C horizons of even a strongly developed soil are likely to contain approximately equal percentages of total potassium. Between different soils, however, these percentages range from less than 0.1 percent potassium by weight to over 4 percent. The average percentage, as shown in Table 12-1, is near 1.5 percent potassium.

An average hectare-furrow slice contains approximately 30,000 kg of potassium. Exchangeable potassium usually constitutes between 100 and 400 kg per hectare-furrow slice, and dissolved potassium in the soil solution is nearly always less than 5 kg per hectare-furrow slice. The nonexchangeable potassium is probably about 99 percent of the total potassium in an average soil. Exchangeable potassium is about 1 percent of the total, and the potassium in solution usually constitutes only about 0.01 percent or less of the total potassium in the soil.

Sources of Potassium

The principal soil minerals containing potassium are muscovite mica, biotite mica, and orthoclase feldspar. Nonexchangeable potassium from within the mineral structures is released as they weather. Milford and Jackson (1966) reported a very high correlation between the surface area of clay-size mica in the soil and the rate of conversion of nonexchangeable to exchangeable potassium. Their results do not imply that all the newly exchangeable potassium came from the mica but only that other sources, such as the feldspar present, were also proportional to the mica.

$$\text{Nonexchangeable } K^+ \underset{}{\overset{\text{(Slow)}}{\rightleftharpoons}} \text{Exchangeable } K^+ \underset{}{\overset{\text{(Rapid)}}{\rightleftharpoons}} \text{Dissolved } K^+$$

Figure 12-1 Equilibrium relationships among the three divisions of potassium ions in soils.

Table 12-1   Total Potassium Contents of the A, B, and C Horizons of Several Kinds of Soil

Soil type	Soil order (see Chapter 16)	% K		
		A horizon	B horizon	C horizon
Granada silt loam	Alfisol	1.7	1.9	1.9
Miami silt loam	Alfisol	1.7	1.7	1.5
Putnam silt loam	Alfisol	1.4	1.6	1.8
Mohave loam	Aridisol	2.3	1.5	1.8
Vernon clay loam	Inceptisol	2.1	1.8	2.4
Williamson silt loam	Inceptisol	1.7	2.5	2.6
Barnes silt loam	Mollisol	1.6	1.7	1.3
Marshall silt loam	Mollisol	1.8	1.7	1.6
Nipe clay	Oxisol	0.1	0.1	0.1
Becket fine sandy loam	Spodosol	1.7	2.9	3.1
Greenville fine sandy loam	Ultisol	0.3	0.3	0.2
Iwo sandy loam	Ultisol	0.7	1.9	3.7
Norfolk sandy loam	Ultisol	0.1	0.1	0.2
Sassafras sandy loam	Ultisol	1.3	1.6	1.7
Tatum silt loam	Ultisol	1.3	1.6	3.5

Source: Calculated from various sources.

Munn, Wilding, and McLean (1976) showed that sand, silt, and clay fractions can all release potassium but that the smaller particles release the most potassium per unit weight.

The importance of the release of nonexchangeable potassium may be realized by considering that a good crop of alfalfa will withdraw about 150 kg of potassium per hectare from the soil each year. This amount compared with the 100 to 400 kg of readily available potassium in a hectare-furrow slice reveals that such withdrawals cannot long continue without a source of replenishment. The supply of readily available potassium, even including subsoil potassium, is quite limited.

Most crops use less potassium than alfalfa (Table 12-2), but they may still need more during the growing season than the soil has available at any one time. Talati, Mathur, and Attri (1974) found the uptake of potassium by wheat growing on an unfertilized soil exceeded the available potassium content of the soil. The available potassium content actually increased during the early stages of growth from 140 to 255 kg/ha in mid-season, then declined to 240 kg/ha by harvest time. The release of nonexchangeable potassium was vital to the growth of the crop. In fact, Talati, Mathur, and Attri showed a significant correlation between the amounts of available and nonexchangeable potassium in the soil. The more nonexchangeable potassium there is to draw on, the higher is the available potassium level maintained in the soil.

Potassium removal by crop growth and the loss caused by leaching are partially balanced when plant and animal residues are returned to the soil. But, there is a net loss that must be made up from some other source. The shortage can be offset for a time as weathering releases nonexchangeable potassium. But, long continued reliance on nonexchangeable potassium can deplete it, too. Potassium fertilization

Table 12–2 Average Potassium Contents of Selected Crops

Crop	Part of crop	Yield, tons/ha	% K	K, kg/ha
Alfalfa	Hay	9	1.87	168
Barley	Grain	3	0.45	13
	Straw	2.5	1.25	31
Corn	Grain	7	0.37	26
	Stover	8	1.27	102
Cotton	Lint, seed	1.5	0.80	12
	Stalks, leaves	1.5	2.00	30
Cowpeas	Hay	5	1.88	94
Lespedeza	Hay	8	0.97	78
Oats	Grain	3	0.50	15
	Straw	3	1.16	35
Peanuts	Nuts	2.5	0.40	10
	Vines	8	0.55	44
Potatoes	Tubers	20	0.53	106
Red clover	Hay	5	1.45	72
Soybeans	Grain	2.5	2.22	55
	Straw	3	0.68	20
Sugar beets	Roots	40	0.15	60
Sweet clover	Hay	12	1.37	164
Sweet potatoes	Roots	20	3.44	69
Timothy	Hay	4	1.23	49
Tobacco	Leaves	1.5	4.40	66
Tomatoes	Fruit	25	0.33	83
Wheat	Grain	3	0.50	15
	Straw	3	0.68	20

Source: Calculated from various sources.

becomes necessary when the soil is no longer able to maintain an adequate level of available potassium.

The most concentrated natural sources of potassium are evaporite deposits of the mineral sylvite (KCl) formed in arid climates. Sylvite precipitates where the conditions are right as a body of briny water evaporates. Sylvite is mined and serves as a source of fertilizer (Figure 12-2). Frequently the sylvite is mixed with other salts such as halite (sodium chloride) in natural deposits. The mixture of sylvite and halite is called sylvinite.

Factors Affecting the Release of Potassium

Potassium is said to be *released* when it changes from a nonexchangeable to an exchangeable form. Release involves weathering processes that break apart, or at least open, the mineral structure. The potassium ions are relatively large cations, nearly as large as oxygen ions. They fit between the layers in micas by penetrating into the six-sided openings in tetrahedral sheets on both sides of them (Chapter 6). Potassium ions in these openings are ionically bonded to both adjoining tetrahedral sheets. The potassium-to-oxygen bonds must be overcome for the layers to separate

Figure 12-2 Mechanical loading of raw potash salts into conveyor cars in a potash mine near Carlsbad, N. Mex. *(American Potash Inst.)*

enough for potassium to escape. Potassium ions in feldspars occur in spaces surrounded by tetrahedra but in a three-dimensional framework rather than in layers. Potassium is released when weathering breaks off fragments from the framework of a potassium feldspar.

The relative resistance to release of potassium from the three principal potassium-containing minerals is as follows: orthoclase feldspar, most resistance; muscovite mica, medium resistance; biotite mica, least resistance.

Minerals release potassium most rapidly when there are environmental changes to promote weathering. Physical effects such as wetting and drying, warming and cooling, and freezing and thawing cause strains in mineral lattice structures. Potassium escapes more readily while the structures are strained. Graham and Lopez (1969) showed that freezing and thawing cause the potassium supply to shift toward an equilibrium. Potassium is released if the supply in solution has been depleted, but potassium can reenter mineral structures if the solution concentration is high. The weathering processes act to speed up the rate of the equilibria shown in Figure 12-1.

Movement of potassium ions from solution into mineral structures is called *potassium fixation.* Potassium fertilization often results in a high enough concentration of potassium to result in fixation. Hanotiaux and Felipe-Morales (1972) measured fixation capacities up to 4 mg of K/100 g of soil (80 kg of K per hectare-furrow slice) in Belgian soils. They concluded that the type of clay mineral is the most important factor in potassium fixation. Potassium is fixed by expanding lattice minerals—montmorillonite and some illite. The fixed potassium should be regarded as stored rather than lost. Most of it is in the most accessible potassium sites of the illite structure, whereas the remaining native potassium is mostly in the least accessible sites. Therefore the fixed potassium is more readily released than the native potassium in illite. Still, the fixed potassium is rather tightly held and may remain in its place for years.

Page et al. (1963) found that the cation-exchange capacity of the soil is reduced by an amount equal to the amount of potassium fixation. The electrical charges are still there but are not measured in the cation-exchange determination because the potassium ions are held too tightly.

Weathering opens layers and breaks linkages most readily at the outer surfaces of the mineral particles. The bulk of a large particle is therefore resistant to weathering. Potassium is most readily released from clay-size particles of minerals such as biotite, but the supply of weatherable particles becomes depleted. The rate of release becomes slower as the soil becomes older because the potassium, as well as other nutrients, must come from either large particles or more resistant minerals.

The importance of particle size and of the degree of protection of the potassium within the particle may be illustrated by a comparison of muscovite and illite. Muscovite is a primary mineral that resists decomposition and therefore releases its potassium very slowly. Illite may be considered as clay-sized, partially weathered particles of muscovite. Potassium is released more readily from the weakened mica structure and smaller particles of illite.

Leaching of Potassium

Potassium fixation and the cation-exchange capacity of soils are so effective for storing potassium that potassium is difficult to leach. Paterson and Richer (1966) found that 77 years of potassium applications on a Hagerstown silt loam soil had significantly increased the supply of exchangeable K^+ but only in the upper part of the soil. The increase was noted in the top 60 cm of plots that had received 177 kg of KCl (93 kg of K) per hectare each year and only in the top 30 cm of plots that had received annual applications of 13 to 22 tons of manure per hectare.

Vittum, Lathwell, and Gibbs (1968) demonstrated that potassium can be leached but that the process is slow. They supplemented New York precipitation with 10 cm of irrigation water per year. Thirteen years of this leaching reduced available potassium in the furrow slice from 109 to 97 kg/ha. The loss by leaching caused by the irrigation was approximately equalled by the gain from fertilization when 54 kg/ha of potassium fertilizer was applied per year. Actually, much of the "leached" potassium was still within the soil profile but at a depth below the furrow slice.

Leaching occurring in a more humid climate is probably similar to that resulting from supplemental irrigation, but the inherent variability of soils makes the climate effect difficult to evaluate quantitatively. It can be noted, however, that in the United States most of the potassium fertilizer is used in the eastern half where the climate is humid; little is used in the western half where the climate is more arid. Exceptions occur where irrigation has been practiced for many years. Irrigation water leaches potassium from the soil and increases crop needs for potassium through increased yield potential.

Soils of humid regions are likely to contain less total and available potassium than those of arid regions. The difference in available potassium is often proportionately greater than the difference in total potassium. Most of the potassium remaining in the humid-region soils is in sand-size particles of minerals like feldspars and muscovite which are slow to weather. Otherwise it would all be gone. In addition, potassium fixation is less common in the more weathered soils. Montmorillonite clay is common in the less weathered soils, and its 2:1 lattice structure with a relatively high charge density on its layers can fix potassium. Kaolinite clay is common where weathering is more intense, but its 1:1 lattice structure does not fix potassium. Moreover, the lower cation-exchange capacity of kaolinite provides fewer storage sites for potassium and other cations. Tropical soils high in oxide clays have very little cation-exchange capacity and store even less potassium than soils dominated by kaolinite.

Sodium versus Potassium in Soils

The chemistry of sodium is quite similar to that of potassium, but its behavior in the soil is somewhat different. Sodium occurs in feldspars but not in micas. The sodium feldspars weather a little more rapidly than the potassium feldspars.

Sodium ions that have been released to the soil solution are not subject to fixation and are less tightly held to cation-exchange sites than potassium, magnesium, or calcium. Sodium is therefore the easiest basic cation to leach from the soil. This accounts for the sodium content of soils gradually decreasing with time while the potassium content remains nearly constant. Much of the sodium eventually reaches the ocean where sodium chloride is the most abundant salt.

Sodium salts sometimes become concentrated in soils of arid regions because of lack of leaching. This can lead to the formation of sodic soils, as discussed in Chapter 7. The dispersed colloids and high pH of sodic soils make them nearly barren of plant growth.

POTASSIUM AVAILABILITY TO PLANTS

Most soils contain between 1 and 5 kg/ha of potassium dissolved in the soil solution. Dissolved potassium can reach plant roots by diffusion, by mass flow of water being absorbed by the plant roots, or by root elongation. Electrical balance can be maintained in either of two ways when the K^+ is absorbed. The root may exchange another cation such as H^+ for K^+, or it may absorb an anion such as NO_3^- or $H_2PO_4^-$ along with the K^+. Dissolved potassium is very important because it is so readily and completely available, but the amount of it present is too small to meet all plant needs.

Exchangeable K^+ ions cannot move to the plant root unless they are replaced by other cations on the exchange sites. Plant roots that come close enough to such K^+ ions may exchange H^+ ions for them. This process is known as *contact absorption*. Kauffman and Bouldin (1967) found that most of the potassium uptake is accomplished by root extensions that are less than a week old and that almost all the absorbed K^+ comes from the zone within 4 to 5 mm of the root. Exchangeable potassium was reduced as much as 25 percent in that zone during a single week. Farr, Vaidyanathan, and Nye (1969) found an even greater concentration gradient with exchangeable K^+ equal to 1.6 meq/100 g of soil beyond 8 mm from the roots reduced to 0.6 meq next to the roots in 4 days. The concentration next to the roots was back up to 0.9 meq/100 g after 8 days, and there was some depletion out as far as 12 mm. Some exchangeable potassium ions may have been absorbed by contact absorption. Others would have entered the soil solution through an exchange reaction before being absorbed. Exchangeable potassium is an important source for replenishment of a dissolved potassium supply that has been depleted by plant growth.

Nonexchangeable potassium must be released before it can be utilized. Even so, it is an important reserve. Potassium ions from this reserve slowly replenish the supply of dissolved and exchangeable ions when these are depleted. Fraps (1929) grew corn followed by either kafir or sorghum in 72 soils in pots in a greenhouse. He checked the exchangeable potassium supply before and after cropping (Table 12-3) and found that the potassium used by the crop was roughly proportional to the amount of exchangeable potassium in the soil. The exchangeable potassium utilized was about half replenished from nonexchangeable sources even while the crops were growing. Much of the remainder would normally be replenished from plant residues and nonexchangeable reserves during winter periods or other times of dormancy.

The net result of the year-round release of nonexchangeable potassium is a cyclic variation in the available potassium supply. Available potassium is highest in the early part of the growing season before the plant needs are very high. Available potassium declines during the rapid growth phase and reaches a low near the end of the growing season. A gradual increase occurs as potassium is released while the plants are dormant.

The cyclic pattern of exchangeable potassium supply can occur at either high or low levels of potassium. Thomas and Hipp (1968) analyzed the results obtained

Table 12–3   Relation between Exchangeable Potassium in Soil and Uptake of Potassium by Crops

Number of soils	Potassium removed by crops, kg/ha		Exchangeable potassium in soil, ppm		
	Group range	Group average	Before cropping	After cropping	Lost by cropping
11	0–46	39	100	83	17
20	46–93	73	110	85	25
31	93–186	135	167	116	51
1	186–280	221	153	108	45
1	373–466	428	432	202	230
3	466–559	491	548	290	258
4	559–652	595	467	284	183
1	652–745	685	639	315	324

Source: Calculated from data of Fraps, 1929.

by a number of researchers and concluded that the release of nonexchangeable potassium is roughly proportional to the amount of weatherable mica in the soil. Young soils high in mica have a high rate of release and therefore replace the exchangeable potassium nearly as fast as it is used, thus maintaining it at a high level. Such soils may be cropped for many years without need of potassium fertilizer.

Weathering gradually removes the potassium minerals from older soils and the rate of release of nonexchangeable potassium gradually declines. The equilibrium level of exchangeable potassium declines also to the point where cropping can soon deplete the soil of potassium. The same is true of some young sandy soils that had low initial supplies of potassium minerals. Fertilizer potassium is vital for sustained cropping of soils in which potassium release from nonexchangeable sources is slow.

The Effect of Liming on Potassium Availability

Other factors being equal, acid soils are more likely to be deficient in available potassium than neutral soils. The acid soils have more H^+ ions and less of all other cations including K^+ on exchange sites and in solution than neutral soils have. Generally the acid soils are more highly weathered and therefore release potassium more slowly than the neutral soils.

Soil acidity can be neutralized by liming, which adds Ca^{++} and often some Mg^{++} ions to the soil but not K^+. There may be an indirect effect on available potassium, however, and this has been the subject of much investigation with conflicting results. It appears that under some circumstances the available potassium supply may be increased for a time by liming but that the reverse is more often true. Potassium fertilizer is likely to be needed in either case because liming increases plant growth and thereby increases the demand for potassium.

A review by Reitemeier (1951) cites investigations in Hawaii, New York, and Alabama in which the addition of Ca^{++} to acid soils improved the availability of K^+. The Ca^{++} ion is more strongly attracted by cation exchange than the K^+ ion; adding lime can therefore drive exchangeable K^+ out into the soil solution. Liming could thus cause a short-term increase in potassium availability if the clay present is kaolinite (potassium fixation absorbs the K^+ ions if montmorillonite clay is present).

Several investigators have found that potassium release is faster at a low pH than at a higher pH. The data in Table 12-4 were obtained by replacing all exchangeable bases in samples of Clyde silt loam with H^+ alone, Ca^{++} alone, or H^+ plus Ca^{++}. The H^+-saturated sample replenished about half of the exchangeable potassium in 1 month and at the end of a year was practically back to the exchangeable potassium content of the original soil. The soils containing Ca^{++} released much less K^+ from nonexchangeable to exchangeable positions.

Haagsma and Miller (1963) reported results showing that Na- or Ca-saturated resins would not draw nonexchangeable potassium from a Haldimond clay but that H-saturated resins did cause the clay to release potassium. Evidently the H^+ ion has an effect on certain mineral structures that Na^+ and Ca^{++} ions do not have.

Rich and Black (1964) found that NH_4^+ is also a relatively effective ion for extracting nonexchangeable potassium from soil. The effectiveness of NH_4^+ results

Table 12–4   Rate of Potassium Release in Clyde Silt Loam

Soil treatment	Initial pH	Initial exchange-able K*	Exchangeable K* after incubation periods of				K released as percent of original
			30 days	60 days	120 days	360 days	
Original soil	6.25	273	269	273	275	323	18
H⁺-saturated	4.00	—	134	182	247	262	96
Ca⁺⁺ H⁺-saturated	6.25	—	79	107	126	163	60
Ca⁺⁺-saturated	7.42	—	37	71	84	104	38

* Expressed in kilograms per hectare-furrow slice.
Source: Unpublished data from Kirk Lawton, Iowa State University.

from its having the same charge and size as the K^+ ion. The NH_4^+ ion can fit wherever the K^+ ion can without changing the mineral structure.

Several things could happen to the exchangeable potassium that is replaced by calcium when a soil is limed. It might enter the soil solution and be absorbed by growing plants, or it might be moved to a lower depth or lost by leaching. Lysimeter experiments, however, indicate that potassium leaching losses are reduced when the soil is limed (Reitemeier, 1951). It appears that much of the formerly exchangeable potassium becomes fixed after a soil is limed and that the rate of release of nonexchangeable potassium is reduced by the presence of more calcium ions. Such fixation, of course, is limited to soils containing 2:1 lattice clays such as montmorillonite and illite.

The Calcium-to-Potassium Ratio

Plants contain about twice as much potassium as calcium as a result of preferential absorption in spite of the abundance ratio in the soil. Potassium generally represents only about 1 to 3 percent of the exchangeable bases. Usually 75 to 85 percent of the exchangeable bases in soils containing montmorillonite clay are calcium, and 12 to 18 percent are magnesium. The ratio of exchangeable calcium to potassium in soils containing montmorillonite clay is commonly 50:1 and can be as wide as 100:1 while still providing adequate potassium nutrition for growing plants.

The ratio of calcium to potassium in solution is much narrower than the exchangeable cation ratio because calcium is held more strongly than potassium on cation-exchange sites. An exchangeable calcium-to-potassium ratio of 100:1 on montmorillonite clay may be in equilibrium with a dissolved calcium-to-potassium ratio of only 5:1 in the soil solution. Even so, plants containing twice as much potassium as calcium are exhibiting a tenfold preference for absorbing potassium over calcium.

Kaolinite clays generally have narrower exchangeable calcium-to-potassium ratios than montmorillonite. The cation-exchange capacity of kaolinite is only about one-tenth as high as that of montmorillonite, and the percent base saturation of kaolinite is generally lower. Any given amount of exchangeable potassium is therefore a much larger percentage of the total bases if the dominant clay is kaolinite instead of montmorillonite. Furthermore, Ca^{++} ions are more readily dissociated from kaolinite than from montmorillonite. The difference between the exchangeable-calcium-to-potassium ratio of kaolinite and the ratio of these cations in solution is therefore much smaller than it is with montmorillonite.

Bear, Prince, and Malcolm (1945) have suggested that for New Jersey soils it is desirable to maintain a ratio of about 13:1 for exchangeable Ca^{++} to exchangeable K^+. This ratio appears to be quite logical for kaolinite soils because it will give nearly the same calcium-to-potassium ratio in solution as found where the clay is montmorillonite and the exchangeable-calcium-to-potassium ratio is 100:1. A kaolinite soil with exchangeable bases of 80 percent Ca^{++}, 10 percent Mg^{++}, and 6 percent K^+ provides a satisfactory balance of these macronutrient cations. Such a combination in a montmorillonitic soil with a high cation-exchange capacity would probably result in "luxury consumption" of potassium.

Potassium Availability in Calcareous Soils

Naturally neutral soils usually have adequate supplies of all bases including potassium. Most neutral soils are either young or too dry to have lost much of their nutrient-supplying power. The same is true of many calcareous soils. Crops grown on calcareous soils of subhumid, semiarid, and arid regions rarely respond to potassium fertilization.

Certain high-lime soils of Illinois and Iowa are quite different from the usual calcareous soils. These are young soils with poor natural drainage. They contain illite and montmorillonite clays with a high capacity for fixing potassium. Potassium ions are so loosely attracted to the micelles that they are mostly replaced by the more abundant and more strongly held calcium ions. Much of the potassium is fixed by the clays, with the result that these soils contain much less available potassium than adjoining noncalcareous soils (Table 12-5). Crops grown on this kind of high-lime soil respond well to potassium fertilization.

Influence of Water and Air on Potassium Availability

An increase in the water content of the soil increases the supply of dissolved nutrients, but the potassium increase is overshadowed by other factors. More water to contain dissolved plant nutrients is only one reason plants grow better on moist soil than on dry soil. It would be hard to detect whether an increased uptake of potassium from a moist soil was caused by increased availability or simply by more vigorous plant growth.

Too much water in the soil is more limiting to potassium uptake than too little water. Aeration is limited when there is too much water in the soil, and root respiration is therefore limited. As previously noted, plant roots take up potassium preferentially. They continue to absorb potassium even when its concentration inside the root is considerably higher than the concentration in the soil solution. This absorption against a concentration gradient requires energy, which comes from root respiration. Plant uptake of potassium is therefore severely hampered when excess water in the soil inhibits respiration.

Table 12–5   Relation between Exchangeable Potassium and Yield of Corn on Normal and High-lime Areas of Webster Soils

Field number	Area sampled	Exchangeable potassium, kg/ha	Corn yield (1940), q/ha
1	Normal	430	35.2
	High-lime	165	20.8
2	Normal	347	46.4
	High-lime	193	25.0
3	Normal	303	39.5
	High-lime	130	14.2
4	Normal	455	44.8
	High-lime	130	10.6

Source: Calculated from data of Stanford, Kelly, and Pierre, 1941.

Lawton (1945) found that decreased aeration resulting from soil compaction caused a decrease in the uptake of potassium. Conversely, he found that forced aeration of soils increased potassium uptake by plants. Chang and Loomis (1945) found that bubbling CO_2 through nutrient solutions decreased potassium absorption by wheat and caused corn to lose potassium to the solution. Removal of the CO_2 end product of respiration is as important for potassium absorption as maintaining an adequate supply of oxygen.

Bower, Browning, and Norton (1944) studied the effect of tillage practices on plant nutrient deficiencies. They noted that corn growing on soils which had not been plowed showed signs of inadequate potassium, while corn growing on adjacent similar soils which had been plowed appeared to have adequate potassium. Plowing had increased aeration in comparison with shallower tillage, and the increased aeration favored the uptake of potassium.

POTASSIUM IN PLANTS

Potassium is present in plants in the form of organic and inorganic salts. It does not form an integral part of the structure of any known organic compound in plants.

Nearly all potassium salts are soluble and highly ionized in solution. The K^+ ions are mobile within the plant but will not leach out of healthy living plant tissue. They are readily leached from dead plant tissue.

Potassium aids in the uptake of other nutrients and in their movement within the plant. For example, K^+ and NO_3^- may move together. The presence of potassium and other ions in solution helps to maintain the osmotic concentration necessary to keep the cells turgid. Potassium is also important in metabolism in the formation of carbohydrates and proteins.

Crop Response to Potassium and Sodium

Potassium is a macronutrient needed in relatively large amounts by all plants. Sodium is chemically very similar to potassium and is needed by animals in large amounts but is a nonessential element for plant growth. Even so, the two elements

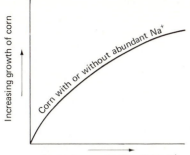

Figure 12-3 The response of corn to available potassium. The amount of sodium present makes little or no difference to the growth of corn.

are very similar, and some of the functions ascribed to potassium are so nonspecific that sodium could be expected to substitute for potassium to some degree. The degree of substitution appears to vary with different plants. Investigators have found three different types of responses by plants supplied varying amounts of potassium and sodium. Corn, cotton, and safflower will serve as examples.

Corn responds to variations in potassium supply as shown in Figure 12-3. This curve is a typical plant response to variations in the supply of any essential growth factor. Furthermore, the response to potassium is the same whether sodium is nearly absent or is present in considerable amounts. If sodium makes any difference in corn growth, it must be at extremely low concentrations. Such low concentrations of sodium never occur in nature and are difficult to produce artificially.

The response of cotton to variations in potassium with and without the addition of sodium is shown in Figure 12-4. The response of cotton to potassium supply when sodium is virtually absent is similar to the response of corn to potassium. But sodium improves the growth of cotton when potassium is deficient. Johan and Amin (1965) found that sodium could greatly delay and reduce, though not always eliminate, potassium deficiency symptoms in cotton. Safflower responds as shown in Figure 12-5 to variations in the supply of potassium and sodium (Aslam, 1975). The potassium response curve with hardly any sodium present is also similar to the curve for corn response to potassium. The curve with adequate sodium supplied is 40 to 50 percent above the other curve at all levels of potassium. It appears that safflower

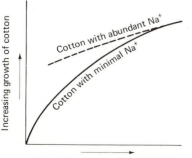

Figure 12-4 The response of cotton to available potassium with and without abundant sodium present. Sodium can partially substitute for potassium in cotton when potassium is deficient.

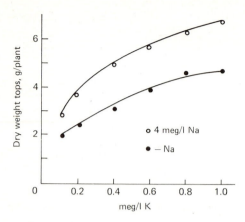

Figure 12-5 The effect of K and Na treatments on the dry weight of safflower seedlings. Sodium increases the growth of safflower at all levels of potassium availability. (*Aslam*, 1975.)

needs both potassium and sodium in macro amounts for maximum rate of growth. The two elements can only partially substitute for each other. Sugar beets have similarly been shown to benefit from sodium along with potassium (Adams, 1961).

Rubidium versus Potassium

Rubidium is the next heavier element that forms monovalent cations. Hafez and Rains (1972) reported that rubidium can partially replace potassium in plant nutrition. Cotton and barley roots were tested and shown to retain Rb^+ preferentially over K^+ ions.

Neither rubidium nor sodium can fully replace potassium. Potassium is the only one of the three that has been shown to be essential for plants to grow and reproduce themselves. Sodium and rubidium should be considered as functional rather than as essential elements.

Potassium Deficiency Symptoms

Potassium deficiencies, like many others, cause plants to be stunted. Stunting, however, tells little by itself because it can have other causes and because it is difficult to detect without a healthy plant for comparison. Better indications can be had by examining the plant tissue.

Potassium moves readily within the plant and tends to be translocated to areas of growth. Deficiency symptoms therefore show first on the older parts of the plant. The potassium moves through the veins and tends to be more concentrated in the veins than elsewhere. First the edges of the older leaves and then areas between veins turn yellow and then brown. Small, brown, necrotic (dead) spots develop while the veins are still green.

A slice across a potassium-deficient cornstalk reveals another deficiency symptom. The vascular bundles are discolored by a dark precipitate. Also the lower part of the cornstalk and its brace roots are weakened, and lodging is increased (Liebhardt and Murdock, 1965). Small-grain crops deficient in potassium also lodge because their straws are weakened.

Soil Tests for Potassium

Many investigators have shown a high correlation between the level of exchangeable K^+ and crop yield. This correlation seems to hold for soils that have not been exposed to too much leaching. Most states in arid and subhumid climates use a neutral ammonium acetate solution to extract exchangeable K^+ and find a good correlation between this test and plant uptake. Several states along the Atlantic coast, however, use acid extractants to remove some nonexchangeable as well as exchangeable K^+ from the more strongly weathered soils that occur there. Both methods are attempts to estimate that elusive quantity called available potassium.

The graph in Figure 12-6 was taken from the work of Bray in Illinois. Bray found a good correlation between crop response to potassium fertilization and the amount of exchangeable K^+ in the soil up to about 150 kg of exchangeable K^+ per hectare-furrow slice. Beyond 150 kg the curve became too flat for reliable predictions; the crop was already yielding 70 or 80 percent of the probable yield attainable with 300 kg or more of available potassium.

Under Iowa and Illinois conditions it is quite certain that 100 kg or less of exchangeable K^+ will prevent the attainment of maximum crop yields unless supplemental potassium is added in fertilizer. Also, crop response to fertilizer potassium is usually small if the soil contains over 200 kg of exchangeable K^+ per hectare-furrow slice. The problem is that many soils contain between 100 and 200 kg of exchangeable K^+ per hectare-furrow slice and only about half of these will produce a significantly larger crop with potassium fertilizer than without it. The test for exchangeable K^+ is useful but should be supplemented with other information, especially in the 100- to 200-kg range. A knowledge of other soil characteristics is necessary in deciding which soils need potassium fertilizer.

Soil Conditions that Influence Potassium Needs

Degrees of weathering and leaching influence potassium needs. Generally a soil that contains a Cca horizon will not need potassium fertilizer because it has not been

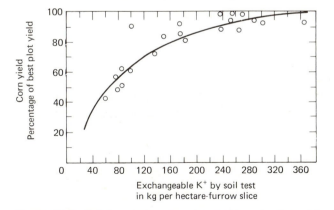

Figure 12-6 Relations between exchangeable potassium and corn yields on plots receiving no potassium fertilizer. (*Bray*, 1944.)

thoroughly leached. In fact, the general line delineating the area where potassium fertilizer is often needed passes through a zone somewhat too humid for the soils to contain Cca horizons. In the United States most of the soils that need potassium fertilization lie east of a line passing from Minnesota to eastern Texas.

The potassium fertilizer line passes directly across the state of Iowa. This state may be divided into three parts. In the eastern third over half of the land in crops needs potassium fertilizer. In the center third about one-third of the land is low in available potassium, and in the western third only a few areas need potassium fertilizer.

The following generalities were developed from experience in Iowa and probably apply to a much larger area of soils having mostly montmorillonite and illite clay minerals.

Naturally neutral (pH 6.5 to 7.5) well-drained soils likely do not need potassium fertilizer. Soils that have not been leached enough to need lime probably still contain adequate potassium. Good drainage makes it possible for the plants to make effective use of the available potassium.

Plants grown on naturally neutral, imperfectly drained soils are likely to give a moderate response to additions of potassium fertilizer. Many of these soils remain wetter than naturally well-drained soils even though a drainage system has been installed. The problem lies in the reduced ability of the plants to absorb available potassium because aeration and root respiration are restricted. The available potassium level should be raised higher in these soils than in well-drained soils.

Well-drained, naturally acid soils that have been limed are likely to contain less available potassium than their neutral counterparts. Crops are likely to show a moderate response to potassium fertilizer applied to raise the available potassium supply to a normal level.

Crops grown on soils that have required both liming and draining are likely to be highly responsive to potassium fertilizer. Not only is the native supply of available potassium likely to be low, but also the needed level for maximum production is likely to be high.

High lime contents in soils often accentuate drainage-related differences in potassium needs. Well-drained soils that are calcareous because of erosion or because of arid climate are likely to be well supplied with available potassium. However, the poorly drained calcareous soils of Iowa and Illinois generally show high crop response to potassium fertilizer. The calcium in these soils causes part of the potassium to become fixed by the clay and competes with the remainder for plant absorption. These soils are difficult to drain completely; the resulting poor aeration restricts selective absorption of potassium by the plant roots. The available potassium supply in these soils should be raised higher than for even the other wet soils of the region.

Soil texture is less important in predicting potassium needs than might be supposed. It is true that sandy soils generally contain less available potassium than other soils for at least two reasons. The sand particles weather very slowly and therefore are slow to release potassium. The sandy soils have low cation-exchange capacities and therefore have small amounts of exchangeable potassium. Sandy soils often contain less than 100 kg of exchangeable potassium per hectare-furrow slice. However, sandy soils are generally well aerated. The plants are able to make effective use of the available potassium that is present.

Differences in texture within the soil profile may be more important than the textures themselves. Soils with strongly developed B horizons may restrict water movement enough to limit aeration not only in the B horizon but also in the A horizon, at least temporarily. Additional potassium is needed to compensate for reduced aeration.

Weather is an important variable related to potassium fertilizer response. It should not be overlooked even if it is uncontrollable. The greatest response to potassium is expected during cool, wet seasons. A wet season means poor aeration, particularly on soils with poor natural drainage.

Response of Crops to Potassium Fertilizer

The response of crops to potassium is related to at least two important plant characteristics: (1) Some plants require a higher ratio of potassium to calcium than others, and (2) some plants need a larger proportion of their potassium requirement within a short period of time.

Albrecht (1940) suggested that plants that are notably high in carbohydrate production and low in protein production require more potassium relative to calcium than do plants with high protein contents. The data in Table 12-6 support Albrecht's suggestion. Legumes are shown to contain approximately equal percentages of potassium and calcium, whereas other plants contain several times as much potassium as calcium.

Crops that grow slowly over a long period of time are less likely to respond to potassium fertilizer than are fast-growing crops. Nonexchangeable potassium may be released rapidly enough to meet the needs of a cool-weather crop like oats and yet be inadequate for a warm-weather crop like corn. Nightingale (1943) points out

Table 12–6 Average Potassium and Calcium Contents of Several Crops

Crop	K, %	Ca, %	K/Ca
Alfalfa (half bloom)	2.22	2.01	1.1
Red clover (various stages of maturity)	1.62	1.53	1.1
Sweet clover (various stages of maturity)	1.30	1.33	1.0
Soybean hay (various stages of maturity)	1.24	1.12	1.1
Oats (aboveground parts in milk stage)	0.96	0.22	4.4
Corn			
60 cm high	9.80	0.68	14.5
Before bloom	6.24	0.55	11.3
Beginning to bloom	7.12	0.66	10.8
Milk stage	2.62	0.50	5.2
Corn stover (mature)	0.74	0.32	2.3
Corn grain	0.40	0.015	26.7

Source: Beeson, 1941.

that plants need an abundant supply of potassium during the time when the photo-synthetic rate and the uptake of nitrates are high. A deficiency of potassium during the period of rapid growth is reflected in low carbohydrate production and in nitrate accumulation in the plant.

Corn responds more to potassium fertilizer than most other field crops. Not only is it a warm-weather crop, but also it requires a large amount of potassium relative to calcium for its high carbohydrate production.

Rotations involving both legume and nonlegume crops may increase potassium fertilizer needs. Frequently the soils are limed to improve the growth of the legume by increasing the availability of phosphorus and several other nutrients. But liming reduces the availability of potassium and at the same time increases the demand for potassium by increasing plant growth. In short, liming to increase legume growth usually calls for potassium fertilization of corn, cotton, or other nonlegumes in the rotation.

Legumes have a lower ratio of potassium to calcium than most crops but will respond to added potassium if its availability in the soil is low. For example, soybean response occurs if the exchangeable potassium drops below about 100 kg per hectare-furrow slice. MacLeod (1965) found that he could increase alfalfa regrowth in an alfalfa-grass mixture by applying potassium, whereas nitrogen favored regrowth of the grass in the mixture.

Most of the potassium in a crop is contained in the leaves and stems. If the crop is harvested for hay or silage, a much larger amount of potassium is removed than if only the seed or grain is harvested. Crop removal has a strong influence on the amount of potassium left for the next crop.

Van Ruymbeke and Ossemerct (1972) determined that the amount of potassium removed by crops depends greatly on rainfall during the last 3 weeks before harvest. They tested cereal crops at various stages of growth and found maximum K contents of 166 to 208 kg/ha at the flowering stage. Leaching removed 79 to 145 kg of K from the plants before harvest. Potassium leaches readily from dead plant tissue because it cannot be bonded into organic molecules. Potassium loss is one of the reasons that hay quality deteriorates when it is rained on during the curing process.

The amount of potassium that a crop can profitably utilize depends on the plant and the climate. Vicente-Chandler et al. (1962) found that grasses growing in the humid tropics gave strong yield response to yearly potassium applications up to 350 kg/ha and some additional response to still higher applications.

Luxury Consumption of Potassium

Adding a potassium fertilizer to a potassium-deficient soil will increase the crop yield. At first the amount of potassium absorbed will remain nearly proportional to the yield. But when the yield begins to level off, the amount of potassium absorbed continues to increase (Figure 12-7) with the result that the percent potassium in the plant increases. Uptake of more potassium than the plant needs to produce maximum yield is known as *luxury consumption.* This tendency to take up more than needed occurs to a degree with any nutrient but is especially evident with potassium.

The significance of luxury consumption of potassium depends considerably on when it occurs and what is done with the crop. Sometimes luxury consumption occurs in a young plant that will fully utilize the potassium by the time it is

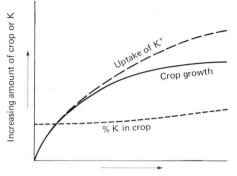

Figure 12-7 Potassium absorption and plant growth at various levels of available potassium illustrating luxury consumption at higher levels.

harvested. Luxury consumption then may have prevented the potassium from being leached or fixed by the soil. Sometimes luxury consumption occurs in a crop that is harvested for grain or seed. Most of the potassium then remains in the plant residues. The potassium soon leaches from these residues and again becomes available potassium in the soil. But when the whole plant is harvested for hay, silage, or some other use, any excess potassium that it has absorbed by luxury consumption is carried away and may never return to the field.

POTASSIUM FERTILIZERS

Potassium fertilizer needs may be identified by deficiency symptoms, plant tissue tests, or soil tests. Soil conditions such as wetness and need for lime also help to identify potassium needs. The amount of potassium to be applied is determined by an economic analysis as discussed in Chapter 9. The next question is, What form of potassium should be applied?

An extract from wood ashes (pot ashes) was an early form of potassium fertilizer. The term *potash* came from this source and has come to be the general term applied to all potassium fertilizers. The potash content of a fertilizer has traditionally been expressed as percent K_2O, just as phosphorus has been expressed as percent P_2O_5. Neither of these oxides is actually contained in fertilizers, and there is a movement to change the method of expressing these nutrients to an elemental basis. Percent P and percent K are used in this book except where the oxide form is specified.

The largest known deposits and the source of most of the potassium in international trade are in Germany and in southern Saskatchewan, Canada. Major U.S. deposits occur in New Mexico, Utah, and California.

Potassium Chloride

Most of the potassium fertilizer in use today is potassium chloride (KCl), commonly known as *muriate of potash*. This compound occurs naturally as the mineral sylvite and, more commonly, as sylvinite (a mixture of KCl and NaCl) and other mixtures. The California sources being utilized are briny lakes; most of the others are buried deposits that presumably represent former briny lakes. Potassium chloride has the advantage of requiring a minimal amount of processing and of being a relatively high-analysis potassium fertilizer. It is partially purified and sold as a fertilizer containing about 50 percent K (about 60 percent if expressed as K_2O).

Potassium Sulfate

Potassium sulfate (K_2SO_4), commonly known as *sulfate of potash,* is used in preference to potassium chloride where the sulfur is also wanted or where chlorine might be detrimental. The sulfate form is equally as effective a source of potassium as the chloride form (Laughlin and Restad, 1964), but the sulfate form is more expensive and has a lower analysis (about 43 percent K or 52 percent K_2O). Sulfur is also a macronutrient, and there are places where both potassium and sulfur are needed. Potassium sulfate is also favored for fertilizing tobacco and potatoes. Excess chlorine can cause watery potatoes and poor-quality tobacco.

Potassium Nitrate

Potassium nitrate (KNO_3) has not been an important fertilizer because of cost, but new methods of producing it are being developed. In nearly pure form it contains about 38 percent K and 13 percent N. Combining these two important nutrients into one compound is obviously advantageous.

All forms of potassium fertilizer are soluble in water, and so there is no problem of unavailability. Water solubility, however, can cause salt damage from high osmotic concentration if too much potassium fertilizer is placed near the seed. The same is true of nitrogen fertilizers. Southwest Potash Corporation (1969) points out that potassium nitrate produces less osmotic concentration than would be caused by equivalent amounts of potassium and nitrogen applied in other fertilizers. Lower osmotic concentration results from providing both soluble nutrients in the same compound rather than including the additional ions that would be present if separate compounds were used.

Placement of Potassium Fertilizer

Potassium does not move far in the soil and should therefore be placed where the plant roots will reach it. Top-dressed applications are satisfactory for established permanent vegetation that has roots near the surface (Smith, 1975).

Potassium fertilizer for tilled crops should be placed within the soil. It may be plowed under or disked into many soils, but there is considerable advantage to a band application to minimize soil contact where potassium fixation is a problem. The fixation capacity of a small volume of soil can be saturated, and the remaining potassium will be exchangeable or in solution. Welch et al. (1966) found that banded potassium was 1⅛ to 4 times as effective as broadcast potassium for corn. Potassium mobility within plants is high enough that a small part of the root system can absorb all the K^+ the plant needs.

REFERENCES

Adams, S. N., 1961, The Effect of Sodium and Potassium Fertilizer on the Mineral Composition of Sugar Beet, *J. Agr. Sci.* **56**:383.

Albrecht, W. A., 1940, Calcium-potassium-phosphorus Relation as a Possible Factor in Ecological Array of Plants, *J. Am. Soc. Agron.* **32**:411–418.

Aslam, M., 1975, Potassium and Sodium Interrelations in Growth and Alkali Cation Content of Safflower, *Agron. J.* **67**:262–264.

Bear, F. E., A. L. Prince, and J. L. Malcolm, 1945, Potassium Needs of New Jersey Soils, *New Jersey Agric. Exp. Sta. Bull.* 721.

Beeson, K. C., 1941, The Mineral Composition of Crops with Particular Reference to the Soils in Which They Were Grown, *USDA Misc. Pub.* 369.

Bower, C. A., G. M. Browning, and R. A. Norton, 1944, Comparative Effects of Plowing and Other Methods of Seedbed Preparation on Nutrient Element Deficiencies in Corn, *Soil Sci. Soc. Am. Proc.* **9**:142–146.

Bray, R. H., 1944, Soil-Plant Relations. I. The Quantitative Relation of Exchangeable Potassium to Crop Yields and to Crop Response to Potash Additions, *Soil Sci.* **58**:305–324.

Chang, H. T., and W. E. Loomis, 1945, Effect of Carbon Dioxide on Absorption of Water and Nutrients by Roots, *Plant Physiol.* **20**:221–232.

Farr, E., L. V. Vaidyanathan, and P. H. Nye, 1969, Measurement of Ionic Concentration Gradients in Soil near Roots, *Soil Sci.* **107**:385–391.

Fraps, G. S., 1929, Relation of the Water-soluble Potash, the Replaceable, and Acid-soluble Potash to the Potash Removed by Crops in Pot Experiments, *Texas Agric. Exp. Sta. Bull.* 391.

Graham, E. R., and P. L. Lopez, 1969, Freezing and Thawing as a Factor in the Release and Fixation of Soil Potassium as Demonstrated by Isotopic Exchange and Calcium Exchange Equilibria, *Soil Sci.* **108**:143–147.

Haagsma, T., and M. H. Miller, 1963, The Release of Nonexchangeable Soil Potassium to Cation-exchange Resins as Influenced by Temperature, Moisture and Exchanging Ion, *Soil Sci. Soc. Am. Proc.* **27**:153–156.

Hafez, A., and D. W. Rains, 1972, Use of Rubidium as a Chemical Tracer for Potassium in Long-term Experiments in Cotton and Barley, *Agron. J.* **64**:413–417.

Hanotiaux, G., and C. Felipe-Morales, 1972, Economie du Sol en Potassium, *Pedologie* **22**:127–147.

Johan, H. E., and J. V. Amin, 1965, Role of Sodium in the Potassium Nutrition of Cotton, *Soil Sci.* **99**:220–226.

Kauffman, M. D., and D. R. Bouldin, 1967, Relationships of Exchangeable and Nonexchangeable Potassium in Soils Adjacent to Cation-exchange Resins and Plant Roots, *Soil Sci.* **104**:145–150.

Laughlin, W. M., and S. H. Restad, 1964, Effect of Potassium Rate and Source on Yield and Composition of Bromegrass in Alaska, *Agron. J.* **56**:484–487.

Lawton, Kirk, 1945, The Influence of Soil Aeration on the Growth and Absorption of Nutrients by Corn Plants, *Soil Sci. Soc. Am. Proc.* **10**:263–268.

Liebhardt, W. C., and J. T. Murdock, 1965, Effect of Potassium on Morphology and Lodging of Corn, *Agron. J.* **57**:325–328.

MacLeod, L. B., 1965, Effect of Nitrogen and Potassium Fertilization on the Yield, Regrowth, and Carbohydrate Content of the Storage Organs of Alfalfa and Grasses, *Agron. J.* **57**:345–350.

Marbut, C. F., 1935, Soils of the United States, Pt. III, *Atlas of American Agriculture,* USDA, Washington, D.C.

Milford, M. H., and M. L. Jackson, 1966, Exchangeable Potassium as Affected by Mica Specific Surface in Some Soils of North Central United States, *Soil Sci. Soc. Am. Proc.* **30**:735–739.

Munn, D. A., L. P. Wilding, and E. O. McLean, 1976, Potassium Release from Sand, Silt, and Clay Soil Separates, *Soil Sci. Soc. Am. J.* **40**:364–366.

Nightingale, G. T., 1943, Physiological Chemical Functions of Potassium in Crop Growth, *Soil Sci.* **55**:73–78.

Paterson, J. W., and A. C. Richer, 1966, Effect of Long-term Fertilizer Application on Exchangeable and Acid-soluble Potassium, *Agron. J.* **58**:589–591.

Reitemeier, R. F., 1951, Soil Potassium, *Adv. Agron.* **3**:113–164.

Rich, C. I., and W. R. Black, 1964, Potassium Exchange as Affected by Cation Size, pH, and Mineral Structure, *Soil Sci.* **97**:384–390.

Smith, Dale, 1975, Effects of Potassium Topdressing a Low Fertility Silt Loam Soil on Alfalfa Herbage Yields and Composition and on Soil K Values, *Agron. J.* **67**:60–64.

Southwest Potash Corp., 1969, K-Nite—the Modern Low Salt Fertilizer, *Farm Technol.* **25**:26–27.

Stanford, George, J. B. Kelly, and W. H. Pierre, 1941, Cation Balance in Corn Grown on High-lime Soils in Relation to Potassium Deficiency, *Soil Sci. Soc. Am. Proc.* **6**:335–341.

Talati, N. R., S. K. Mathur, and S. C. Attri, 1974, Behaviour of Available and Non-exchangeable Potassium in Soil and Its Uptake in Wheat Crop, *J. Indian Soc. Soil Sci.* **22**:139–144.

Thomas, G. W., and B. W. Hipp, 1968, Soil Factors Affecting Potassium Availability, Chap. 13 in *The Role of Potassium in Agriculture,* Am. Soc. Agron., Crop Sci. Soc. Am. and Soil Sci. Soc. Am., Madison, 509 p.

Van Ruymbeke, R., and C. Ossemerct, 1972, Le Lessivage par les Eaux de Pluie de la Potasse des Vegetaux, *Pedologie* **22**:301–317.

Vicente-Chandler, Jose, R. W. Pearson, Fernando Abruna, and Servando Silva, 1962, Potassium Fertilization of Intensively Managed Grasses under Humid Tropical Conditions, *Agron. J.* **54**:450–453.

Vittum, M. T., D. J. Lathwell, and G. H. Gibbs, 1968, Cumulative Effects of Irrigation and Fertilizer on Soil Fertility, *Agron. J.* **60**:563–565.

Welch, L. F., P. E. Johnson, G. E. McKibben, L. V. Boone, and J. W. Pendleton, 1966, Relative Efficiency of Broadcast versus Banded Potassium for Corn, *Agron. J.* **58**:618–621.

Chapter 13

Calcium, Magnesium, and Sulfur

Calcium, magnesium, and sulfur are macronutrients in plant nutrition. They are abundant elements and are usually present in soils in large-enough amounts to meet the needs of plants. These elements are usually not considered as fertilizer elements and therefore receive less notice than nitrogen, phosphorus, and potassium. Even so, large amounts of calcium, magnesium, and sulfur are added to the soil in fertilizers and in soil amendments.

The use of calcium and magnesium in the form of lime to raise the pH of acid soils was discussed in Chapter 8. The use of sulfur and sulfur compounds to lower pH was also discussed in Chapter 8. The presence of calcium and sulfur in nitrogen, phosphorus, and potassium fertilizers was pointed out in Chapters 10 to 12. These applications of calcium, magnesium, and sulfur for other purposes undoubtedly eliminate many deficiencies that would otherwise require fertilization with these elements. Deficiencies may become more prevalent in the future because the percentages of calcium and sulfur in fertilizers have declined steadily in recent years, as shown in Figure 13-1. The trend has been to increase the analyses of N, P, and K, and to decrease the amounts of other elements present even if they are also plant nutrients.

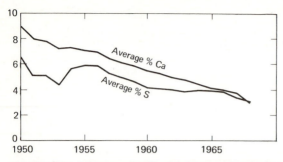

Figure 13-1 Average amounts of calcium and sulfur in ammonium sulfate, superphosphate, and triple superphosphate as percentages of the total amount of fertilizer applied in the United States from 1950 to 1968.

CALCIUM

Calcium occurs in soils and plants as the divalent cation Ca^{++}. Some calcium minerals such as calcite, $CaCO_3$, and gypsum, $CaSO_4 \cdot 2H_2O$, have very low solubilities. These minerals are leached out of the sola of humid-region soils but remain in the sola of soils in arid regions.

Calcium also occurs in the minerals apatite, plagioclase, and hornblende. Calcium minerals weather slightly faster than the average soil minerals. There is therefore a tendency for the percent calcium in a soil to gradually decline as weathering and leaching progress. The loss of calcium from soils of humid regions would undoubtedly be much more rapid if Ca^{++} ions were not strongly attracted to cation-exchange sites.

Calcium in Soils

Calcium may constitute more than 5 percent of the weight of a saline soil in an arid region or as little as 0.01 percent of the weight of a soil in the humid tropics. Most soils of humid temperate regions contain about 1 to 2 percent calcium.

Very low calcium contents occur in highly leached soils with low cation-exchange capacities. This includes some tropical soils—those that contain mostly oxide clays. Crop response to calcium as a fertilizer may occur on some tropical soils but has not yet been documented. Such low calcium contents are unlikely to be found in temperate regions except in very acid, sandy soils.

Calcium usually represents 75 to 85 percent of the total exchangeable bases present in temperate region soils. Most soils contain between 200 and 10,000 kg of exchangeable calcium per hectare-furrow slice. The exchangeable calcium is in equilibrium with the few kilograms of dissolved calcium present in the soil solution, even though there is likely to be 1,000 times as much exchangeable calcium as dissolved calcium in the soil.

The great concentration of calcium on cation-exchange sites is a result of the relatively small hydrated size of Ca^{++} ion relative to its +2 charge. The charge concentration causes a preferential adsorption of Ca^{++} ions over other ions that are likely to be present. Preferential adsorption is strong in clays such as montmorillonite that have high cation-exchange capacities and less pronounced in clays such as kaolinite that have low cation-exchange capacities. Calcium adsorption is highest in humus (Naylor and Overstreet, 1969) because humus has a high cation-exchange capacity and because some of the Ca^{++} ions may be chelated by organic compounds present in humus (chelates are organic materials that can enclose metallic cations; they will be dicussed in Chapter 14).

The exchangeable calcium in a soil has an important relation to soil pH and to the availability of several nutrient elements. The amounts of calcium and other basic cations present in a soil decline as a soil becomes more acid and increase as it becomes more alkaline. An excess of calcium causes calcium carbonate to precipitate and buffer the pH to a value near 8. Excess calcium usually results in low solubility of phosphorus, iron, manganese, boron, and zinc, and sometimes causes deficiencies of one or more of these essential plant nutrients.

Calcium, like other nutrients, becomes depleted in the zone immediately surrounding growing plant roots as shown in Figure 13-2. This depletion apparently results from the withdrawal of soil water from the same zone. The concentration of calcium in the soil water remaining near the root actually increases slightly (Brewster and Tinker, 1970). The same is true of magnesium and sodium concentrations in the soil water but not for potassium. Plants absorb potassium so rapidly that its concentration in solution near a growing plant root drops to less than half of the original concentration.

One conclusion that can be reached from the above information is that more than one type of transport mechanism is involved in moving ions to plant roots. Potassium ions move to the root much faster than does the soil water; diffusion therefore appears to be the dominant transport mechanism for potassium. Calcium, magnesium, and sodium ions move to the plant roots slightly slower than does the soil water unless the humidity is high enough to slow the rate of transpiration

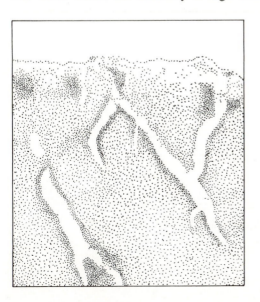

Figure 13-2 Calcium-45 depletion around wheat roots. Dark areas represent calcium-45 in soil or plant. Lighter areas represent regions from which calcium-45 has been depleted by plant growth. *(Drawn from Wilkinson, Loneragan, and Quick, copyright 1968 by American Association for the Advancement of Science.)*

(Elgawhary, Malzer, and Barber, 1972); mass flow therefore appears to be the dominant transport mechanism for them.

Calcium dissolved in the soil solution can move by mass flow and by diffusion, but exchangeable calcium has a very low mobility. Wadleigh (1957) reports that plant roots will not enter soil layers that are devoid of calcium even though other conditions are favorable for growth and calcium is available in other layers. Monovalent ions such as Na^+ and K^+ are more mobile because they are less strongly attracted to cation-exchange sites than Ca^{++} ions.

Calcium in Plants

Calcium is a structural component of cell walls and is therefore vital in the formation of new cells. Furthermore, calcium is so integrated into cell walls that it cannot be removed from old cells to form new cells (Loneragan and Snowball, 1969). Plants deficient in calcium are stunted because they produce fewer and smaller cells. They have weak stems because their cell walls are less than normal thickness. Rangnekar (1974) showed that substitution of potassium for all the calcium in the nutrient solution of 24-day old tomato seedlings began to limit their growth rate after only 4 days and halted growth in about 10 days (Figure 13-3).

A calcium shortage restricts the growth of roots as well as stems, leaves, etc. The inability of calcium-deficient roots to elongate rapidly handicaps the plant for exploiting new portions of the soil volume to obtain water and nutrients. Miller, Peverly, and Koeppe (1972) found that an increased calcium supply stimulated the uptake of phosphorus by corn roots. Restricted root growth could produce or aggravate other nutrient deficiencies as well.

The amount of calcium required to promote good root growth increases under a variety of adverse conditions. Gerard (1971) found that high soil temperatures increased the calcium needs of pea and cotton roots and that the high osmotic concentrations of saline soils also increased calcium needs. A study of cotton roots

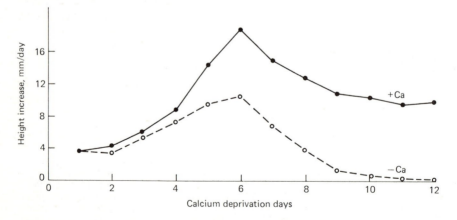

Figure 13-3 The growth rates of tomato seedlings after half of them were deprived of calcium at 24 days of age. (*Ragnekar,* 1974.)

showed that salinity reduced the uptake of calcium enough to require 140 h instead of 70 h to form thick cell walls (Gerard and Hinojosa, 1973).

Calcium supplies are smaller in acid soils than in alkaline soils. Part of the peril of aluminum toxicity is related to calcium deficiency. The amount of calcium needed to promote good root growth was found by Lund (1970) to depend on the concentration of aluminum in solution.

Calcium is normally abundant in plant leaves. A calcium deficiency prevents the growth and unfolding of new leaves. It also prevents the growth of the margins of existent leaves and therefore results in curled leaves.

The amount of calcium utilized by plants differs greatly from one plant to another Table 13-1 shows that legumes use much more calcium and magnesium than do plants of the grass family. Legumes usually respond very well to the liming of acid soils.

It was shown in Table 12-6 that legumes utilize nearly equal amounts of calcium and potassium but corn and oats utilize several times as much potassium as calcium. The data indicate that divalent cations (Ca^{++} and Mg^{++}) carry most of the positive charges absorbed by legumes and monovalent cations (K^+ and Na^+) carry most of the positive charges absorbed by other plants.

Calcium in Fertilizers

Most of the calcium in fertilizers is contained in phosphorus compounds and is applied for the benefit to be derived from the phosphorus. Apatite and ordinary superphosphate contain over twice as much calcium as phosphorus. Triple superphosphate, however, contains only about two-thirds as much calcium as phosphorus, and the ammonium phosphates contain no calcium at all. An increasing reliance on these latter materials for phosphorus needs has greatly reduced the calcium content of the average fertilizer in the United States (as was shown in Figure 13-1).

Lime ($CaCO_3$) and gypsum ($CaSO_4 \cdot 2H_2O$) are the two materials most likely to be considered if and when specific needs for calcium fertilizer are identified.

Table 13-1 Average Amounts of Calcium and Magnesium in Legumes and Nonlegumes*

Crop	Calcium	Magnesium	Ca:Mg ratio
Legumes			
Alfalfa	13.91	3.55	3.8:1
Red clover	11.42	2.70	4.2:1
Soybeans	12.29	3.88	3.2:1
Nonlegumes			
Corn	2.24	0.86	2.6:1
Oats	1.65	0.98	1.7:1
Wheat	1.45	0.87	1.7:1

*In kilograms per 1,000 kg dry plant materials.

Ordinary superphosphate could also serve; 70 percent of its calcium is in the form of gypsum. Lime would be used if it were advisable to raise the pH. Gypsum would be used if a neutral salt were desired. The neutral effect of gypsum might very well be desired for a tropical soil with a low buffer capacity resulting from a low cation-exchange capacity.

MAGNESIUM

The magnesium ion, Mg^{++}, is chemically similar to the calcium ion, Ca^{++}. Nevertheless, there are important differences in the behavior of these ions in minerals and plants.

The nonhydrated Mg^{++} ion is small enough to fit into octahedral spaces in mineral structures, whereas the Ca^{++} ion requires larger spaces. Magnesium in rocks is closely associated with the iron-containing (ferromagnesium) minerals such as olivene, various inosilicates, and biotite mica. The most important exception to this is the mineral dolomite ($CaCO_3 \cdot MgCO_3$).

None of the magnesium minerals mentioned are as resistant to weathering as are the feldspars, quartz, and hydrous oxides. Soils tend to be depleted of magnesium minerals somewhat sooner than they are depleted of the more resistant potassium, sodium, and calcium minerals. The association with iron ends as the iron is oxidized from Fe^{++} to insoluble Fe^{+++} compounds while the magnesium is leached away.

Magnesium in Soils

Finer particles contain more magnesium than coarser particles. Mokwunye and Melsted (1973a) tested nine temperate and tropical soils and found the following distribution of magnesium in the soil separates: clay 51 to 70 percent of the total magnesium present, silt 22 to 42 percent, and sand 0.1 to 11 percent. They found that severe weathering, soil erosion, and clay eluviation all tend to reduce the magnesium content of surface soil horizons. Magnesium can be released equally well from either the silt or the clay fraction and even the interlayer magnesium and lattice structure magnesium are partly available to plants (Christenson and Doll, 1973).

Exchangeable magnesium is the largest source of available magnesium in soils. Magnesium ions behave more like calcium ions when they are in the soil-solution cation-exchange complex than when they are in minerals or plants. From 12 to 18 percent of the exchangeable bases are normally Mg^{++} ions, an amount that is second only to the 75 to 85 percent represented by Ca^{++} ions. Magnesium excess is indicated when exchangeable Mg^{++} represents more than 40 to 60 percent of the cation-exchange capacity; magnesium deficiency is indicated by less than 3 to 8 percent exchangeable Mg^{++} (Martin and Page, 1969). Some highly leached soils are near the deficiency level and require magnesium fertilization. Lombin and Fayemi (1975) anticipate serious magnesium deficiencies in Nigeria in the near future.

The percentages of Ca^{++} and Mg^{++} ions in the soil solution are much lower than on the cation-exchange sites. Preferential adsorption of the double-charged ions is so strong that their concentrations in solution are reduced to about the same level as the concentrations of K^+ and Na^+ ions. Preferential adsorption of Mg^{++} by cation exchange combined with preferential absorption of K^+ by plants can result in magnesium deficiencies in plants and animals when potassium fertilizer is used on low-magnesium soils. Claasen and Wilcox (1974) found that fertilizer applications of either K^+ or NH_4^+ reduced the uptake of both Ca^{++} and Mg^{++} by corn. High Ca^{++} levels resulting from lime applications can also aggravate a magnesium deficiency if the lime used is low in $MgCO_3$.

Magnesium occupies a unique position in vermiculite. The layers of this clay mineral are sufficiently expanded to accommodate two layers of water and enough Mg^{++} ions to neutralize the excess negative charges in the layers. Vermiculite provides an excellent source of Mg^{++} ions as it weathers. It is possible that some vermiculite might form from montmorillonite if large amounts of Mg^{++} ions are added to a soil. This would be beneficial in terms of maintaining a satisfactory level of Mg^{++} availability over a number of years.

Magnesium in Plants

Magnesium is vital to the production of chlorophyl because every molecule of chlorophyl contains a magnesium ion at the core of its complex structure. It is, in fact, the only metallic element contained in chlorophyl. Most of the magnesium in plants is found in either chlorophyl or seeds, though a lesser amount is distributed through other parts of the plant. Part of the distributed magnesium functions in the enzyme system involved in carbohydrate metabolism.

The magnesium content of several different crops is shown in Table 13-1. It may be noted from the table that plants commonly contain between one-fourth and two-thirds as much magnesium as calcium. Legumes contain considerably more magnesium and calcium than do most other plants.

The behavior of magnesium inside the plant resembles the behavior of potassium much more than that of calcium. Magnesium is readily translocated from one part of the plant to another. Deficiencies show first on the lower leaves with a loss of green color moving inward from the margins and tips of the leaves. Magnesium deficiency in corn produces a distinct striping with yellow or even white streaks running full length of the leaves parallel to the veins.

The magnesium supply may be adequate to prevent deficiency symptoms from showing in plants and yet be inadequate for the nutrition of animals that eat the plants. Magnesium deficiency causes grass tetany in ruminants. Mayland and Grunes (1974) recommend that the magnesium concentration in forage should be about 0.2 percent to avoid grass tetany. Mayland, Grunes, and Lazar (1976) report that wheat forage is more likely than wheatgrass to cause grass tetany because wheat contains more potassium but less magnesium and calcium.

Magnesium Applications to Soil

Most of the magnesium applied to soil is contained in dolomitic limestone. Pure dolomite is a 1:1 ratio of $CaCO_3 \cdot MgCO_3$, but most dolomitic limestone is a mixture of dolomite and calcite. A ratio of six calcium ions to one magnesium ion would approximate normal soil proportions. But, many limestones are highly calcitic and contain much less magnesium than the 6:1 ratio. Use of calcitic lime can produce a need for magnesium fertilizer.

Magnesium fertilizers in the United States are most used on truck crops grown near the Atlantic coast where the soils formed from rocks low in magnesium. High rainfall has resulted in strongly leached soils deficient in magnesium. Liming and fertilization for intensive truck cropping have increased the magnesium requirement. Some crops such as potatoes and tobacco grown in the area are particularly sensitive to magnesium deficiency.

The principal magnesium fertilizer is magnesium sulfate ($MgSO_4$), sometimes known as epsom salts. Magnesium sulfate is soluble in water and can be applied to either soil or plants. Spraying it on the leaves gives the quickest response. Soil applications are longer lasting and require higher rates. Gallaher, Harris, Anderson, and Dobson (1975) tested rates as high as 68 kg/ha Mg and raised the Mg content of sorghum leaves from 0.06 up to 0.22 percent. Fertilization increased the level of residual soil magnesium available for future crops.

SULFUR

Sulfur behaves like nitrogen in many respects. Both of these elements are held in organic compounds by covalent bonds with carbon. Each exists in several different oxidation states and tends to be oxidized to its highest oxidation state (SO_4^{--} or NO_3^-) in well-drained soils. Sulfur and nitrogen are both unavailable to plants in their elemental states and are most available when in their most oxidized states. Sulfur, like nitrogen, forms soluble anions that can be lost by leaching when its concentration builds up in the soil solution. Sulfur solubility, however, is limited when calcium is abundant. Gypsum ($CaSO_4 \cdot 2H_2O$) is only slightly soluble and may precipitate.

Occurrence of Sulfur in Soils

Soil parent materials average about 0.06 percent sulfur. Weathering, leaching, and plant use deplete the supply of mineral sulfur to low levels so that most of the sulfur in the solum occurs in organic combination covalent bonded into complex molecules. A hectare-furrow slice in a humid region usually contains between 200 and 1,000 kg of sulfur; nearly all of it is organic (Figure 13-4). The B horizon of a humid-region soil has a relatively low concentration of sulfur because most of the inorganic sulfur has been lost and there is not enough organic sulfur to replace it.

Soils of arid regions commonly contain about the same amount of sulfur as those in humid regions but more of it is in mineral form. Half or more of the sulfur in the A horizon may still be inorganic (Joshi, Choudhari, and Jain, 1973). The B horizon is usually low in sulfur in arid regions as well as in humid regions.

Organic sulfur occurs in soil in several forms including carbon-bonded sulfhydryl (–SH) groups and sulfate (SO_4^{--}) groups with one of their oxygens bonded to a carbon. About one-third of the organic sulfur in some soils is present in the amino acids cystine and methionine (Freney, Melville, and Williams, 1975).

Soils usually contain a few kilograms per hectare of sulfate sulfur dissolved in the soil solution and attracted to positively charged sites on clay or humus. Sulfates

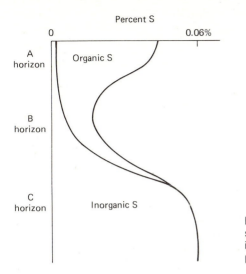

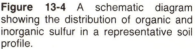

Figure 13-4 A schematic diagram showing the distribution of organic and inorganic sulfur in a representative soil profile.

are available to plants but the supply is usually small and must be replenished from other sources. Soils high in clay store more sulfate sulfur than sandy soils. Many tropical soils have more positively charged colloids than other soils and therefore store more adsorbed sulfate ions. Hague and Walmsley (1973) attribute the sulfate adsorption capacity to hydrated iron and aluminum oxides. Soils of arid regions contain the highest concentrations of dissolved sulfates and precipitated gypsum because of low leaching intensity. Saline soils often contain large amounts of sulfates throughout their sola. Other soils of arid and semi-arid regions may contain gypsum accumulations near the bottom of their sola.

Some very poorly drained soils contain sulfur in the sulfide form, most commonly combined with iron as ferrous sulfide (FeS). Sulfides are so insoluble that a large amount of FeS and other sulfides can be present in a soil with no ill effects as long as the soil is waterlogged. Such soils should not be drained because the sulfides are then oxidized to sulfates and the pH drops to about 2—too acid to support plant growth. Many areas that have been strip-mined for coal are barren from acidification because mine spoils are often high in sulfides. Nothing will grow on such spoil heaps, and pools of water nearby are usually too acid for fish to live in them.

The Sulfur Cycle

Sulfur goes through a cycle of oxidation in the soil and reduction in the plant much like the nitrogen cycle (Figure 13-5). Gains to and losses from this cycle take several forms. Sulfur from soil parent materials is gradually released as the minerals weather. Leaching losses of sulfur depend on climate as do those of other soluble materials. Average annual leaching losses of sulfur are about 45 kg/ha in the eastern United States (Lipman and Conybeare, 1936). Significant but variable losses of sulfur occur by erosion and by removal in crops and in forages consumed by animals.

An exchange between the sulfur of soil and plants and atmospheric sulfur dioxide is an important part of the sulfur cycle. Plant and animal residues contain

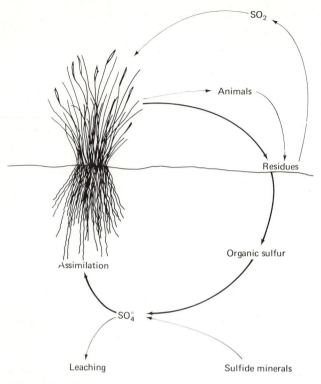

Figure 13-5   The sulfur cycle.

sulfur that may be released to the atmosphere as sulfur dioxide when they decay. Coal and petroleum also contain sulfur and produce sulfur dioxide when they burn. Industrial use of fossil fuels causes the sulfur content of the atmosphere near manufacturing centers to be much higher than elsewhere. The SO_2 concentration in the atmosphere is normally about 0.05 ppm but can rise above 1 ppm near smelters if no pollution control devices are used. Concentrations of 1 ppm are toxic to most plant growth and make the rainfall acid and corrosive.

Sulfur dioxide dissolves in water, and rainfall carries it back to the soil. Table 13-2 shows that rainfall adds about 30 kg of sulfur per hectare per year to the soils of most parts of the state of Indiana. This is more than enough sulfur to replace that removed by field crops in Indiana (similar additions occur in many parts of the world). Even so, the soils near Gary, Ind. (an industrial city) receive nearly five times as much sulfur in precipitation as do the soils in nonindustrial areas. Many soils would surely contain excess sulfur if it were not removed by leaching.

Masclet, Nagy, and Trocme (1971) analyzed the sulfur content of precipitation for several years at Isneauville, France, and found that the precipitation carried down more sulfur in wet years than in dry years. Proximity to the sea also increased the content of sulfur and of other elements in the precipitation. Slutskaya (1972) reported that sulfur in the precipitation at 13 sites in Russia ranged from 5 to 45 kg/ha.

Table 13–2   Sulfur in Precipitation at 11 Locations in Indiana in 1 year, Sept. 1, 1946, to Aug. 31, 1947

Location	Total sulfur, kg/ha	Total rainfall, mm
Nashville	22	993
Gary (industrial city)	143	811
Monterey	31	869
Orland	29	942
Indianapolis	37	930
Rising Sun	33	1,034
Evansville	32	1,031
Larwill	26	906
Branchville	34	1,054
Lafayette	24	963
Montgomery	31	1,069

Source: Calculated from data of Bertramson, Fried, and Tisdale, 1950.

Many different microbes can release soluble sulfur by decomposing organic matter. A simple version of this part of the sulfur cycle suggests that sulfur is released initially as sulfides and subsequently oxidized to the sulfate form. Organic acids like sulfinic acid ($R-SO_2H$) may also be released as intermediate compounds and oxidized to sulfate. No matter what intermediate compounds are formed, sulfates are the end products and the principal inorganic form of sulfur in soils under aerobic conditions.

Sulfur as a Plant Nutrient

Plants absorb sulfur from the soil as the sulfate (SO_4^{--}) ion or from the air as sulfur dioxide (SO_2). Plant metabolism reduces SO_4^{--} and SO_2 to forms that can be built into organic molecules. Sulfur is a vital part of all plant proteins and of some plant hormones.

Westermann (1975) showed that alfalfa is quite sensitive to sulfur deficiency and requires 0.15 to 0.20 percent sulfur in the hay to produce maximum yield (Figure 13-6). Sulfur not needed for synthesis of organic compounds often accumulates as sulfates in plants. Thomas, Hendricks, and Hill (1950) found that the organic sulfur contents of nearly 1,000 leaf samples all fell within a narrow range (0.2 to 0.4 percent). Nevertheless, the total sulfur contents of these leaves varied widely because of differences in their inorganic sulfur contents, especially the sulfate form.

The sulfur and phosphorus contents of several crops are compared in Table 13-3. Plants use approximately as much sulfur as phosphorus; nitrogen needs are usually 4 to 10 times larger than those of either sulfur or phosphorus. Sulfur deficiencies slow down protein synthesis for two reasons. The sulfur-containing amino acids are vital constituents of proteins. Other amino acids may accumulate if sulfur is deficient. Futhermore, sulfur is essential for the action of enzymes involved in nitrate reduction, and a sulfur deficiency slows the formation of all amino

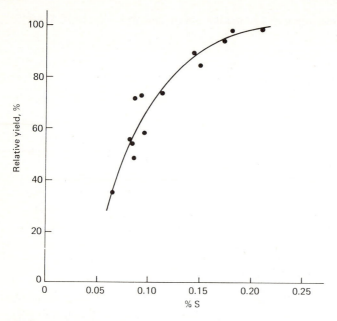

Figure 13-6 The relationship between relative yield and percent total S in alfalfa. (*Westermann,* 1975.)

acids. Sulfur-deficient plants therefore tend to accumulate nitrate nitrogen in their tissues.

Sulfur deficiency symptoms resemble nitrogen deficiency symptoms because both are related to protein and chlorophyl deficiencies. However, the pale green color from nitrogen deficiency tends to be most apparent in older leaves, whereas paleness is likely to show on new growth as well when sulfur is deficient.

Legumes are often particularly sensitive to sulfur deficiencies because sulfur is required for nitrogen fixation (Tisdale, 1974). Yields of alfalfa and forage crops are frequently doubled or tripled by additions of sulfur to sulfur-deficient soils (Conrad, 1950).

Sulfur in Fertilizers

Ammonium sulfate contains about 24 percent S (and only about 21 percent N). Ordinary superphosphate contains about 11 percent S in the form of gypsum produced by treating the rock phosphate with sulfuric acid (its P content is about 9 percent). Potassium sulfate contains about 18 percent S (and 43 percent K). These particular fertilizers are usually not the cheapest sources of N, P, or K alone, but they may be the best choices if sulfur is needed in addition to one of these other macronutrients.

The modern trend is toward higher-analysis fertilizers that contain more N, P, and K but less sulfur. High yields obtained by large applications of N, P, and K fertilizers can increase the demand for sulfur above the supply available in many soils. It may be wise to go back to some of the fertilizers mentioned above when

Table 13-3   Amounts of Phosphorus and Sulfur Contained in Certain Crops

Crop	Assumed yield, tons/ha	S, kg/ha	P, kg/ha
Alfalfa	9	26	21
Corn	7	20	19
Cotton	1.5	25	8
Potatoes	20	28	12
Tobacco	1.5	3	4
Wheat	3	10	11

Source: Calculated from various sources.

sulfur deficiencies occur. An alternative is to apply either gypsum (about 18 percent S) or elemental sulfur. Both of these materials have been shown to be effective (Hoeft and Walsh, 1975). Gypsum and sulfur are also used to lower soil pH as discussed in Chapter 8, but they are applied at rates between 10 and 100 kg/ha of sulfur when used as fertilizers. Soil amendment application rates are several times as large as the fertilizer rates.

REFERENCES

Adams, W. E., A. W. White, Jr., and R. N. Dawson, 1967, Influence of Lime Sources and Rates on "Coastal" Bermudagrass Production, Soil Profile Reaction, Exchangeable Ca and Mg, *Agron. J.* **59**:147–149.

Bertramson, B. R., Maurice Fried, and S. L. Tisdale, 1950, Sulfur Studies of Indiana Soils and Crops, *Soil Sci.* **70**:27–42.

Brewster, J. L., and P. B. Tinker, 1970, Nutrient Cation Flows in Soil around Plant Roots, *Soil Sci. Soc. Am. Proc.* **34**:421–426.

Buyanovskiy, G. A., 1973, Distribution of Bacteria Involved in the Sulfur Cycle and Their Role in Soil Processes, *Sov. Soil Sci.* **5**:179–186.

Christenson, D. R., and E. C. Doll, 1973, Release of Magnesium from Soil Clay and Silt Fractions during Cropping, *Soil Sci.* **116**:59–63.

Claasen, M. E., and G. E. Wilcox, 1974, Comparative Reduction of Calcium and Magnesium Composition of Corn Tissue by NH_4–N and K Fertilization, *Agron. J.* **66**:521–522.

Conrad, John P., 1950, Sulfur Fertilization in California and Some Related Factors, *Soil Sci.* **70**:43–54.

Cressman, H. K., and J. F. Davis, 1962, Sources of Sulfur for Crop Plants in Michigan and Effect of Sulfur Fertilization on Plant Growth and Composition, *Agron. J.* **54**:341–344.

Elgawhary, S. M., G. L. Malzer, and S. A. Barber, 1972, Calcium and Strontium Transport to Plant Roots, *Soil Sci. Soc. Am. Proc.* **36**:794–799.

Freney, J. R., G. E. Melville, and C. H. Williams, 1975, Soil Organic Matter Fractions as Sources of Plant-available Sulphur, *Soil Biol. Biochem.* **7**:217–221.

Gallaher, R. N., H. B. Harris, O. E. Anderson, and J. W. Dobson, Jr., 1975, Hybrid Grain Sorghum Response to Magnesium Fertilization, *Agron. J.* **67**:297–300.

Gerard, C. S., 1971, Influence of Osmotic Potential, Temperature, and Calcium on Growth of Plant Roots, *Agron. J.* **63**:555–558.

Gerard, C. J., and E. Hinojosa, 1973, Cell Wall Properties of Cotton Roots as Influenced by Calcium and Salinity, *Agron. J.* **65**:556–560.

Gupta, Umesh C., and R. L. Veinot, 1974, Response of Crops to Sulfur under Greenhouse Conditions, *Soil Sci. Soc. Am. Proc.* **38**:785–788.

Hague, I., and D. Walmsley, 1973, Adsorption and Desorption of Sulphate in Some Soils of the West Indies, *Geoderma* **9**:269–278.

Hoeft, R. G., and L. M. Walsh, 1975, Effect of Carrier, Rate, and Time of Application of S on the Yield, and S and N Content of Alfalfa, *Agron. J.* **67**:427–430.

Joshi, D. C., J. S. Choudhari, and S. V. Jain, 1973, Distribution of Sulphur Fractions in Relation to Forms of Phosphorus in Soils of Rajasthan, *J. Indian Soc. Soil Sci.* **21**: 289–294.

Kang, B. T., and O. A. Osiname, 1976, Sulfur Response of Maize in Western Nigeria, *Agron. J.* **68**:333–336.

Lipman, J. G., and A. B. Conybeare, 1936, Preliminary Note on the Inventory and Balance Sheet of Plant Nutrients in the United States, *New Jersey Agric. Exp. Sta. Bull.* 607.

Lombin, L. G., and A. A. A. Fayemi, 1975, Critical Level of Mg in Western Nigerian Soils as Estimated under Greenhouse Conditions, *Agron. J.* **67**:272–275.

Loneragan, J. F., and K. Snowball, 1969, Calcium Requirements of Plants, *Aust. J. Agric. Res.* **20**:465–478.

Lund, Z. F., 1970, The Effect of Calcium and Its Relation to Several Cations in Soybean Root Growth, *Soil Sci. Soc. Am. Proc.* **34**:456–459.

McCart, G. D., and E. J. Kamprath, 1965, Supplying Calcium and Magnesium for Cotton on Sandy, Low Cation Exchange Capacity Soils, *Agron, J.* **57**:404–406.

MacLean, A. J., J. J. Jasmin, and R. L. Halstead, 1967, Effect of Lime on Potato Crop and on Properties of a Sphagnum Peat Soil, *Can. J. Soil Sci.* **47**:89–94.

Martin, J. P., and A. L. Page, 1969, Influence of Exchangeable Ca and Mg and of Percentage Base Saturation on Growth of Citrus Plants, *Soil Sci.* **107**:39–46.

Masclet, A., C. Nagy, and S. Trocme, 1971, Apports de Divers Elements par les Precipitations, *Bull. de L'Assoc. Francaise pour L'Etude der Sol 1971* (2):3–8.

Mayland, H. F., and D. L. Grunes, 1974, Magnesium Concentration in *Agropyron desertorum* Fertilized with Mg and N, *Agron. J.* **66**:79–82.

Mayland, H. F., D. L. Grunes, and V. A. Lazar, 1976, Grass Tetany Hazard of Cereal Forages Based upon Chemical Composition, *Agron. J.* **68**:665–667.

Miller, R. J., J. H. Peverly, and D. E. Koeppe, 1972, Calcium-stimulated [32]P Accumulation by Corn Roots, *Agron. J.* **64**:262–266.

Mokwunye, A. U., and S. W. Melsted, 1973a, Interrelationships between Soil Magnesium Forms, *Commun. Soil Sci. Plant Anal.* **4**:397–405.

Mokwunye, A. U., and S. W. Melsted, 1973b, Magnesium Fixation and Release in Soils of Temperate and Tropical Origins, *Soil Sci.* **116**:359–362.

Naylor, D. V., and Roy Overstreet, 1969, Sodium-calcium Exchange Behavior in Organic Soils, *Soil Sci. Soc. Am. Proc.* **33**:848–851.

Rangnekar, P. V., 1974, Effect of Calcium Deficiency on the Translocation and Utilization of C[14]–photosynthate in Tomato Plants, *Plant Soil* **41**:589–600.

Rhue, R. D., and E. J. Kamprath, 1973, Leaching Losses of Sulfur during Winter Months when Applied as Gypsum, Elemental S or Prilled S, *Agron. J.* **65**:603–605.

Slutskaya, L. D., 1972, Sulfur as Fertilizer, *Sov. Soil Sci.* **4**:52–72.

Starkey, R. L., 1950, Relations of Microorganisms to Transformations of Sulfur in Soils, *Soil Sci.* **70**:55–66.

Thomas, M. D., R. H. Hendricks, and G. R. Hill, 1950, Sulfur Content of Vegetation, *Soil Sci.* **70**:9–18.

Tisdale, S. L., 1974. Sulphur: Part 1—An Introduction. *Fert. Solutions* **18**(6):8–16.

Wadleigh, C. H., 1957, Growth of Plants, *Soil,* 1957 Yearbook of Agriculture, USDA, pp. 38–49.

Westermann, D. T., 1975, Indexes of Sulfur Deficiency in Alfalfa. II. Plant Analyses, *Agron. J.* **67**:265–268.

Wilkinson, H. F., J. F. Loneragan, and J. P. Quirk, 1968, Calcium Supply to Plant Roots, *Science.* **161**:1245–1246.

Chapter 14

The Micronutrients

A micronutrient is an element that plants must have to complete their life cycles but need only in a small amount. These elements have often been called *trace elements* or *minor elements,* but micronutrient is the preferable term. "Minor elements" is inappropriate because these elements are essential to plant growth and that which is essential is not minor. The term *trace elements* is less objectionable but it emphasizes the wrong aspect. Some micronutrients are present in both plants and soil in large amounts, but only small amounts are actually required as nutrients.

The micronutrients are often incidental components of fertilizers, being present as impurities. Such incidental inclusions are decreasing, however, because manufacturers and buyers now favor fertilizers with higher analyses of the principal fertilizer elements. Common fertilizers often supplied adequate amounts of one or more micronutrients where they were needed in years past. The newer high-analysis fertilizers, however, contain fewer impurities and do not carry the micronutrient bonuses unless they are specially added. Elimination of incidental inclusions will require in the future that more fertilizers have deliberate additions of the micronutrients.

New high-analysis fertilizers are one of several reasons that micronutrient deficiencies are becoming more common. Another factor is the shift many farmers have made from manure to inorganic fertilizers. Manure contains some of every essential element and therefore helps prevent micronutrient deficiencies. Modern farmers, in addition, are obtaining much higher yields than were common years ago.

Higher yields require more plant nutrients and can deplete the supply of available nutrients in the soil. Succeeding crops may therefore suffer from micronutrient deficiencies. This is especially true of some new varieties that require larger amounts of certain nutrients to produce their full yield potential.

The use of micronutrient fertilizers in the United States is shown in Table 14-1. The amounts used are much smaller than the millions of tons of macronutrients used (Chapter 9) because micronutrient deficiencies are less common and application rates are smaller than for macronutrients. Micronutrient fertilizer use is most common on acid soils, alkaline soils, and organic soils used for growing crops sensitive to a particular deficiency. Soil and crop distribution combine to give regional patterns to micronutrient fertilizer requirements.

DISCOVERY OF THE MICRONUTRIENTS

Cyril G. Hopkins published a classic textbook, *Soil Fertility and Permanent Agriculture,* in 1910. At that time 10 elements were recognized as essential to plants. The following expression became popular in Hopkins' time to help students remember the essential elements:

C HOPKNS CaFe Mg (See Hopkins Cafe, mighty good)

The only micronutrient in the 1910 list was iron. More iron is required than any of the other micronutrients but still considerably less than any of the macronutrients. It was known that plants contained many elements besides those listed, but the others had not been proven essential.

Even the "chemically pure" compounds available in 1910 were likely to contain trace amounts of micronutrients which were adequate for plant growth. Distilled water may dissolve enough copper, zinc, or boron from the condensing tube to keep plants from showing deficiency symptoms. Even the micronutrient supply contained

Table 14–1   Use of Micronutrients as Fertilizers in the United States in Year Ending June 30, 1975

Micronutrient	Metric tons of element	Areas of largest use
Boron	450	Pacific coast and north central states
Copper	475	Atlantic coast and north central states
Iron	1,930	Pacific coast, mountain, and north central states
Manganese	10,600	Atlantic coast and north central states
Molybdenum	130	Central and south Atlantic states
Zinc	11,850	Central and Pacific coast states

Source: USDA Crop Reporting Board.

in the seed or caught from the dust in the air can provide enough micronutrients to grow a plant.

Careful work has added several more elements to the list of essential plant nutrients. Some of these had been suspected earlier but had to await proof. For example, boron had been known to stimulate plant growth for several years, but it was 1915 before it was widely accepted as essential for plants to complete their life cycles. Warrington proved the essentiality of boron in 1923. Manganese and zinc were added to the list at about the same time as boron. Copper was proved essential in 1931 and molybdenum in 1942.

Broyer, Carlton, Johnson, and Stout added chlorine to the list of essential micronutrients in 1954. Hopkins' list of 10 essential elements was thereby enlarged to 16. Other beneficial elements have significant influence on plant growth even though they have not been shown to be vital for completing the life cycle. Such elements as silicon, sodium, rubidium, cobalt, and vanadium may be called functional or beneficial rather than essential. Further research, of course, may add some of them to the list of essential micronutrients.

Silicon forms crystals of plant opal as mentioned in Chapter 7. These crystals are commonly elongated and serve to strengthen the stems of plants. Plant opal crystals occur in many forms unique to various species of plants and endure long after the organic material has disintegrated, thus providing a means for identifying types of vegetation that have long ago vanished.

The effects of sodium and rubidium on plant growth were discussed in Chapter 12 along with their relationship to potassium. Both sodium and rubidium are beneficial to some plants. Some workers have suggested that sodium is essential in small quantities, but most do not list it. Practically, it does not matter whether sodium is essential or not because sodium is too abundant to ever be deficient in nature.

Cobalt is required for nitrogen fixation by Rhizobium bacteria in their symbiotic relationship with nodulating legumes. Cobalt is not listed as an essential element because legumes as well as other plants can use nitrogen from other sources and do not have to rely on nitrogen fixation by Rhizobium. But, legumes lose the advantage of symbiotic nitrogen fixation if cobalt is lacking.

Vanadium is able to partially replace molybdenum in nonsymbiotic nitrogen fixation by some microbes. It has also been indicated as being needed for some vegetable and grain crops to make their best growth.

ROLES OF MICRONUTRIENTS

The potency of the micronutrients is impressive. Each plant requires only a tiny amount of each micronutrient, yet it must have that tiny amount. Clearly they cannot serve as building blocks of major plant components; the amounts involved are inadequate for such a role.

Some micronutrients function in the enzyme systems of plants. Anion-forming elements such as boron and molybdenum can be part of the structure of enzyme molecules. A very small amount of such an element is all that is required to make enough enzyme to catalyze an essential plant process. Cation-forming elements such as copper are more likely to serve as coenzymes that activate an enzyme but are not an integral part of the molecule.

Some micronutrients function in oxidation-reduction processes of plant metabolism. Elements such as iron, copper, and manganese can change valences and thus enter into oxidation-reduction reactions.

Several of the micronutrients are involved in the production of chlorophyl. Deficiencies of these elements can appear very similar to the pale green condition characteristic of nitrogen deficiency.

Applying Micronutrient Fertilizers

Care must be taken in the application of micronutrient fertilizers, especially those containing elements such as boron that can be toxic to plants. The right amount of the nutrient element must be applied, and it must be uniformly distributed over the area involved. Spotty distribution can leave small areas where the element is still deficient and result in toxic concentrations in nearby areas. The usual solution to this problem is to apply the micronutrients along with the macronutrients in a carefully mixed fertilizer.

Unfortunately, shortages of micronutrients can reduce yield and quality of crops without being severe enough to produce deficiency symptoms. Anderson and Boswell (1968) found that 2 to 4 kg/ha of manganese and 0.45 to 0.9 kg/ha of boron increased cotton yields even though there were no deficiency symptoms. Limitations like this are hard to detect and often are uncovered only by trial and error plus "educated guesses."

BORON

Boron was shown to be toxic to plants before it was shown to be an essential nutrient. Cases of boron toxicity were identified during the first decade of this century. Boron toxicity was again observed during World War II when European sources of potassium fertilizer were replaced with materials from the western United States. The United States sources of potassium contained enough boron impurities to cause toxicity problems.

Boron compounds are sufficiently soluble in water to make the water toxic to plant growth. Boron content is therefore one of the factors considered in determining water quality for irrigation. Recognition of boron as an essential element led to the realization that several problems that had been considered nonparasitic diseases are actually boron deficiencies. Some of these "diseases" are top sickness of tobacco, heart rot of beets, corky core of apples, brown rot of cauliflower, snakehead of walnuts, and cracked stem of celery. Boron deficiency also causes brown heart of turnips, internal brown spot of sweet potatoes, split roots in carrots, blossom blast of pears, and yellow terminal leaves of alfalfa.

Boron in Soils

Whetstone, Robinson, and Byers (1942) tested over 300 soil samples and found that total boron contents were between 9 and 198 kg per hectare-furrow slice. The available boron ranged from 0.9 to 145 kg per hectare-furrow slice and was generally near half of the total boron present. The available boron averaged 38 kg per hectare-furrow slice. Most of the available boron in soils is associated with organic matter (Sparr, Jordan, and Turner, 1968).

Boron deficiencies are widespread in humid regions. In the United States they occur most commonly in three general areas: the uplands near the Atlantic coastal plain; a region around the Great Lakes; and the Pacific Northwest (especially the Pacific coastal region) (Figure 14-1).

Several different forms of boron in soils were identified by Il'in and Anikina (1974) including water-soluble boron, boron bound to organic matter, boron in clay minerals, and borosilicates. They indicated that saline soils in arid zones in Russia contain high concentrations of boron that are often toxic to plants and animals. Boron toxicity also occurs in certain irrigated areas of the western United States and would probably be more prevalent in arid regions if it were not for decreased solubility of boron compounds in alkaline soils. Kubota and Allaway (1972) suggest that boron toxicity is more dependent on high boron contents in irrigation water than in soil.

Boron is subject to leaching losses in humid regions. Leaching losses coupled with crop removal will gradually cause more and more of the soils of humid regions to need boron fertilizer. The lowest boron contents are in acid, leached soils, coarse sandy soils, and organic soils. Overliming an acid soil can reduce the availability of the boron that is present and thus induce a boron deficiency.

Boron in Plants

Normal plant leaves usually contain between 25 and 100 ppm boron. Eaton (1944) reported boron concentrations much higher than normal (3,875 ppm in muskmelon, 3,080 ppm in zinnia, and 2,245 ppm in sweet clover), but these were toxic accumulations. A normal crop usually contains between 50 and 500 g/ha of boron. For example, 9 tons of alfalfa with 40 ppm boron would contain 360 g of boron.

Alfalfa is a good indicator crop for boron deficiency. A yellowing of terminal alfalfa leaves indicates a possible boron deficiency. The symptoms show mostly on young leaves because boron is relatively nonmobile in plants. Application of borax to small areas will show whether boron actually is deficient.

Boron deficiencies have long been known to retard the uptake of calcium (Branchley and Warrington, 1927). Since boron availability decreases as the pH rises above neutral, it is possible to create a boron deficiency by overliming a soil. In rare cases, the excess lime might reduce the uptake of calcium because of the boron deficiency. Another macronutrient interaction with boron was pointed out by Jones and Scarseth (1944). They observed that high potassium uptake caused higher boron uptake, lower calcium uptake, and lower limits of boron toxicity.

Boron Fertilizer

An extensive source of boron occurs in the form of borax ($Na_2B_4O_7 \cdot 10H_2O$) in the desert area of California, particularly in Death Valley. This area serves as a source of boron for fertilizer and other boron products. Borax is soluble in water and therefore available to plants without requiring any chemical treatment.

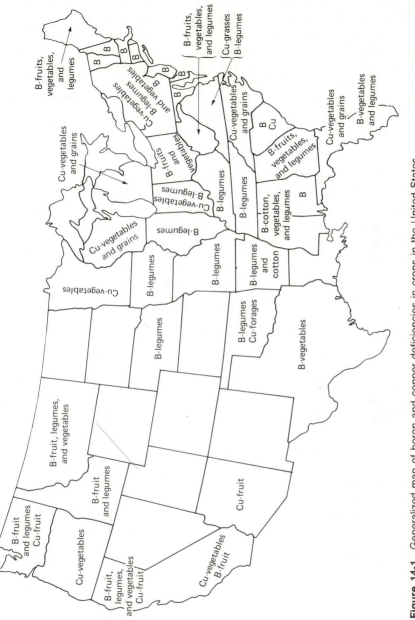

Figure 14-1 Generalized map of boron and copper deficiencies in crops in the United States. (Compiled from various sources.)

Some commercial fertilizers are advertised as having boron added to them in areas where boron deficiencies are known to exist. Such fertilizers are appropriate where the deficiencies exist, but they must be used cautiously to avoid boron toxicity. The boron application rate should be between 1 and 5 kg/ha (not over 50 kg of borax). Soils with low cation-exchange capacities and low calcium saturation should receive less than 25 kg/ha of borax.

Fly ash collected from smokestacks has been suggested as a source of boron (Plank and Martens, 1974). The boron content is only 100 to 1,000 g per ton but millions of tons of fly ash are collected each year. The ash also carries significant quantities of potassium and other nutrients.

CHLORINE

Chlorine and sodium have often been cited as elements that are needed by animals but not by plants. Current evidence suggests that chlorine is essential for both plants and animals. It is a micronutrient for plants but a macronutrient for animals. Even this statement must be understood to apply to the amounts needed rather than to amount absorbed. Plants usually contain large amounts of chlorine even though small amounts will suffice to meet plant needs.

Chlorine in Soils

Chlorine ions (Cl^-) are the same charge and about the same size as hydroxyl ions. Chlorine therefore substitutes for hydroxyl ions in mineral structures.

Considerable amounts of chlorine are bonded onto organic compounds. The chlorine atoms must always be at the edge of such compounds and never in the middle of a chain because chlorine forms only one covalent bond.

Chlorides (metallic ions associated with Cl^-) are highly soluble compounds. Solubility accounts for the ease with which chlorine is leached from well-drained soils of humid regions. Soils that contain soluble salts usually contain large amounts of chlorides in solution. Such chlorides are highly mobile in the soil. They move downward when it rains and upward when water evaporates from the soil surface.

Upward movement of water produces saline soils in arid regions. Evaporation leaves enough salts (including chlorides) at the soil surface to form white crusts during dry periods. Extreme cases of salinity occur in such low areas as the floor of Death Valley or the flats along the Great Salt Lake. The "soil" in these areas is almost pure sodium chloride.

Chlorine in Plants

Broyer, Carlton, Johnson, and Stout showed in 1954 that chlorine was essential to the growth of tomato plants. They have shown since 1954 that chlorine is also essential for several other plants. Production of a chlorine deficiency required that the water used be purified by treating it with activated charcoal and then distilling it. Gaseous chlorine had to be cleaned from the air because several plant species were able to absorb enough chlorine through their leaves to meet their needs. Great care had to be exercised to avoid introducing chlorine through dust, sweat, or impure chemicals.

Plants actually absorb much more chlorine than they require. It undoubtedly serves in part as a balancing anion that is absorbed along with cations such as Ca^{++}, Mg^{++}, K^+, and NH_4^+. This function can, of course, be filled by other anions such as NO_3^-, $H_2PO_4^-$, and SO_4^{--}. Unique functions which can be served only by chlorine ions are yet unidentified. A very small amount of chlorine (less than 2 ppm in solution) will suffice for these unique functions, whatever they are.

Chlorine is sufficiently abundant that excesses are far more common than deficiencies. Excess chlorine can cause watery potatoes or poor burning quality in tobacco. Such problems can often be avoided by using potassium sulfate as a fertilizer rather than potassium chloride.

The abundance of chlorine makes it difficult to prove that chlorine is an essential plant nutrient. A small amount of chlorine usually finds its way into an experiment by way of water, other chemicals used, dust, perspiration from workers' hands, or even in the seeds that are planted. It is therefore much easier to prove that chlorine is beneficial than that it is essential. The data in Table 14-2 are good evidence that chlorine is beneficial to plants. Other plants which have shown a less marked response to chlorine are barley, buckwheat, corn, and beans. Cotton and tobacco have also shown marked response to chlorine additions when the supply was very low.

COPPER

Stimulation of plant growth by copper was recognized around 1900 in areas where bordeaux mixture (copper sulfate and lime) was used as a fungicidal spray on trees and truck crops. Several years later, Floyd (1917) recommended the use of bordeaux mixture as a cure for dieback in citrus on the basis of several years of experimentation. Still later it was found that onions and lettuce could be grown on previously unproductive peat soils after the addition of copper sulfate.

Copper in Soils

Copper occurs as Cu^{++} ions in most soils and as Cu^+ ions where the oxidation level is low. Copper ions are held so tightly by cation-exchange sites that they are even less mobile than Ca^{++}. The concentration in solution is only a few parts per million. Copper is most soluble in acid soils, and its solubility decreases as the pH rises.

Table 14–2 Experimental Results Showing that Chlorine Is Beneficial to Plant Growth

Plants studied	Yield without chlorine (% of yield with chlorine)
Lettuce	30
Tomatoes	35
Cabbage	42
Sugar Beets	49
Subterranean clover	50
Alfalfa	68

Source: Private communication from Dr. T. C. Broyer.

Copper deficiencies are most common in organic soils, especially those formed from sphagnum moss. Organic soils lack weatherable minerals to replenish the supply of copper as it is gradually lost. Some acid sandy and gravelly soils also lack weatherable copper-containing minerals and are deficient. Copper deficiencies are among the least common deficiencies in the United States but have been identified in Washington, California, Florida, South Carolina, and the Great Lakes region (Beeson, 1945). States where copper deficiency has been identified are shown in Figure 14-1.

Copper in Plants

Copper is important as a coenzyme that is needed to activate several plant enzymes. It is also involved in chlorophyll formation. Copper uptake seems to be inversely related to iron uptake. Too little copper causes iron to accumulate in plants. Too much copper causes chlorotic symptoms similar to those of iron deficiency. Copper deficiency symptoms occur mostly on new growth because copper is relatively nonmobile in plants.

Robinson and Edgington (1945) showed a wide range in the copper content of plants. They cited unpublished data of R. S. Holmes showing that 25 samples of wheat grain from the Great Plains and Portugal contained no more than 7.5 ppm copper. They also cited the work of Lehman (1896) showing that certain nonagricultural plants accumulated from 223 to 560 ppm copper. Somewhere between 5 and 50 ppm copper seems to be representative of most plants. An average crop removes only about 100 g of copper from a hectare of soil.

Copper Fertilizers

The most common copper fertilizer is hydrated copper sulfate ($CuSO_4 \cdot 5H_2O$). This form contains about 25 percent copper and is soluble. It is the ingredient that makes bordeaux mixture effective as a copper fertilizer as well as a fungicide.

The amount of copper sulfate to apply to copper-deficient soils depends primarily on the cation-exchange capacity. The amounts vary from 5 kg of copper sulfate (about 1¼ kg of copper) on soils with extremely low cation-exchange capacities to 300 kg (75 kg of copper) on Florida Everglades soils with high cation-exchange capacities. The higher rates are required to compensate for the low mobility of exchangeable Cu^{++}.

Copper sulfate is an effective fertilizer when applied to neutral or acid soils, but precipitation of insoluble compounds reduces its effectiveness on alkaline soils. However, there are two good alternatives. One is to apply a copper chelate to the soil (chelates are discussed in the latter part of this chapter). The chelate form is soluble but very little ionized so it neither precipitates in insoluble forms nor is it immobilized by adsorption on exchange sites. Wallace and Mueller (1973) found that beans grown in a calcareous loam soil in a greenhouse absorbed more copper from soil fertilized with 5 ppm of chelated copper than from 50 ppm of copper applied as copper sulfate.

The other alternative is to spray copper on the plants instead of applying it to the deficient soil. Nutrients can be absorbed through leaves and other plant parts as well as by roots. Either copper sulfate or a copper chelate can be used for this purpose. The soil must, of course, provide enough copper to satisfy the plant needs until it is large enough to spray.

IRON

Iron was recognized as an essential element as early as 1845. It is an abundant element in rocks and soils, but it is also one of the most commonly deficient micronutrients. The problem is the extremely insoluble nature of certain compounds of ferric (Fe^{+++}) iron. Ferric compounds accumulate in highly weathered soils and are major constituents of the red soils of tropical regions. Fossil remnants of some ancient soils contain enough iron to serve as iron ore, but the iron is too insoluble to meet plant needs even for a micronutrient.

Iron in Soils

Most of the iron contained in igneous rocks is in the ferrous (Fe^{++}) form. The iron in waterlogged soils tends to remain in this form and contributes to the bluish-gray colors that indicate wetness. Many compounds of ferrous iron have low solubilities, but ferric iron compounds are even less soluble. Iron deficiencies are limited to alkaline soils if much ferrous iron is present.

Imperfectly drained soils usually contain rust mottles, formed by the somewhat mobile ferrous iron moving to points where it is oxidized and precipitated as ferric iron. Ferric compounds become concentrated enough in spots in some subsoils to form small hardened masses called *iron concretions.* In extreme conditions a continuous layer of soil is cemented together by precipitated iron.

Much of the iron in well-drained soils is in the ferric form and is associated with humus in coatings on mineral particles. The reddish color of ferric oxide (Fe_2O_3) mixed with humus and other soil components produces many shades of brown in soils. Such iron is mostly unavailable to plants. Iron deficiencies may result if the soil minerals do not gradually release ferrous iron to replace that being oxidized year by year to ferric iron.

The solubilities of ferric and ferrous iron are much lower at high pH than at low pH. Both $Fe(OH)_3$ and $Fe(OH)_2$ have low solubilities and can be precipitated at high pH because OH^- ions become more abundant when the pH rises. Other iron compounds also become less soluble at higher pH. Precipitation of previously available iron is one of the hazards of overliming a soil. The resulting iron deficiency is known as *lime-induced chlorosis.*

The low solubilities of iron phosphates make it possible for either iron or phosphorus to contribute to a deficiency of the other. The adsorption and precipitation of phosphorus by iron compounds in acid soils was discussed in Chapter 11. High phosphorus levels in neutral or alkaline soils can similarly reduce iron availability and contribute to iron deficiency (Brown and Jones, 1975).

Iron deficiencies also can result from an excess of manganese (Lee, 1972) and possibly from excess copper. Manganese and copper can serve as oxidizing agents and convert ferrous iron to the more insoluble ferric form. Iron deficiencies caused by manganese toxicity occur in acid soils that otherwise would probably supply

Figure 14-2 Generalized map of iron and manganese deficiencies in the United States. (Compiled from various sources.)

adequate iron for plant growth. Some areas where iron deficiencies have been identified are shown in Figure 14-2.

Iron in Plants

Iron is absorbed by plants as the ferrous ion. The amount needed is not large enough
for iron to be considered a macronutrient. Iron is necessary for the formation of
chlorophyl and functions in some of the .enzymes of the respiratory system
(Schneider, Chesnin, and Jones, 1968). An iron deficiency results in the younger
leaves being small and pale green or yellow in color. This shortage of chlorophyl is
called *chlorosis.* The younger leaves are more affected than the older leaves because
iron is relatively immobile inside the plant. Often the veins remain green while the
areas between veins turn yellow from iron chlorosis (Figure 14-3). Iron-deficient
corn and sorghum leaves show pale stripes between veins.

Annual plants may show iron deficiencies early in the season but recover later
as the soil warms up. Larger root systems and increased solubility of iron at higher
temperatures help the plants obtain more iron in the summer than in the spring.

Plants differ in their susceptibility to iron chlorosis. Soybeans become chlorotic
on certain high-lime soils in Iowa (Figure 14-4) where corn, oats, clover, and alfalfa
show no chlorosis. Peach trees become chlorotic on the Houston clay soils in Texas,
but cotton, corn, clover, and small grains grown on these soils show no chlorosis.
Most plants obtain adequate iron for their needs from neutral or acid soils, but
rhododendrons and azaleas are likely to become chlorotic if the soil pH is above 6.0.

Figure 14-3 An iron-deficient soybean leaf is much smaller and lighter colored between the veins
than a normal leaf.

Figure 14-4   A soybean field in central Iowa showing iron chlorosis. The light-colored soybeans are yellow because the high-lime soil reduces the solubility of iron.

Brown and Jones (1975) indicate that root exudates are one reason for differences in susceptibility to iron deficiency. Resistant plants release an exudate that reduces Fe^{+++} to Fe^{++}.

Iron Fertilizers

Ferrous sulfate has been used for many years as a treatment for iron deficiency. This salt is soluble in water and can be applied to either the plant or the soil as the circumstance demands. Florists sometimes include a small package of ferrous sulfate with azaleas or other potted plants which require appreciable quantities of soluble iron.

Application of ferrous sulfate to calcareous soils is generally ineffective. These soils already contain many times more iron than would ever be applied as a fertilizer. The problem is low availability resulting from the high pH. The added iron will soon react to form ferrous hydroxide or other insoluble compounds like those already in the soil. Adding a form of soluble iron that can ionize and react in the soil is therefore not effective.

Foliar sprays of ferrous sulfate have proven effective when the iron requirement was not too high. A 1 percent solution of ferrous sulfate ($FeSO_4 \cdot 7H_2O$) is recommended as a spray applied directly on the leaves of affected plants (Webb, 1965). The spray may be repeated in 7 to 10 days if needed.

In Florida, where almost half the citrus trees have been affected by lack of iron, the use of ferrous sulfate has not proven satisfactory. Most of the soils used for Florida citrus are acid, but other ions, especially copper, interfere with the availability of iron. An improved treatment of iron chlorosis using chelates was developed at the Florida Citrus Experiment Station at Lake Alfred. Applications of 10 to 20 g of actual iron per tree in a chelated form have been effective in correcting chlorosis for as long as 2 years. Chelates are discussed in the last section of this chapter.

MANGANESE

A small amount of manganese is essential, but a large amount is toxic to plants. Manganese can exist in several different oxidation states, but research indicates that most of the manganese in the soil solution is present as Mn^{++} (Geering, Hodgson, and Sdano, 1969). This form is also found in rocks. The Mn^{++} ion can fill the same role in mineral structures as Fe^{++} and Mg^{++} ions, but manganese is less abundant than either iron or magnesium.

Manganese in Soils

Manganese leaches from well-drained acid soils because oxidation and acidity increase its solubility. Dissolved manganese moves to wetter and/or more alkaline positions where it precipitates in small, dark-colored, hardened bodies called *concretions* or *nodules*. Manganese concretions occur in the lower horizons of some soils of semiarid regions because the pH is high in those horizons. Concretions occur in humid regions in soil horizons that are wet enough from seepage water to have reducing conditions during some seasons. Certain footslope soils in Israel were found to have 50 to 80 percent increases in their total manganese contents (Yaalon, Jungreis, and Koyumdjisky, 1972). The coarser fractions contained as much as 5,000 ppm of manganese where the clay fraction contained less than 1,000 ppm. Exchangeable manganese is a smaller quantity, usually between 10 and 100 ppm.

Manganese deficiencies occur more often and for different reasons in humid regions than in arid regions. Acid sandy soils are low in manganese because most of their supply has been lost by leaching. Many muck soils are deficient in manganese (and other cations, too) because the organic materials release and lose cations more readily than anions. Calcareous and overlimed soils may be deficient in manganese because of low solubility even though considerable total manganese is present.

Manganese deficiencies have been identified in southern California soils, along the Atlantic coastal plain (especially in Florida), and in muck soils of the Great Lakes region (Beeson, 1945). Manganese deficiencies in Florida have been found in sandy soils, alkaline soils, and mucks (Wischhusen, 1948). Some manganese deficient areas are shown in Figure 14-2.

Manganese toxicity can occur in acid soils that have not been excessively leached. Lowering soil pH by addition of sulfur can induce a manganese toxicity (Morris, 1948). Liming to raise the pH is an effective means of overcoming the toxicity. Increasing the Fe^{++} and Zn^{++} supplies can also help to avert manganese toxicity because these ions reduce plant uptake of Mn^{++}.

Manganese in Plants

The manganese content of plants varies greatly according to the availability of manganese. Morris (1948) found that the manganese content of sweet clover grown in greenhouse pots varied from 29 to 858 ppm depending on the pH to which the soil was adjusted. Robinson and Edgington (1945) found manganese contents of oak leaves in the eastern United States ranging from 425 to 4,500 ppm. Brown, Hills, and Krantz (1968) found that sugar beets could tolerate up to 5,590 ppm manganese in their leaves. These high values are associated with high manganese concentrations in the soil. Robinson and Edgington state that "exchangeable manganese in the organic layers of Podzol soils [Spodosols] sometimes equals and occasionally exceeds the exchangeable calcium." The usual range of manganese concentrations in plants is from 20 to 400 ppm. Manganese contents of crops at different levels of sufficiency are shown in Table 14-3.

Manganese functions in chlorophyl development and in the enzyme systems of plants. Its various valences make it possible for manganese to be either a metallic coenzyme or a part of an organic molecule.

Manganese, like iron, is a relatively immobile element in plants. Deficiency symptoms appear on the younger leaves of the plant first. The symptoms vary with the plant but include a pale color similar to iron chlorosis between the veins of broad-leaved plants.

The similarity between manganese and iron causes a form of competition between the two elements. Symptoms of iron toxicity correspond to symptoms of manganese deficiency, and symptoms of manganese toxicity correspond to those of iron deficiency.

Manganese Fertilizers

Manganese deficiency can be corrected by spraying a 1 percent solution of manganese sulfate directly on the affected plants. Manganese sulfate can also be applied as a fertilizer on organic soils or leached mineral soils. Application of manganese sulfate to alkaline soils is generally not successful because of the low solubility of manganese in alkaline conditions.

Table 14-3   Manganese Contents of Various Crops

Crop	Part	Classification of nutrient level, ppm			
		Deficient	Marginal	Sufficient	Toxic
Alfalfa	Top 6 in. at bud	0–20	21–35	35–200	400+
Apples	Leaves	0–20	21–80	81–150	250+
Beans (snap and white)	Leaves at bloom	0–25	26–60	61–200	475+
Citrus	Bloom cycle leaves	0–17	18–25	26–100	200+
Corn	Sixth leaf	0–10	11–20	21–100	?
Cotton	Leaves at bloom	0–17	18–30	31–200	600+
Peas	Top 1/3 at bloom	0–25	26–40	41–70	150+
Peaches	Leaves	0–20	21–45	46–200	?
Pears	Leaves	0–25	26–40	41–150	?
Potatoes	Leaves at bloom	0–30	31–50	51–200	500+
Soybeans	Leaves at bloom	0–15	16–22	23–100	170+
Sugar beets	Leaves	0–20	21–90	91–350	1,100+
Sugarcane	Leaves	0–10	11–32	33–100	?
Tobacco	Third leaf from top	0–100	101–170	175–500	3,000+
Tomatoes	Leaves at bloom	0–20	21–65	66–120	210+
Wheat, oats, and barley	Tops at boot stage	0–12	13–18	19–100	700+

Source: Stangel, 1969.

Manganese chelates are also suitable forms of manganese fertilizer. The chelate form is more expensive than manganese sulfate but would be much preferred for applications on alkaline soils.

Both chelated and nonchelated manganese fertilizers were found by Mortvedt and Giordano (1975) to have enhanced effectiveness when applied in combination with phosphate fertilizers. The most effective method was band application of Mn-containing phosphate fertilizers.

Manure usually contains enough manganese to prevent manganese deficiency where it is applied regularly. Organic manganese from manure will remain available in alkaline soils much longer than manganese applied as manganese sulfate. The amount of manganese in the manure will, of course, be partially dependent on the amount of manganese in the soil where the animal feed was grown.

MOLYBDENUM

The molybdenum requirement of plants is the smallest of any of the micronutrients. Some investigators have reported plant analyses in which no molybdenum was found. Probably there was some, but the amounts were too small to be measured by the techniques used. Plants containing more than 10 to 20 ppm Mo are toxic to ruminants. The toxicity can be overcome by supplying larger than usual amounts of copper to the animals (Kubota and Allaway, 1972).

The chemistry of molybdenum somewhat resembles that of phosphorus. Phosphate minerals usually contain some molybdenum as impurities.

Molybdenum in Soils

Molybdenum concentrations in soils are lower than those of any other essential element. An average soil probably contains about 4 kg of molybdenum per hectare-furrow slice. Molybdenum tends to accumulate in A horizons and to increase as the organic-matter percentage increases (Kereszteny, 1973). Most of the molybdenum is locked up in organic and mineral structures or adsorbed on positively charged exchange sites so that the amount available to plants at any one time is no more than a few hundred grams per hectare.

Molybdenum can have several different valences, but the prevalent available form is the molybdate ion (MoO_4^{--}). The solubility of this ion increases as the pH rises. Molybdenum deficiencies are therefore most likely to occur in acid soils. Most of the molybdenum deficiencies reported in the United States have been in states bordering either the Great Lakes, the Atlantic Ocean, or the Pacific Ocean (Figure 14-5). Molybdenum deficiencies in arid or semiarid regions are rare except in Australia where millions of hectares of grazing lands need to be fertilized with molybdenum. The addition of 70 g of Mo per hectare to the phosphorus fertilizer raised the carrying capacity of pasture in the Melbourne area from 0.6 sheep to 7 to 10 sheep per hectare (Stout, 1972).

Molybdenum in Plants

Molybdenum is essential in symbiotic nitrogen fixation and in the reduction of nitrate nitrogen to the amine form. Thus a molybdenum deficiency can cause a

Figure 14-5 Generalized map of molybdenum and zinc deficiencies in crops in the United States. (Compiled from various sources.)

nitrogen deficiency in the plant (Hagstrom, 1968). The first deficiency symptoms to show are likely to be those of the nitrogen deficiency.

Molybdenum contents of peanut plant parts were found to range from 0.07 to 5.1 ppm by Welch and Anderson (1962). The higher molybdenum contents were associated with either a higher pH or with applications of molybdenum fertilizer.

The crops most likely to suffer from molybdenum deficiencies are cauliflower, citrus, and legumes (Turner, 1964). "Whiptail" of cauliflower is one of the most striking results of molybdenum deficiency.

Citrus trees in Florida are frequently subject to "yellow spot." This problem was identified as a molybdenum deficiency by Stewart and Leonard (1953). The symptom appears as large, interveinal chlorotic spots and has been known by the name "yellow spot" for years.

Molybdenum Fertilizers

"Yellow spot" in citrus can be corrected with a foliar spray applied at the rate of 3 g of sodium molybdate per mature tree. Allen and Laughlin (1967) found a foliar spray effective for treating molybdenum deficiency in cauliflower also.

Molybdenum requirements are so small that the necessary amount can sometimes be supplied by treating the seed. Applications of 100 g of Mo per hectare on the seed have been effective remedies for deficiencies in soybeans (Webb, 1965).

Molybdenum fertilizers applied to soils have given variable results. As little as 5 g of Mo per hectare as sodium molybdate or ammonium molybdate has produced satisfactory plant growth on molybdenum-deficient soils (Stout, 1972) but molybdenum fertilizers applied to the soil have not been a satisfactory treatment for "yellow spot" in citrus. Gupta (1969) showed that it is often necessary to apply both molybdenum fertilizer and lime to overcome a molybdenum deficiency. Added molybdenum did little good for the crops on soils at pH 5.0.

Phosphorus fertilizers usually contain small amounts of molybdenum that help to satisfy molybdenum needs where they are applied.

ZINC

Zinc is a widely distributed element that occurs in small but adequate amounts in most soils and plants. Zinc deficiencies result in some soils because of too little total zinc being present in their parent materials and in others because of unfavorable soil reaction. Zinc deficiencies occur in small, widely scattered areas. Some of these areas are shown in Figure 14-5.

Zinc in Soils

Most mineral soils contain between 20 and 600 kg of Zn per hectare-furrow slice, but the solubility of zinc compounds in water is very low. The solubility of zinc hydroxide indicates that about a gram of Zn^{++} would be contained in solution in the hectare-furrow slice of neutral soil. The solubility increases in an acid soil but decreases in an alkaline soil. Zinc deficiencies resulting from low solubility may occur at any pH above 6.0, depending largely on how fast the supply in solution can be replenished from other sources.

The zinc ion (Zn^{++}) is strongly adsorbed on the cation-exchange sites of silicate clays. Carbonates, and perhaps some other soil minerals, also adsorb zinc. Zinc adsorption is one of the factors that limits the concentration of zinc in solution, but it is also a source of replenishment when the zinc supply in solution has been depleted. As another source, some soil minerals contain small amounts of zinc within their crystal structures and release it gradually as they are weathered.

Zinc deficiencies in acid soils generally indicate a very low total zinc content and are most likely to occur in very sandy soils where leaching has removed what little available zinc they once contained.

Several workers have noted that high concentrations of phosphorus in soils tend to reduce the availability of zinc. Reduced phosphorus uptake sometimes results when the zinc content of the soil is increased (Sharma, Krantz, Brown, and Quick, 1968). Zinc phosphate has a low solubility, but this fact alone seems insufficient to account for the apparent antagonism between phosphorus and zinc.

Ward, Langin, Olson, and Stukenholtz (1963) found that increased soil compaction and wetter soils reduced zinc uptake by corn. This effect appears similar to the reduced ability of plants to absorb potassium in poorly aerated soils. Root respiration is too restricted in such circumstances to provide enough energy to absorb these cations in adequate amounts when the soil solution is very dilute.

Wet conditions, however, favor the uptake of zinc by rice, probably because the rice plant is able to translocate oxygen inside its roots. Giordano and Mortvedt (1972) found that zinc uptake by rice plants growing on flooded soil was as much as five times as great as on moist soil.

Low soil temperature reduces zinc uptake and accentuates a zinc deficiency. Joham and Rowe (1975) reported that the optimum level of zinc for cotton decreased from 25 to 1 ppm when the temperature warmed from 15 to 23°C.

Zinc in Plants

Robinson and Edgington (1945) cite several studies showing that plants normally contain a few hundred parts per million of zinc. Some plants contained over 1,000 ppm zinc, but these were grown in soils or solution cultures unusually high in zinc. Near the other extreme, the data of Rogers, Gall, and Barnette (1939) show that crotalaria grown on a soil low in zinc contained only 8 ppm zinc.

The difference between plants in their ability to absorb zinc is marked. Weeds growing on the same soil as the crotalaria contained 140 ppm zinc. Camp (1945) suggested that the use of green manure crops able to absorb zinc would ensure an adequate supply of zinc for the next crop through organic-matter decomposition.

Zinc is needed for protein metabolism and appears to be involved somehow in the production of chlorophyl. A characteristic zinc deficiency symptom in citrus and in corn is a green midrib and veins in the leaves with white areas between the veins.

Boawn, Rasmussen, and Brown (1969) found that zinc contents less than 15 ppm in growing field beans delayed maturity up to 30 days. Zinc contents of 20 ppm or more produced normal growth. Lo and Reisenauer (1968) found alfalfa to be unusually tolerant of low levels of zinc. Highest yields of alfalfa could be produced with as little as 6 ppm zinc in the leaves.

Fruit crops, especially peaches and citrus, are sensitive to zinc deficiency. A number of field crops including corn, soybeans, cotton, and potatoes are also sensitive to shortages of zinc. Significant zinc contents of several crops are given in Table 14-4.

Zinc Fertilizers

Zinc sulfate is the principal zinc fertilizer in use. It may be applied at rates of 2 to 20 kg of zinc per hectare. Shukla and Morris (1967) found it to be slightly more effective than other forms such as zinc oxide or zinc chelate. Much depends on the need, however. Sometimes a more slowly available form is preferable because of its long-lasting effect. Sharpee, Ludwick, and Attoe (1969) found that zinc oxide fused with sulfur supplied zinc over a period of months or even years.

Zinc sulfate can be applied to either soil or plants. The soil application is likely to be the most satisfactory if the soil is acid and not too sandy. The zinc will be adsorbed and become unavailable in alkaline soils and may be leached from the sandy soils. A slow-release fertilizer or a plant spray is preferable with either the sandy or the alkaline soils. Boehle and Lindsay (1969) recommend ¼ to 1 percent Zn in a foliar spray for emergency applications. A wetting agent should be included in the spray when maximum absorption of zinc is required.

Manures and plant residues returned to the soil contain zinc; if these are returned in adequate amounts, deficiencies may be averted. Zinc and other fertilizers are more likely to be needed when few residues are returned to the soil.

Table 14-4 Zinc Contents of Various Crops

Crop	Plant part	Deficient	Low	Sufficient	High
		Classification of nutrient level, ppm			
Alfalfa	Vegetative	0–15	16–20	21–70	71+
Apples	Leaves	0–15	16–20	21–50	50+
Citrus	Leaves	0–15	16–25	26–80	81–200
Corn	Leaves	0–10	11–20	21–70	71–150
Cotton	Vegetative	—	—	20–30	—
Grapes	Petioles	—	0–30	31–50	50+
Grass	Vegetative	—	—	15–80	—
Peaches	Leaves	0–16	17–20	21–50	50+
Pears	Leaves	0–10	11–16	17–40	41+
Potatoes		—	0–16	17–40	—
Soybeans	Vegetative	0–10	11–20	21–70	71–150
Sugar beets	Vegetative	0–10	11–20	21–70	70+
Tobacco	Vegetative	—	0–20	21+	—
Tomatoes	Leaves	0–10	11–20	21–120	120+
Wheat, oats, barley	3–12 in. growth	0–10	11–20	21–40	41–50

Source: Boehle and Lindsay, 1969.

THE USE OF CHELATES WITH MICRONUTRIENTS

The term *chelate* (pronounced kē'late) is derived from the Greek word meaning *claw.* Chelates are water-soluble organic compounds that can immobilize metallic cations. The cations are exchangeable with other cations because they are very slightly ionized from the organic chelating agent.

The advantage of chelated micronutrients can be illustrated with iron. Iron supplied in the form of ferrous sulfate is soluble in water and is readily available to plants, but as the ferrous sulfate dissolves it also ionizes. The ferrous ion is soon oxidized and precipitated as ferric hydroxide or some other equally insoluble compound. On the other hand, iron chelate is also soluble in water, but it does not ionize. The iron is held in a soluble form and is readily available for root absorption.

One of the best-known chelating compounds is ethylene diamine tetraacetic acid (EDTA). The ionizable H^+ from the acetic acid portion of the EDTA molecule can be replaced by metallic cations. Such replacements are symbolized with a prefix indicating the chelated element such as Fe-EDTA or Zn-EDTA. The structure of Zn-EDTA is diagrammed in Figure 14-6.

Fe-EDTA is usually applied to the soil rather than as a foliar spray because it has been known to cause damage to leaves. Soil applications of Fe-EDTA amounting to 10 to 20 g of actual iron per tree have proved effective in correcting chlorosis in

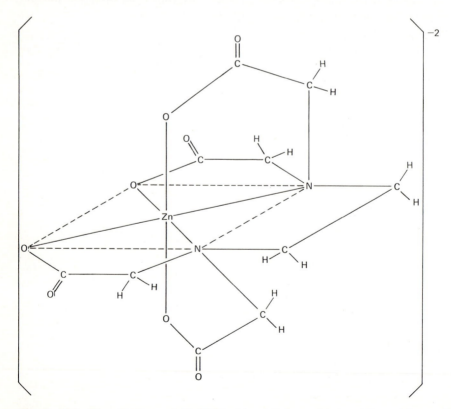

Figure 14-6 The structure of Zn-EDTA (zinc ethylene diamine tetraacetate or $[C_2H_4N_2 (CH_2COO)_4Zn]^{--}$). (*After Geigy,* 1967.)

citrus. Some radioactive-tracer studies have indicated that the entire Fe-EDTA molecule is absorbed by plant roots. Other studies have indicated that the metallic cation may sometimes be separated from the molecule before it is absorbed. The slight tendency of chelates to ionize would make the separation possible.

The fate of the Fe-EDTA compound in the soil is a matter of interest. It is resistant to both microbial attack and hydrolysis. One application can prevent chlorosis for as long as 2 years. During this time some of the compound is absorbed by plant roots. Some of the iron may be replaced by other cations, and the EDTA may sometimes be adsorbed on clays. Some of the compound is undoubtedly lost by leaching inasmuch as it is water soluble.

The stability of Fe-EDTA and its effectiveness in controlling chlorosis are much greater in acid soils than in calcareous soils. Another material, hydroxyethyl ethylene diamine triacetic acid (EDTA-OH), is much more stable in calcareous soils. Several other chelates are also available and are adapted to particular circumstances. Some chelation undoubtedly occurs naturally in the soil by organic components complexing metallic cations.

Chelates are used more with iron than with other micronutrients, but copper, manganese, and zinc can also be chelated. These chelates are less used because foliar sprays of these elements are cheaper and are usually adequate to meet plant needs. Iron can also be applied to some plants as a foliar spray of ferrous sulfate. Such a spray is considerably less expensive than a soil application of Fe-EDTA-OH for soybeans growing on a calcareous soil.

Calcium and magnesium can also be chelated, but this is not a practical means of supplying these macronutrients. The cost would be much too great, and the application of inorganic compounds is more satisfactory for the large amounts required for macronutrients.

REFERENCES

Allen, Z. D., and W. M. Laughlin, 1967, Molybdenum Deficiency of Cauliflower Influenced by Soil Moisture and Fertilizer Rate, *Agron. J.* **59**:505–506.

Anderson, O. E., and F. C. Boswell, 1968, Boron and Manganese Effects on Cotton Yield, Lint Quality, and Earliness of Harvest. *Agron. J.* **60**:488–492.

Beauchamp, E. G., and G. Lean, 1973, Evaluation of Surfactants for Zinc Absorption by Soybean Leaf Tissues, *Commun. Soil Sci. Plant Anal.* **4**:1–7.

Beeson, K. C., 1945, The Occurrence of Mineral Nutritional Diseases of Plants and Animals in the United States, *Soil Sci.* **60**:9–14.

Boawn, L.C., P. E. Rasmussen, and J. W. Brown, 1969, Relationship between Tissue Zinc Levels and Maturity Period of Field Beans, *Agron. J.* **61**:49–51.

Boehle, John, Jr., and W. L. Lindsay, 1969, "Micronutrients—The Fertilizer Shoe-Nails." Pt. 6, In the Limelight—Zinc, *Fert. Solutions* **13**(1):6–12.

Branchley, W. E., and K. Warrington, 1927, The Role of Boron in the Growth of Plants, *Ann. Bota.* **41**:167–187.

Brown, A. L., F. J. Hills, and B. A. Krantz, 1968, Lime, P, K, and Mn Interactions in Sugar Beets and Sweet Corn, *Agron. J.* **60**:427–429.

Brown, J. C., and W. E. Jones, 1975, Phosphorus Efficiency as Related to Iron Inefficiency in Sorghum, *Agron. J.* **67**:468–472.

Broyer, T. C., A. B. Carlton, C. M. Johnson, and P. R. Stout, 1954, Chlorine—A Micronutrient Element for Higher Plants, *Plant Physiol.* **29**:526–532.

Camp, A. F., 1945, Zinc as a Nutrient in Plant Growth, *Soil Sci.* **60**:157–164.

Darst, B. C., and J. H. Reeves, 1968, Micronutrients—The "Fertilizer Shoe-Nails." Pt. 3, A Closer Look at Copper, *Fert. Solutions* **12**(3):26–31.

Eaton, F. M., 1944, Deficiency, Toxicity, and Accumulation of Boron in Plants, *J. Agric. Res.* **69**:237–277.

Floyd, B. F., 1917, Dieback, or Exanthema of Citrus Trees, *Fla. Agric. Exp. Sta. Bull.* 140.

Forno, D. A., S. Yoshida, and C. J. Asher, 1975, Zinc Deficiency in Rice. I. Soil Factors Associated with the Deficiency, *Plant Soil* **42**:537–550.

Geering, H. R., J. F. Hodgson, and Caroline Sdano, 1969, Micronutrient Cation Complexes in Soil Solution. IV. The Chemical State of Manganese in Soil Solution, *Soil Sci. Soc. Am. Proc.* **33**:81–85.

Geigy, 1967, *Micronutrients for Row Crops,* Geigy Chemical Corp., Ardsley, New York, 42 pp.

Giordano, P. M., and J. J. Mortvedt, 1972, Rice Response to Zn in Flooded and Nonflooded Soil, *Agron. J.* **64**:521–524.

Gupta, U. C., 1969, Effect and Interaction of Molybdenum and Limestone on Growth and Molybdenum Content of Cauliflower, Alfalfa, and Bromegrass on Acid Soils, *Soil Sci. Soc. Am. Proc.* **33**:929–932.

Hagstrom, G. H., 1968, Micronutrients—The "Fertilizer Shoe-Nails." Pt. 5, A Closer Look at Molybdenum, *Fert. Solutions* **12**(5):26–33.

Il'in, V. B., and A. P. Anikina, 1974, Boron Salinization of Soils, *Sov. Soil Sci.* **6**:68–75.

Joham, H. E. , and Viola Rowe, 1975, Temperature and Zinc Interactions on Cotton Growth, *Agron. J.* **67**:313–317.

Jones, H. E., and G. D. Scarseth, 1944, The Calcium-Boron Balance in Plants as Related to Boron Needs, *Soil Sci.* **57**:15–24.

Kereszteny, B., 1973, Distribution of Total B, Cu, Mn and Mo Contents in the Profiles of Some Soil Types in the Little Plain, and Its Relationship to Certain Soil Characteristics, *Acta Agron.* **22**:115–130.

Kubota, Joe, and W. H. Allaway, 1972, Geographic Distribution of Trace Element Problems, Chap. 21 in *Micronutrients in Agriculture,* Soil Sci. Soc. Am., pp. 525–554.

Lee, C. R., 1972, Interrelationships of Aluminum and Manganese on the Potato Plant, *Agron. J.* **64**:546–549.

Lehman, D. S., 1963, Some Basic Principles of Chelation Chemistry, *Soil Sci. Soc. Am. Proc.* **27**:167–170.

Lo, S. L., and H. M. Reisenauer, 1968, Zinc Nutrition of Alfalfa, *Agron. J.* **60**:464–466.

MacKay, D. C., E. W. Chipman, and U. C. Gupta, 1966, Copper and Molybdenum Nutrition of Crops Grown on Acid Sphagnum Peat Soil, *Soil Sci. Soc. Am. Proc.* **30**:755–759.

Morris, H. D., 1948, The Soluble Manganese Content of Acid Soils and Its Relation to the Growth and Manganese Content of Sweet Clover and Lespedeza, *Soil Sci. Soc. Am. Proc.* **13**:362–371.

Mortvedt, J. J., and P. M. Giordano, 1970, Application of Micronutrients Alone or with Macronutrient Fertilizers, *Commun. Soil Sci. Plant Anal.* **1**:273–286.

Mortvedt, J. J., and P. M. Giordano, 1975, Crop Response to Manganese Sources Applied with Ortho- and Polyphosphate Fertilizers, *Soil Sci. Soc. Am. Proc.* **39**:782–787.

Norvell, W. A., and W. L. Lindsay, 1969, Reactions of EDTA Complexes of Fe, Zn, Mn, and Cu with Soils, *Soil Sci. Soc. Am. Proc.* **33**:86–91.

Olomu, M. O., G. J. Racz, and C. M. Cho, 1973, Effect of Flooding on the Eh, pH, and Concentrations of Fe and Mn in Several Manitoba Soils, *Soil Sci. Soc. Am. Proc.* **37**: 220–224.

Plank, C. O., and D. C. Martens, 1974, Boron Availability as Influenced by Application of Fly Ash to Soil, *Soil Sci. Soc. Am. Proc.* **38**:974–977.

Robinson, W. O., and Glen Edgington, 1945, Minor Elements in Plants, and Some Accumulator Plants, *Soil Sci.* **60**:15–28.

Rogers, L. H., O. E. Gall, and R. M. Barnette, 1939, The Zinc Content of Weeds and Volunteer Grasses and Planted Land Covers, *Soil Sci.* **47**:237–243.

Schneider, E. O., Leon Chesnin, and R. M. Jones, 1968, Micronutrients—The "Fertilizer Shoe-Nails." Pt. 4, The Elusive Nutrient—Iron, *Fert. Solutions* **12**(4):18–24.

Scott, H. D., S. D. Beasley, and L. F. Thompson, 1975, Effect of Lime on Boron Transport to and Uptake by Cotton, *Soil Sci. Soc. Am. Proc.* **39**:1116–1121.

Sharma, K. C., B. A. Krantz, A. L. Brown, and James Quick, 1968, Interaction of Zn and P in Top and Root of Corn and Tomato, *Agron. J.* **60**:453–456.

Sharpee, K. W., A. E. Ludwick, and O. J. Attoe, 1969, Availability of Zinc, Copper, and Iron in Fusions with Sulfur, *Agron. J.* **61**:746–749.

Shukla, U. C., and H. D. Morris, 1967, Relative Efficiency of Several Zinc Sources for Corn (*Zea mays* L.), *Agron. J.* **59**:200–202.

Siman, A., F. W. Cradock, P. J. Nichols, and H. C. Kirton, 1971, Effects of Calcium Carbonate and Ammonium Sulphate on Manganese Toxicity in an Acid Soil, *Aust. J. Agric. Res.* **22**:201–214.

Sparr, M. C., C. W. Jordan, and J. R. Turner, 1968, "Micronutrients—The Fertilizer Shoe-Nails." Pt. 2, A Closer Look at Boron, *Fert. Solutions* **12**(2):22–32.

Stangel, P. J., 1969, Micronutrients—The "Fertilizer Shoe-Nails." Pt. 7, A Closer Look at Manganese, *Fert. Solutions* **13**(2):38–44.

Stewart, Ivan, and C. D. Leonard, 1953, Correction of Molybdenum Deficiency in Florida Citrus, *Proc. Am. Soc. Hortic. Sci.* **62**:111–115.

Stout, P. R., 1972, Introduction to Micronutrients in Agriculture, *Soil Sci. Soc. Am.,* pp. 1–5.

Turner, Jim, 1964, Micronutrients . . . a Growing Need in the South, *Better Crops Plant Food* **48**(1):34–36.

Wallace, A., and G. V. Alexander, 1973, Manganese in Plants as Influenced by Manganese and Iron Chelates, *Commun. Soil Sci. Plant Anal.* **4**:51–56.

Wallace, A., and R. T. Mueller, 1973, Effects of Chelated and Nonchelated Cobalt and Copper on Yields and Microelement Composition of Bush Beans Grown on Calcareous Soil in a Glasshouse, *Soil Sci. Soc. Am. Proc.* **37**:907–908.

Ward, R. C., E. J. Langin, R. A. Olson, and D. D. Stukenholtz, 1963, Factors Responsible for Poor Response of Corn and Grain Sorghum to Phosphorus Fertilization: III. Effects of Soil Compaction, Moisture Level, and Other Properties on P-Zn Relations, *Soil Sci. Soc. Am. Proc.* **27**:326–330.

Warrington, Katherine, 1923, The Effect of Boric Acid and Borax on the Broad Bean and Certain Other Plants, *Ann. Bot.* **37**:629–672.

Webb, J. R., 1965, Do You Recommend Micronutrients? *Fert. Agric. Chem. Dealers Conf. Proc. 17th,* Des Moines, Iowa, Jan. 19, pp. 22–26.

Welch, L. F., and O. E. Anderson, 1962, Molybdenum Content of Peanut Leaves and Kernels as Affected by Soil pH and Added Molybdenum, *Agron. J.* **54**:215–217.

Whetstone, R. R., W. O. Robinson, and H. G. Byers, 1942, Boron Distribution in Soils and Related Data, *USDA Tech. Bull.* 797.

Wischhusen, J. F., 1948, The Story of Fertilizer Manganese, *Fert. Rev.* **23**(4):10.

Yaalon, D. H., Chava Jungreis, and Hanna Koyumdjisky, 1972, Distribution and Reorganization of Manganese in Three Catenas of Mediterranean Soils, *Geoderma* **7**:71–78.

Chapter 15

Variations in Plant Composition

The amount of each element contained within a plant varies with the type of plant, its stage of maturity, and environmental conditions. Part of the variation is dependent on the supply of each element available in the soil. Available nutrient supplies varying within certain limits may change plant composition without much change in yield. Variations beyond these limits will reduce yield.

CATION-ANION BALANCE

The need for electric balance in both soil and plant places restrictions on nutrient uptake. There must be a balance of cations and anions in the soil and in the plant at all times. The absorption of plant nutrients must therefore be balanced. When a plant absorbs a cation, it must also absorb an anion or else release another cation in exchange for the one absorbed. More precisely, the net numbers of positive charges absorbed as cations and negative charges absorbed as anions must be equal. Otherwise the plant and the soil would develop different electric potentials and an electric current would flow between the two.

Inorganic anions are utilized by the plant in forming proteins and other organic molecules. Cation charges that were balanced by charges of the inorganic anions must then be balanced by charges of newly formed organic anions. The charges remain balanced, but the proportions of organic and inorganic ions vary with the plant and the plant part. Supporting tissues such as stems and stalks are usually high in cations such as Ca^{++} balanced by organic anions. Seeds are high in elements such as nitrogen, phosphorus, and sulfur whose form has been changed from inorganic anions into organic molecules.

Harvesting grain or some other plant part high in protein leaves a crop residue that is not balanced on the basis of inorganic cations and anions alone (Pierre, Meisinger, and Birchett, 1970). Protein contains nitrogen that was probably absorbed in the nitrate form. Some of the cations that were absorbed along with the nitrate anions are contained in the crop residues. These residues usually tend to raise the soil pH because they return to the soil more inorganic cations than inorganic anions (the organic ions disappear as the residue decomposes). Plant residues returned to the soil therefore partly offset the acidifying effects of leaching and of fertilizers containing the ammonium form of nitrogen (Chapter 8).

Absorption of Cations

The positive electric charges of cations in the soil are balanced partly by anions in solution but mostly by the negative charges of clay and humus micelles. The micelles are not absorbed by plants; therefore, any absorbed cations that were associated with the micelles must be replaced with other cations. Replacement cations may come from the soil solution or from the plant. Most of the cations released by growing plants are H^+ ions; these, of course, tend to make the soil more acid. Replacement cations from the soil solution must in turn be replaced from other sources such as organic-matter decomposition, mineral weathering, or carbonic acid. Carbonic acid is always available in soil when plants are growing because it forms from CO_2 evolved by root respiration and microbial activity.

Absorption of Anions

Most of the anions absorbed by plants come from the soil solution. They are combined with carbohydrates in the synthesis of complex organic compounds. Elements such as nitrogen, phosphorus, and sulfur which unite with oxygen to form complex anions are firmly bonded to carbon in organic matter. They become integral parts of organic compounds and no longer function as separate inorganic anions. The charge balance in the plant would be upset if organic anions, hydroxyl ions, and bicarbonate ions were not formed in the process. Some changes involve complex oxidation-reduction reactions that take place within the plant. An excess of hydroxyl or bicarbonate ions formed by such reactions may move from the plant into the soil in exchange for other anions being absorbed or may be accompanied by surplus H^+ or other cations.

Most of the absorbed cations remain in water-soluble form in the plant, whereas most of the anions are incorporated into organic molecules. This difference between the utilization of cations and anions in plants results from the ability of anion-forming elements to enter into covalent bonds. Monovalent cations such as K^+ and Na^+ remain independent and are not tied into the organic structures. Most of the divalent cations also remain as independent ions in solution but not all of them. Some

Mg^{++} ions are combined in chlorophyl, and some may become parts of organic compounds in seeds. Some Ca^{++} ions are incorporated into organic compounds in cell walls.

ELEMENTAL COMPOSITION OF PLANTS

Plants absorb some of every ion available to them. They may contain traces of such unneeded elements as gold and silver. Even the potential toxicity of arsenic, selenium, or mercury does not prevent plants from absorbing ions of these elements. Even so, there is some selectivity in the absorption of elements. Most plants will absorb proportionately more K^+ ions relative to the amount available than they will of Ca^{++}, Mg^{++}, or Na^+ ions. Of the various anions, plants absorb more NO_3^- than any other type.

The average elemental composition of several kinds of plants is shown in Table 15-1. Trace amounts of many more elements would also be present but at very low levels. It may be noted from this table that oxygen, carbon, and hydrogen make up most of the plant weight. Potassium, calcium, magnesium, and sodium make up about 80 to 90 percent of the mineral cation content of plants. Similarly, nitrogen, phosphorus, sulfur, and chlorine constitute over 90 percent of the anion-forming elements (excluding carbon, hydrogen, and oxygen from consideration). These eight elements are compared on a weight percentage basis and on a milliequivalent basis

Table 15-1 Average Elemental Composition of Indicated Plant Materials in kg/10,000 kg of Dry Weight

Element	Corn silage	Corn grain	Wheat grain	Soybean seed	Alfalfa hay	Bluegrass hay
O	4,500					
C	4,400					
H	630					
N	130	144	211	606	245	131
Si	120					
K	90	29	42	150	197	167
Ca	25	2	4	25	147	40
P	16	27	39	59	24	27
Mg	16	10	14	28	31	19
S	15	12	20	22	29	12
Cl	15	4	8	3	28	55
Al	11					
Na	3	1	6	22	15	10
Fe	0.9	0.3	0.6	0.8	2.7	1.5
Mn	0.6	0.05	0.4	0.3	0.6	0.8
Zn	0.3	0.10	0.05		0.5	
B	0.1					
Cu	0.05	0.03	0.08	0.15	0.18	0.09
Mo	0.01					

Source: Compiled from various sources.

in Table 15-2. On the milliequivalent basis, the anion-forming elements shown in Table 15-2 outnumber the cation-forming elements by 96 to 80 for the grasses and by 202 to 144 for the legumes. The difference results from the way nitrogen, phosphorus, sulfur, and chlorine are combined in organic compounds. The negative charges they carried as anions are only partly replaced by organic anions.

Nitrogen is by far the predominating anion-forming element taken up by plants even if an allowance is made for the nitrogen absorbed as the ammonium cation rather than in the anion form. The data show that more potassium is taken up by plants than any other cation on a weight basis when analyses of a large number of plants are averaged together. But if legumes are analyzed separately from grasses, it is usually found that more calcium than potassium is taken up as shown in Table 15-2. The higher content of calcium in legumes is related to the charge balance at the surfaces of plant roots. Legume roots have higher cation-exchange capacities than grass roots have and therefore attract divalent cations more strongly.

One important point not shown by the preceding tables is that plants which contain a high content of total anions also contain a high content of total cations. The milliequivalents of anion-forming elements will usually outnumber the milliequivalents of cations by a ratio of about 4:3. The charge balance is maintained during absorption by part of the nitrogen entering as NH_4^+ and by H^+ and OH^- ions moving in either direction between plant and soil.

EFFECT OF ONE ANION ON ANOTHER

Additions of one anion may cause either a decrease or an increase in the absorption of other anions. A decrease is most likely to result if the two elements are chemically similar. For example, an increase in sulfur uptake by a plant can decrease the concentration of selenium. Sulfur and selenium are so similar that they probably

Table 15–2 Average Chemical Composition of Grasses and Legumes

Element	% by weight		meq/100g	
	Grasses	Legumes	Grasses	Legumes
K	1.54	1.13	39	29
Ca	0.33	1.47	16	73
Mg	0.21	0.38	17	32
Na	0.18	0.24	8	10
Total meq			80	144
N	0.99	2.38	71	170
P	0.20	0.21	6	7
S	0.15	0.22	9	14
Cl	0.37	0.38	10	11
Total meq			96	202

Source: Compiled from various sources.

compete for the same entries into the plant root. Phosphorus and arsenic also compete with each other because of chemical similarity.

Information about the uptake of arsenic and selenium is valuable even though neither is essential to plant growth. Animals need small amounts of selenium but can be poisoned by excessive amounts of either arsenic or selenium. Westermann and Robbins (1974) reported that sulfur fertilization on a sulfur-deficient soil reduced the selenium concentration of alfalfa from low to deficient for animal nutrition. The reduced concentration was a dilution effect from increasing the alfalfa yield from 1.81 to 5.56 tons per hectare. The total selenium uptake showed a small increase but not nearly proportional to the increased yield.

Some studies have shown actual decreases in total uptake of selenium when sulfur was applied or of arsenic when phosphorus was applied. Thus, fertilization might be used as part of a program to decrease selenium or arsenic toxicity. Soil acidification and control of certain plants that accumulate selenium are also helpful for reducing selenium toxicity.

UPTAKE OF ESSENTIAL MACRONUTRIENT ANIONS

Nitrogen, phosphorus, and sulfur are all essential constituents of protein. They are built into organic structures in relatively fixed proportions so that a deficiency of any one of these elements limits the utilization of the other two. A severe deficiency of nitrogen, for example, can cause phosphorus and sulfur to accumulate in solution in plant sap because they are not being made into protein. The accumulation in solution reaches all the way through the plant and makes it more difficult for the roots to absorb more of these elements from the soil. Thus, a deficiency of nitrogen can reduce the uptake as well as the utilization of phosphorus and sulfur. Similar effects can be noted when either phosphorus or sulfur is deficient.

Plants typically use about ten parts of nitrogen to one part each of phosphorus and sulfur. Figure 15-1 illustrates a close relationship between phosphorus and protein in several plants. Based on an average of 16 percent nitrogen in protein, the N:P ratio of the line in Figure 15-1 is about 9:1.

Not all sets of nitrogen and phosphorus data show such close correlation as the data in Figure 15-1. The ratio between these two elements may vary even in the same variety of plant because of differences in the availability of nitrogen and phosphorus. Hoagland and Arnon (1941) report that plants grown in moist soils have higher phosphorus contents relative to nitrogen than plants grown in relatively dry soils.

There is considerable evidence (Sheets, 1944, and Smith, Capp, and Potts, 1949) that nitrogen fertilizers can decrease the percentage of other mineral elements in crops. The plants grow so much more vigorously with nitrogen fertilizer that the percentages of other elements are reduced in spite of increased total absorption of each element. Reduction of magnesium concentration to a level that resulted in grass tetany in livestock was reported by Mayland, Grunes, and Stuart (1974).

Interactions of N, P, and K

The data from Hanway and Dumenil shown in Table 15-3 illustrate interactions between nitrogen, phosphorus, and potassium fertilizers. Nitrogen was the most

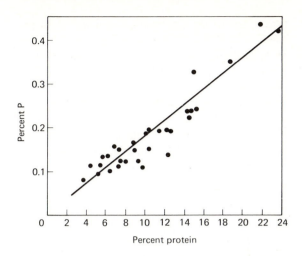

Figure 15-1 The correlation of protein and phosphorus in samples of Kentucky bluegrass, Canada bluegrass, poverty grass, bent grass, timothy, and redtop grown in New York pastures. (*Johnstone-Wallace,* 1937.)

limiting element in the experiment; nitrogen fertilizer increased yield by 40 percent over the check and increased the concentration of both nitrogen and phosphorus in the corn leaves. The increase in phosphorus must have resulted from more vigorous root growth.

Phosphorus was the second limiting element and resulted in significant yield increase when applied along with nitrogen. Treatments combining nitrogen and phosphorus had much higher phosphorus concentrations in the corn leaves than any other combination.

Nitrogen and phosphorus together increased yield enough to have a diluting effect on potassium. The reduced potassium concentration became a limiting factor

Table 15-3   Effects of N, P, and K Fertilizers on Yields and Leaf Composition of Corn Grown on Nesset Silt Loam in Iowa

Fertilizer treatment*	Leaf analyses			Corn yield q/ha
	% N	% P	% K	
None	1.9	0.20	1.6	29
N	2.6	0.22	1.5	41
P	1.8	0.22	1.5	30
NP	2.6	0.28	1.3	48
K	1.9	0.20	1.8	30
NK	2.6	0.22	1.8	41
PK	1.6	0.20	1.9	24
NPK	2.5	0.27	1.6	55

*Fertilizer rates per hectare were 67 kg of N, 30 kg of P, and 55 kg of K, broadcast and plowed under.
Source: Hanway and Dumenil, 1965, *Plant Food Rev.*

so it was finally beneficial to add potassium to the fertilizer. The final treatment combining all three elements increased yield by 90 percent over the check.

The fertilizer rates used in Hanway and Dumenil's data were too low to reach maximum yield. It would have been profitable to have doubled or tripled the amount of nitrogen and to have increased phosphorus and potassium by lesser amounts.

The interaction between nitrogen and phosphorus shows clearly in Figure 15-2. Phosphorus fertilizer is much more effective at increasing the P content of plant tissue when nitrogen is abundant than when nitrogen is deficient. The nitrogen content declines with added phosphorus if nitrogen is deficient but increases if nitrogen is abundant.

EFFECT OF ONE CATION ON ANOTHER

An increase in one anion often increases the uptake of other essential anions, but it has long been established that an increased uptake of one cation causes a decreased uptake of other cations. This effect is known as *cation competition.* Allaway and Pierre (1939) and Bower and Pierre (1944) showed that an increased uptake of calcium and magnesium is accompanied by a corresponding decrease in uptake of potassium.

Part of the cation needs are specific, such as magnesium for chlorophyl and calcium for cell walls, but most of the cations are needed only to balance the charges of anions. Thus, for each essential cation there is a certain minimum amount necessary for plant growth. Once this minimum amount has been met, the additional amount of each cation absorbed serves mainly to maintain a neutral charge balance. Plants can show a great deal of variation between the ratios of different cations and still have normal growth. Hunter, Toth, and Bear (1943) found that alfalfa could have a ratio of calcium to potassium anywhere between 1:1 and 5:1 and still have normal growth. They found that a depression in yield occurred if the ratio was

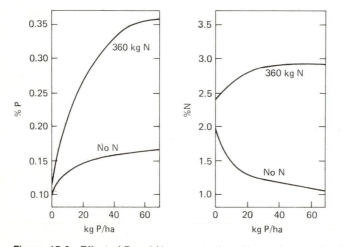

Figure 15-2 Effect of P and N concentrations in corn leaves of various rates of P fertilizer with and without 360 kg of N per hectare. (*Redrawn from Hanway and Dumenil,* 1965.)

Table 15–4 Chemical Composition of Plants as Affected by High and Low Concentrations of Calcium

Crop	Solution	Relative ion equivalents				
		Ca	K	Mg	N	P
Barley plants	Control	8	20	5	50	17
Pea plants	Control	9	21	6	47	17
Barley plants	Low Ca	5	26	5	50	14
Pea plants	Low Ca	5	27	4	51	13

Source: Newton, 1923.

between 5:1 and 8:1; potassium deficiency signs appeared if there was more than eight times as much calcium as potassium in the alfalfa.

Corn normally contains about twice as much potassium as calcium plus magnesium. However, Stanford, Kelly, and Pierre (1941) found that corn may still have normal growth with a ratio of 3½ times as much calcium plus magnesium as potassium.

The largest variations in cation ratios are in the calcium-to-potassium relation. These two cations appear to be strongly competitive with each other. An increased uptake of one almost invariably causes a depressed uptake of the other. This is also true of magnesium and calcium, and magnesium and potassium, but to a lesser extent. The data in Table 15-4, from solution-culture work, show a marked tendency for calcium and potassium to replace each other.

The data in Table 15-5 show that the calcium-to-potassium ratio can be changed by altering the potassium supply as well as by altering the calcium supply. Increasing amounts of potassium fertilizer increased the potassium content and decreased the calcium content of soybean hay.

Competition versus Antagonism

Competition between dissimilar cations is probably a direct result of their positive charges. The more abundant a cation is, the larger its share of the absorbed charges will be. Other cations will therefore have smaller shares.

Table 15–5 Percentages of K and Ca in Soybean Hay as Affected by Different Levels of Potassium Fertilization

K in fertilizer, kg/ha	Percentage in hay (air-dry basis)	
	K	Ca
0	0.65	1.21
11	0.91	1.08
22	1.23	0.97
34	1.63	0.85
45	1.88	0.76
56	1.86	0.70

Source: Calculated from data of Adams, Bogg, and Roller, 1937.

Antagonism refers to exceptionally strong competition between pairs of similar cations. The specific nature of antagonism indicates that the ions involved enter the root the same way whereas other ions may have other entries. Most plant physiologists postulate the existence of several kinds of carrier molecules that selectively transport ions into the inner space of roots. Similar ions must enter by the same carrier molecules. They are therefore so strongly competitive for entry that they are "antagonistic" to each other. Pairs of similar anions likewise exhibit antagonism and are believed to share carrier molecules.

UPTAKE OF MICRONUTRIENTS

Micronutrients interact in various ways with each other, with the macronutrients, and with the environment. For example, the uptake of zinc can be reduced by high concentrations of phosphorus or of manganese, by lack of oxygen, or by cool soil temperatures. The phosphorus influence results from the low solubility of zinc phosphate reducing the concentration of zinc in the soil solution. Manganese and zinc are similar enough that they probably compete for entrance to the plant root through the same carrier molecules and are therefore antagonistic to each other. Ohki (1975) reported that a high level of zinc in manganese-deficient cotton plants was reduced to a barely adequate level of 25 ppm when manganese was supplied. Singh and Steenbert (1974) showed that supplying zinc could reduce the manganese content of corn by 40 percent.

Zinc uptake in wet soil may be reduced both by cool temperature and by lack of oxygen. Cool, cloudy weather is often associated with zinc deficiency in corn (Edwards and Kamprath, 1974). Low temperature reduces solubilities and slows the rates of chemical and biological processes, thus reducing zinc uptake. Lack of oxygen slows respiration and thus deprives the plant of energy needed to absorb nutrients against a concentration gradient.

Iron, like zinc, has multiple relationships that influence its availability: Iron phosphates have low solubility; iron and manganese are antagonistic to each other (Ohki, 1975); oxidation changes iron to the less soluble ferric form; and high pH reduces the solubility of both the ferrous and the ferric forms. Positive interactions also exist, e.g., sulfur can be very helpful for the absorption of iron. Any of several forms of sulfur sprayed on cowpeas increased their yield between 30 and 40 percent by helping to overcome iron deficiency (Bansal and Singh, 1975).

Many other micronutrient interactions occur that cannot be fully explored here. Some nutrients such as nitrogen may influence most any other nutrient (Walker and Peck, 1973).

PLANT GROWTH RESPONSE TO A DEFICIENT ELEMENT

Justus von Liebig (1803–1873) analyzed plant samples and developed the basis for modern concepts of plant nutrition. Liebig proposed a *law of the minimum* stating that plant growth is proportional to the amount available of the most limiting plant nutrient. Expansion of this concept has led to the inclusion of water, temperature, and soil conditions as possible limiting factors along with the plant nutrients. This theory suggests that increases in plant growth are obtained by supplying more of the one most limiting factor. This will, of course, cause that factor to become adequate and some other factor to become most limiting. Maximum yield is obtained when all factors are supplied in adequate or optimum amounts.

Liebig's theories marked an important advance in the understanding of plant nutrition. His ideas still have merit even though they have been found imprecise in some regards. For example, the law of the minimum neglects the possibility of interactions between nutrients (positive and negative interactions of fertilizer elements were discussed in Chapter 9). Experiments have shown that two or more growth factors may interact to limit growth more severely than the one most limiting factor. If there are three limiting factors present, respectively, at levels that individually should limit growth to 70, 80, and 90 percent of its potential, the actual growth will be approximately 0.70 X 0.80 X 0.90, or about 50 percent rather than 70 percent of normal. Improved growth could be achieved by adding any one of the three factors. Maximum growth will, of course, require that all needed factors be adequately supplied.

A German scientist named Eilhardt Mitscherlich concluded that plant growth response to a limiting element is not proportional as Liebig proposed but rather follows a law of diminishing returns. Mitscherlich developed an equation to express this law mathematically:

$$dy/dx = k(A-y)$$

where dy is the yield increase resulting from a small addition, dx, of the limiting factor; k is a constant for a particular crop and growth factor; A is the maximum possible yield; and y is the yield under the actual conditions.

Perhaps the simplest way to interpret Mitscherlich's equation is to say that the yield improvement resulting from any small additional unit of fertilizer is proportional to the remaining possible improvement until the maximum yield is attained. Each additional unit of fertilizer gives a slightly smaller benefit than the previous unit. The benefit from applying many units of fertilizer is, of course, the sum of the benefits obtained from each successive unit.

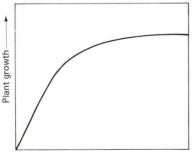

Figure 15-3   A typical growth response curve (schematic).

A typical growth response curve is shown in Figure 15-3. Mitscherlich's equation usually fits the central portion of the response curve fairly well. This is often the only significant part of the curve because extremely high or low nutrient levels seldom occur. The soil usually supplies enough of each nutrient to mask the lower part of the curve. Fertilizer applications in excess of the amount required for maximum growth are obviously wasteful. Furthermore, excessively large fertilizer applications can cause pollution of groundwater and streams.

Critical Percentage of an Element in a Plant

Macy (1936) studied plant analyses in an effort to clarify and apply the theories of Liebig and Mitscherlich. He observed that when a small addition of a severely deficient element is made, the increased uptake of the deficient element causes a proportional increase in growth. The content of that element in the plant remains at a fixed minimum percentage. With increasing additions of the element, however, the growth rate increases by an ever-smaller amount in accord with the law of diminishing returns, and the percentage composition of the element in the plant increases. Finally, still larger additions of the element cause it to be absorbed in greater quantities giving a further increase in percent composition but no increase in yield. Macy termed the increase in percentage composition as *poverty adjustment* up to the level required for maximum yield; he called the increased absorption beyond that level *luxury consumption.* The *critical percentage* was defined as the percentage of the limiting element in the plant above which there is luxury consumption and below which there is poverty adjustment. Macy's ideas are illustrated in Figure 15-4.

Macy pointed out that Mitscherlich's equation fits that portion of the yield curve corresponding to poverty adjustment and that Liebig's law of the minimum fit the portion of the yield curve corresponding to the minimum percentage of the

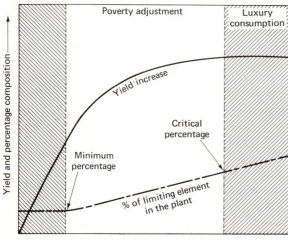

Figure 15-4 A schematic diagram illustrating the theories of Macy.

Table 15-6 Effects of Different Amounts of Superphosphate on Bermuda and Rhodes Grass

Superphosphate, mg	Total yield of dry matter, g	Phosphorus in dry matter, %
Bermuda grass		
0	19.5	0.08
200	52.5	0.08
400	59.5	0.10
800	72.8	0.14
1,600	76.4	0.21
Rhodes grass		
0	26.1	0.07
200	59.4	0.09
400	81.3	0.10
800	91.9	0.14
1,600	95.0	0.26

Source: Fudge and Fraps, 1945.

element. In many sets of data the deficiency studied is not severe enough to include the minimum percentage part of the total growth response curve.

Macy suggested that the theory of critical percentage in plants is applicable to anions but not to cations. One cation can replace another to a certain extent, but anions do not appear to be mutually replaceable in plants.

The data in Tables 15-6 to 15-8 illustrate the theories of Liebig, Mitscherlich, and Macy. The first addition of phosphorus to Bermuda grass (Table 15-6) increased yield but did not increase the percentage composition. The response was following Liebig's law of the minimum. The minimum percentage of phosphorus in Bermuda grass appears in these data to be 0.08 percent. Succeeding increases in phosphorus caused a percentage increase in phosphorus and a growth response approximating Mitscherlich's law of diminishing returns. Maximum yield and the critical percentage were probably reached somewhere between the 800- and 1,600-mg rates of superphosphate application. The data for Rhodes grass in Table 15-6 show results similar to those for Bermuda grass except that the minimum percentage zone is not detected.

Table 15-7 Effects of Different Amounts of Superphosphate on Oat Grain

Superphosphate, kg/ha	Gose farm Yield, kg/ha	P, %	Smith farm Yield, kg/ha	P, %	Purdie farm Yield, kg/ha	P, %
0	140	0.29	196	0.33	165	0.35
168	204	0.31	222	0.37	161	0.37
337	230	0.35	233	0.39	172	0.41
505	237	0.36	230	0.43	183	0.44
673	197	0.42	233	0.43	176	0.44

Source: Unpublished data, Iowa Agricultural Experiment Station.

Table 15–8   Effects of Different Amounts of Superphosphate on Alfalfa

P kg/ha	Gose farm Yield, tons/ha	P, %	Smith farm Yield, tons/ha	P, %	Purdie farm Yield, tons/ha	P, %
0	1.7	0.20	1.5	0.18	2.8	0.22
15	2.5	0.21	2.2	0.20	3.1	0.22
29	2.7	0.23	2.3	0.22	4.1	0.24
59	3.1	0.27	3.5	0.26	4.6	0.29
118	3.7	0.33	3.5	0.32	4.8	0.33

Source: Unpublished data, Iowa Agricultural Experiment Station.

The data in Table 15-7 indicate that the critical percentage of phosphorus in oat grain is about 0.35 percent. Adding more phosphorus to a soil producing grain with 0.35 percent phosphorus would not be expected to increase yields. The Purdie farm oat field appears to have provided adequate phosphorus even without fertilizer. The oats on the Smith farm needed a small application of phosphorus, and those on the Gose farm needed a moderate application for maximum yield.

Apparently all the fields shown in Table 15-8 were deficient in phosphorus. There was no indication of 100 percent sufficiency of phosphorus on any of these fields. Beeson (1941) showed that alfalfa may have as much as 0.51 percent phosphorus and as little as 0.15 percent phosphorus.

Optimum Level of an Element in a Plant

The critical percentage of an element is the most appropriate level to have if maximum growth is desired and fertilizer cost is a negligible factor. Maximum profit from truck crops and maximum satisfaction from ornamental plants might well be achieved by providing nutrient elements up to the critical percentages. The fertilizer cost for achieving critical percentages in field crops, however, is likely to be excessive, relative to the value of the crop.

Hanway and Dumenil (1965) suggest that the *optimum level* is equally as important as the critical percentage. The optimum level is taken as the percentage of a nutrient element present in a crop when it is producing the maximum possible profit. The optimum level can be evaluated by an economic interpretation as shown in Figure 15-5. The optimum level of an element can never exceed the critical percentage of that element in the same plant tissue. Hanway and Dumenil indicate that the optimum level for corn in Iowa will give about 97 percent of maximum yield. At this level, the corn leaf opposite and below the ear shoot will contain about 3.0 percent N, 0.30 percent P, and 1.5 percent K when 75 percent of the silks have emerged.

RELATION OF AGE OF A PLANT TO ITS CHEMICAL COMPOSITION

Annual plants may absorb 75 percent or more of their total nutrients by the time they have made 50 percent of their growth. There is a definite lag in dry-matter

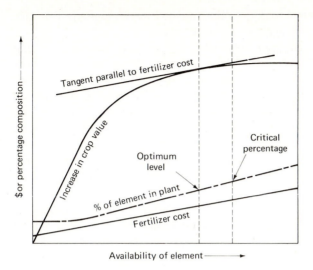

Figure 15-5 A schematic diagram illustrating the optimum level of an element in plant tissue. (*As defined by Hanway and Dumenil,* 1965.)

accumulation as compared with nutrient accumulation, especially during the early phases of growth. The nutrients must be absorbed before the dry matter can be produced.

The lag in dry-matter accumulation causes young plants to contain higher percentages of essential mineral elements than they will contain when they are older. Pasture plants are therefore most nutritious when they are young. Table 15-9 shows that the percentage of protein declines as grasses become more mature. The fiber content of the grass increases considerably with maturity and makes the forage

Table 15-9 Chemical Composition of Grasses at Different Stages of Maturity

Grass stage	Protein %	Fiber %	% N	% P	% K	% Ca
Bluegrass						
Pasture	5.5	7.6	0.88	0.13	0.59	0.16
Headed out	4.8	10.6	0.77	0.10	0.73	0.09
In seed	4.0	14.7	0.64	0.13	0.87	0.08
Orchardgrass						
Young	4.4	5.6	0.70	0.12	0.63	0.13
Heading	3.5	8.1	0.56	0.08	—	0.07
In bloom	2.6	10.7	0.42	0.07	—	0.07
Crested wheatgrass						
Very young	6.6	5.2	1.06	0.08	—	0.12
Young	4.0	8.9	0.64	0.07	—	0.14
Mature	3.3	23.2	0.53	0.08	—	0.15

Source: Morrison, *Feeds and Feeding,* Abridged, 9th ed., 1961.

tougher and harder for an animal to digest. Most nutrient percentages decline with time, though some may increase again as the seed ripens.

Figure 15-6 shows that small-grain crops had taken up 45 percent of their total nitrogen and potassium by the time they had produced 20 percent of their dry matter. The phosphorus uptake was also ahead of the dry-matter production but by a lesser amount. Similar relations for corn are shown in Figure 15-7. Corn takes up phosphorus slightly ahead of dry-matter production, nitrogen considerably ahead, and potassium absorption is ahead of the other nutrients as well as being far ahead of dry-matter production.

The rate of potassium uptake may be very high during the early stages of growth without injurious effect. The data for corn in Figure 15-7 show that it contained more potassium in August than at the end of the season. Apparently some potassium was lost from the plant as it matured. The loss may have resulted from potassium being leached from dead leaves, from excretion to leaf surfaces whence it was washed off by rain, or by diffusion back into the soil through the root system.

Relations to Fertilizer Programs

The ability of plants to take up an excess of potassium early in growth without injurious effect is important to the agronomist. Enough potassium to supply the entire crop needs may be placed in the soil at planting time or before. Neither the seed nor the plant will be injured so long as the osmotic concentration of the soil solution is not excessive.

Some fertilizer is often placed within 5 to 8 cm. of the seed at planting time as a "starter" for vigorous seedling growth. The amounts of potassium and nitrogen

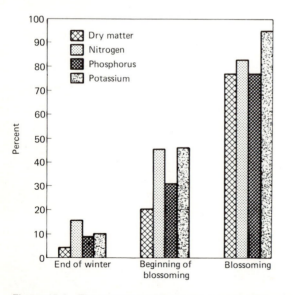

Figure 15-6 The percentage of the total dry matter, nitrogen, phosphorus, and potassium accumulated at different stages of growth. Average of winter wheat, winter rye, and barley grown in Germany. (*Remy,* 1938.)

in starter fertilizer must be limited so that the osmotic concentration of the soil solution near the seed will not be excessive. Larger amounts of fertilizer may be applied before seeding as plowdown or during early stages of plant growth as side-dressing (fertilizer placement is discussed in Chapter 9). Nitrogen is the element most often supplied as side-dressing, partly as a matter of convenience and because side-dressing delays the time of fertilizer application. The delay may permit better evaluation of growing conditions and fertilizer needs as well as reduce leaching losses of nitrate nitrogen from porous soils.

The amount of nitrogen needed per day by a corn crop can be calculated by the use of the curve in Figure 15-7 by allowing 3.5 kg of nitrogen for each 100 kg of corn. A 7,000-kg crop of corn would need about 245 kg of nitrogen during the growing season. The curve indicates that about 60 kg of this nitrogen is needed during the first 20 days and about 100 kg during the next 25 days. In other words, the average nitrogen needs for a 7,000-kg corn crop are about 3 kg per day for the first 20 days and about 4 kg per day for the next 25 days. Fertilizer applications

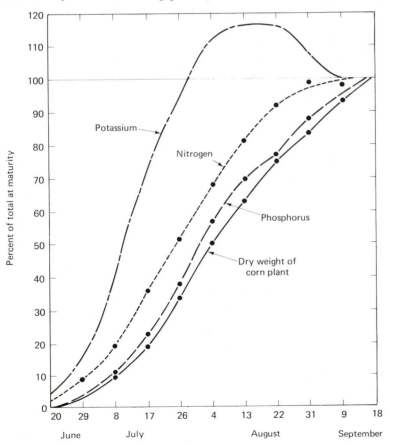

Figure 15-7   The uptake of nutrients compared with the growth rate of corn. (*Calculated from data of Sayre*, 1948.)

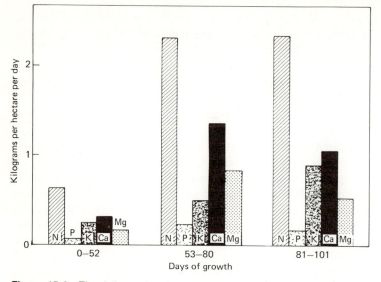

Figure 15-8   The daily uptake of nutrients by soybeans growing on Webster silt loam. (*Prepared from data of Hammond*, 1947.)

should be adjusted to meet crop needs at this time with due allowance for the rate of release of nitrogen from organic-matter decomposition.

The relatively low nitrogen requirement of corn during the first 2 to 3 weeks of growth results from the small size of the plant at that time. A pattern similar to that for corn, though not identical, is followed by other annual plants. Hammond (1947) found that soybeans took up an average of less than a kilogram of nitrogen per hectare per day during the first 50 days of growth, but they took up an average of nearly 3 kg of nitrogen per hectare per day after the first 50 days. Hammond's data are illustrated in Figure 15-8. These data show that the uptake of other nutrients is also slower at first than it will be later.

REFERENCES

Adams, J. E., H. M. Bogg, and E. M. Roller, 1937, Effect of Fertilizers on Composition of Soybean Hay and Seed and of Crop Management on Carbon, Nitrogen, and Reaction of Norfolk Sand, *USDA Tech. Bull.* 586.

Allaway, W. H., and W. H. Pierre, 1939, Availability, Fixation, and Liberation of Potassium in High-lime Soils, *J. Amer. Soc. Agron.* **31**:940–953.

Baker, J. M., and B. B. Tucker, 1973, Critical N, P and K Levels in Winter Wheat, *Commun. Soil Sci. Plant Anal.* **4**:347–358.

Bansal, K. N., and H. G. Singh, 1975, Interrelationship between Sulfur and Iron in the Prevention of Iron Chlorosis in Cowpea, *Soil Sci.* **120**:20–24.

Banwart, W. L., and W. H. Pierre, 1975, Cation-Anion Balance of Field-grown Crops. II. Effect of P and K Fertilization and Soil pH, *Agron. J.* **67**:20–25.

Bear, F. E., 1953, *Soils and Fertilizers,* 4th ed., Wiley, New York, 402 pp.

Beeson, K. C., 1941, The Mineral Composition of Crops with Particular Reference to the Soils in which They Were Grown, *USDA Misc. Pub.* 369.

Bower, C. A., and W. H. Pierre, 1944, Potassium Response of Various Crops on High-lime Soils in Relation to Their Content of Potassium, Calcium, Magnesium, and Sodium, *J. Am. Soc. Agron.* **26**:608–614.

Drover, D. P., 1972, Cation Exchange in Plant Roots, *Commun. Soil Sci. Plant Anal.* **3**:207–209.

Edwards, J. H., and E. J. Kamprath, 1974, Zinc Accumulation by Corn Seedlings as Influenced by Phosphorus, Temperature, and Light Intensity, *Agron. J.* **66**:479–482.

Fudge, J. F., and G. S. Fraps, 1945, The Value of Different Phosphate for Various Texas Soils and Grasses, as Indicated by Pot Experiments, *Texas Agric. Exp. Sta. Bull.* 672.

Hammond, L. C., 1947, *Rate of Nutrient Uptake by Soybeans on Two Iowa Soils,* master's thesis, Iowa State University, Ames.

Hanway, J. J., and L. C. Dumenil, 1965, Corn Leaf Analysis, *Plant Food Rev.* **11**(2):5–8.

Hoagland, D. R., and D. I. Arnon, 1941, Physiological Aspects of Availability of Nutrients for Plant Growth, *Soil Sci.* **51**:431–444.

Hunter, A. S., S. J. Toth, and F. E. Bear, 1943, Calcium-potassium Ratio for Alfalfa, *Soil Sci.* **55**:61–72.

Johnstone-Wallace, D. B., 1937, The Influence of Grazing Management and Plant Associations on the Chemical Composition of Pasture Plants, *J. Am. Soc. Agron.* **29**:441–455.

Jungk, A., and S. A. Barber, 1974, Phosphate Uptake Rate of Corn Roots as Related to the Proportion of the Roots Exposed to Phosphate, *Agron. J.* **66**:554–557.

Laughlin, W. M., P. F. Martin, and G. R. Smith, 1973, Potassium Rate and Source Influences on Yield and Composition of Bromegrass Forage, *Agron. J.* **65**:85–87.

Macy, Paul, 1936, The Quantitative Mineral Nutrient Requirement of Plants, *Plant Physiol.* **11**:749–764.

Mayland, H. F., D. L. Grunes, and D. M. Stuart, 1974, Chemical Composition of *Agropyron desertorum* as Related to Grass Tetany, *Agron. J.* **66**:441–446.

Melsted, S. W., H. L. Motto, and T. R. Peck, 1969, Critical Plant Nutrient Composition Values Useful in Interpreting Plant Analysis Data, *Agron. J.* **61**:17–20.

Ohki, K., 1975, Mn and B Effects on Micronutrients and P in Cotton, *Agron. J.* **67**:204–207.

Pierre, W. H., John Meisinger, and J. R. Birchett, 1970, Cation-Anion Balance in Crops as a Factor in Determining the Effect of Nitrogen Fertilizers on Soil Acidity, *Agron. J.* **62**:106–112.

Remy, T., 1938, Fertilization in Its Relationship to the Course of Nutrient Absorption by Plants, *Soil Sci.* **46**:187–200.

Russell, J. S., 1972, A Theoretical Approach to Plant Nutrient Response under Conditions of Variable Maximum Yield, *Soil Sci.* **114**:387–394.

Sayre, J. D., 1948, Mineral Accumulation in Corn, *Plant Physiol.* **23**:267–281.

Sheets, O. A., 1944, Effect of Fertilizer, Soil Composition, and Certain Climatological Conditions on the Calcium and Phosphorus Content of Turnip Greens, *J. Agric. Res.* **68**:145–190.

Singh, B. R., and K. Steenberg, 1974, Plant Response to Micronutrients. III. Interaction between Manganese and Zinc in Maize and Barley Plants, *Plant Soil* **40**:655–667.

Smith, J. C., L. C. Capp, and R. C. Potts, 1949, The Effects of Fertilizer Treatment upon Yield and Composition of Wheat Forage, *Soil Sci. Soc. Am. Proc.* **14**:241–245.

Stanford, George, J. B. Kelly, and W. H. Pierre, 1941, Cation Balance in Corn Grown on High-lime Soils in Relation to Potassium Deficiency, *Soil Sci. Soc. Am. Proc.* **6**:335–341.

Terman, G. L., S. E. Allen, and B. N. Bradford, 1975, Nutrient Dilution-Antagonism Effects in Corn and Snap Beans in Relation to Rate and Source of Applied Potassium, *Soil Sci. Soc. Am. Proc.* **39**:680–685.

Westermann, D. T., and C. W. Robbins, 1974, Effect of SO_4-S Fertilization on Se Concentration of Alfalfa (*Medicago sativa L.*), *Agron. J.* **66**:207–208.

Chapter 16

Soil Classification and Survey

There are thousands of different kinds of soils. This fact was mentioned in Chapter 1 and needs to be emphasized again at this point. Furthermore, a great deal of information can be gathered about each soil—descriptions of its physical and chemical properties, of its various uses, of its response to management, etc. Classification systems are designed to simplify this mass of information and make it manageable.

Many characteristics of a soil can be recorded by describing the physical, chemical, and organic properties discussed in the early chapters of this book. Such descriptions are good references to have on record, but they are too cumbersome for many uses. Brevity is achieved by assigning names to each soil. A name conveys a long list of properties to persons acquainted with the soil. Classification systems are essential for assigning names in an orderly manner.

The usefulness of a set of soil names depends on a common understanding and consistency among the users. A particular soil name must always mean that the soil has certain specified properties; but enough variation is allowed for a name to be given to extensive soil areas in spite of differences in some properties. Indeed, there probably are no two soils that are identical in all respects. The name means only that specified soil properties are within stated limits.

The defined properties of a soil are termed *differentiating characteristics* (Cline, 1949). Additional *accessory characteristics* may also be associated with most members of the particular group of soils, but the soil name would apply to a soil whose differentiating characteristics fit the definition even if it did not have the usual accessory characteristics. Still other *accidental characteristics* would show no consistency in relation to the soil name. For example, soil color and clay content might be among the differentiating characteristics for a particular soil. On the basis of these characteristics, the organic-matter content and the cation-exchange capacity could probably be predicted with reasonable accuracy and would therefore be accessory characteristics. The supply of available nitrogen, however, might vary greatly in response to management practices and would be an accidental characteristic of the soil.

PURPOSE OF SOIL CLASSIFICATION

A classification system must organize knowledge in a meaningful way. It must emphasize important points and ignore irrelevant detail. Unfortunately, facts that are irrelevant to one user may be important to another. Each significantly different group of users is therefore likely to need a special classification system. Economists, engineers, and pedologists may all classify soils but in very different manners.

A "natural" system of soil classification has been the goal of many persons. Actually, a classification system is not a natural thing that can be discovered. Rather, it is the product of humanity's creative mind. As such, it can and must be devised for particular uses. The viewpoints and understandings of the persons who devise and use the classification system determine its nature. Several different systems of classifying soil will therefore be discussed in this chapter.

NATURE OF CLASSIFICATION

A classification system consists of a set of classes. All members of the population to be classified must fit into definable classes in the system. The definitions should make it clear that a certain individual belongs to one particular class and not to any of the other similar classes. The definitions must be specific enough that any qualified user would place the same individual in the same class.

Classification systems necessarily generalize. Most systems provide means of generalizing to more than one degree by the use of categories[1] (levels of generalization). Individuals are placed in classes at the lowest categorical level. These classes are further classified into broader classes at higher categorical levels. Classes at the lower levels convey the most information, but their use requires rather detailed knowledge of the individuals classified. Such classes serve well for detailed local use, but the classes at higher categorical levels serve better for more general purposes.

Figure 16-1 illustrates a hypothetical soil classification system with three categorical levels. A class at the lowest level in this system tells three things about the soil (depth, texture, and reaction), whereas a class at the highest level specifies only one thing (texture). The additional information specified at the lower levels neces-

---

[1]The word *category* is used here in a technical sense as a level of generalization and will not be considered synonymous with class.

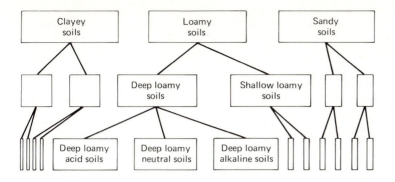

Figure 16-1 A hypothetical soil classification system with three categorical levels. Each box represents a class consisting of many individual soils. (An actual system would provide descriptions for all the boxes, but most are left empty here for simplicity.)

arily makes more divisions and requires that more classes be provided in the lower categories than in the higher categories. The higher categories are designed for use where broad generalizations are needed; the lower categories serve more specific needs, usually on a more local basis.

The criteria for differentiating characteristics in a classification system must be carefully chosen. These criteria and the definitions and names that go with them determine the nature and the value of the system. Differentiating characteristics should be simple to use and yet relate to important soil properties. They need to be assigned to appropriate categorical levels. The importance of categorical level can be illustrated by the system in Figure 16-1. No single class in that system contains all the acid soils. Texture and depth were placed at higher categorical levels than reaction, with the result that acid soils were distributed through several classes.

Usually the criteria chosen for classification purposes are much more complex than those outlined in Figure 16-1. This is especially true of the criteria for the higher categories. Pedologists usually choose properties that are thought to indicate the action of the soil-forming factors. For instance, C. F. Marbut (1935) proposed that all "normal soils" be divided into two groups, Pedocals and Pedalfers. The differentiating criterion was the presence or absence of a layer of calcium carbonate accumulation at the base of the solum. The presence of such a layer usually indicates that the soil formed in a subhumid or drier climate. Its absence often indicates a humid climate. Unfortunately for this criterion, certain parent materials and certain topographic positions cause exceptions to the desired results.

Soil Pedons

Individual plants, animals, and objects of many kinds are easily identified and counted. Not so with soils. Who can say how many individual soils there are in a field where they all merge into one mass? This problem was ignored for a long time. *Pedons* have since been defined to serve as the smallest classifiable soil units (Soil Survey Staff, 1960). Pedons are intended to be large enough to contain the entire root system of an average-sized plant. They also should be of a size suitable for field examination, description, and sampling.

The tops and bottoms of pedons are the same as the solum. The lateral boundaries are arbitrarily set to give a surface area of about 1 m² except in special circumstances. Soils with large variations in their properties within short horizontal distances are allowed to have pedons as large as 10 m² (about 3½ m of width) so that their properties may be fully represented. The pedon size reverts to 1 m² and two soils are named where the variations cannot be covered within a 10-m² pedon.

The term *polypedon* is used to designate a soil area composed of many similar pedons (Figure 16-2) that are all given the same soil name (Soil Survey Staff, 1975). Ideally, polypedons should be shown on detailed soil maps. Practically, the polypedons can only be approximated on soil maps.

SOIL CLASSIFICATION SYSTEMS

The forerunners of modern systems of soil classification can be traced to a Russian named Dokuchaiev who published his first classification of soils in 1879 (Basinski, 1959). The Russian work was largely unknown to the rest of the world until Glinka translated it into German in 1914 and Marbut translated the German into English in 1927. The field of soil classification has developed rapidly since that time.

The United States has made extensive use of two systems of soil classification. The first of these was published in the 1938 Yearbook of Agriculture (Baldwin, Kellogg, and Thorp) under the heading Classification of Soils on the Basis of Their Characteristics. This system will be referred to as the "1938 System" for want of a better title.

The 1938 System

The 1938 System of soil classification became widely known and used in the United States and abroad. Many other countries based their own systems upon it and used similar terminology and criteria. Soil survey programs were keyed to it and soils literature contains frequent reference to it and use of its terms. Significant features of the 1938 System will be discussed later in this chapter.

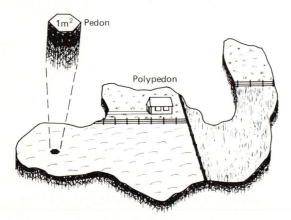

Figure 16-2   A polypedon includes all of the contiguous pedons belonging to the same soil series.

Use of the 1938 System revealed weaknesses as well as strengths. Among the more serious defects are the following: (1) some soils do not fit well in any of the classes at one or more categorical levels; (2) definitions are not always precise enough to make a soil fit only one class in each category; (3) some of the categories have not been very useful; and (4) some of the soil groupings appear illogical.

Soil Taxonomy

The present system of soil classification in the United States was developed in an effort to overcome weaknesses of the older systems. It went through a series of stages, called *approximations,* beginning in the 1950s. A nearly complete version was published by the Soil Survey Staff (1960) as *Soil Classification, a Comprehensive System, 7th Approximation.* The usage increased with each succeeding approximation until the 7th Approximation was officially adopted by the Soil Conservation Service for nationwide use in 1965. Developments since the publication of the 7th Approximation have adapted it to several types of soils that are better represented in other countries than in the United States. The completed system was published as *Soil Taxonomy* in 1975 by the Soil Survey Staff. Soil Taxonomy is intended to be usable for soils anywhere in the world.

The 1938 System classified soil profiles into soil types, series, families, etc. Soil Taxonomy gives profiles an areal extent by defining pedons and polypedons for classification into series and other categories. Diagnostic horizons are another feature that the 1938 System lacked.

Diagnostic Horizons   The use of diagnostic horizons helps make Soil Taxonomy more precise than the older systems. Diagnostic horizons are used to define soil classes at various categorical levels. Diagnostic horizons that necessarily occur at or near the soil surface are called *epipedons.* Simplified definitions of epipedons are given in Table 16-1.

Epipedons include but are not limited to A1 horizons. Underlying A2, A3, B1, and even B2 horizons that qualify by the definition are included as part of the epipedon.

The mollic epipedon is considered by many to be the heart of the Soil Taxonomy system of classification. It is intended to include soils that have high natural fertility and productivity. Large areas of soils devoted to grain production have mollic epipedons.

Diagnostic Subsurface Horizons   Several different horizons form beneath other horizons. Erosion may remove the overlying horizons and expose a subsurface horizon, but its properties are still influenced by the former presence of the eroded horizons. For example, a horizon containing illuvial clay must have formed beneath an eluvial horizon.

The diagnostic horizon that is roughly equivalent to an A2 horizon is called an *albic horizon.* Albic horizons may occur at any position in the profile where there is a strong leaching action or a sufficiently impermeable underlying layer resulting

Table 16–1 Simplified Definitions of Epipedons

Epipedon	Definition
Mollic	The relatively thick, dark-colored upper horizon usually associated with grassland soils. It must contain at least 0.6 percent organic carbon and have a base saturation of 50 percent or more. The minimum thickness ranges from 10 cm over bedrock through one-third of the solum to 25 cm in a deep soil.
Umbric	A thick, dark-colored upper horizon similar to the mollic epipedon but having a base saturation less than 50 percent.
Ochric	An upper horizon that is too thin, too light in color, or too low in organic carbon to be a mollic epipedon.
Histic	A layer of peat or muck too thin to constitute a Histosol (organic soil). The thickness ranges from 20 to 60 cm and the organic carbon must be at least 12 to 18 percent, depending on the clay content as explained in Chapter 5.
Anthropic	An upper horizon that has accumulated a high concentration of available phosphorus (more than 250 ppm) because of human habitation or has acquired the characteristics of a mollic epipedon because of long-continued irrigation.
Plaggen	An artificial epipedon at least 50 cm thick produced by adding manure and bedding materials to fields.

Source: Condensed from Soil Survey Staff, 1975.

in strong eluviation. Eluviation removes clay, iron oxides, organic matter, and other materials from albic horizons. The resulting color is therefore the color of the sand and silt grains, usually a light gray.

Another group of diagnostic horizons normally occur in the subsoil or B horizon position. The principal characteristics of these horizons are given in Table 16-2. The argillic horizon is the most common type of B horizon in soils. The spodic horizon occurs in Spodosols, and the oxic horizon in Oxisols.

A few diagnostic horizons represent materials that precipitate from the soil water and may occur anywhere in a soil profile. They often occur within one of the other diagnostic horizons. These are the *calcic horizon* (more than 15 percent $CaCO_3$ and at least 5 percent more $CaCO_3$ than the C horizon); the *gypsic horizon* (at least 5 percent more gypsum than the C horizon); and the *salic horizon* (at least 2 percent by weight of soluble salts). Each of these horizons has a minimum thickness of 15 cm.

Two diagnostic horizons are hard layers that occur at the base of the solum. These are called *duripans* and *fragipans*. Duripans form in arid to subhumid climates and fragipans in humid climates.

Duripans are cemented by silica (SiO_2) precipitated when water evaporates into dry substrata. Duripans are especially common in arid regions where a porous material underlies the solum. Calcium carbonate and other salts are often deposited along with the silica in arid regions. Duripans are hard whether wet or dry and prevent the penetration of plant roots.

Fragipans are layers with high bulk density and very low permeability. They form in humid climates in loamy materials containing little or no calcium carbonate. Fragipans are hard and brittle when dry but soften considerably when wet. Few plant roots penetrate fragipans and those that do are concentrated in vertical cracks bounding large polygons that commonly occur in fragipans.

Plinthite (formerly called laterite) is an iron-rich product of long, intense tropical weathering. It forms in certain subsurface horizons where all but the most resistant materials are weathered away. Quartz sand and kaolinite clay remain and iron accumulates as dark-red mottles. Plinthite typically forms in soils in terrace or level upland positions. The horizons that become plinthite are saturated for part of the year but are well enough drained for the water to escape and thoroughly leach basic cations out of the soil.

Plinthite is hard when dry and hardens irreversibly if it is baked. Pieces can be carved from the soil and baked in the sun to make brick. Several drying periods that reduce the soil to the wilting point can cause plinthite to harden in place and form *ironstone.* Ironstone can occur as segregated bodies intermixed with soil material or as a continuous rocklike layer. Formation of ironstone involves bonding between positively charged iron oxides and negatively charged kaolinite clays. Drying causes enough shrinkage to reduce the bond length to a stable position that is not broken by rewetting. Ironstone layers are a serious impediment to root growth and cause extensive barren areas where the ironstone is near the surface.

Categories in Soil Taxonomy The Soil Taxonomy system classification has six categorical levels, named *orders, suborders, great groups, subgroups, families,* and *series.* Only the series is interchangeable between Soil Taxonomy and the 1938 System, and it occurs at a different level. The new system is so designed that any

Table 16–2 Simplified Definitions of Diagnostic Subsoil Horizons

Horizon	Definition
Argillic	An illuvial horizon at least one-tenth as thick as the overlying soil and containing at least 3 to 8 percent more clay than the A horizon (see Chapter 3 for details) including clay films on soil peds, in pores, and/or on sand grains.
Agric	An illuvial horizon formed below the plow layer in cultivated soils by the accumulation of silt, clay, and humus transported through the large pores formed by tillage.
Natric	An argillic horizon that contains more than 15 percent exchangeable Na^+ and has columnar or prismatic structure.
Spodic	A soil horizon that has accumulated illuvial organic matter and iron and aluminum oxides. The colors are bright (high chroma), especially near the middle of the horizon. Most spodic horizons form in sandy parent materials under strongly acid leaching.
Cambic	A soil horizon formed by weathering of minerals within the horizon rather than by illuviation. Cambic horizons have subsoil structures such as blocky or prismatic, and/or bright colors, but are too weakly developed to qualify as argillic or spodic horizons.
Oxic	A very highly weathered soil horizon composed of iron and aluminum oxides, kaolinite clay, and highly resistant minerals. Common in tropical soils. Most oxic horizons have less than 10 meq of exchangeable bases per 100 g of clay.

Source: Soil Survey Staff, 1975.

of the four highest categories can be used for making generalized soil maps. This is a marked improvement over the 1938 System.

A systematic nomenclature has been adopted for the four highest categories of Soil Taxonomy. The names used in these four categories are coined words that help to define some of the soil properties. The nomenclature has generated a great deal of debate because some of the names have peculiar sounds. Nevertheless, it is a useful system because the categorical level and some important soil properties can be readily identified from the name. The system of nomenclature is outlined in Table 16-3. An analysis of the example in Table 16-3 reveals that the Clarion soils have mollic epipedons (are Mollisols), occur in humid regions (Udolls combines the *ud* from *udus,* meaning *humid,* with *oll* from Mollisols), and have relatively simple profiles without excess horizons (the *hapl* in Hapludolls comes from *haplous* meaning *simple*). The adjective *typic* in the subgroup name indicates that they are typical Hapludolls. The family designation indicates a subsoil clay content between 18 and 35 percent with no one kind of soil mineral predominating and specifies an average annual soil temperature between 8 and 15°C. Any soil name above the family level automatically specifies the names of all the higher categories. For example, Duric Albaqualf is the name of a subgroup. It belongs to the Albaqualf great group (drop the adjective), the Aqualf suborder (drop the first syllable), and the Alfisol order (the order name containing the syllable *alf*).

Orders in the Soil Taxonomy System of Classification   There are 10 orders in Soil Taxonomy. These 10 orders and their suborders are discussed in the following paragraphs. A schematic arrangement of several of these orders according to climate is shown in Figure 16-3.

Alfisols obtain their name from the chemical symbols Al and Fe. Many of the forested soils of the northeastern United States as well as much of Europe and northeastern China are Alfisols. A typical Alfisol has an ochric epipedon and an albic horizon; it must have an argillic horizon. Fragipans are common. The soil pH is usually slightly acid, but the base saturation is above 50 percent.

Table 16–3   The System of Nomenclature for the Six Categories in the New System of Soil Classification

Category	Method of naming	Example
Order	Names end with -sol	Molli*sol*
Suborder	Names end with a syllable from the order name	U*doll*
Great group	Names end with the suborder name	Hapl*udoll*
Subgroup	Adjective added to great group name	Typic Hapludoll
Family	Divisions of subgroups based on texture, mineralogy, etc., or Divisions based on local name	Fine loamy, mixed mesic typic Hapludoll, or Clarion family
Series	Usually geographic names	Clarion

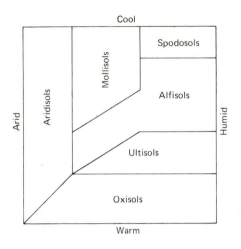

Figure 16-3 A schematic illustration of the relations between climate and some of the orders of the new system of soil classification. Names of dominantly grassland soils are written vertically, and names of dominantly forested soils are written horizontally.

Suborders of Alfisols are recognized on the basis of wetness *(Aqualfs)*, cold climates *(Boralfs)*, humid climates *(Udalfs)*, moderately dry seasons *(Ustalfs)*, or lengthy dry seasons *(Xeralfs)*. Udalfs often occur in a broad zone between Mollisols and Spodosols. Udalf A1 horizons are not as thick and dark colored as those of Mollisols, and their A2 horizons are not nearly as prominent as those of Spodosols.

Ustalfs and Xeralfs occur in warm climates and usually have reddish colors whereas the other Alfisols have mostly grayish-brown colors. Ustalfs occur in belts between Aridisols and Ultisols or Oxisols. Xeralfs occur mostly in areas with Mediterranean-type climates including parts of Australia, South Africa, Chile, and California. The natural vegetation usually includes woody shrubs and grasses instead of the trees that predominate on other Alfisols.

Alfisols are important agricultural soils occupying considerable areas in temperate regions. They are commonly used for producing corn, soybeans, small grains, and many other crops. Their natural fertility is lower than that of Mollisols because of the more intense leaching they have undergone. Most of them require liming and fertilizing with nitrogen, phosphorus, and potassium to make them highly productive. Xeralfs and some Ustalfs require irrigation for the production of many crops.

Aridisols are soils of arid regions. Much of the western United States, Australia, northern Africa, Arabia, northern China, Chile, and other arid areas are included. Seasons when the soil is both warm enough and moist enough for plant growth are less than 3 months long. The natural vegetation is a mixture of grasses, small shrubs, cacti, and various flowering plants, often sparsely spaced with considerable open area between plants. The plants are either annuals that grow and mature rapidly or perennials that can remain dormant for months. Aridisols have ochric epipedons and can have argillic or natric horizons. Calcic, gypsic, and salic horizons and duripans are common. Many aridisols are shallow over bedrock, duripans, or loose material. Aridisols are divided into two suborders on the basis of the presence of argillic or natric horizons *(Argids)* or the lack of significant clay accumulations *(Orthids)*.

Aridisols are commonly used as grazing lands, especially for spring and fall pasture. Livestock are often moved to more humid areas at higher elevations for the summer and down to irrigated valleys for the winter. Some Aridisols will produce low yields of small grains by the summer-fallow system. High yields of almost any crop suited to the temperature conditions are produced in irrigated areas of Aridisols. The soils normally have high base saturation so lime and potassium fertilizer are seldom needed. The phosphorus supply is usually good also because of little leaching. But the low organic-matter contents of Aridisols lead to a need for nitrogen fertilizer, especially under irrigation. Some elements such as boron, arsenic, and selenium have accumulated in a few areas of Aridisols to levels that are toxic to plants or animals.

Entisols are very young soils (from rec*ent*) that have little or no horizon differentiation. Most Entisols are formed either in fresh deposits such as alluvium along a stream or in actively eroding steep slopes. The only Entisols on old stable surfaces are those formed in parent materials that are very resistant to weathering, such as an old beach sand that is nearly pure quartz.

Entisols may show a small accumulation of organic matter and can even have a histic epipedon. They may also have had a slight loss of carbonates from the upper several centimeters. Entisols in alluvium and colluvium may have several buried horizons showing organic matter accumulations and loss of carbonates. The soils that might be considered most characteristic of Entisols are classified as Orthents. Other subgroups are provided for wet Entisols *(Aquents),* Entisols formed where people have mixed soils deeply enough to destroy their natural horizons *(Arents),* Entisols with floodplain layering *(Fluvents),* and very sandy Entisols *(Psamments).*

Entisols can occur almost anywhere, and their use and management depend greatly on where they are. The soil fertility of Entisols depends greatly on the parent material because these soils are little changed from their parent materials. Psamments are generally unproductive but many other Entisols are among the most productive soils of their vicinities. Aquents and Fluvents are likely to need artificial drainage and flood control.

Histosols are soils dominated by organic materials. More than half of the upper 80 cm of soil contains at least 12 to 18 percent organic carbon, depending on the clay content as explained in Chapter 5. The organic materials are of low enough bulk density to constitute well over half the soil volume even at 12 percent organic carbon by weight. Most Histosols are formed in lakes, marshes, and swamps and some are only an organic mat floating on water. Thin Histosols blanket some landscapes in Alaska and Siberia but these, too, are usually saturated with water in their lower parts. A few Histosols are recognized as having formed under well-drained conditions where leaves and twigs have accumulated directly over bedrock or gravel until the organic layer constitutes a soil able to support vegetation.

Small areas of Histosols occur in many places. Larger areas occur in Siberia, Alaska, Florida, California, and Michigan. Suborders are provided for Histosols formed under well-drained conditions *(Folists),* and for Histosols dominated by fibrous materials *(Fibrists),* by mixed fibrous and decomposed materials *(Hemists),* and by well-decomposed organic materials *(Saprists).*

Folists are used mostly for woodland and wildlife. Other Histosols are also used mostly for wildlife unless they are drained. Deep Histosols often become highly productive cropland after drainage. Those located near population centers are often used for truck crops because the organic soil is porous, easy to work, and does not restrict root development. Most cropped Histosols are highly responsive to fertilization with complete fertilizers. Even the micronutrients may be deficient because there is little mineral matter to weather and release fresh supplies of essential elements. Unfortunately, drainage of a Histosol leads to its eventual destruction. The entrance of air into the organic soil causes decomposition at a rate higher than that at which the organic matter can be replaced. Furthermore, wind readily wafts away lightweight organic particles from exposed areas of cropped Histosols. The value of the area after the Histosol is gone depends on the underlying mineral matter.

Inceptisols are soils in an early stage of development; the name is derived from the Latin word *(inceptus)* meaning beginning. Their most common diagnostic horizons are ochric or umbric epipedons, cambic horizons, fragipans, and duripans. Inceptisols do not have strong enough development to include argillic, natric, spodic, or oxic horizons. They do not have gypsic or salic horizons, either, because these would indicate a drier climate. Any soil with a plaggen epipedon (at least 50 cm of artificial soil produced by manuring) is considered to be an Inceptisol.

Inceptisols are common in mountains and in humid or subhumid areas, especially in those that were covered by glaciers within the past 20,000 years including the northeastern United States, eastern and northern Canada, and parts of Europe. Suborders are provided for Inceptisols that have high contents of allophane *(Andepts)*, evidence of wetness in their profiles *(Aquepts)*, light-colored or shallow epipedons *(Ochrepts)*, plaggen epipedons *(Plaggepts)*, tropical climate *(Tropepts)*, and dark-colored epipedons *(Umbrepts)*.

Inceptisols are likely to be used in the same manner as their more strongly developed neighbors (often Alfisols, Ultisols, Spodosols, or Oxisols). Many of them have acid reactions and need both lime and fertilizer to be productive. Fragipans and duripans, where they occur, restrict water and root penetration. Pans increase the effects of excessive wetness in the spring and dryness in the summer in many Inceptisols.

Mollisols have mollic epipedons. They may be wet or dry, old or young, but they all have mollic epipedons. Some of them have argillic or natric horizons. They can have albic horizons, calcic horizons, or duripans. Most Mollisols developed under grass vegetation in temperate climates between Aridisols on the dry side and Alfisols or Spodosols on the humid side. Mollisols are very common in the plains of north central United States and the adjoining part of Canada, a strip through the Ukraine eastward across the plains of southern Russia, Mongolia, and northern China, and in Uruguay and adjoining areas of Argentina, as well as in other areas with prairie vegetation (Figure 16-4).

Mollisols are divided into suborders according to whether they have albic horizons *(Albolls)*, evidence of wetness in their profiles *(Aquolls)*, cold climates *(Borolls)*, over 40 percent calcium carbonate in the horizon below the mollic epipe-

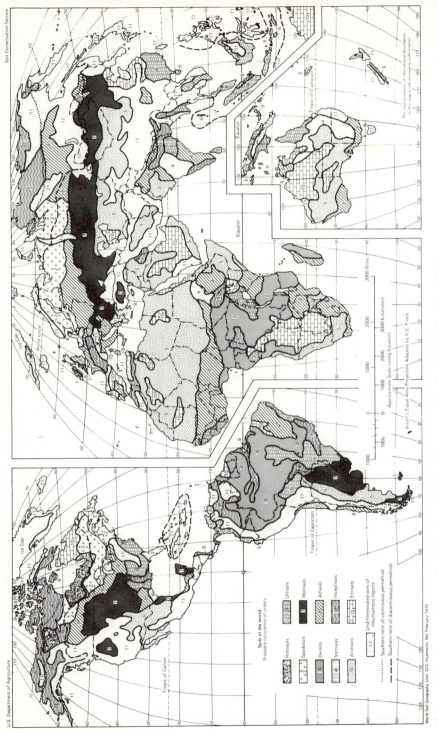

Figure 16-4 World distribution of the soil orders of the new system of soil classification. *(Soil Conservation Service, USDA.)*

Soils of the world

Probable occurrence of orders

Histosols	Ultisols	
Spodosols	Mollisols	
Oxisols	Alfisols	
Vertisols	Inceptisols	
Aridsols	Entisols	

Undifferentiated soils of mountainous regions

⋯⋯⋯ Southern limit of continuous permafrost

▬ ▬ Southern limit of discontinuous permafrost

Aitoff's Equal Area Projection Adapted by V.C. Finch

Approximate Scale (along Equator)

World Soil Geography Unit, SCS. Hyattsville, Md. February 1970

U.S. Department of Agriculture

Soil Conservation Service

don *(Rendolls),* humid climates *(Udolls),* moderate dry seasons *(Ustolls),* or extensive dry seasons *(Xerolls).*

Most Mollisols are highly productive for grain and other crops. A large part of the world's wheat and other small grains is grown on Ustolls and Xerolls. The dry summers common in these areas come about the time the small grains are maturing, and the ensuing harvest is made easier by the lack of rain. Udolls and associated Aquolls are generally the most productive soils in the world for growing corn and soybeans. Much of the area of Aquolls has had tile drainage installed to make crop production possible. Albolls, Udolls, and some Aquolls need lime and potassium fertilizer. Nitrogen and phosphorus fertilizer are widely used on all kinds of Mollisols.

Oxisols are soils that have oxic horizons or contain a continuous layer of plinthite within 30 cm of the surface. Most Oxisols occur in the tropics including much of Brazil, Central America, equatorial Africa, southeast Asia, and part of India, but not all tropical soils are Oxisols. Ultisols are common in the tropics and Alfisols, Mollisols, Entisols, and Inceptisols also occur there. Oxisols occur on the more stable, older surfaces and are therefore deeper and more highly weathered than the other soils. Intense weathering and leaching produce deep sola and high clay contents in Oxisols but remove most of the soil fertility. Most of the clay is of the oxide and kaolinitic types so the cation-exchange capacities are less than 10 meq/100 g of clay. Vegetation may grow rapidly in spite of low fertility because the climate is favorable. Organic matter decomposes and releases its nutrients rapidly in the warm, moist conditions.

Oxisols generally lack the strong soil structure and distinct horizon boundaries that are common in several other soil orders. They are considered not to have argillic horizons even though many of them have considerable variation in clay content. The distinction is made on the basis of lack of clay films in the oxic horizon and boundaries so gradual that more than 30 cm are required to change from a horizon of one clay content to another with 1.2 times as much clay. Suborders of Oxisols are based largely on moisture regimes with provisions being made for Oxisols that are very wet and/or contain plinthite *(Aquox),* nearly always moist and relatively high in organic matter *(Humox),* dry for at least 90 days and moist for at least 90 consecutive days during the year *(Ustox),* too dry for cultivation unless irrigated *(Torrox),* or representative of the central concept of Oxisols *(Orthox).*

The natural vegetation of Oxisols ranges from tropical rain forest to desert, although the soils are so old that the drier Oxisols may have formed under a different climate. Extensive areas with ironstone near the surface are now barren in spite of favorable moisture regimes. A wide variety of tropical crops such as sugarcane, pineapple, bananas, and coffee are grown on the Orthox and on favorable areas of Aquox, Humox, and Ustox. Some areas of Ustox and Torrox are irrigated. The soil fertility is low in all Oxisols and response to fertilizers has sometimes been disappointing because too much is lost by leaching. Several small applications of fertilizer each year may be used to reduce leaching losses. The climate is warm and some areas produce two or three crops per year. Much work is needed to develop really satisfactory ways of managing Oxisols. Many areas have been cropped under a shifting slash-and-burn cultivation, but increasing population makes this method impractical.

Spodosols are soils with spodic horizons. They form under intensely acid leaching conditions that cause humus, iron, and aluminum to move from the A to the B horizon. They are most common in sandy parent materials with somewhat poor drainage under evergreen trees in a cool humid climate. A typical profile includes a litter layer above the soil, little or no A1 horizon, a prominent ashy-gray A2 horizon, a black Bh horizon containing illuvial humus and aluminum, a reddish-brown Bir horizon containing illuvial iron, and a B3 horizon in which the iron color fades with depth. The spodic horizon includes the Bh, Bir, and upper B3 horizons. Thin cemented layers may occur within the spodic horizon. Other layers such as a fragipan, a duripan, or an argillic horizon may occur beneath the spodic horizon.

Spodosols combine bright colors with horizons that are irregular in thickness and depth (Figure 16-5) or even intermittent. In addition to the most typical Spodosols *(Orthods),* there are suborders based on wetness *(Aquods),* high iron contents *(Ferrods),* or high contents of humus in the spodic horizon *(Humods).*

Spodosols are colorful, but they are not very fertile. The fertility and high acidity cause most of them to be left in forest vegetation. Much of the developed area of Spodosols produces pasture or hay, though some areas are cropped. Good management with adequate fertilization is required for use as cropland. Cultivation alters the profile by mixing the upper horizons. Use of lime and fertilizer may lead to degradation of the spodic horizon over a period of several decades.

Ultisols are strongly weathered soils. The name comes from the same root as *ultimate* but is somewhat misleading because the Oxisols are more strongly weathered than the Ultisols. But Ultisols are more strongly weathered than most other soils. They resemble Alfisols in always having an argillic horizon and not having a mollic epipedon but are more weathered and leached and brighter colored than Alfisols. Kaolinite is the dominant clay type so its cation-exchange capacities are relatively low. Ultisols are the dominant soils of warm, humid regions such as the southeastern United States (Figure 16-6), southern Brazil, eastern India, and southeastern China. Ultisols are also common in association with Oxisols in tropical areas.

Ultisols generally form under forest vegetation. They have base saturations below 35 percent in some parts of their profiles, but the reactions are much less acid than they would be if the clay were mostly montmorillonite instead of kaolinite. Suborders of Ultisols are based on wetness *(Aquults),* more than 0.9 percent organic carbon in the upper 15 cm of the argillic horizon *(Humults),* humid climates *(Udults),* moderate dry seasons *(Ustults),* or very dry summers *(Xerults).*

Ultisols have relatively low fertility as a consequence of the strong weathering and leaching they have undergone. Sandy loam A horizons over clay B horizons are common in Ultisols. The warm, humid climate is favorable for crop production except for the dry summers of the Xerults. Good soil management, complete fertilizers, and moderate amounts of lime are needed for good yields. Cotton, corn, peanuts, and tobacco are among the important crops grown on Ultisols. Row crops combined with high rainfall intensities have caused severe erosion on many Ultisols.

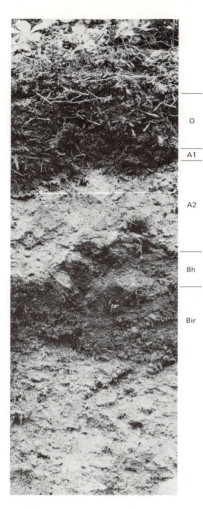

O

A1

A2

Bh

Bir

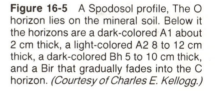

Figure 16-5 A Spodosol profile, The O horizon lies on the mineral soil. Below it the horizons are a dark-colored A1 about 2 cm thick, a light-colored A2 8 to 12 cm thick, a dark-colored Bh 5 to 10 cm thick, and a Bir that gradually fades into the C horizon. *(Courtesy of Charles E. Kellogg.)*

Vertisols are soils that develop cracks at least 1 cm wide and 50 cm deep during dry seasons. They contain at least 30 percent clay, and most of it is of the montmorillonite type so the soil is able to shrink and swell. Vertisols have a unique "self-swallowing" action in which surface soil falls into the cracks during dry seasons; when it rains, the soil absorbs water through the cracks as well as the surface. As the soil swells it also shifts to accommodate the extra soil that fell down the cracks. The mixing action that results from repeated wet and dry seasons prevents the development of strong horizonation and often produces a hummocky topography known as *gilgai.*

Vertisols and Histosols are the least extensive of the 10 soil orders. The largest areas of Vertisols are in India, central Africa (Sudan and Chad areas), eastern Australia, Uruguay, the Yucatan Peninsula, and Texas. Most of them had grass for their native vegetation. Suborders are based on the frequency and duration of dry

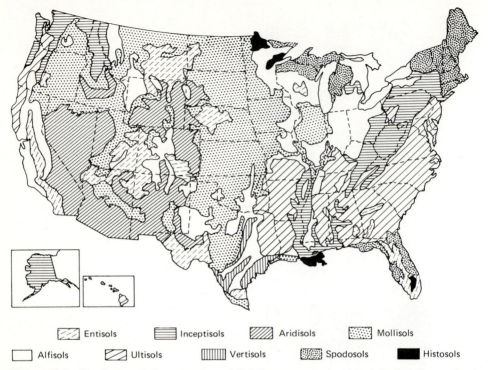

Legend:
Entisols Inceptisols Aridisols Mollisols
Alfisols Ultisols Vertisols Spodosols Histosols

Figure 16-6 Distribution of the soil orders of Soil Taxonomy in the United States. (Generalized from *Soil Conservation Services, USDA.*)

periods that produce cracking. There are Vertisols that are cracked open less than 90 days per year *(Uderts)*, that have open cracks for more than 90 days per year but are moist for at least 60 consecutive days during the growing season *(Usterts)*, that are cracked open during most of the summer *(Xererts)*, or are cracked open nearly all the time *(Torrerts)*.

Most Vertisols have high fertility because they have high cation-exchange capacities and the mixing action helps maintain high base saturations; but their use is complicated by soil movement. Trees and fences may tilt, building foundations and pavement may shift and crack, pipelines may break, etc. Irrigation systems are needed to use the Torrerts and Xererts to grow crops but the shifting soil makes the systems difficult to maintain. The high clay contents make Vertisols very sticky when wet. Still, much cotton, small grains, and other crops are grown on Vertisols. Most of the uncultivated area is used as grazing land.

Names of Great Groups and Subgroups The suborders of the new system of soil classification are divided into great groups by attaching prefixes to the suborder name. Some of the most used prefixes are listed in Table 16-4. This table can be used to interpret soil properties from the names of great groups.

Great groups are divided into subgroups by adding an adjective to the name. The adjectives specify that the soils in the great group are either typic (typical) or have some special property that is named. Many of the prefix syllables used in naming suborders and great groups are made into subgroup adjectives by adding *-ic* (for example, there are Albic Natraqualfs and Aquic Fragiudults). Some of the suborder and great group names are also used as adjectives (such as Tropeptic, Haplorthox, and Udorthentic Haplustolls). Some additional words are needed for conditions not covered by any of the preceding (such as Lithic Haplargids where *lithic* means less than 50 cm deep to bedrock).

Families Soil families are identified by adding descriptive adjectives to the subgroup name. For example, the Sassafras series belongs to the fine-loamy, siliceous, mesic family of Typic Hapludults. Fine loamy restricts the particle size range to 18 to 34 percent clay and at least 15 percent material between 0.1 and 75 mm in diameter. Siliceous means the mineral material is over 90 percent quartz or other SiO_2 minerals. Mesic means the average annual soil temperature is between 8 and 15°C. Other terms can be used to specify slope class, depth of root zone, soil reaction, consistence, cracks, and coatings. Most of the distinguishing properties are applied to the *control section* of the soil. The control section coincides with the upper 50 cm of the argillic horizon or the 25- to 100-cm zone of soils that do not have an argillic horizon. Detailed definitions are given in Soil Taxonomy (Soil Survey Staff, 1975).

Table 16–4 Some of the Prefixes Used to Make Great Group Names from the Suborders of the New System of Soil Classification

Prefix	Related word	Meaning
Alb-	Albino	Has an albic horizon
Arg-	Argillite	Has an argillic horizon
Bor-	Boreal	Cool climate
Calc-	Calcium	Has a calcic horizon
Camb-	Cambic	Has a cambic horizon
Cry-	Crystal	Cold climate
Dur-	Durable	Has a duripan
Eutr-	Eutrophic	High base saturation
Frag-	Fragile	Has a fragipan
Hapl-	Haploid	Minimum development
Hum-	Humus	Presence of humus
Natr-	Natrium	Has a natric horizon
Pale-	Paleosol	Old development
Plinth-	Plinthite	Presence of plinthite
Quartz-	Quartz	High quartz content
Rhod-	Rhododendron	Dark red color
Sal-	Saline	Has a salic horizon
Sulf-	Sulfur	Presence of sulfides or sulfates
Torr-	Torrid	Arid climate
Trop-	Tropical	Continually warm
Ud-	Udometer	Humid climate
Umbr-	Umbrella	Has an umbric epipedon
Ust-	Combustion	Dry climate with hot summers
Xer-	Xerophyte	Annual dry season

Source: Soil Survey Staff, 1975.

The properties chosen for a particular family name are intended to make the soils in the family fairly uniform in use and management needs. The descriptive name can be replaced with the name of a soil series for more practical use on a local basis. Miami family, for example, can be used for the fine-loamy, mixed, mesic family of Typic Hapludalfs.

Soil Series   Soil series are the only classes that are designed to fit into both Soil Taxonomy and the 1938 System of soil classification. Soil series are subdivided on the basis of texture to form soil types in the 1938 System. However, the soil series name conveys all the other information about the classification of a soil and is the real primary unit of soil classification. This fact was recognized by making the soil series the lowest categorical level in Soil Taxonomy.

The process of soil classification usually begins with the assignment of a soil series name to a group of soil pedons. The pedons must resemble one another on the basis of the kind and sequence of horizons present. Presumably they resemble each other physically because of similarity in the action of the five soil-forming factors that produced their characteristics. There are many thousands of soil series, each with its own set of soil properties resulting from a particular combination of effects of the soil-forming factors. Soil correlation to ensure that soil series names are defined and used consistently is performed cooperatively in the United States by the Soil Conservation Service of the USDA and many of the state agricultural colleges.

Soil series names are usually chosen from some geographical feature of the vicinity where the series is first established. A complete soil profile description plus specification of the range of characteristics to be included in that series is required for the establishment of a new soil series. The soil-forming factors involved and the differences between the new series and other similar series must also be included.

The 1938 System of Soil Classification

The 1938 System was the official soil classification system in the United States until the new system was adopted in 1965. Many of its terms are frequently encountered in soils literature. The names of the great soil groups are especially well known so their characteristics will be summarized here. Approximate relationships between the great soil groups of the 1938 System and the orders of Soil Taxonomy are shown in Table 16-5. The relationships are not exact because different criteria are used in the two systems, but they are close enough to be useful for translating information from one system to the other.

Details of the 1938 System are contained in the Yearbook of Agriculture (Baldwin, Kellogg, and Thorp). Most of the official modifications made in the system were published in a special issue of *Soil Science,* February, 1949. These modifications are included in the discussion that follows.

The 1938 System has six categories: orders, suborders, great soil groups, families, series, and types. The system has 3 orders, 11 suborders, and 36 official great soil groups (some unofficial great soil groups were also used). These three highest

Table 16–5   A General Classification of the Great Soil Groups of the 1938 System into the Orders of Soil Taxonomy

Great Soil Groups 1938 System (as revised)	Orders in Soil Taxonomy including most of the soils
Alluvial soils	Entisols, Inceptisols, and Mollisols
Alpine Meadow soils	Spodosols
Ando soils	Inceptisols
Bog soils	Histosols
Brown soils	Aridisols and Mollisols
Brown Forest soils	Inceptisols
Brown Podzolic soils	Spodosols and Inceptisols
Brunizems (Prairie soils)	Mollisols
Chernozems	Mollisols
Chestnut soils	Mollisols
Degraded Chernozems	Alfisols and Mollisols
Desert soils	Aridisols
Gray-Brown Podzolic soils	Alfisols
Gray Wooded soils	Alfisols
Groundwater Laterite soils	Ultisols
Groundwater Podzols	Spodosols
Grumusols	Vertisols
Half-Bog soils	Inceptisols
Humic Glei soils	Mollisols, Inceptisols, and Ultisols
Laterite soils	Oxisols
Latosols	Ultisols and Inceptisols
Lithosols	Entisols
Low Humic Glei soils	Entisols, Inceptisols, etc.
Noncalcic Brown soils	Alfisols
Planosols	Alfisols and Mollisols
Podzols	Spodosols
Red Desert soils	Aridisols
Reddish Brown soils	Aridisols and Alfisols
Reddish Chestnut soils	Mollisols and Alfisols
Reddish Prairie soils	Mollisols
Red-Yellow Podzolic soils	Ultisols
Regosols	Entisols and Inceptisols
Rendzinas	Mollisols
Sierozems	Aridisols
Solonchaks	Aridisols
Solonetz soils	Aridisols and Mollisols
Soloth soils	Alfisols and Mollisols
Sols Bruns Acides	Inceptisols
Tundra soils	Inceptisols

levels are intended to contain all the soils of the world. The numbers of classes required in the three lower levels are considerably larger and somewhat indefinite. There are over 10,000 soil series in the United States alone.

Three of the categories in the 1938 System, the orders, great soil groups, and series, proved to be of greater significance to soil classification than the other three. The suborders and families received relatively little use; the type is a subdivision of the series based only on texture of the surface soil.

Orders   The three orders in the 1938 System are named *Zonal, Intrazonal,* and *Azonal.* These are closely related to the particular factors of soil formation that have had dominant influence on the soil. Zonal soils are those whose characteristics are determined mostly by the two active factors of soil formation, climate and vegetation. Zonal soils are more extensive than the other two orders and generally receive the most attention.

Intrazonal soils have distinct soil profiles that reflect a dominant influence of topography or unusual parent material. Most Intrazonal soils occur on nearly level or slightly concave topography and have poorer drainage than the Zonal soils (parent materials are a less common cause of Intrazonal soils). Land-use problems associated with wetness and with saline and alkaline soils are common in Intrazonal soils.

Azonal soils lack distinct profile characteristics. The cause may be extreme youth, a high rate of erosion, or parent material that is highly resistant to change. Extremely young Azonal soils may be very productive, but the others are likely to be unproductive.

Zonal, Intrazonal, and Azonal soils occur in mixed patterns all over the earth. The concept of zonality helped to explain how soils form, but it has been of limited value for defining units to be used in soil mapping. This is a weakness of the 1938 System. Ideally, the classes of the highest orders should be useful for preparing maps of whole countries or perhaps of the entire world. The intermixing of soils of the three orders prevents such use. The suborders of the 1938 System are better adapted for such maps but have received little use. A major reason for not using the suborders is their long names such as "Light-colored podzolized soils of the timbered regions." Most large-scale maps have therefore been based on the great soil groups.

Great Soil Groups   All soils belonging to any one great soil group have certain broad similarities in their soil profiles. Podzols have acid reactions and bright colors and contain A2, Bh, and Bir horizons. Sierozems are light colored, only slightly leached, and relatively shallow because they form in arid climates. Chernozems have thick, black, A1 horizons, etc. Sometimes the characteristics a soil does not have are as significant as those it does have. For example, the soils of the Gray-Brown Podzolic great soil group have grayish-brown A1 horizons, gray A2 horizons, and brown B2 horizons, but they do not have B3ca (nor any other ca) horizons.

Gray-Brown Podzolic soils represent one of the 20 Zonal great soil groups. The relations of these 20 great soil groups to climate and vegetation are illustrated in Figure 16-7. Most of the Zonal great soil groups occur on the North American

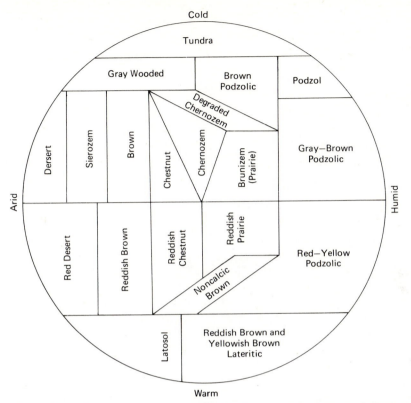

Figure 16-7 A schematic diagram relating the Zonal great soil groups to climate and vegetation. Names printed vertically are grassland and desert soils; horizontal lettering indicates podzolic and lateritic soils (forested soils) and tundra (shrub vegetation); printed at an angle are names of soils bearing the influence of both grass and trees.

continent in geographical positions approximating the pattern shown in Figure 16-7, but the pattern is distorted by mountains, floodplains, etc., giving rise to the Azonal and Intrazonal soils. The resulting general distribution in the United States is shown in Figure 16-8.

Colors are mentioned in many of the Zonal great soil groups, especially when the Russian words *Sierozem* (gray soil), *Brunizem* (brown soil), and *Chernozem* (black soil) are translated. Colors are among the most obvious and generally consistent soil properties related to climatic and vegetation zones.

In addition to the 20 Zonal great soil groups there are 13 Intrazonal and 3 Azonal great soil groups in the 1938 System. Some feature or features of parent material, topography, or time control the most significant properties of soils in the Intrazonal and Azonal great soil groups.

Characteristics of Zonal Great Soil Groups   The *Desert* soils are light gray in color and low in organic matter (usually only a fraction of 1 percent of organic matter is present in any part of a Desert soil). Their sola are commonly only 20 to

General Distribution of the Great Soil Groups

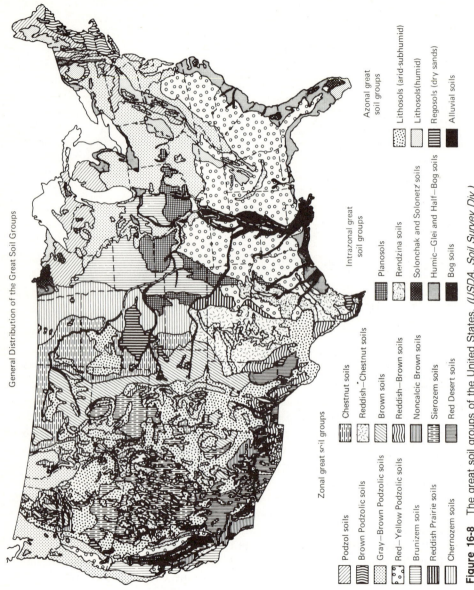

Zonal great soil groups

Podzol soils
Brown Podzolic soils
Gray–Brown Podzolic soils
Red–Yellow Podzolic soils
Brunizem soils
Reddish Prairie soils
Chernozem soils

Chestnut soils
Reddish–Chestnut soils
Brown soils
Reddish–Brown soils
Noncalcic Brown soils
Sierozem soils
Red Desert soils

Intrazonal great
soil groups

Planosols
Rendzina soils
Solonchak and Solonetz soils
Humic–Glei and Half–Bog soils
Bog soils

Azonal great
soil groups

Lithosols (arid-subhumid)
Lithosols (humid)
Regosols (dry sands)
Alluvial soils

Figure 16-8 The great soil groups of the United States. *(USDA, Soil Survey Div.)*

40 cm deep to underlying Cca horizons. The natural vegetation of Desert soils is a sparse mixture of grasses, small shrubs, cacti, and various flowering plants. The vegetation grows rapidly during a moist period (often in the spring), then lies dormant for months. Much of the desert is used for grazing purposes, usually with the result that the edible plants die, leaving the cacti and shrubs as the only vegetation.

The *Sierozem, Brown, Chestnut,* and *Chernozem* soils are a climatic sequence reaching from the deserts to the borders of humid regions. The colors of the A horizons range progressively from gray (Sierozems) to black (Chernozems). Increasing precipitation makes the more adequately watered soils in this sequence progressively deeper, more productive under natural conditions, and higher in organic matter than the drier soils. The degree of profile development and the horizon thicknesses tend to increase with more precipitation, but the main horizons (A1, A3, B2, and Cca) are the same in mature soils of all these great soil groups.

The *Red Desert, Reddish-Brown,* and *Reddish-Chestnut* soils are warmer-climate equivalents of the grayer soils with similar names. The redder colors result from more intense weathering that releases more iron and decomposes organic matter faster than in cooler climates. The frost-free seasons are longer in the red-soil regions, but the effective growing seasons are generally about equal to those of the cooler regions with similar soils. Plants are naturally dormant through the warm, dry summers unless they receive irrigation water.

The *Brunizem (Prairie)* and *Reddish-Prairie* soils also form under grass vegetation and closely resemble the Chestnut and Reddish-Chestnut soils. However, the Brunizems occur in humid climates and are therefore more thoroughly leached than other grassland soils. Brunizems do not have Cca horizons because deep percolation carries the calcium carbonate from the solum to the groundwater.

Degraded Chernozem and *Noncalcic Brown* soils bear the influence of both grass and trees. The Degraded Chernozems were formed under tall grass vegetation but have since been overgrown by forest. They still have a very dark A1 horizon, but it is thinner than it originally was because its lower part has been converted into an A2 horizon. The solum as a whole is generally more acid and more strongly developed than most grassland soils.

The Noncalcic Brown soils form under a mixed forest and grass or brush and grass vegetation. They occur in warmer climates than the Degraded Chernozems and have browner (sometimes reddish-brown) colors. The Noncalcic Brown soils pass through wet and dry periods because of markedly seasonal precipitation.

Tundra soils form in climates so cold that they have permafrost beneath the soil. Every summer the soil thaws enough for some flowering plants and shrubs to grow during the long arctic days. The organic matter thus produced is preserved by the cold, and a peaty soil layer is produced above the gray mineral horizons. Yellowish-brown mottles indicate the imperfect drainage resulting from the lack of water penetration through the underlying permafrost.

Podzols are the most colorful soils of any of the great soil groups. Figure 16-5 shows a Podzol profile. Most Podzol profiles are low in clay and are highly acid. The horizons are irregular in thickness and depth and are sometimes intermittent. Cemented layers are common in Podzol B horizons.

Podzols may be considered as end members of the acid-leaching sequence. Other soils that bear the effects of acid leaching are said to be *podzolized* and are called *podzolics.* The Podzols occur in cool, humid climates. Most typically, they form under evergreen forest in sandy parent materials with imperfect drainage.

Brown Podzolics, Gray-Brown Podzolics, Red-Yellow Podzolics and Gray-Wooded soils all have acid reactions and A2 horizons resulting from the acid-leaching process known as *podzolization.* All these soils develop under forest vegetation but under varying climates. The Brown Podzolics are quite similar to the Podzols but have less prominent A2 horizons and are somewhat less leached. All the other podzolic soils have noteworthy clay accumulations in their B horizons.

Gray-Wooded (Gray Podzolic) soils border the Tundra but lack permafrost and are therefore better drained than Tundra soils. They occur in the driest climates of any of the podzolics, but the profiles are leached and acid because the cool temperatures make the precipitation highly effective relative to its amount. Their profiles are grayer but otherwise quite similar to those of the Gray-Brown Podzolics.

Gray-Brown Podzolics occupy an area between Podzols and Brunizems. They form mostly under deciduous forest vegetation in cool, humid climates. They have A1 horizons, but these are not as thick and dark colored as those of the Brunizems. The Gray-Brown Podzolics have A2 horizons but not nearly such prominent ones as the Podzols have. They generally have stronger profile development and more clay accumulation in their B2 horizons than the Brunizems have. The reactions are acid but less so than those of the Podzols.

Red-Yellow Podzolics occur in warm, humid climates and are more strongly weathered than any of the preceding great soil groups. They resemble the Gray-Brown Podzolics in their horizon sequences, but the Red-Yellow Podzolics are brighter colored, more leached, and often somewhat deeper and stronger developed than the Gray-Brown Podzolics. The fertility level of the Red-Yellow Podzolics is relatively low because they generally are low in organic matter and have sandy A horizons.

Latosols and *lateritic* soils occur in tropical climates. They are the best-known soils of the tropics, though extensive areas of Red-Yellow Podzolics and even some black soils also occur there. The name *laterite* comes from the Latin word *later,* which means *brick.* This name is used because some of these soils have layers that harden irreversibly when dried. Brick-sized pieces are cut from the moist soil, dried in the sun, and used for building purposes.

The lateritic soils are the most weathered of all the great soil groups. Some of these soils contain as much as 80 percent oxide clays in their mineral matter. Lateritic soils generally have very deep profiles. They have reddish and sometimes yellowish colors. The climate is warm year-round and humid all or much of the year. The native vegetation is tropical forest.

Latosols form under savannah vegetation (nonforested tropical areas of grasses and flowering plants). These soils have distinct wet and dry seasons. Irrigation is required if crops are to be grown during the dry season.

Characteristics of Intrazonal Great Soil Groups   The 13 Intrazonal and 3 Azonal great soil groups are listed in Table 16-6. Most of the Intrazonal great soil groups show some influence from excess water. Some of them have been influenced by excessive salts; some are formed under bog conditions; and some are very strongly differentiated.

Solonchak, Solonetz, and *Soloth* soils have been influenced by excess salts, usually because of a water table within 1.5 m of the soil surface. These soils occur mostly in arid or semiarid regions. The Solonchak soils contain large quantities of soluble salts because the natural precipitation is inadequate to counteract upward movement from the water table. A white salt crust forms on the surface of Solonchak soils during dry periods and causes them to be called *white alkali.* The salt concentration may or may not be enough to prevent plant growth.

Solonetz soils are leached of excess salts, but they contain enough exchangeable sodium to disperse their clay particles and cause a very low permeability. Clay movement is relatively rapid in the dispersed condition, and the soils therefore develop strong profiles rather rapidly. Plant growth is very poor because of the high pH values and the low permeability. What little organic matter the Solonetz soils contain tends to dissolve, move to the surface in capillary water, and form thin, black flakes as the water evaporates. This has led to the name *black alkali* for these soils. As was discussed in Chapter 8, these are some of the most difficult soils there are to reclaim.

Soloth soils presumably were once Solonetz soils. Their natural drainage has improved sufficiently for leaching to occur. The A horizons of Soloth soils are generally neutral or even acid in reaction, though the B horizons may still be alkaline. The A horizons are darkening from the top down but are still light colored below; the B horizons are high in clay and generally have a columnar structure. Plant growth on Soloth soils is usually fair to poor but definitely better than on Solonetz soils.

Alpine Meadow, Bog, Half Bog, Humic Glei, and *Low Humic Glei* soils all form under conditions of poor or very poor drainage. Each of these soils has a much higher organic-matter content than its better-drained neighbors. Excess salts are not a problem because enough water is present to keep them dissolved. Water-loving

Table 16–6   The Intrazonal and Azonal Great Soil Groups of the 1938 System of Soil Classification (as Revised in 1949)

Intrazonal great soil groups				Azonal great soil groups
Salt- and sodium-affected soils	Bog soils	Strongly differentiated soils	Soils high in CaCO$_3$	
Solonchak	Alpine Meadow	Groundwater	Brown Forest	Alluvial
Solonetz	Bog	Laterite	Rendzina	Lithosol
Soloth	Half Bog	Groundwater		Regosol
	Humic Glei	Podzol		
	Low Humic Glei	Planosol		

plants such as reeds, sedges, mosses, ferns, swamp trees (cypress, etc.) grow very well in the marshes and swamps occupied by these soils. Peats and mucks are often intermixed with mineral soils in these areas. Frequently a layer of peat or muck overlies a bluish-gray mineral soil with bright-colored rust mottles.

Alpine Meadow soils form in cool to frigid regions and are likely to be associated with Tundra soils. Of the other four great soil groups included here, the Bog soils contain the most organic matter. They are deep peats or muck over peat and often constitute the wettest part of a swamp or marsh. The Half Bogs are shallower peats or mucks over mineral soil. The Humic Gleis are dark-colored mineral soils with usually between 10 and 30 percent organic matter. The Low Humic Gleis have only a thin dark-colored mineral soil high in organic matter over a rust-mottled, bluish-gray mineral horizon.

Groundwater Laterites, Groundwater Podzols, and *Planosols* are imperfectly to poorly drained soils. The wetness in these soils has had the effect of accentuating the normal soil-forming processes. All these soils have horizons with low permeabilities overlain by leached A2 horizons. The soil reactions are acid throughout the profile in most of these soils. Plant growth is limited by the impermeable layers, poor drainage, and acid reactions.

Groundwater Laterites occur in flat areas in the tropics. Surface drainage is poor, and many of them have only 10 or 20 cm of infertile soil material over a layer of hardened laterite. Productivity is very low.

Groundwater Podzols form in temperate regions in poorly drained sandy materials that contain iron minerals. Ferrous (Fe^{++}) iron moves through the water in the saturated part of the soil and is oxidized to ferric iron (Fe^{+++}) at the base of the aerated part of the soil. The ferric iron precipitates and cements the subsoil into a hardpan (sometimes called *ortstein*).

Planosols occur in flat areas with either grass or forest vegetation. They are similar to Gray-Brown or Red-Yellow Podzolic soils but more strongly developed. Their B horizons are claypans that have low permeability. Their A2 horizons are formed immediately above the B2 horizons because these horizons are saturated during wet periods.

Brown Forest and *Rendzina* soils owe their unique properties to a high content of calcium carbonate in their parent materials and to being relatively young soils. Little illuviation has occurred in the soils belonging to these two great soil groups. The surface soils are dark, and the color fades gradually with depth into the calcareous parent materials. These soils are often only moderately deep or even shallow, but they are quite fertile. Usually the native vegetation is deciduous trees on the Brown Forest soils and mostly grass on the Rendzinas.

Characteristics of Azonal Great Soil Groups   Alluvial soils, Lithosols, and Regosols are Azonal soils. They show very little soil profile development because of either extreme youth or lack of weatherable minerals.

Alluvial soils occur in relatively fresh alluvium. They are often among the most fertile and productive soils of their vicinity. The topography is nearly level and generally favorable for most uses. Textures tend to be variable because the parent materials are usually stratified and streaked by former stream channels. Soil drainage ranges from poor to good. Usually the only apparent profile development is a darkening of the surface soil by organic-matter accumulation. Buried surface layers can sometimes be detected by these same dark colors.

Lithosols are shallow soils forming on imperfectly weathered hard rocks. Usually they are droughty, and they often contain rock fragments. They are most common on steep topography and in arid climates.

Regosols form in unconsolidated materials such as loess, wind-blown sand, glacial till, or old alluvium that is being eroded. Usually they are on steep topography where erosion prevents much soil horizon differentiation. The type of vegetation depends on the climate of the area, but the vegetative growth on Regosols is usually sparse in comparison with that on other soils of the same vicinity.

SOIL SURVEYS

Soil maps are prepared by several different agencies to show what kinds of soil occur in each part of the area mapped. The most comprehensive soil survey program is that carried on by the Soil Conservation Service in cooperation with the state agricultural colleges.

Soil survey publications are usually prepared on a county basis. Early maps were published on a single sheet, usually at a scale of 1 to 126,720 or 1 to 63,360 (½ in. or 1 in. per mile). Colors were used along with letter symbols to identify the soils. After a few years the scale was changed to 2 in. per mile so that more detail could be shown on the map. Most current maps are published at a scale of either 1 to 20,000 or 1 to 15,840 (4 in. per mile). Aerial photographs are now used as base maps both for the field work and for the published maps. An example of a modern soil map is shown in Figure 16-9.

Soil survey reports contain soil descriptions, classifications, and interpretations for various uses along with the soil maps. Soil mapping units are currently classified according to soil series that fit into both Soil Taxonomy and the 1938 System of soil classification.

Soil-mapping Units

Each area delineated on a soil map such as the one in Figure 16-9 is designated as a *soil-mapping unit.* Most soil-mapping units shown on detailed soil maps are phases of soil series. Some other kinds of mapping units are used for special situations such as soils too intermixed to be separated on the map.

Soil *phases* are made by dividing soil series (or sometimes other classes) into more specific units that are significant for practical use even though they are not needed for purposes of soil classification. For example, most soil series occur on a range of slopes that is wide enough to influence the use and management of the soil. Soil series are therefore frequently divided into slope phases. Texture of the plow layer is a phase criterion in Soil Taxonomy, though it had its own special category (soil type) in the 1938 System. Other common phase criteria are degrees of erosion or of stoniness. Most mapping unit names include some phase designations. "Everly

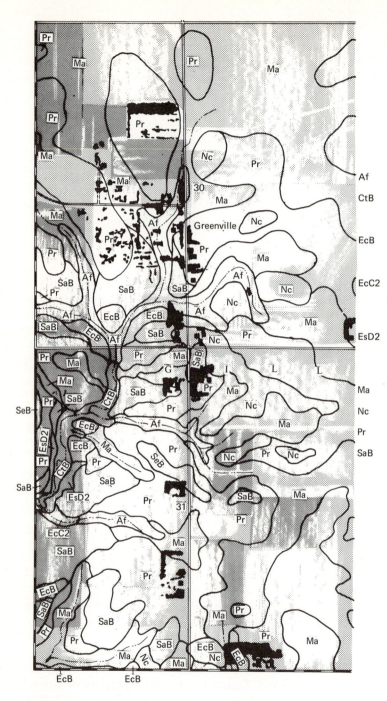

The map legend reads:

Af — Afton silty clay loam

CtB — Colo-Terrill Complex, 2 to 5 percent slopes

EcB — Everly clay loam, 2 to 5 percent slopes

EcC2 — Everly clay loam, 5 to 9 percent slopes, moderately eroded

EsD2 — Everly-Storden Complex, 9 to 15 percent slopes, moderately eroded

Ma — Marcus silty clay loam

Nc — Nicollet clay loam

Pr — Primghar silty clay loam

SaB — Sac silty clay 2 to 5 percent slopes

Figure 16-9 A 2-mi² portion of the soil survey of Clay County, Iowa, published in 1968. The scale of the published map is 1 to 15,840 (4 in. equals 1 mi).

clay loam, 5 to 9 percent slopes, moderately eroded" is an example from Figure 16-9. Everything except the series name (Everly) is a phase designation according to the new system.

Not all the area designated on a map as Everly clay loam, 5 to 9 percent slopes, moderately eroded will actually fit that description. Perhaps part of the area is a different texture; part may have only a 4 percent slope. Some parts may belong to a soil series other than Everly. Such deviations are inevitable because there are too many tiny soil areas to be shown on most soil maps. Small soil areas that differ from the name given to the mapping unit in which they occur are called *inclusions*. Mapping units that contain no more than 15 percent inclusions are usually considered acceptable.

Soil *complexes* are used as mapping units where two or more soils are too intermixed to be mapped separately at the map scale used. The Colo-Terrill Complex, 2 to 5 percent slopes in Figure 16-9, is an example of a phase of a soil complex. Colo and Terrill are two different soil series, but they are intermixed in some landscapes.

Undifferentiated units are used where two or more soils could be separated but are mapped together because their differences are not important. For example, the differences among extremely stony areas of several different soils may be unimportant if they can be used only as grazing land.

Soil associations are groups of soils that occur in repeating patterns on the landscape. Often the hilltops are one series, the hillsides another, and the valley floors another. The separate areas may be large enough to be mapped on a detailed map but not on a generalized map. Soil associations are therefore used for making small maps of a large area such as a county or a state. Figure 16-10 is a portion of the soil association map of Gem County, Idaho. Figure 16-11 shows how some of the Gem County soil associations fit together on the landscape.

The soil series that occur together in a soil association may be quite different from one another. They may have different vegetation, different soil age, and different parent materials. Usually there are differences in topography. The main requirement is that the soils appear together repeatedly in an identifiable pattern within an area that can be shown on a map. It is important that the pattern be well defined so that the map users can usually identify the individual soils correctly in the field.

A *toposequence* is a small soil association with differences related to topography. All the soils in a toposequence form in the same climatic zone under similar vegetation. The soil parent material may vary within a toposequence if the variation is related to topographic position.

A soil *catena* is a toposequence formed entirely in one kind of parent material. Several different catenas having different parent materials and vegetation may occur in the same soil association.

INTERPRETIVE CLASSIFICATION OF SOILS

Soil maps contain information that is useful for many different purposes, but this information must be interpreted before it can be used. An arrangement of lines and

Figure 16-10 A portion of the soil association map of Gem County, Idaho. *(From Troeh et al., 1965.)*

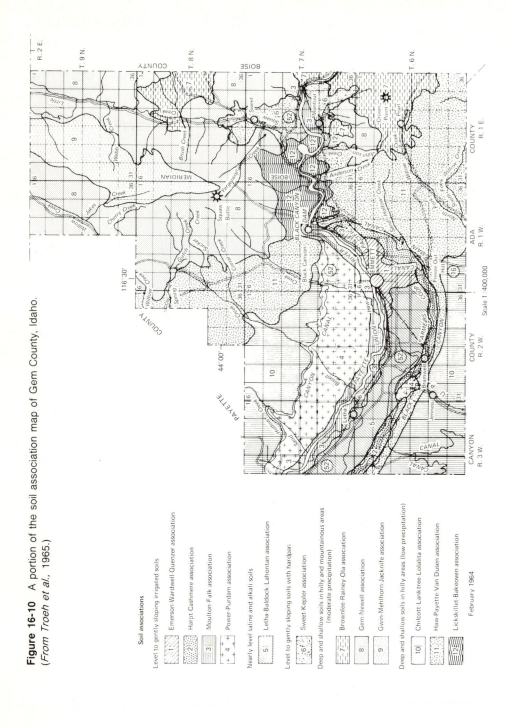

Soil associations

Level to gently sloping irrigated soils

- Emerson-Wardwell-Quenzer association
- Harpt-Cashmere association
- 3 Moulton-Falk association
- 4 Power-Purdam association

Nearly level saline and alkali soils

- 5 Letha-Baldock-Lahontan association

Level to gently sloping soils with hardpan

- 6 Sweet-Kepler association

Deep and shallow soils in hilly and mountainous areas (moderate precipitation)

- 7 Brownlee-Rainey-Ola association
- 8 Gem-Newell association
- 9 Gwin-Mehlhorn-Jacknife association

Deep and shallow soils in hilly areas (low precipitation)

- 10 Chilcott-Lanktree-Lolalita association
- 11 Haw-Payette-Van Dusen association
- 12 Lickskillet-Bakeoven association

February 1964

392

Figure 16-11 A view from Freezeout Hill looking north across the Emmett Valley portion of Gem County, Idaho. This view shows several of the soil associations of Figure 16-10 including no. 11 (right foreground), nos. 8 and 9 around Squaw Butte (right background), nos. 10 and 11 (left and center background), and nos. 1 to 5 (in the irrigated land).

symbols is meaningless without an interpretation. Furthermore, different uses require different interpretations of the same soil map.

Suitability

Interpretive classifications are often made so that suitability maps can be prepared from soil maps. A suitability map can be prepared for any type of land use. Some of the most common types show the soil suitability for building sites, septic tank drain fields, irrigation by surface or sprinkler methods, or the growing of some specified crop.

Suitability maps are made by first evaluating all soil-mapping units in the area according to suitability for the specified use. Symbols representing the suitability evaluation of each soil are then placed on a map. The map used is usually a copy or a tracing of the soil map. Suitability maps are often colored to make the classes easy to see.

Interpretations for Engineering

Interpretive classifications of soils for engineering purposes are often made into tables rather than maps. The engineer may be interested in placing a building, a dam, or a road at a particular site. The soil map will tell the engineer what kind of soil is there, and the interpretive table will tell what problems are likely to be encountered. Interpretive tables for engineering purposes usually include an engineering classification of soil texture, soil strength characteristics, permeability and drainage characteristics of the soil, and pertinent special features.

Economic Classification of Soils

The capacity of the soil to produce crops can be evaluated monetarily and made into an economic classification. Economists are interested in performance characteristics rather than in physical characteristics of soil. They want to know what yields can be obtained and what the costs will be to obtain various yield levels from each soil. They can use this information to calculate probable profit (or loss) and land value.

Rural property tax valuations are an important use of economic land classifications. Many counties base their tax assessments partially or entirely on such classifications. The number of acres of each kind of soil in each landholding is measured and mutiplied by its economic value factor. These products are then summed and multiplied by a scale factor to determine assessed valuations.

Table 16–7   Land-Use Capability Classes

	Land suitable for cultivation
Class I	Land of good productivity practically free of erosion and suitable for cultivation without special practices; some areas may need to be fitted for cultivation, as by clearing or simple drainage.
Class II	Land of moderate to good productivity suitable for cultivation with ordinary or simple practices to prevent erosion or effect satisfactory drainage, as by contouring, growing protective cover crops, and carrying out rather easy drainage operations, as with small ditches, where needed.
Class III	Land of moderate to good productivity suitable for cultivation with intensive practices, such as terracing, strip-cropping, heavy fertilization, and installation of extensive drainage facilities.
	Land suitable for limited cultivation
Class IV	Land of moderate productivity suitable chiefly for pasture and hay because of steepness of slope, with occasional use for row crops. When it is cultivated, intensive erosion-prevention practices usually are required.
	Land not suitable for cultivation
Class V	Land not suitable for cultivation but useful for grazing or forest with normal precautions to ensure sustained use.
Class VI	Land not suitable for cultivation but suitable for grazing or the growing of trees with strict precautions for sustained use.
Class VII	Land not suitable for cultivation but suitable for grazing or forestry when used with extreme care to prevent erosion.
Class VIII	Land not suitable for cultivation, grazing, or forestry, although it may have some value for wildlife. This land ordinarily is extremely rough, stony, sandy, wet, or susceptible to severe erosion.

Source: Bennett, 1947.

Figure 16-12 An illustration of the eight land-use capability classes. *(Soil Conservation Service, USDA.)*

Land-use Capabilities

The land-use capability system has had the widest use of any interpretive classification. It is an evaluation of land according to its limitations for agricultural use and employs three categorical levels known as classes, subclasses, and units.

There are eight classes in the highest category of the land-use capability system. The eight classes are based on limitations or hazards in use ranging from class I with no special limitations through progressively greater hazards or limitations to class VIII which has very restricted use (see Table 16-7 and Figure 16-12). Subclasses are formed from classes II through VIII by adding one of four letters to designate what kind of problem limits the use of the land. The letters are:

e = erosion hazard
w = wetness problem
s = soil limitation
c = climate limitation

Land-use capability units are formed by adding numbers to the subclasses. For example, capability units IIe1, IIe2, and IIe3 would represent several class II soils with erosion hazards stemming from different causes. Some might be on slopes subject to runoff, others might be sandy and subject to wind erosion, and still others might be subject to occasional damaging overflow from a nearby stream.

The Soil Conservation Service uses the land-use capability system for the preparation of conservation plans for its farmer-cooperators. The soil maps provided in these plans are colored according to a standard color scheme with I colored green; II, yellow; III, red; IV, blue; V, dark green or white; VI, orange; VII, brown; and VIII, purple.

REFERENCES

Arkley, R. J., 1971, Factor Analysis and Numerical Taxonomy of Soils, *Soil Sci. Soc. Am. Proc.* **35**:312–315.

Baldwin, Mark, C. E. Kellogg, and James Thorp, 1938, Soil Classification, *Soils and Men,* USDA Yearbook, pp. 979–1001.

Basinski, J. J., 1959, The Russian Approach to Soil Classification and Its Recent Development, *J. Soil Sci.* **10**:14–26.

Bennett, H. H., 1947, *Elements of Soil Conservation,* McGraw-Hill, New York, 406 p.

Bridges, E. M., 1970, *World Soils,* Cambridge Univ. Press, 89 p.

Buol, S. W., F. D. Hole, and R. J. McCracken, 1973, *Soil Genesis and Classification,* Iowa State Univ. Press, 360 p.

Cline, M. G., 1949, Basic Principles of Soil Classification, *Soil Sci.* **67**:81–91.

Dekker, L. W., and M. D. deWeerd, 1973, The Value of Soil Survey for Archaeology, *Geoderma* **10**:169–178.

Gerasimov, I. P., 1973, Chernozems, Buried Soils, and Loesses of the Russian Plain: Their Age and Genesis, *Soil Sci.* **116**:202–210.

Huddleston, J. H., and G. W. Olson, 1967, Soil Survey Interpretation for Sub-surface Sewage Disposal, *Soil Sci.* **104**:401–409.

Malo, D. D., B. K. Worcester, D. K. Cassel, and K. D. Matzdorf, 1974, Soil-Landscape Relationships in a Closed Drainage System, *Soil Sci. Soc. Am. Proc.* **38**:813–818.

Marbut, C. F., 1935, The Soils of the United States, *Atlas of American Agriculture,* USDA, Washington, D.C.

Olson, G. W., 1964, Application of Soil Survey to Problems of Health, Sanitation, and Engineering, *Cornell Univ. Agr. Exp. Sta. Memoir* 387, 75 p.

Schelling, J., 1970, Soil Genesis, Soil Classification and Soil Survey, *Geoderma* **4**:165–193.

Shrader, W. D., and F. F. Riecken, 1964, Soil Survey Use in Iowa: Crop Yield Data Help Farmers, Tax Assessors Benefit from Maps Covering All Counties, *Soil Conserv.* **30**(5):109–110.

Simonson, R. W., 1962, Soil Classification in the United States, *Science* **137**:1027–1034.

Soil Survey Staff, 1960, Soil Classification, A Comprehensive System, 7th Approximation, *Soil Cons. Ser., USDA,* Washington, D.C., 265 p.

Soil Survey Staff, 1975, *Soil Taxonomy,* Soil Cons. Ser., USDA Handbook No. 436, U.S. Government Printing Office, Washington, D.C., 754 p.

Thornburn, T. H., 1966, The Use of Agricultural Soil Surveys in the Planning and Construction of Highways, *Soil Surveys and Land Use Planning, Amer. Soc. Agron.* and *Soil Sci. Soc. Amer.,* pp. 87–103.

Troeh, F. R., J. C. Chugg, G. H. Logan, C. W. Case, and Virgil Coulson, 1965, *Soil Survey of Gem County Area, Idaho,* U.S. Government Printing Office, Washington, D.C., 196 p.

Walker, P. H., G. F. Hall, and R. Protz, 1968, Soil Trends and Variability across Selected Landscapes in Iowa, *Soil Sci. Soc. Am. Proc.* **32**:97–101.

Wertz, W. A., 1966, Interpretation of Soil Surveys for Wildlife Management, *Amer. Midland Naturalist* **75**:221–231.

Westin, F. C, and C. J. Frazee, 1976, LANDSAT Data, Its Use in a Soil Survey Program, *Soil Sci. Soc. Amer. J.* **40**:81–89.

Chapter 17

Land Use and Soil Management

Land has too often been considered a limitless resource because new areas were still available for development. Such a viewpoint leads to exploitation, waste, and ruin of land resources. The idea that land "wears out" has been generated under such conditions even though it was more wasted than worn.

Population increases force revision of concepts of boundless land resources. Calculations have been made relating standards of living to population densities. An area of about 1 hectare of agricultural land per person has been suggested as a minimum for maintaining the American standard of living. The idea is valid whether the area is exactly 1 hectare or not. An increasing population living on a fixed land area must use its land with an increasing degree of efficiency or its standard of living will decline. Eventually a limit must be reached beyond which efficiency can no longer increase.

Usually there is a degree of competition for the use of land, especially for highly desirable areas. Part of this competition arises from several persons having similar desires. For example, several different farmers might want to own the same farm. This type of competition is normally resolved by the land going to the highest bidder.

There is also competition among the different types of land use for which a particular piece of land may be suitable. Land that is level and fertile is best for many different uses. The ability of a potential user to pay for such land is not always the sole consideration for determining what should be done with it. The use that is chosen may affect the use and value of other parcels of land either favorably or adversely. Also, society may decide that certain areas either should or should not be used for particular purposes. Such decisions may be written into zoning laws and other regulations that restrict the use of land. Furthermore, certain agencies and organizations are granted the power of eminent domain whereby they can preempt land for their purposes and force owners to sell even if they do not wish to relinquish it.

LAND USE IN THE UNITED STATES

The 50 states encompass about 937 million hectares of land and water. The relative amounts of land utilized for forest, pasture, cropland, miscellaneous, special uses, and construction purposes are shown in Figure 17-1. Any of these broad uses can, of course, be divided into a myriad of more specific uses.

About 60 percent of the land in the United States is privately owned. Virtually 100 percent of it is utilized in one way or another. Much land is used in more than one way. Cropland, for example, may also be used for grazing purposes or for wildlife. The particular use or uses for which an area of land is suitable depend on such diverse factors as the nature of the soil, topography, and climate of the area. The availability of irrigation water becomes a decisive factor in determining land use in arid regions. Considerable influence may also be exerted by the location relative to roads and population centers, the presence or absence of buildings, and the personal preferences of the people managing it. The concept of land encompasses these various facets of the setting of the area. The term *land* therefore has a much broader meaning than *soil*.

People are dependent on land for a large proportion of their needs and wants. Farmers use land to grow crops that provide not only food but also raw materials

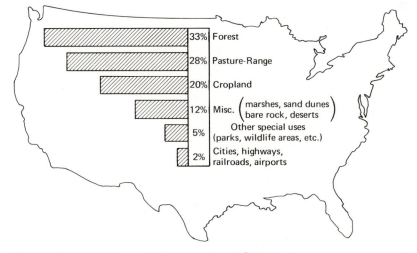

Figure 17-1 Major uses of land in the United States.

for many uses. Livestock graze on grass that comes from the land. Forests also depend on the land, and so do gardens, parks, and anything else that involves growing plants. Homeowners with a lawn, a flower garden, and a vegetable garden are as dependent on their plots of land to support plant growth as farmers are dependent on theirs.

PLANNING LAND USE

Land-use plans are made for areas of various sizes. A small-area plan might call for backyards to be subdivided for use as garden plots, shade trees, barbecue areas, and game areas. Different parts of a lot may lend themselves better to one use than to another. Some games require a level area of certain minimum dimensions. A natural updraft is an asset for an outdoor fireplace. An area sheltered from the wind is advantageous for a picnic table. A swale may be too poorly drained to grow a tree, a strip next to the sunny side of the house may be too warm and dry for a flower garden, etc. Good land-use planning takes all these factors into account.

Homeowners may find that certain parts of even a small plot of ground need special treatment. A steep area needs protection from erosion. Lawns and gardens usually need to be fertilized, but the kind and amount of fertilizer needed varies. The vegetation also helps to determine fertilizer needs. For example, a pin oak tree may require an application of iron that most other plants would not need. The interaction of soil and plants is very important in making plans and putting them into practice.

Regional Land-Use Plans

Larger areas also need to be planned. An airport needs a large, level area; certain types of industry need large supplies of water; most users need access to their land by road and some need railroads; housing areas need either sewers or septic tanks. The functioning of these facilities depends greatly on soil and topography. Regional planners need good maps on which to base decisions; they find topographic maps and soil survey maps invaluable.

A regional land-use plan helps to harmonize community needs and available land resources. Areas are designated for housing development, industrial development, parks, forests, and agricultural land. Roads, railroads, and airports are included in the plan. Various limitations such as flood hazards on bottomlands and the possibility of landslides on steep land are vital considerations in planning. Environmental pollution must be limited to acceptable levels. Unplanned areas are very likely to violate important environmental considerations and not put land to its most appropriate use.

City Planning

Much of what has been said of regional planning applies to city planning as well. The city may first be divided into sectors such as residential areas, shopping areas, recreational areas, industrial areas, etc. Plans may be made for each sector as well as for the city as a whole. The plans will include through streets and residential streets, fuel, water, and sewer systems, etc. All components need to be fitted to the topography and soils of the area (Figure 17-2).

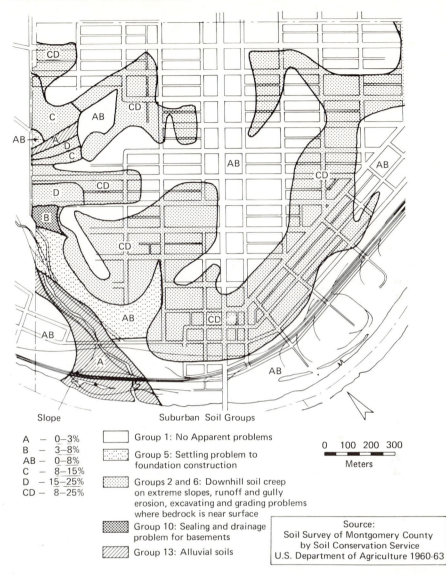

Source:
Soil Survey of Montgomery County
by Soil Conservation Service
U.S. Department of Agriculture 1960-63

Slope | Suburban Soil Groups

A – 0–3%
B – 3–8%
AB – 0–8%
C – 8–15%
D – 15–25%
CD – 8–25%

☐ Group 1: No Apparent problems

Group 5: Settling problem to foundation construction

Groups 2 and 6: Downhill soil creep on extreme slopes, runoff and gully erosion, excavating and grading problems where bedrock is near surface

Group 10: Sealing and drainage problem for basements

Group 13: Alluvial soils

0   100 200 300
Meters

Figure 17-2 Map of Conshohocken Borough, Montgomery County, New York, showing some soil problems important to urban planning. *(From Witwer, copyright 1966 by American Society of Agronomy and Soil Science Society of America.)*

More detailed planning will include design of public buildings and landscaping their surroundings. Flowers, shrubbery, and trees along streets, in malls, and in parks can do much to beautify a city. Many cities employ specialists to plan ways to make their environments both attractive and functional.

Land Use Near the City

Cities influence the use of land that lies beyond their borders. Many people earn their living in the city but live in the suburbs or other nearby areas. Many of them have a tract of land up to a few hectares in size. Such tracts are more than enough for a lawn and garden, but they are not large enough to constitute a farm. This type of land use has grown markedly in the United States in recent decades. The land area involved in small tracts near cities now commonly exceeds the area of the cities.

Some small tracts are large enough to include a field. Sometimes these fields are rented to a nearby farmer, who often uses them to pasture a few horses, cows, or other livestock. These uses are normally sidelines of less concern to the landowners than some of their other activities. Many people enjoy owning some land but do not know much about how to manage its use. Nor can they afford to make large investments in time, money, and equipment to work the land. The unique needs of owners of small tracts have become a matter of concern to several agencies that were once directed only toward farm problems. These include the agricultural extension services of the state universities, the Farmer's Home Administration, and the Soil Conservation Service.

A nearby city may cause agricultural land to be used more intensively than it would otherwise be used. There are two main reasons for this. One is that land taxes are usually high near a city and only an intensive use will give enough profit to pay the taxes. The other is that the city is a market for a number of perishable but profitable crops. Various vegetables, fruits, and specialty crops are more likely to be grown near a city than far away. Truck gardens, nurseries, flower producers, etc., need both the market area and the labor supply of the city. Some of these uses have very particular soil needs related to such properties as texture, pH, and drainage. Some people can make highly profitable use of small areas of especially suitable soil.

Highway Planning

Society invests tremendous sums in transportation systems. Adequate planning for these systems can help to make them serve their purposes well and reduce costs as compared with poorly planned systems. An unnecessary road is a very costly investment—not to have a road can be even more costly to people who need it. Alternate routes vary in construction and maintenance costs. Relocation of a road may save millions of dollars if it thereby avoids a peat bog, a landslide area, or some other unstable condition. Aerial photographs with stereoscopic coverage (Figure 17-3), soil surveys, and other maps help the highway planner immensely. Savings on highway construction costs can easily exceed the cost of a soil survey of an entire county through which the highway passes.

Highway planners need to know the strength and drainage characteristics of the soils where a road is to be built, they need to know of excavation problems where cuts are to be made, and they need to know the type, extent, and location of potential sources of gravel and other road-building materials. This type of information is as vital as the information on traffic volume and loads to the design of roadbeds and the selection of paving materials. Some soil and substratum materials are so unstable that it is expedient to replace them with other materials if the road cannot be relocated to avoid them.

Highway builders must provide for both surface and subsurface drainage. Runoff water from the highway must have some place to go, and all the natural

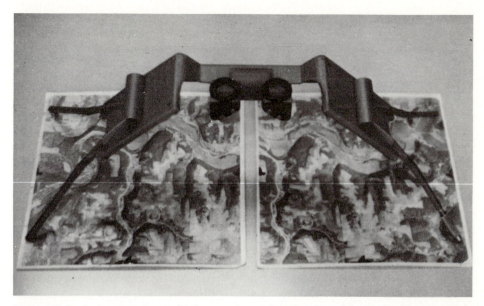

Figure 17-3 A stereoscope permits the user to see the elevation of the land surface by viewing two different photographs of the same area simultaneously.

drainageways that cross the highway must still be able to flow. The subsurface flow of water must not be so restricted as to cause poor drainage in an adjoining field or under the roadbed. The roadbed drainage has a marked effect on the strength of the roadbed and the suitability of different kinds of construction materials. Susceptibility to frost heaving in cold climates is also a much greater hazard where subsurface drainage is poor.

Even such farm-type problems as soil fertility and erosion control are concerns of highway planners and builders. Many roadbanks are steep and highly susceptible to erosion. A good vegetative cover will usually solve the problem if it can be established soon enough. A layer of topsoil is often placed over all areas where soil has been moved except for the roadbed itself. These areas are then seeded, and lime, fertilizer, and mulches are provided where needed to establish the vegetative cover.

Parks

Public parks ranging in size from the smallest city park to Yellowstone National Park need to be planned and managed. Decisions regarding what facilities will be provided and developed are important. The placement of facilities in appropriate locations with a usable system of access roads and paths can be critical. The vehicular and pedestrian traffic that a well-used park receives create special problems in establishing and maintaining vegetative cover. Compaction can prevent the growth of roots and thereby kill plants. Ball games and other activities may "wear out" even a good stand of grass. Fire is always a hazard to dry grass and trees. Denuded areas tend to become either "mudholes" or erosion channels; either of these conditions reduces the recreational value of the park.

Public cooperation is especially important to park management. Facilities break down if they are misused. Litter can spoil the appearance of the landscape and kill vegetation. Traffic must be kept away from newly planted areas until the vegetation is well established. Parks can provide much more public enjoyment if they also receive public cooperation. Careful planning of the facilities helps to meet public needs and make it easy for the public to cooperate. Good planning also keeps trails from crossing wet, seepy spots and avoids mismatch of plant needs to soil pH, soil depth, and other factors that influence plant adaptation.

Waste Disposal

Spoiling land for other use by dumping trash on top of it was once a common practice. Such methods of disposal are unpopular now and are forbidden by law in many places. Land utilized in such a way is wasted and usually becomes an eyesore. Auto wrecking yards are coming under increasing condemnation as a form of sight pollution.

Some areas are devoted to waste-disposal purposes for a limited time only. For example, a landfill trash-disposal area may be completely filled, covered, and utilized for some other purpose. Some landfills have been converted into parks that show no hint of their former use. Others may revert to cropland or pasture. Buildings should not generally be placed on such sites because settling may occur. Suitable uses usually require that some kind of vegetation be established. Therefore, the fertility of the soil placed on top of the fill is important.

Sewage Disposal   Sewage treatment requires a significant land area no matter what form it takes. Usually it begins with a basin where most of the suspended solids are removed by allowing them to settle to the bottom or be skimmed off the top. This process, known as *primary treatment,* requires a large-enough area to permit the sewage to remain in the tank for 2 h or longer. Primary treatment can remove 60 percent or more of the solids and perhaps a third of the biological oxygen demand (BOD) of the sewage (Foster, 1959). About half of all sewage treatment plants in the United States provide only primary treatment before emptying the effluent into a stream.

Secondary sewage treatment is by bacterial action in either a trickling filter (a deep bed of stones about 10 cm in diameter covered with bacteria) or an activated sludge tank (where air is blown through to promote bacterial growth). Either of these methods requires less land area than the sand beds that were used originally. Secondary sewage treatment should remove 90 percent or more of the solid material and of the BOD of the sewage. The effluent is usually returned to a stream after secondary treatment, as shown in Figure 17-4.

Tertiary treatment is necessary if dissolved materials must be removed from effluent. Two principal methods have been developed. The first, a chemical treatment to precipitate the offending solutes, is quite expensive. Increasing attention has therefore been given recently to the second method, that of using the effluent to irrigate a field. Some kind of vegetation should be growing on the field to utilize

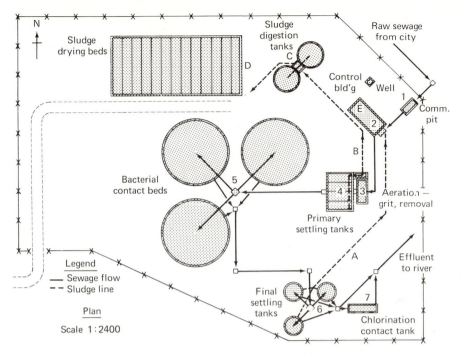

Figure 17-4 Plan of the sewage treatment plant at Ames, Iowa.

plant nutrients from the sewage. The vegetation should preferably be removed as a forage crop to prevent the buildup of high concentrations of plant nutrients in the soil. Tile drainage is usually needed to remove the purified water from beneath the soil.

Individual homes often use a septic tank and a tile drain to replace the sewage treatment plant. The septic tank serves as a settling basin and an area for microbial action. The tile drain distributes the effluent to the soil by subirrigation. The natural drainage and permeability of the soil are obviously important factors in determining whether the tile drain will function effectively. Soils with low permeability and poor drainage seriously detract from many otherwise desirable building sites.

Waste disposal often requires a reversal of principles usually followed in growing plants. Much more water and nutrients than the plants need can be added to the soil, and excess water will be thoroughly filtered before it reaches the groundwater table. Most of the solutes can be absorbed by plants or otherwise eliminated. The idea is for the plants to absorb ions (whether nutrients or not, as long as they are harmless) in large amounts. Luxury consumption is an advantage rather than a disadvantage. Denitrification is an ally that may be promoted by periods of waterlogging. Precipitation of insoluble phosphates may be considered a good thing for waste disposal. Such principles are especially important where large quantities of waste accumulate. Feedlots that handle tens of thousands of cattle have to dispose of mountains of manure. Large swine and poultry operations have similar problems. Most of them do not have nearly enough land to make profitable use of the manure produced by the livestock.

Each Use Is Unique

The uses discussed in the preceding pages are illustrative of the many uses of land. Each use has its own particular needs and adaptations. Level, fertile land is best for many but not all uses. A perfectly level golf course, for example, would be much less interesting and challenging than one built on rolling topography.

The variety of demands that may be placed upon land can be amplified by a few more brief examples. Lakeshore real estate is in high demand for recreational purposes, but there are only a limited number of lakes. Cemeteries will be needed as long as people bury their dead. Most cemeteries are built on hills so that the burial sites will be well drained. Railroad switchyards should have just enough slope for the cars to roll to their designated positions. The list could go on, but these examples should suffice to illustrate that land-use planning is not a simple matter.

AGRICULTURAL USE OF LAND

Tree crops such as orchards and Christmas trees should be included with cropland and pasture as agricultural uses of land. Forest is sometimes considered an agricultural use as well and will be included here. The growth of trees depends on the same basic principles as the growth of other plants.

Forest Land Use

The influence of forest vegetation on soil properties has been indicated in several parts of this book. Forest soils tend to be more acid and often are more strongly developed than grassland soils. The presence of an A2 horizon is usually associated with forest vegetation. Most forest soils have O1 and many O2 horizons on their surfaces. All these factors are related to the influence of the forest vegetation and of the climatic and other factors that are associated with the forest.

The growth of trees is as strongly influenced by the soil as the growth of other plants. Shallow soil, heavy B horizons, or low fertility can limit the growth of trees and even exclude some species that would otherwise grow in the area. Poorly drained soils often produce different trees (or perhaps grass instead of trees) than do well-drained soils. A cypress tree, for example, can grow in a swamp, but an oak or a maple needs better drainage. A wet spot in a pine and fir forest is likely to produce an aspen grove. The trees growing in an area and the soil that occurs beneath them are related in many ways.

The U.S. Forest Service uses a soil survey program to gather soil information for forest management. These surveys are conducted in the same general manner as those made by the Soil Conservation Service and the agricultural colleges. The information is used for predicting growth rates of trees, planning forest land use, laying out routes for forest roads, etc.

Forest planning often relates to when and how the trees should be harvested. Production is maximized by harvesting near the time when the trees have completed their period of faster-than-average growth (Figure 17-5). The forest-site index can be used to estimate when the optimum harvest age will be reached. The numerical value of a site index is the average height of a group of trees when they are at a specified index age. A forester can use a set of standard curves to estimate the site index of a tree stand whose age differs from the index age. The same set of standard curves permits prediction of future growth rates, optimum time of harvest, and amount of lumber or other wood products that will be produced.

Studies of environmental factors and of other plants growing with the trees can also be used as guides for the most appropriate age or size of trees to be harvested (Jones, 1969). Other plant species growing on a site give a good indication of the available moisture supply and therefore of how fast the trees can grow. Information on soil type, steepness and direction of slope, and depth to water table can be used in a similar way.

Harvesting Forests

Forest harvest may be either partial or complete. Clear cutting removes all the trees. Selective cutting leaves the more rapidly growing trees to continue growing and to reseed the area. Selection of the trees to be left to reseed the area may be more important than selection of the trees to be cut.

Provision may be made to replant clear-cut areas. The new trees may or may not be the same species as the old. This situation is an opportunity to plan the forest more completely than otherwise. Species to be planted should be chosen on the basis of climate, soil, topography, and other environmental factors as well as marketability.

Careful consideration of how to control erosion is needed when a forest is harvested. Logging operations to remove trees can cause severe erosion on steep forested land. Building roads, dragging trees across the ground, and leaving openings in the forest cover all increase the erosion potential. However, erosion can usually be controlled by good management. Roads can be built nearly enough on the contour to avoid high water velocities or they can be placed on ridges above the accumulation of runoff water. Harvested trees can be lifted and carried to the loading point rather than dragged across the ground. Carrying helps preserve the litter layer and thus leaves the soil protected even where the forest canopy is broken.

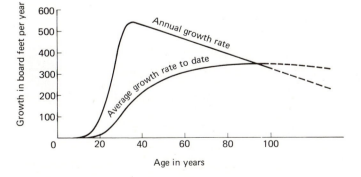

Figure 17-5   Growth rate of upland oak on an excellent site. The suggested age of harvest is where the two curves cross, between 90 and 100 years. (*Plotted from data of Schnur,* 1937.)

Clear cutting as a method of forest harvest has been criticized because it exposes large areas to increased erosion, but loggers like the efficiency of clear cutting. One important lumber species, Douglas fir, will not reproduce under shaded conditions. Douglas fir stands are established naturally where forest fires cleared the land. Carefully managed clear cutting of areas where the erosion hazard is not too severe also provides areas for establishing Douglas fir.

Tree Crops

Many trees are grown outside of forests. Some are intended for shade and ornamental purposes around homes and businesses rather than for crops; windbreaks are another noncrop use for trees. However, there are several different ways in which trees are grown as crops.

Some farmers plant areas to trees that will produce marketable lumber or other wood products. Walnut, oak, and many other species are sometimes planted in areas that are not being farmed. Many years are required to produce a crop, but the value of the land gradually increases as the trees grow. A farmer may receive a profit even if he sells the land before a single tree is harvested.

Many farmers have found it profitable to grow Christmas trees on their land. Figure 17-6 shows Christmas trees growing inside the city of Los Angeles, Calif. The site is near an electric power station and is traversed by too many power lines to be used for building purposes. Christmas trees are a profitable crop adapted to the site.

Large tracts of land are planted to orchards in some parts of the country. California and Florida are noted for oranges and other citrus fruit. The northern United States is too cool for citrus, but states such as Washington grow apples,

Figure 17-6 Christmas trees growing beneath power lines in Los Angeles, Calif.

cherries, peaches, etc. Small orchards are common throughout most of the nation except Alaska.

Orchards involve long-term commitment of the land. No harvest is obtained from the trees for the first few years (another crop is sometimes interplanted between the trees during these years). Orchards generally need good drainage and a fairly high fertility level. Sloping areas are usually preferred to low-lying flat areas because the low areas are more likely to have frost damage as well as poor drainage. Cover crops are often used in orchards to reduce erosion and help protect the soil from compaction under traffic. The fertility needs of the cover crop should be added to those of the trees when a fertilizer program is prepared for an orchard.

Trees are sometimes used in tropical climates as soil-improving crops to rejuvenate cropland. The fertility of such soils is very low and declines to nearly nothing under cropping in the tropics. Most of the soil fertility is associated with the soil organic-matter content, and this decomposes rapidly in the warm, moist climate. Allowing the tropical forest to return for several years rebuilds the supply of organic matter and permits the soil to regain some fertility.

Forage Crops

Pasture and hay crops help promote or maintain good soil structure. Legumes used for hay or pasture can add to the nitrogen supply in the soil. These favorable effects can be anticipated when large quantities of organic materials in the form of animal manure, green manure, or other residues are returned to the soil. Improvements in soil physical condition and in yields of succeeding crops have led to the popular concept that forage crops are soil-improving crops. But there are exceptions. Removing hay from a field without returning manure, for example, probably removes more nutrients from the soil than almost any other crop. The soil is depleted of plant nutrients rather than improved when forages are removed from the area.

Most forage crops are close-growing grasses and legumes that provide good erosion control. This fact is often of more significance than the effect on soil fertility, especially where fertilizer is inexpensive. Soil permeability is often improved by the growth of forage crops because their extensive root systems open new soil pores and favorably influence soil structure. Improved permeability reduces both runoff and erosion. Favorable effects may persist for a year or two after the forage crop is removed and another crop planted.

Forage crops may be grown for either a short time or a long time in a particular place. The shortest time would be for a few months that are off-season for the other crop or crops grown on the land. These crops provide vegetative cover for the land and are therefore called *cover crops*. Often they are plowed under as green-manure crops instead of being utilized as forage crops.

Forage crops are often grown for 1 to 3 years at a time in rotation with other crops. The forage is normally utilized as pasture, hay, or silage. Soil conservationists recommend that the last growth of such forage crops be incorporated into the soil as green manure.

Much land is permanently devoted to the production of forage crops for pasture or range use, often land that is not suitable for plowing because of dryness, wetness, stoniness, or steepness. Grazing management is critical on pastures or rangeland because it may be difficult or impossible to reestablish desirable vegetation that has been lost by poor management. Overgrazing or grazing when the land is too wet can seriously damage both the vegetative cover and the soil structure.

Permanent grazing land in arid regions is known as *rangeland*. The forage plants on rangeland need an opportunity to reseed themselves periodically. A system known as *rotational deferred grazing* is designed to permit reseeding at least 1 year and often 2 years out of every 3. The system requires the use of at least three separate grazing areas. The livestock are placed in one grazing area as soon after spring growth begins as the soil is dry enough to avoid trampling damage. They remain in the first area for 2 to 3 weeks and then are transferred to the second area. The forage in the second area should already be growing vigorously before the livestock arrive there. The livestock are left in the second area for most of the grazing season.

Transfer to the third area comes after the forage plants have matured and set seed. The livestock can then eat the dried forage with little damage to the plants. The first area, too, may have recovered enough to set some seed. The second area was used the hardest of any but it will be left until last to reseed itself the next year (Table 17-1).

Rotational deferred grazing has proven to be of great value on rangeland in the western United States. Many ranchers in arid areas provide summer grazing either in more humid mountainous areas or in irrigated pastures. The dry rangeland is used for spring and fall grazing.

Pastures in more humid regions also give their best production through controlled grazing systems. Some dairy farmers use electric fences to provide their milk cows with a small area of tall forage to graze every day. Enough cows can be placed in the pasture to eat practically all the growth in this small area in one day. The forage makes optimum growth when 20 to 30 such pasture areas are provided so that each one has 3 to 4 weeks to recover before being grazed again. Forages such as alfalfa that can cause grazing animals to bloat are much safer to graze under this type of controlled grazing.

Small Grains

Wheat, barley, oats, and rye are very widely adapted and extensively grown crops. These small-grain crops are relatively fast-growing annuals. They are either drilled in rows (usually about 15 cm apart) or scattered in broadcast fashion to produce a similar stand. The close spacing and rapid growth of small grains provide better erosion control than intertilled row crops but not as good as forage crops.

Table 17-1   A Rotational Deferred Grazing System Plan Utilizing Three Pastures

	Area 1	Area 2	Area 3
First year	Early grazing	Main grazing season	Late grazing
Second year	Main grazing season	Late grazing	Early grazing
Third year	Late grazing	Early grazing	Main grazing season
Fourth year	Same as first year		

Small grains are grown as cash crops in areas where most row crops are not adapted. For example, wheat does well in the Great Plains area of the United States where the summers are too dry for the corn and soybean crops that are grown farther east. The small grains grow during the cool weather of late fall and early spring when many other plants are dormant. They are already maturing by the time the soil becomes dry in the summer. Their range is expanded into still drier zones by means of summer fallow. Nothing is planted during the summer-fallow year, and weeds are killed so that soil water is not lost by transpiration. A significant part of the rainfall for the summer-fallow year is stored in the soil for use by the next year's crop as illustrated in Chapter 18.

Small grains are also grown in rotation with row crops and forage crops. A small-grain crop and a forage crop are often planted simultaneously as companion crops. The grain crop provides cover and erosion control long before the forage crop and is harvested while the forage crop is still small. The growth of the forage crop is somewhat slowed (not "nursed" as is often suggested), but this disadvantage is more than offset by the market value of the grain.

Small grains are quite responsive to nitrogen fertilizer. Additions of nitrogen are necessarily limited in dry climates to the amounts needed for the crop yields that can be produced with the available water supply. Too much nitrogen may cause vegetative growth to predominate and reduce the amount of grain produced. Furthermore, the plants may grow so tall that they will lodge (fall over) and make the crop difficult to harvest. Nitrogen fertilizer is also limited or withheld from companion crops. The forage crop may suffer too much from the competition if the small-grain crop does too well.

Rice is an important small-grain crop in warm climates. Some types of rice require flooding; other types are grown more nearly like other small grains. Level areas of soils that will hold water are needed for the flooded varieties. Careful land leveling and diking are essential for attaining highest yields. Research in California has shown the optimum depth of water on flooded rice to be about 5 cm.

Row Crops

Corn, cotton, potatoes, soybeans, and many other crops are planted in rows and cultivated periodically to control weeds. These are often the highest-value crops grown by farmers and therefore the focal point of their efforts. They are grown on the best land. The fertilizer program is geared to the responsiveness of row crops. Drainage and possibly irrigation are provided where needed.

Most farmers who grow row crops would like to maximize the acreage of these crops. This acreage may be limited by any of several factors including labor, equipment, marketing quotas, and erosion hazard or other soil limitations. The erosion hazard is particularly serious with row crops that have a wide area of bare soil between rows. This soil is loosened every time the crop is cultivated. The damage to severely eroded land can often be traced to its having been used too much for intertilled row crops. Some land is suited to frequent or even continuous row crop, some to row crop in rotation with other crops, and some should not be used at all for row crops.

CROPPING SYSTEMS

There are many different kinds of cropping systems. Probably the most significant distinction among them is whether the same crop is grown on the same land year after year or if a crop rotation (two or more crops in a recurring sequence) is used instead. The relative merits of these two approaches have long been argued. Both are practiced with varying degrees of success.

The single-crop approach is advocated by some persons on the basis of matching soils and crops. For example, row crops may be grown continuously on level land, and the steeper land may be used only for forage crops. Furthermore, each area may be treated specially for the crop it produces. Perhaps the soil fertility is maintained at a high level for the row crop and at a lower level for the pasture. The soil may be limed to a higher pH for one crop than for another, or some particular plant nutrient may be emphasized more for one particular crop. The effects of such treatments carry over from one year to another and can probably best be utilized by growing the same crop repeatedly.

Crop rotations also have strong support. Sloping land may suffer excessive soil loss if row crops or small-grain crops are grown on it for too many years in succession. Yet, the soil loss may remain within tolerable limits if those same crops are grown on the same land in rotation with forage crops. Organic-matter content may decline and soil structure deteriorate under row crops but improve under forage crops in the rotation. Mazurak and Ramig (1962) reported that increasing the number of years of grass (forage) crops in the rotation increased aggregate stability and permeability to air and water. Carreker et al. (1968) reached similar conclusions and also found that crust strength was reduced by growing grass.

Some reasons for using crop rotations have decreased in importance in recent years. The increased use of fertilizer may improve the growth of a row crop enough to offer more soil protection and to maintain a higher level of organic matter than was previously possible with continuous row crops. Also, some crops that were formerly grown in 100-cm rows are now being grown with 50- or 75-cm row spacing. The narrower rows combined with more vigorous growth reduce the rate of erosion associated with the row crop. Furthermore, some of the weeds and insect pests that once became serious problems if one crop were grown continuously can now be chemically controlled.

Rotations are still an important part of agriculture, however, even though their importance has declined. Forage crops are still needed to maintain many soils in satisfactory physical condition and to limit erosion to acceptable levels. Some pests are best controlled by alternating types of crops on the land.

Crop rotations require more planning than does planting the same crop year after year. It is usually desirable to produce consistent amounts of each crop each year even though the crops shift to different fields. A simple but essential principle makes this possible. There must be the same number of fields or groups of fields as there are years in the rotation. Equal numbers of fields and years make possible a block plan such as that shown in Table 17-2. This plan will give constant production

Table 17–2   A Plan Showing How a 5-year Rotation Is Applied to Five Fields (or Five Groups of Fields): the Sequence Repeats Beginning with the Sixth Year

Fields	Years*				
	1	2	3	4	5
1	R	R	G	F	F
2	R	G	F	F	R
3	G	F	F	R	R
4	F	F	R	R	G
5	F	R	R	G	F

*R stands for row crop; G for small grain; and F for forage crop.

of each crop each year if the fields are all equal in production capacity. Separate plans and sets of fields should be made if more than one rotation is to be used.

Evaluating a Cropping System

Several criteria are used to evaluate cropping systems. A good system must maintain an adequate level of production and net income, or the farmer is likely to discontinue it. Crop yield is therefore an important measure of success. But it is possible for yield to increase because of improved varieties, fertilizer, and management techniques even though the natural yield potential of the soil is declining. The system might appear satisfactory by a yield criterion in spite of serious soil deterioration.

Percent Organic Matter   Maintenance of a specified organic-matter content is sometimes used as a criterion for a good cropping system. Organic-matter content is related to soil fertility, soil structure, and permeability. More organic matter usually means a more productive soil. This relation has cause and effect working both ways. Organic matter is a storehouse for plant nutrients and therefore helps promote plant growth. Vigorous plant growth produces more plant residues to replenish the organic-matter supply.

There is, however, no magic number that can be stated as the percentage of organic matter that soils need to contain. Plants can be grown in materials that initially contain no organic matter. The natural organic-matter contents of mineral soil A1 horizons are mostly between 1 and 6 percent, but wet soils and soils of cold climates may range up toward 100 percent organic matter. An organic-matter content of 3 percent might represent a serious decline if the soil were a Mollisol that originally contained 6 percent, but an organic-matter content of 3 percent in an irrigated Aridisol would probably represent an increase from about 1 percent in the virgin soil.

A satisfactory organic-matter content must be related to other factors. It should be stabilized at a level that maintains good soil structure and productivity. Exactly what this level may be depends on soil texture, type of clay, and other factors.

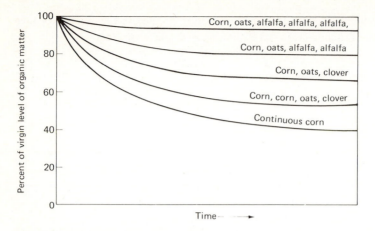

Figure 17-7 A schematic diagram showing the effect of different cropping systems on soil organic matter, assuming use of enough mineral fertilizers for maintenance of mineral nutrients.

The effect of several different cropping systems on soil organic matter in a corn belt soil is illustrated schematically in Figure 17-7. The critical part of each of the curves in Figure 17-7 is the stability level on the right-hand side of the figure. The cropping system evaluation must be based on this level whether or not it has been reached yet. Any other basis invites serious damage to the soil resource.

The level of production of the crops grown on a field influence its organic-matter content almost as much as what the crops are. The Jordan plots in Pennsylvania (the second oldest plots in America) will serve to illustrate this principle. The land was farmed from 1838 to 1868 and used for miscellaneous plot work from 1869 to 1880. The plots were changed to a 4-year rotation of corn-oats-wheat-mixed hay in 1881. Fertilizer and manure treatments were applied whenever corn or wheat

Table 17–3 Average Crop Yields and Soil-organic-matter Contents of the Jordan Plots in Pennsylvania under a Corn-Oats-Wheat-Mixed Hay Rotation

Fertilizer treatment on corn and wheat since 1881	Average yield of all crops 1882–1930, kg/ha year	Organic-matter contents after 72 years of plot work (in 1940)	
		kg/ha	% of grass borders
Untreated	1,406	56,100	61
P–K*	2,351	65,800	72
N–P–K†	2,631	65,700	71
13 tons manure/ha + lime	2,594	77,400	84
22 tons manure/ha	2,605	81,800	89

* P–K = 23 kg P and 93 kg K per hectare per rotation.
† N–P–K = 80 to 112 kg N per hectare per rotation plus P–K.
Source: Based on data of White, Holden, and Richer, 1945.

crops were grown on designated plots. The resulting average yields and soil organic-matter contents are shown in Table 17-3.

Yields on the unfertilized plots gradually declined, but yields on the fertilized plots remained nearly constant throughout the period covered by the data presented by White, Holden, and Richer (1945). The yields after 1900 from the N—P—K and the manure treatments were over twice as high as those from the untreated plots. The N—P—K treatment provided about the same amount of nitrogen as 22 tons of manure and produced a comparable yield but did not raise the organic-matter content as much as manure did.

None of the plots contained as much organic matter when tested in 1940 as did the grass borders between plots. The grass had returned large amounts of organic matter to the soil in the borders. Decomposition of organic matter was relatively slow there because the borders were not cultivated. The soil in these borders should have had about the same organic-matter content as it had before it was farmed. The soil in the untreated plots contained 39 percent less organic matter than the soil in the grass borders. The fertilized soil contained 28 or 29 percent less and the manured soils contained 11 and 16 percent less organic matter than the grass borders. Crop harvest removed a large part of the organic matter that would return to the soil in nature. Manure treatments made up for most of this loss. The N—P—K fertilizers helped promote increased crop growth, and this provided more crop residues than the untreated plots had but did not match the organic matter added to the manured plots.

Aggregate Stability   Soil fertility deficiencies can generally be alleviated by the use of fertilizer. Physical limitations such as soil depth, permeability to water, air, and roots, and tendency to form a crust are not so easily circumvented. Aggregate stability is sometimes used as an indicator of the physical condition of the soil. Aggregate stability is related to organic-matter content, but it also depends partly on the type of organic matter present and on soil chemistry.

Mazurak and Ramig (1962) found that the average size of water-stable aggregates (those that will withstand shaking in water) increased curvilinearly and approached a maximum as grass stands became older. Similar results were reported by Carreker, Bertrand, Elkins, and Adams (1968). These workers found that infiltration rates and air permeabilities improved along with aggregate size and that crust strength declined as aggregate stability increased. Benoit, Willits, and Hanna (1962) found that a rye winter cover crop improved aggregate stability in soils used to produce row crops.

Aggregate stability is more sensitive to year-by-year changes in soil treatment than is organic-matter content. Aggregate stability is a good means of comparing the effects of various treatments on a specified type of soil. Comparisons with other soils of similar texture are probably valid, but there are inherent relations between texture and aggregate stability that must not be overlooked. Sandy soils may have very few stable aggregates because their clay contents are low. Fortunately, the sandy soils usually have enough large pores to permit air and water to move readily even without aggregates. The significance of aggregate stability increases as the clay and silt contents of the soil increase.

Management Practices

Large differences in crop growth and in soil condition can be caused by varying management practices. A good farmer learns to perform tillage, seeding, fertilizing, pest control, and harvest operations at or near optimum times and in appropriate ways. Frequently the farmer has to weigh the advantages and disadvantages of alternative methods.

Tillage   Conventional tillage for grain crops and row crops has long involved plowing, disking, and harrowing to prepare ground for seeding. Plowing turns the soil over and covers crop residues, usually producing a rough, cloddy surface. Disking breaks the clods into smaller pieces, and harrowing smooths and compacts the surface to form a seedbed. Reduced soil-organic-matter contents and increased compaction are predictable results of all this tillage. Ways have therefore been sought to grow crops with less tillage or even with no tillage. Farmers often find that their fields do not need to be made as smooth as they once made them. Sometimes two or more operations can be combined by using special implements or attaching more than one implement to the tractor at the same time. Fewer trips across the field will save the farmer money, destroy less organic matter, and cause less soil compaction than conventional tillage.

Wheel-track planting was one of the early methods of reducing tillage for row crops. Most of the disking and harrowing that would ordinarily be used to prepare a seedbed is eliminated. Instead, the seed is planted in the narrow strip that is crushed and compacted by the tractor wheel as the field is plowed. The area between rows is left rough and cloddy. The rough surface is favorable for water infiltration and erosion control.

Till planters (Figure 17-8) are a more recent design for performing the entire tillage and planting operation at one time. Combining these operations into one trip is favorable for erosion control because it leaves plant residues on the land until planting time. Tractor time and traffic are reduced to a low value, yet the seeds are planted in a tilled seedbed.

The ultimate in tillage reduction eliminates all tillage except planting. Jones et al. (1968) planted corn directly into chemically killed sod or crop residues and reported yields ranging from 100 to 139 percent of yields obtained by conventional methods. Other workers have reported variable results with no-tillage and minimum-tillage practices including yield decreases as well as increases. Reducing tillage usually reduces costs, however, so profits may increase even with yield decreases.

Reduced tillage can reduce soil loss to a small fraction of the amount suffered under conventional tillage (Chapter 19). Water is also conserved because surface cover reduces both evaporation and runoff. Blevins et al. (1971) attributed increased corn yields in Kentucky averaging about 10 percent to water conservation providing reserve moisture to last through short droughts. Griffith et al. (1973) reported that residues on the soil surface reduced soil temperature as much as 3.8°C.

Figure 17-8 A till planter performing tillage, applying fertilizer and pesticides, and planting corn, all in one trip across the field near Ida Grove, Iowa. *(Soil Conservation Service, USDA.)*

Tillage reduction methods have disadvantages as well as advantages. They usually work satisfactorily on well-drained soils but are poorly suited for use on wet, heavy soils. The slower evaporation makes a wet soil stay wet longer. Seed germination and plant growth are retarded by low soil temperatures in the spring. Weed control is totally dependent on chemicals and often becomes less effective after a few years without tillage. Multipurpose equipment such as that shown in Figure 17-8 reduces the flexibility of farming operations and increases the amount of work to be done at planting time.

There has been some concern about how to apply fertilizer and lime with reduced tillage. Of course, it is possible to use occasional tillage to incorporate lime into the soil. The same could be done for phosphorus and potassium fertilizers but may not be necessary. Several workers have reported that surface applications of P and K are satisfactory. Reduced erosion keeps the fertilizer from being lost. Roots are able to grow near the surface and obtain the fertilizer when there is no cultivation.

Mulching Crop residues can be either an asset or a problem. Burning grain stubble was once common practice because it was difficult to plow under. Usually the next year's yield was higher on burned fields than on unburned ones. The problem was primarily an inadequate nitrogen supply to decompose the residues and produce a large crop at the same time. This problem can be alleviated by using adequate nitrogen fertilizer to meet all needs. The same problem occurs with the residues of many nonlegume crops and the same remedy usually applies. Burning wastes both organic matter and nutrients. It should be avoided because of the waste and because it contributes to air pollution.

Some farmers plow under their crop residues, others use a chisel plow or a disc to leave all or part of the residues on the soil surface. Residues left or placed on the surface are known as *mulches;* they help keep the soil permeable by reducing the tendency to crust. They also help to control both wind and water erosion. Mulches shade and insulate the soil and thus lower the soil temperature. Temperature reduction is a good thing on a hot summer day, but it is a detriment when it slows plant growth in the spring or delays field work because the soil stays wet too long.

Crop residues should eventually return to the soil and add to its organic-matter content. Timing this return can be important. It should be delayed until the residues are no longer needed as mulch. Furthermore, the possibility of competition between microbes decomposing the residues and the needs of a new crop must be considered. Residue decomposition can take from a few weeks to several months. The decomposition rate is fastest when the residues are mixed into the upper part of a warm, moist, well-aerated, fertile soil. Residues that are plowed under in a layer will persist much longer than residues that are disked into the surface of the same soil.

Mulches are often used in gardens and on newly seeded lawns or roadbanks as well as on fields. Straw, chopped cornstalks, wood shavings, sawdust, and other plant residues are among the materials used. Nitrogen fertilizer should be added if the material has a wide carbon-to-nitrogen ratio (as explained in Chapter 5).

Pest Control   Crops cannot produce maximum yields if they are ravaged by weeds, insects, or disease. Various methods of pest control are used with varying results. Cultivation kills weeds in row crops. Crop rotations help to keep some disease problems from building up but are ineffective against others. A few types of insects have been controlled by biological methods using diseases, natural predators, or sterile males of the insect species. Crop-breeding techniques have produced varieties that are resistant to various pests. Unfortunately, new pests seem to develop about as fast as old ones are brought under control.

Many pests are controlled by means of chemicals—weeds are treated with herbicides and insects with insecticides. Some of these chemicals have become controversial because of pollution problems. The rate of decomposition of the chemicals and the end products are important factors. Some pesticides will decompose into harmless materials within a few weeks, whereas others require many years.

Varieties, Rates, and Dates   Many factors are involved in crop production and must be matched together in some workable combination. Some factors, such as soil type and climate, must be regarded as fixed. Crop varieties, seeding rates, timing of operations, and other variables must be matched to the fixed factors. For example, farmers in a corn-belt state choose corn varieties to match the growing season in their part of the state. A few extra growing days permit them to choose varieties with higher yield potentials than farmers in a cooler climate can use. They must, of course, adjust their planting and harvesting dates to make use of these extra days of spring and fall growth.

Wheat producers face a different kind of climatic limitation. Much wheat is grown in areas with dry summers. Stored soil moisture can extend the growing season for perhaps 2 weeks beyond the rainy season. The crop must mature by the time the available soil moisture is exhausted. The next crop cannot be planted until such time as its moisture needs will be met.

Optimum planting rates depend not only on the crop but also on moisture and fertility factors. Too many plants compete excessively with each other and do not produce well; plants that are seeded too sparsely may grow well, but the total yield is limited by the number of plants. The optimum seeding rate provides for an optimum plant density with due allowance for percent germination and seedling mortality rates.

Timing of operations involves both yearly and seasonal considerations. Should this crop be planted on this field this year or next year? Should plowing be done in the fall or in the spring for a spring-planted crop? The fall plowing may produce a mellower seedbed and avoid spring wetness, but fall plowing may result in much more erosion than spring plowing. Timing of cultivation may be critical for weed control. Timing of fertilization may have much to do with crop response and possible loss of nutrients by leaching or erosion. Nutrient loss is not only a cost to the farmer but also contributes to eutrophication of streams. Loss of fertilizer nutrients is largely avoidable by proper application, timing, and erosion control techniques.

REFERENCES

Allemeier, K. A., 1973, Application of Pedological Soil Surveys to Highway Engineering in Michigan, *Geoderma* **10**:87–98.

Amemiya, Minoru, 1970, Tillage Alternatives for Iowa, *Iowa State Univ. Coop. Ext. Ser.* Pm-488, 8 p.

Bauer, K. W., 1973, The Use of Soils Data in Regional Planning, *Geoderma* **10**:1–26.

Barnett, A. P., E. G. Diseker, and E. C. Richardson, 1967, Evaluation of Mulching Methods for Erosion Control on Newly Prepared and Seeded Highway Backslopes, *Agron. J.* **59**:83–85.

Benoit, R. E., N. A. Willits, and W. J. Hanna, 1962, Effect of Rye Winter Cover Crop on Soil Structure, *Agron. J.* **54**:419–420.

Blevins, R. C., Doyce Cook, S. H. Phillips, and R. E. Phillips, 1971, Influence of No-tillage on Soil Moisture, *Agron. J.* **63**:593–596.

Carreker, J. R., A. R. Bertrand, C. B. Elkins, Jr., and W. E. Adams, 1968, Effect of Cropping Systems on Soil Physical Properties and Irrigation Requirements, *Agron. J.* **60**: 299–302.

Davidson, J. M., Fenton Gray, and D. I. Pinson, 1967, Changes in Organic Matter and Bulk Density with Depth under Two Cropping Systems, *Agron. J.* **59**:375–378.

Doyle, R. H., 1966, Soil Surveys and the Regional Land Use Plan, *Soil Surveys and Land Use Planning, Soil Sci. Soc. Amer.* and *Amer. Soc. Agron.*, Madison, Wis., pp. 8–14.

Eckbo, Garrett, 1950, *Landscape for Living,* McGraw-Hill, New York.

Fink, R. J., and Dean Wesley, 1974, Corn Yield as Affected by Fertilization and Tillage System, *Agron. J.* **66**:70–71.

Foster, W. S., 1959, The Disposal of Wastes, *The Book of Popular Science* **5**:326–336.

Green, D. E., E. L. Pinnell, L. E. Cavanah, and L. R. Williams, 1965, Effect of Planting Date and Maturity Date on Soybean Seed Quality, *Agron. J.* **57**:165–168.

Griffith, D. R., J. V. Mannering, H. M. Galloway, S. D. Parsons, and C. B. Richey, 1973, Effect of Eight Tillage-planting Systems on Soil Temperature, Percent Stand, Plant Growth, and Yield of Corn on Five Indiana Soils, *Agron. J.* **65**:321–325.

Hartwig, N.L., 1974, Crownvetch Makes a Good Sod for No-till Corn, *Crops Soils* **27** (2):16–17.

Huddleston, J. H., and G. W. Olson, 1967, Soil Survey Interpretation for Subsurface Sewage Disposal, *Soil Sci.* **104**:401–409.

Jones, J. N., J. E. Moody, G. M. Shear, W. W. Moschler, and J. H. Lillard, 1968, No-tillage System for Corn (*Zea mays* L.), *Agron. J.* **60**:17–20.

Jones, J. R., 1969, Review and Comparison of Site Evaluation Methods, *USDA Forest Serv. Res. Paper* RM-51, 27 p.

Larson, W. E., 1972, Tillage during the Past 25 Years, *Crops Soils* **25**(3):5–6.

Mazurak, A. P., and R. E. Ramig, 1962, Aggregation and Air-Water Permeabilities in a Chernozem Soil Cropped to Perennial Grasses and Fallow-grain, *Soil Sci.* **94**:151–157.

McCormack, D. E., 1974, Soil Potentials: A Positive Approach to Urban Planning, *J. Soil Water Cons.* **29**:258–262.

Miller, D. E., and W. D. Kemper, 1962, Water Stability of Aggregates of Two Soils as Influenced by Incorporation of Alfalfa, *Agron. J.* **54**:494–496.

Northey, R. D., 1973, Insurance Claims from Earthquake Damage in Relation to Soil Pattern, *Geoderma* **10**:151–159.

Pendleton, J. W., and D. B. Egli, 1969, Potential Yield of Corn as Affected by Planting Date, *Agron. J.* **61**:70–71.

Phillips, S. H., and H. M. Young, Jr., 1973, *No-tillage Farming,* Reiman Assoc., Milwaukee, 224 p.

Schaller, F. M., D. H. Sims, and R. L. Hine, 1968, *Making Rural and Urban Land Use Decisions,* Soil Conservation Society of America, Ankeny, Iowa, 40 p.

Schnur, G. L., 1937, Yield, Stand, and Volume Tables for Even-aged Upland Oak Forests, *USDA Tech. Bull.* 560, 88 p.

Shear, G. M., and W. W. Moschler, 1969, Continuous Corn by the No-tillage and Conventional Tillage Methods: A Six-year Comparison, *Agron. J.* **61**:524–526.

Smika, D. E., A. L. Black, and B. W. Greb, 1969, Soil Nitrate, Soil Water, and Grain Yields in a Wheat-fallow Rotation in the Great Plains as Influenced by Straw Mulch, *Agron. J.* **61**:785–787.

Van Doren, D. M., Jr., G. B. Triplett, Jr., and J. E. Henry, 1975, No-till is Profitable on Many Soil Types, *Crops Soils* **27**(9):7–9.

Westerveld, G. J. W., and J. A. van den Hurk, 1973, Application of Soil and Interpretive Maps to Nonagricultural Land Use in the Netherlands, *Geoderma* **10**:47–65.

White, J. W., F. J. Holden, and A. C. Richer, 1945, Maintenance Level of Nitrogen and Organic Matter in Grassland and Cultivated Soils over Periods of 54 and 72 Years, *J. Amer. Soc. Agron.* **37**:21–31.

Witwer, D. B., 1966, Soils and Their Role in Planning a Suburban County, *Soil Surveys and Land Use Planning,* Soil Sci. Soc. Amer. and Amer. Soc. Agron., Madison, Wis., pp. 15–30.

Chapter 18

Water Management

Plant growth is restricted if the soil contains either too little or too much water for an extended period of time. Too little water causes plants to wilt and eventually to die. Too much water causes a lack of oxygen for root respiration. Some fluctuation in soil water content is desirable, but extremes can be disastrous.

The soil water must be replenished periodically. Unfortunately, the precipitation needed to replenish water supplies is controlled by factors other than how much water is needed. People must either adjust the plants they grow and the way they grow them to the water supply that nature provides or find ways to manipulate that water supply to better fulfill their needs. Possible manipulations include (1) conditioning the soil to absorb more (or, rarely, less) of the natural precipitation, (2) mulching to reduce water loss by evaporation, (3) irrigation to provide water when precipitation is inadequate, and (4) drainage to remove excess water.

WEATHER RELATIONS

Most crops are grown with natural precipitation and are therefore dependent on the weather for their water supply. People can do little to change the weather, but they can respond to changes in weather and plan programs in anticipation of certain weather patterns. They can select crops that can be expected to grow reasonably well and to compete economically with other crops in the vicinity.

Improvements in the science of long-range weather forecasting will produce benefits in many fields. For example, the water level in reservoirs can be allowed to drop for flood control purposes when a wet season is coming; the water level in these same reservoirs can be held as high as possible when a drought is anticipated.

Agriculture could benefit greatly from accurate long-range weather forecasts. Some of the adjustments that farmers can make include selection of crop varieties to be grown, time and rate of planting, and the amount of fertilizer to be applied.

Rainfall Distribution

Rainfall comes at different times in different places. Each kind of plant has its own particular moisture needs. For example, wheat and other small grains mature early in the summer and can therefore be grown in areas that have winter-spring precipitation but dry summers. Corn, soybeans, and other warm-season crops require summer precipitation or irrigation. Cotton, sugarcane, and many other crops require warm temperatures as well as adequate amounts of water.

Water needs of plants vary from one month to another. Seedlings do not transpire as much water per day as do larger plants. Furthermore, transpiration increases with warmer temperatures. A crop that grows vigorously through the summer months, therefore, has a much higher water requirement in summer than it has in either spring or fall. Corn is a good example of such a crop. The corn belt in the United States receives an average of about 900 to 1,000 mm of precipitation per year with reasonably equal amounts to be anticipated each month. Even so, the corn belt usually has a surplus of moisture early and late in the season and a deficit during the summer, as shown in Figure 18-1.

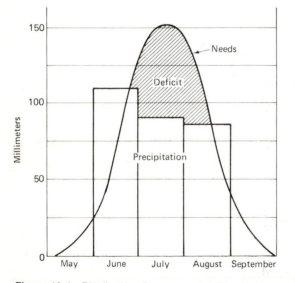

Figure 18-1   Distribution of summer rainfall in the United States corn belt in relation to the needs of corn.

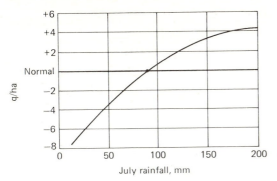

Figure 18-2 Response of corn to July rainfall in the United States corn belt.

The water requirement in July shown in Figure 18-1, allowing for some runoff and evaporation, is about twice the available rainfall. The deficiency in July and August rainfall is normally balanced by soil moisture stored in the preseason period if the soil has a large-enough water-holding capacity. Rain at a critical time during the July-August period can mean the difference between failure and a successful crop.

Figure 18-2 shows how corn responds to July rainfall. Normal July rainfall averages 90 mm in the United States corn belt. There is a marked decrease in yield as the rainfall decreases below normal. The response is curvilinear and results in maximum yields at about twice normal July rainfall. The response to water is similar to the response to applications of fertilizer.

Temperature as a Weather Factor

There is a strong interaction between soil water and air temperature which is often overlooked. Below-normal rainfall may result in normal yield if the air temperature remains below normal, because the water requirement of the crop declines as the temperature drops. Yields are often limited by precipitation and soil water-holding capacity rather than by fertility when the summer is unusually warm.

The response of corn to June, July, and August temperatures in the corn belt is shown in Figure 18-3. The optimum daily average temperature for corn is about

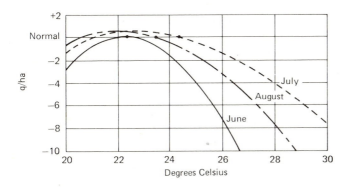

Figure 18-3 Response of corn to June, July, and August temperatures in the United States corn belt.

22°C. The normal July and August temperatures are higher than this. The optimum daily range is from 16°C at night to 30°C in the daytime. Temperatures below about 10°C stop the growth of corn (though not of all plants). Temperatures above 32°C adversely affect yields because of the higher water requirement at high temperatures. High temperatures are especially damaging at the time of seed production; they increase the rate of respiration relative to photosynthesis. The yield is lower because less of the products of photosynthesis are stored in the seed.

Figure 18-4 shows the fluctuations of July-August temperatures in the corn belt from 1900 to 1975. The graph shows evidence of cyclical weather patterns in the central United States; weather cycles have been confirmed in earlier periods by tree ring studies (Weakly, 1943, 1962). Cyclical weather patterns have significance to meteorologists (Willett, 1964), climatologists (Newman, 1970), agriculturalists (Thompson, 1973), and others.

There is a popular belief that summer temperatures are becoming cooler because air pollutants are screening out the sun's energy. However, the period from 1902 to 1911 was as cool as the decade of the 1960s. There was a warming trend in the corn belt from 1902 to 1936 and a cooling trend from 1936 to the late 1960s. The July-August temperatures of the 1930s were all above normal. The July-August temperatures in the 1960s were all below normal. Neither of these decades proves that the normal temperatures have changed; an alternation of warm and cool periods is normal.

Periods of warmer-than-normal weather are usually associated with below-normal rainfall. High temperatures are a manifestation of dryness. Much of the heat from the sun is dissipated by evaporation when surface soil moisture is high, but more of the sun's energy is left to heat the atmosphere when the soil is drier.

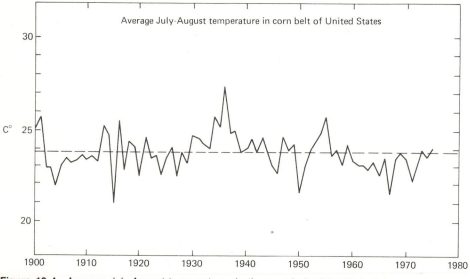

Figure 18-4 Average July-August temperatures in the corn belt of the United States.

Effect of Weather on Technology

The cooler-than-normal weather and the above-normal precipitation of the 1960s were favorable for the production of corn in the corn belt. With favorable weather, fertility tended to limit yields more than water. Consequently, there was a rapid increase in the use of fertilizer, especially nitrogen (Figure 18-5) during this decade. The average corn population increased from 34,000 stalks per hectare in 1960 to 44,000 in 1970. Corn yields increased rapidly under the combined effects of favorable weather and technology (Figure 18-6). A decade of unfavorable weather could make it much more difficult to obtain such high yields. Both plant populations and fertilization rates would have to be reduced in accord with the amount of water available for producing a crop. Farmers failing to make such adjustments would suffer from high costs and low yields.

Drier areas can be affected even more dramatically than the corn belt by weather fluctuations. Wheat production in the Great Plains region of the United States, for example, expanded into drier regions when there was above-normal precipitation. Some of these areas became "dust bowls" during the warmer- and drier-than-normal years of the 1930s and 1950s. The crops did not grow because the soil was too dry. The unprotected soil then blew away, clouding the atmosphere for hundreds of miles.

WATER CONSERVATION

Conservation practices that make the best use of available water are becoming increasingly important. Erosion control is still important, but the emphasis is changing to include water conservation as a necessity. This change in emphasis has

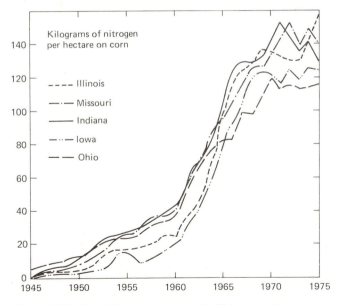

Figure 18-5   Trend in use of nitrogen fertilizer on corn.

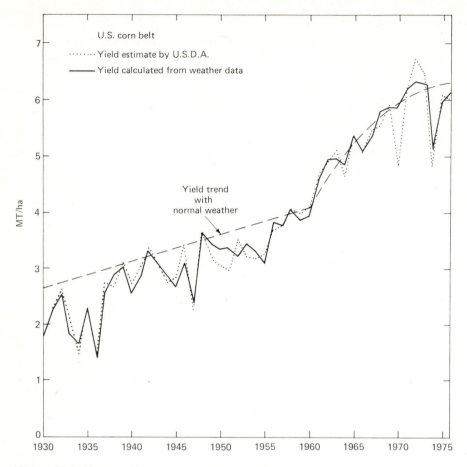

Figure 18-6 Corn yield in the United States corn belt showing trend with normal weather.

resulted largely from improved soil fertility practices. Understandably, the value of a centimeter of water conserved increases as the soil fertility is raised.

Only about half of the precipitation water reaching the soil in humid regions is utilized by crops; the remaining water is lost by evaporation from the soil surface, transpiration by weeds, runoff, and deep percolation. Some of these losses are unavoidable, but some can be controlled.

At conservative average figures, it is estimated that 5 cm of water saved is enough to produce 5 q of wheat, 15 q of corn, or 50 kg of lint cotton per hectare. At times of water shortage the increases would be considerably larger than these estimates, but at times when the soil is filled with water, it would do no good to conserve more water.

It is estimated that good soil and water management could conserve an average of 5 cm of water that is now being wasted. Water conservation therefore has the potential to raise the yield of corn in the United States as much as did the introduction of hybrid seed.

Runoff

Runoff accounts for most of the water lost that could be saved. Runoff occurs because much rainfall comes faster than it can enter the soil. Soils vary greatly in the rates at which they can absorb water. Infiltration rates range from less than 0.2 cm/h to more than 5 cm/h. Rainfall rates may sometimes exceed the infiltration rates of even highly permeable soils and cause runoff.

Rain falling at the rate of 2 cm/h would produce no runoff if the soils on which it fell had infiltration rates of at least 2 cm/h. But if such a rain fell for an hour on a soil with an infiltration rate of 1 cm/h, there would be between 0.5 and 1 cm of runoff (part of the 1-cm excess would be caught by vegetation and by irregularities in the soil surface).

Runoff can be reduced if the infiltration rate of the soil can be increased or if the water can be held on the soil for a longer period of time. Either of these conditions permits more water to enter the soil and less to run off. A good granular soil structure and a rough soil surface promote infiltration and reduce runoff.

The effects of tillage on runoff vary greatly with the type of tillage, the type of soil, topography, and time. Soil covered by a thick stand of vegetation usually has a minimum of runoff. Crop residues continue to give considerable protection from runoff if they are left on the soil surface. Plowing the soil and burying the residues usually produces a rough surface that helps to prevent runoff. But raindrop impact on the bare soil will produce a crust and reduce the infiltration rate after a time (the time required may vary from a few minutes to a few days depending on the stability of the soil structure). Further cultivation may produce temporary increases in infiltration but soon leads to reduced infiltration and increased runoff. Maximum runoff from sloping land will occur where the residues are plowed under and the surface is smoothed and then left without further cultivation.

Mulching Protects the Soil   The impact of raindrops on bare soil tends to break soil aggregates apart and produce a crust of puddled soil. Such crusts have low permeabilities that reduce infiltration and increase runoff. Crust formation is best avoided by maintaining some kind of cover on the soil at all times. The best cover is a thick stand of vegetation that intercepts the raindrops. A mulch of crop residues or other material placed on the soil surface can be used to afford similar protection during times when vegetative cover is lacking. Borst et al. (1945) found that runoff was reduced by one-third in Ohio by plowing a meadow crop only 8 to 10 cm deep instead of the more conventional 15 to 18 cm. The shallow plowing left the sod only partially covered, and it still helped to protect the soil. Other crop residues can also be manipulated to give more than their usual erosion protection. Mannering and Meyer (1961) found that shredded cornstalks left on the soil surface gave enough added protection to reduce soil loss to less than half that where cornstalks were left standing after harvest.

Much research has been done in the semiarid Great Plains to develop a machine that will leave residues on the soil surface as it loosens the soil and kills weeds. Such mulching tools are very helpful for conserving water, but fertility problems have been associated with their use. The mulch system requires a high level of fertility, particularly with respect to the nitrogen status of the soil. Plant nutrients can be tied up until the residues are returned to the soil and decomposed.

The moisture conserved by mulching is obviously important in dry climates. It is also important when dry periods occur in humid climates. Straw mulch applied after the second cultivation in eight experiments in North Carolina increased corn yields under dry conditions an average of 13 q/ha according to Krantz (1949). This mulch was applied at the rate of 7 tons per hectare; other studies have shown that applications of 2 to 4 tons of straw mulch are also effective in conserving water. The importance of the water conserved depends greatly on whether a dry period occurs. Under moist conditions the yields on the above plots were 63 and 67 q/ha, respectively, for the unmulched and mulched conditions.

Crop Rotations Influence Runoff   Grass-legume crops included in crop rotations maintain the soil in a better condition for absorbing water than the continuous culture of a clean-tilled crop. Table 18-1 shows the results of 9 years of study of runoff from plots at Zanesville, Ohio. The continuous corn lost 165 mm more water per year than did the rotation corn. The meadow crops in the rotation had very little runoff.

Grass-legume crops often exhaust the subsoil moisture in the fall to a greater depth than corn does. A dry winter and spring could, therefore, lead to the meadow soil starting the next growing season with less water than a soil that had grown corn the previous year. Usually, however, this difference is more than offset by water infiltration during the fall, winter, and spring months. The grass-legume sod increases infiltration and reduces runoff during the winter as well as the summer months. Studies in Iowa showed that soils in grass-legume cover the previous year had accumulated more moisture to a depth of 1.5 m during the fall, winter, and spring months than adjacent soil that had been in corn the previous year (there was a good supply of winter-spring precipitation during these studies).

Contour Furrows Reduce Runoff   Rough surfaces produced by tillage implements reduce runoff by providing small depressions that can store water until it infiltrates. The volume of storage is higher when furrows are made on the contour than when they are up- and downslope. Contour listing is an especially effective form of contour tillage for storing water. Listers are designed like V-shaped plows that

Table 18-1   Runoff Losses from Continuous Corn and 4-year Rotation Plots, Zanesville, Ohio, 1934-1942

Total rainfall, mm	Annual runoff, mm				
	Continuous corn	Rotation corn	Rotation wheat	Rotation meadow	
				1st year	2nd year
988	399	234	246	15	6

Source: Borst, McCall, and Bell, 1945.

make deep furrows by throwing soil to both sides. A comparison of runoff losses and soil erosion under different crops and tillage methods in western Iowa is shown in Table 18-2. Alfalfa-bromegrass meadow conserved both soil and water and had a very favorable carry-over effect on the contour-listed corn grown in the corn-oats-meadow-meadow rotation. Contouring reduced runoff 29 mm per year, and contour listing saved nearly 45 mm of water from running off when compared with tillage up- and downhill. Soil loss with contour listing was reduced to one-fifth that from conventional tillage up- and downhill.

Planting on the contour has little effect on runoff unless there are ridges to hold the water. Contour tillage was compared with corn rows up- and downhill by the soil conservation experiment station in Ohio. Contour planting with smooth cultivation did not reduce runoff. Contour cultivation that formed ridges, however, reduced runoff by 40 percent.

Terraces Control Runoff   A ridge-type terrace is an effective means of controlling runoff and reducing erosion. It may or may not conserve water. Some terraces are built with tile outlets to dispose of the water and therefore do not conserve water. Other terraces are built to gradually move water around the hill. Some of this water may infiltrate. Still other terraces are built level and have closed ends so that all the water must infiltrate.

Table 18–2 Soil and Water Loss as Affected by Cropping and Tillage Practices on Ida Silt Loam on a 12 Percent Slope

Rotation and tillage method	1948–1957	
	Average soil loss, tons/ha-year	Average water loss, mm/year
Corn		
Corn-oats (sweet clover)		
Uphill and downhill	56.5	82
Contoured	22.6	53
Contour listed	11.0	37
Corn-oats-meadow-meadow		
Contour listed	2.9	14
Oats		
Corn-oats (sweet clover)	6.1	49
Corn-oats-meadow-meadow	6.1	27
Alfalfa-bromegrass		
Corn-oats-meadow-meadow		
First year meadow	0.4	20
Second year meadow	0	14

Source: Moldenhauer and Wischmeier, 1960.

The usefulness of increased water infiltration behind a terrace depends greatly on slope. The water-storage area behind a terrace on steep topography is little wider than the terrace channel. Miller and Shrader (1973) found 2.5 to 5 cm more water stored in the soil of terrace channels throughout the summer than in the soil midway between terraces. The corn in the channel yielded 10 to 13 percent more than the rest of the field. It is usually more advantageous, however, to have the water distributed over more area. Contour cultivation is likely to improve the effectiveness of water use as much as or more than terraces on steep land (Figure 18-7).

Terraces on relatively flat land can be highly effective at conserving water. Nearly level topography (less than ½ percent slope) at Spur, Tex., lost an average of 36 mm of runoff per year over a 13-year period (Dickson, Langley, and Fisher, 1940) where the rows were up- and downslope. A field with closed, level terraces had no runoff and produced 47 percent more cotton than the unterraced field (the average annual rainfall was 475 mm). These terraces were very effective at holding water on the land because a terrace 0.5 m high on a slope of ½ percent can hold water back for a width of 100 m. The corresponding width on a 5 percent slope is only 10 m for a ridge 0.5 m high.

Evaporation Losses

About one-fifth of all precipitation falling on cropland is lost by evaporation, with a high proportion of this loss being unavoidable. Much precipitation comes as light showers that moisten only the top few centimeters of soil. A 10-mm rain will raise an average soil from the wilting point to field capacity to a depth of only 6 cm. Evaporation from the soil surface is usually effective to a depth of about 5 to 8 cm. Thus, most of a 10-mm rain will be lost by evaporation if it is followed by several dry days.

Figure 18-7 Small contour ridges or furrows are especially desirable in reducing runoff and erosion. The water that cannot be held in the furrows is caught above the terrace. Under conditions of high rainfall, the terrace is built with enough grade to conduct the excess water off the field. *(Soil Conservation Service, USDA.)*

Evaporation increases as the humidity decreases and as the temperature increases; wind also hastens evaporation. The rate of evaporation from a moist soil surface is about as fast as that from an open-water surface. The rate of evaporation decreases as the soil surface becomes drier and most of the water being evaporated must move to the surface by capillarity. The rate of capillary movement declines as the distance increases and usually becomes negligible when about 8 cm of soil is air-dry. There is little or no capillary movement to the surface after that even if the soil below 8 cm is at field capacity.

Slow evaporation losses may continue for some time after capillary movement has ceased, the mechanism involved being vapor transport. The relative humidity of soil air is nearly 100 percent until the soil is almost air-dry. Moist air from below exchanges with the drier atmosphere above and thereby permits more water to evaporate. This type of loss is normally small unless large cracks form in the soil.

Evaporation losses can be reduced by making the water penetrate more deeply rather than staying near the surface. Massee and Siddoway (1969) found that fall chiseling increased storage of winter precipitation by 5 to 8 cm. Chisel slots can be made more effective by the use of *vertical mulching* whereby plant material is used to keep the chisel slots open. Fairbourn and Gardner (1973) found vertical mulching to be very effective for reducing evaporation.

Some farmers cultivate their land soon after a rain so that a dust mulch will form. This practice is not recommended as a means of reducing evaporation losses unless it is needed to control weeds. Without cultivation, the dry layer on top of the soil will form and evaporation will decline to a negligible rate.

A bare soil often reaches a temperature several degrees above air temperature in the daytime. The rate of evaporation from such a soil can be quite high during the summer. A thick mulch can reduce the daytime temperature of the surface soil and thereby reduce the rate of evaporation. The soil temperatures in the previously cited North Carolina mulch studies (Krantz, 1949) were reported when the air temperature was 35°C. The soil temperature at 1 cm depth was 27°C in the mulched area and 48°C in the unmulched area. Krantz attributed a part of the increased yield from the mulched plots to water saved through reduced evaporation.

Many different materials can be used for mulching, including almost any crop residue. Truck crops are sometimes grown through a sheet of plastic that serves as a mulch. The Iowa State University Horticulture Department has shown that ground corncobs make an excellent mulch for flower beds. Sawdust, wood shavings, and bark chips are all used for mulch (fresh wood that contains turpentine should usually be avoided). Gravel layers can also be used as mulches.

A mulch not only reduces soil temperature and decreases evaporation losses, it also controls weed growth if applied thickly enough. One general caution applies to the use of moist organic mulches. The soil nitrogen supply should be maintained at a high-enough level to decompose any mulch that is incorporated into the soil (Chapter 10).

Fallowing

Fallowing is a means of storing water in the soil from one year to the next. It permits farmers in regions of low rainfall to increase their yields by cropping their land every second year rather than every year. Often the increased yield is enough to maintain about the same total production over the 2-year period, or at least the same net profit after reduced costs are considered. Furthermore, fallowing greatly reduces the risk of crop failure in semiarid regions.

A wheat field in the 400-mm precipitation zone in South Dakota will illustrate the principle of fallowing. A good crop of wheat needs about 500 mm of water. Fallowing provides that amount in a 400-mm zone in the following manner. Early in the spring of the fallow year, and periodically through most of the growing season, the land is tilled to prevent weed growth, because weeds would use up the water that is to be stored. The tillage should be only as often and as deep as necessary to kill weeds. Water will be lost by evaporation from the tilled soil, but the lack of vegetation results in saving water that would have been lost by transpiration. Greb, Smika, and Black (1970) reported savings ranging from 16 percent of the precipitation falling on bare fallow to 37 percent of the precipitation on heavily mulched fallow. A saving of 25 percent of 400 mm of precipitation stored as illustrated in Figure 18-8 would amount to 100 mm of water; these 100 mm added to the next year's 400 mm make the 500 mm needed by the crop.

Summer fallowing is sometimes used in subhumid regions for weed control rather than for water conservation. The year of fallowing permits large numbers of weeds to germinate but not to set seed. The repeated cultivation also weakens and eventually kills perennial weeds.

Fall plowing is often used in subhumid regions as a means of conserving fall precipitation for the following crop. This practice is common in the western part of the corn belt when corn follows a legume-grass meadow in the rotation. Plowing the meadow in the fall avoids further transpiration losses. The reduced transpiration between the time of fall plowing and spring plowing may make a considerable difference in the corn growth the following summer.

Fall plowing is significant as a water conservation practice in the drier part of the corn belt but not in the more humid part. Most corn-belt soils receive an excess of water in the spring and therefore do not need the increased carry-over from the preceding fall. It should be recognized, too, that fall plowing increases the erosion hazard. Water erosion on sloping land can be considerably worse where the land is

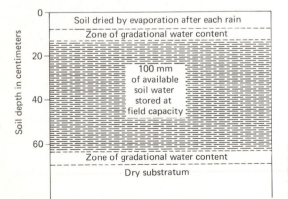

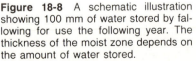

Figure 18-8 A schematic illustration showing 100 mm of water stored by fallowing for use the following year. The thickness of the moist zone depends on the amount of water stored.

fall-plowed than it would be with spring plowing. And fall-plowed land is likely to be subject to wind erosion even if it is level. Consequently, fall plowing should be used only where it is really needed and the erosion hazard is not excessive. Sometimes a compromise is possible. Shallow plowing or other tillage may kill the meadow crop and yet leave the sod at the surface to control erosion. Another approach is to kill the sod chemically without any tillage.

Percolation Losses

Losses of water by deep percolation occur in humid climates and in irrigated areas. The loss is accompanied by a loss of available plant nutrients. The nutrient losses are at least partially controllable, as will be seen later. The water losses in humid regions are largely unavoidable, but their occurrence should be recognized. The amount of percolation loss occurring per year is equal to the amount of infiltration minus the amount of evapotranspiration. Practices that increase infiltration in humid regions or that reduce evaporation and transpiration will therefore increase percolation losses.

Percolation losses resulting from irrigation can be held to a minimum by careful irrigation. Some percolation loss is essential when saline soils are irrigated or when the irrigation water contains significant amounts of soluble salts. In either case, enough irrigation water should be provided to leach the soluble salts from the soil.

Many irrigated soils suffer unnecessary percolation losses resulting from the way the water is applied. Irrigation furrows that are too long result in the upper end of the field being overirrigated by the time the lower end receives adequate irrigation. Uneven irrigation is especially likely on highly permeable soils.

Irregular slopes can cause percolation losses either with or without irrigation. Water runs to the low places and often makes them too wet. The wetness problem can be solved by land leveling if the surface irregularities are minor such as those common in river floodplains and low terraces. Some sloping land can be leveled into a system of terraces that produce the same general effect as level bottomlands. More rolling areas are the most difficult to handle, especially with irrigation. Sprinkler irrigation is often the answer to the problem of uniform water application on rolling lands, especially those that have reasonably high infiltration rates.

Nutrients Lost by Percolation

The loss of plant nutrients in percolating water is usually more serious than the loss of the water. Water losses by percolation are limited to water that is surplus at the time of the loss (though a water shortage might develop at some other season of the year). Water will not percolate beyond the root zone until the soil it passes through is at field capacity.

Measurements of macronutrient losses by leaching in central New York State are shown in Table 18-3. Phosphorus losses are usually negligible because of low solubility; nitrogen losses are the most variable and controllable of the macronutrient losses. Large losses of nitrogen can occur only if the supply of nitrate nitrogen is built up by either fertilization or nitrification processes in the soil (discussed in Chapter 10). Growing plants can utilize large amounts of nitrate nitrogen and

Table 18–3   Annual Loss of Nutrients by Leaching from Bare and Cropped Dunkirk Silty Clay Loam Soils as Measured by Cornell University Lysimeters over a 10-year Period

Treatment	kg/ha-year					
	N	P	K	Ca	Mg	S
Bare	77	Trace	81	447	70	59
Rotation	9	Trace	64	258	49	49
Grass	3	Trace	69	292	56	50

Source: Calculated from Bizzell and Lyon, 1927.

thereby drastically reduce the leaching losses. Reduced nitrogen loss is important not only because of the cost of fertilizer nitrogen but also because high concentrations of nitrate in drinking water are undesirable and can even become toxic to livestock and people. Also, some of the groundwater eventually seeps back to surface streams and can be a factor in causing eutrophication (though nitrates, phosphates, and other nutrients reaching the water via surface routes are usually much more important factors in this process).

Losses of the macronutrient cations, calcium, magnesium, and potassium in percolating water are mostly unavoidable in humid regions. Management programs in humid regions must compensate for these losses. The potassium is generally replaced in the fertilizer program, and the calcium and magnesium are replaced by liming. Soils that do not suffer percolation losses usually have adequate reserves of cations and seldom require either liming or potassium fertilization.

Soils with high levels of fertility generally lose more nutrients by leaching than soils with low fertility levels. Table 18-4 shows the results of lysimeter studies with Marshall silt loam in southwestern Iowa. These studies show that large applications of manure greatly increase the amount of nutrients lost by leaching. Growing crops reduced but did not stop these losses. The effect of plant growth acts two ways in reducing leaching losses. Some of the nutrient ions are absorbed and used by the plants and thereby protected against leaching. The second way is even more important. The growing plants transpire large amounts of water and thereby drastically reduce the amount lost by percolation. Reduced water percolation automatically results in reduced nutrient losses.

Leaching losses in subhumid regions (in temperate climates, those below about 750 mm of precipitation annually) are of little importance. They do occur in areas that receive water that runs off from higher areas, but these spots are a small percentage of the total area. In general, the underlying parent material of subhumid and drier regions is permanently dry. One enterprising person learned this by installing some lysimeters near Moscow, Idaho. The lysimeters were abandoned after a few years of observation. Not a drop of leachate had come through. Some nutrients do leach from the sola in arid regions, but they accumulate in a layer at the usual depth of water penetration. These layers are usually light colored, and some of them are cemented into rocklike masses called *duripans* or *hardpans*.

Table 18–4 Calcium and Magnesium in Percolate from Lysimeters in Marshall Silt Loam[*]

Crop and treatments	1935		1936	
	Ca, kg/ha	Mg, kg/ha	Ca, kg/ha	Mg, kg/ha
Fallow	197	48	45	14
Fallow plus 16 tons of manure	442	118	126	34
Corn	195	42	5	1
Corn plus 16 tons of manure	238	55	57	14

[*]Precipitation: 1935, 822 mm; 1936, 559 mm.
Source: Browning, Norton, McCall, and Bell, 1948.

IRRIGATION

The surest way to avoid a deficiency of water for plant growth is to irrigate. People have been irrigating for thousands of years in arid regions where crops can be grown no other way. Now there is interest in, and some practice of, supplemental irrigation in humid regions where the goal is to increase the crop yield.

Irrigation is costly, but so is crop failure. High-value crops are the ones most likely to be irrigated because they are most likely to repay the cost of irrigation. But many farmers with convenient water sources have found irrigation profitable even with ordinary field crops.

Irrigation systems can be classified as *surface, subsurface,* and *sprinkler* types. Subsurface irrigation is limited in extent because it requires a fortuitous set of conditions that permit a type of controlled drainage system to either add or remove water. These conditions are rare and will receive no further consideration here.

Surface Irrigation

Surface irrigation takes several different forms depending on how the water is controlled. Water may be guided across a pasture, or a field of hay or small grain, by means of small channels called *corrugations.* Row crops usually have somewhat larger channels called *furrows* (Figure 18-9). Both corrugation and furrow irrigation require that the water from a supply ditch be subdivided into many small streams, each with just enough water to supply the area adjacent to its own small channel. This subdivision may be accomplished by various means including a set of small trenches, siphon tubes that lift water over the ditch bank, and gated pipes that have individual openings with adjustable gates for each irrigation row.

Surface irrigation also includes various forms of flooding. The simplest (and most erratic) of these consists of supplying water by means of ditches on the ridges in a field; from there, the water is released to flow toward the low places, usually with some guidance from someone with a shovel. This system, known as *wild flooding,* is mostly restricted to close-growing crops such as pasture, hay, or perhaps small grains.

Figure 18-9 Furrow irrigation in Idaho.

Better control of flood irrigation can be achieved by surrounding a carefully leveled area with ditches and ridges. Sometimes a flat basin is formed. Basin irrigation is readily accomplished by turning in the desired amount of water and allowing it to stand there until it infiltrates.

Border irrigation is the most common type of flood irrigation. Borders have ridges down both sides and a water-supply ditch at the top. The area within a border strip has a slope from end to end but must be level from one side to the other so that water will spread completely across it. The water supply must either be shut off at the proper time for the right amount of irrigation at the bottom of a closed border, or a ditch must be provided for the "waste water" at the bottom end of the field. Waste water also occurs with the corrugation and furrow methods of irrigation. The waste water may not actually be wasted, because it can often be used as part of the irrigation water for another field.

Drip or *trickle irrigation* is a relatively new method of applying water that can reduce evaporation losses (Bucks, Erie, and French, 1974) and permit irrigation with relatively saline water. Trickle irrigation uses pipes and small tubes, called *emitters,* to deliver water directly to the root zone of the plants being irrigated. Water is saved when plants are young because much of the area between plants is left dry. Savings also occur in orchards where each tree is watered by several emitters but the areas between trees are left dry.

Trickle irrigation can use saline water because the water moves away from the plants rather than toward them. Any salt concentrations that develop can be kept outside the root zone where the water stops moving.

Under favorable conditions, trickle irrigation requires only about two-thirds as much water as furrow irrigation (Bernstein and Francois, 1973). With close-growing crops, however, there would be no water savings because there is no area to leave dry. The other main limitation on trickle irrigation is the cost of installing all the pipelines and emitters. The installation can probably best be justified for orchards or other high-value perennial crops grown where water is at a premium.

Sprinkler Irrigation

Sprinkler irrigation is especially important where the land is too rolling for efficient surface irrigation or where only occasional irrigation is needed. Many irrigation farmers spend large sums to "level" their land for surface irrigation (the "leveled" land usually has a slope, but it is made smooth and uniform). This expense is justifiable to obtain a uniform flow of water across suitable fields in arid regions. Land leveling is usually too costly for supplemental irrigation in humid regions. Also, sprinkler irrigation can be accomplished with a set of portable pipes, whereas surface irrigation usually requires a rather permanent ditch system.

Sprinkler lines usually consist of aluminum pipes 6, 9, or 12 m long that can be coupled and uncoupled rapidly. A rotating sprinkler on a riser (Figure 18-10) is mounted near each coupling. The circles covered by the sprinklers must overlap considerably so that the entire area will be irrigated as uniformly as possible. The sprinkler line is disassembled and moved to the next location after several hours of operation.

Figure 18-10 Sprinkler irrigation in an orchard on sloping, sandy soils in Idaho.

Some sprinkler systems operate by gravity pressure from a high-level water supply. Much more commonly the pressure is created by a pump. Water may reach the sprinkler line through a portable main line or a buried main line with risers. Another method, known as *center-pivot irrigation,* is to mount the sprinkler line on wheels so that it can rotate around and around a well or a pipeline riser placed in the center of the field. Many such sprinkler lines are about 400-m long and irrigate all but the corners of square 64-ha fields. One county in Nebraska has 700 of these rotating center-pivot systems (Dolan, 1972).

Design Factors for Irrigation Systems

The design of an irrigation system is based on the infiltration rate of the soil, its available water-holding capacity, and the amount of water needed. Topography and soil erodibility can also pose serious limitations, especially for surface irrigation systems.

Flood irrigation necessarily adds water at the infiltration rate of the soil. Corrugations and furrows have their own infiltration rates that are limited by the wetted surfaces of the channels. Irrigation continues until the soil contains enough water to raise the entire root zone to field capacity. The next irrigation must take place before the soil is dried to the wilting point. Sometimes it takes place sooner so that less water will be needed each time.

Irrigation water should be made to flow full length of corrugations, furrows, or borders within a reasonably short time. The upper end of the field will receive considerably more water than the lower end if there is too much time lag. A rule of thumb is that water should reach the bottom end of the field by the time the irrigation period is one-fourth complete.

The maximum length that an irrigation "run" (distance the water flows) should be is controlled by the maximum size of a nonerosive stream. If, for example, the irrigation period is to last for 8 h, then the maximum length of run should be the distance that the maximum nonerosive stream will flow in 2 h. Longer runs should be subdivided to avoid excessive erosion. Some soils with uniform slopes between 0.3 and 0.5 percent can have runs as long as 400 m without suffering erosion. Flatter slopes must have shorter runs because the water will barely move across them. The maximum length of run becomes progressively shorter as the slope increases above 0.5 percent because the erosion hazard increases with slope. The size of the irrigation streams must be reduced as the slope increases.

Sprinkler irrigation is likely to be chosen where surface irrigation runs would have to be shorter than about 100 m. Sprinkler irrigation normally applies water at rates between 0.5 and 2.5 cm/h. A rate slightly below the infiltration rate of the soil should be chosen so that there will be no runoff. The length of irrigation time and the interval between sprinkler irrigations depend on the available water-holding capacity of the soil, as already explained for surface irrigation.

Supplemental Irrigation

Supplemental irrigation is used where crops can be grown without irrigation. It is used to broaden the choice of crops, to improve the yield potential of crops, or to avoid yield reductions that would otherwise occur in dry years. It is kept on a standby basis to be used when needed and may be left unused during a moist year when a good yield can be produced without irrigation.

Much of the significance of supplemental irrigation stems from the fact that water is especially critical at certain stages of plant growth. For example, it takes about 500 mm of water to produce 60 q of corn. Such a crop requires about 8 mm of water per quintal, but at tasseling time it may take less than half that much water to increase yield by 1 q. Table 18-5 shows data from Virginia where 76 mm of water produced a yield increase of 23.8 q/ha, or nearly 1 q for each 3 mm of added water.

All crops need water, and each has its own critical periods when the water is especially important. The probability of these critical periods coinciding with a dry period varies with the crop and the climate of the area. Of course, the value of any yield increase that may be achieved by irrigation varies with the crop. Few farmers using supplemental irrigation have adequate systems to irrigate all their land. They must therefore choose carefully which fields will give the greatest additional return with irrigation.

High-income crops such as citrus, strawberries, tomatoes, and other fruit and vegetable crops usually receive priority on the water available for irrigation. But many of the high-income crops are also high-cost and high-risk crops. It is often possible to make a smaller but more reliable profit by irrigating ordinary field crops that have stable prices.

Suitable crops growing on sandy soils frequently give the largest yield increases from irrigation. The potential increase is large because the yield without irrigation is likely to be lower on the sandy soils than on heavier soils. In addition, there is always a possibility that too much water will be applied either by an overzealous irrigator (the subsoil water content should be checked occasionally) or by an ill-timed rain shortly after irrigation. Too much water is less likely to be a problem on sandy soils than on heavier soils.

Attaining the high-yield potential made possible by irrigation requires that other good practices be used also. The fertilizer response of an irrigated crop is usually higher than that of a nonirrigated crop, and the possibility of crop failure

Table 18-5   Corn Yields Associated with Three Irrigation Practices and No Irrigation, Blacksburg, Va., 1954

Period of irrigation	Irrigation applied, mm	Yield per hectare, q	Increase per hectare from irrigation, q
No irrigation (with 320 mm of rainfall from May to September)	0	40.6	
Tasseling through milk stage	76	64.4	23.8
Early growth plus tasseling through milk stage	152	74.9	34.3
Continuous through season	269	80.6	40.0

Source: Jones and Moody, 1955.

resulting from heavy fertilization followed by a water shortage is eliminated. Therefore, higher fertilizer rates are usually more profitable on irrigated land than on nonirrigated land. Similarly, good seed, proper tillage practices, and other factors of crop production take on an added significance under an irrigation system of farming because farmers are dealing with a greater investment and a greater return. They can make money faster if they manage well, but they can also lose money faster if they manage poorly.

DRAINAGE

Some of the world's most fertile soils are not cropped or are very limited in their use because of wetness. Poorly drained soils are usually fertile because they are not leached very much—the water cannot escape to carry solutes away. Furthermore, they are usually high in organic matter because decomposition is slow where aeration is poor. The removal of excess water from such soils is a prerequisite to using the land for a desired crop.

Extensive level areas, especially those occurring in low topographic positions, are generally wet enough to at least partially restrict land use. Capillary rise may result in a poorly aerated soil even though the water table is actually 50 cm or more below the soil surface. Any large pores that are present and contain air are likely to have a low percentage of oxygen and a high percentage of carbon dioxide. These conditions create problems in the absorption of water and nutrients by plants (Chang and Loomis, 1945). Plants may actually wilt on dry, hot days in the presence of excessive soil moisture (Kramer, 1949).

Poor aeration reduces the absorption of potassium by some crops, particularly corn. The decomposition of organic matter and the release of nitrogen and other anions are retarded by poor aeration. Ammonium nitrogen cannot be oxidized to nitrate under anaerobic conditions. Furthermore, denitrification may reduce nitrate nitrogen to elemental nitrogen or to nitrous oxide and cause a loss of nitrogen from the soil as a result of waterlogged conditions. The yellowing of corn in poorly drained spots is frequently caused by nitrogen deficiency or, sometimes, by potassium deficiency.

Wet spots in fields are particularly troublesome to farmers because they cannot till or plant them as early as adjacent better-drained soils. The growing season and crop yield are both reduced if farmers wait for the wet spots to dry out. If they try to work across the wet spots, they may get stuck. Going around them is troublesome and costly, and the wet areas often become weedy as the season progresses.

Poor drainage is both common and troublesome in irrigated areas. Most irrigation occurs in valleys; the soils in valleys often have water tables at shallow depths. Bringing in irrigation water is likely to raise the water table and increase the problems associated with wetness.

Wetness causes the same problems in arid regions as in humid regions plus a considerable likelihood of producing saline and sodic soils. High evaporation and low precipitation cause water to move to the soil surface, evaporate, and leave soluble salts behind. Sodium salts are highly soluble and are especially likely to accumulate under such conditions.

A typical irrigated valley in an arid region has two or even three systems of ditches; one carries the irrigation water and distributes it to the various fields; another provides subsurface drainage to keep the water table from rising too high; the third handles any waste water that flows off the bottom ends of the fields (this third system of ditches is often partially integrated with the other two).

Drainage Systems

Drainage systems may be described as random, regular, or interceptor, and as surface or subsurface. Random systems are used where there are scattered wet spots and the drainage system goes from one to another. Regular systems use uniformly spaced ditches or tile lines to drain large, wet areas. Interceptor drains keep water from adjoining areas from running onto or seeping into the area being drained (Figure 18-11). Surface drains carry away excess water before it infiltrates. Subsurface drains remove water from beneath the soil. Subsurface drainage is one practice that can hardly ever be overdone because the water will not drain to a content lower than field capacity.

The choice between surface and subsurface drainage often depends on the infiltration rate of the soil. Slow infiltration rates result in excess water standing on the soil surface long enough to be removed by surface drainage. Removal of water by surface drainage reduces or eliminates leaching and may prevent the soil from ever becoming waterlogged. Both surface and subsurface drainage systems may follow either regular or random patterns, the difference being controlled largely by topography and the resulting wetness pattern.

High infiltration rates and artesian water (water under pressure beneath the soil) create situations that can be solved only by subsurface drainage. High infiltration rates are associated with either coarse-textured soils or soils with good granular structure. The functioning of subsurface drainage in these soils depends greatly on whether the coarse texture or granular structure extends through the subsoil giving it a high permeability. The horizontal permeability is especially important for subsurface drainage because the water must move laterally to reach the drains. Horizontal permeability is often higher than vertical permeability.

Ditch or Tile Drainage

Many people assume that ditches are used for surface drainage and tile for subsurface drainage. This is usually true but not always. Tile lines may have risers and surface inlets so that surface water can flow into them. Deep ditches are at least as

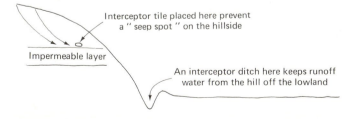

Interceptor tile placed here prevent a " seep spot " on the hillside

Impermeable layer

An interceptor ditch here keeps runoff water from the hill off the lowland

Figure 18-11 Interceptor tile and interceptor ditches prevent water from external areas from reaching the area being drained.

effective as tile lines for subsurface drainage. Then, too, it should be remembered that several other types of drainage exist, including bedding systems (ditches shallow enough and with flat-enough sides to be crossable), mole drainage (a subsurface channel created by pulling a torpedo-shaped device through the subsoil), and various substitutes for tile such as rock drains and wooden boxes.

The choice between ditches and tile is based partly on soil conditions and partly on the balance between convenience and cost. Tile lines cost more to install than open ditches but do not interfere with anyone crossing the land. A drainage ditch placed next to a road, an irrigation ditch, or some other field boundary is usually less objectionable than a ditch in the middle of an open area. Many drainage systems consist of tile lines in the fields that empty into large outlet ditches. The outlet ditches are often constructed by a drainage district or some other cooperative effort rather than by an individual.

Drainage Design Factors

Soil characteristics determine to a large extent the depth and spacing of tile lines or ditches. An extra meter of depth that terminates in layers of lower permeability than those above does little good. An extra few centimeters of depth that reaches a more permeable layer is highly beneficial. Other depth factors that should be considered for tile lines include the usual depth of frost penetration and the depth required to avoid breakage of tile under whatever traffic may cross the line. Typical tile depths are between 1 and 1.5 m, partly because greater depth costs more.

The spacing between tile lines is largely a matter of the horizontal permeability of the soil being drained. Tile lines may have to be only 10 m apart in very tight clay soils, whereas they may be 30 to 50 m apart if they are in a layer of porous sandy material. Spacings of 18 to 30 m are common in loamy soils. The most critical points in the system are usually midway between the lines, as shown in Figure 18-12.

Many tile drainage systems that were effective at the time of installation later become inadequate. The gradual destruction of soil structure under cultivation causes a loss of large pore spaces and a decline in soil permeability. This problem

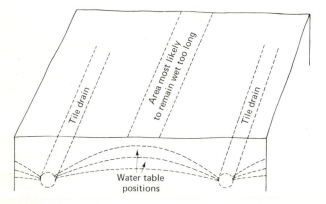

Figure 18-12 Block diagram showing successive positions of a water table as it lowers after a rain on a field with tile drainage.

is most likely to develop in medium- or fine-textured soils under intensive cropping. Inclusion of grasses and legumes in the crop rotation helps to increase the aeration pore space and improve the rate of drainage.

REFERENCES

Bennett, O. L., D. A. Ashley, and B. D. Doss, 1966, Cotton Responses to Black Plastic Mulch and Irrigation, *Agron. J.* **58**:57–60.

Bernstein, Leon, and L. E. Francois, 1973, Comparisons of Drip, Furrow, and Sprinkler Irrigation, *Soil Sci.* **115**:73–86.

Bizzell, J. A., and T. L. Lyon, 1927, Composition of Drainage Waters from Lysimeters at Cornell University, *Proc. Int. Congr. Soil Sci.* **2**:342–349.

Borst, H. L., A. G. McCall, and F. G. Bell, 1945, Investigations in Erosion Control and the Reclamation of Eroded Land, Zanesville, Ohio, 1934–1941, *USDA Tech. Bull.* 888.

Browning, G. M., R. A. Norton, A. G. McCall, and F. G. Bell, 1948, Investigations in Erosion Control and the Reclamation of Eroded Land, *USDA Tech. Bull.* 959.

Bucks, D. A., L. J. Erie, and O. F. French, 1974, Quantity and Frequency of Trickle and Furrow Irrigation for Efficient Cabbage Production, *Agron. J.* **66**:53–57.

Chang, H. T., and W. E. Loomis, 1945, Effect of Carbon Dioxide on Absorption of Water and Nutrients by Roots, *Plant Physiol.* **20**:221–232.

Dickson, R. E., B. C. Langley, and C. E. Fisher, 1940, Water and Soil Conservation Experiments at Spur, Texas, *Texas Agr. Exp. Sta. Bull.* 587.

Dolan, Jeanie, 1972, The Impact of Center Pivot Irrigation on Dryland Farming, *Crops Soils* **25**(3):9–11.

Doss, B. D., and C. E. Scarsbrook, 1969, Effect of Irrigation on Recovery of Applied Nitrogen by Cotton, *Agron. J.* **61**:37–40.

Fairbourn, M. L., and H. R. Gardner, 1973, Increasing Soil Water Storage with a Vertical Mulch, *Crops Soils* **25**(4):10–11.

Fairbourn, M. L., and H. R. Gardner, 1975, Water-repellant Soil Clods and Pellets as Mulch, *Agron. J.* **67**:377–380.

Follett, R. F., E. J. Doering, G. A. Reichman, and L. C. Benz, 1974, Effect of Irrigation and Water-table Depth on Crop Yields, *Agron. J.* **66**:304–308.

Fouss, J. L., 1971, Tomorrow's Drainage Systems Today, *Crops Soils* **23**(7):12–14.

Francois, L. I., 1975, Effects of Frequency of Sprinkling with Saline Waters Compared with Daily Drip Irrigation, *Agron. J.* **67**:185–190.

Gelashvili, F. N., 1973, Effect of Irrigation with Thermal Waters on the Content of Microelements in Some Soils of Georgia, *Sov. Soil Sci.* **5**:212–218.

Greb, B. W., D. E. Smika, and A. L. Black, 1970, Water Conservation with Stubble Mulch Fallow, *J. Soil Water Conserv.* **25**:58–62.

Jones, J. N., Jr., and J. E. Moody, 1955, Quarterly (March) Report on Progress in Soil and Water Conservation Research, *USDA*.

Kramer, P. J., 1949, *Plant and Soil Water Relationships,* McGraw-Hill, New York.

Krantz, B. A., 1949, Fertilize Corn for Higher Yields, *N. Carolina Agr. Exp. Sta. Bull.* 366.

Ligon, J. T., Don Kirkham, and H. P. Johnson, 1964, The Falling Water Table between Open Ditch Drains, *Soil Sci.* **97**:113–118.

Lucey, R. F., and M. B. Tesar, 1965, Frequency and Rate of Irrigation as Factors in Forage Growth and Water Absorption, *Agron. J.* **57**:519–523.

Mannering, J. V., and L. D. Meyer, 1961, The Effects of Different Methods of Cornstalk Residue Management on Runoff and Erosion as Evaluated by Simulated Rainfall, *Soil Sci. Soc. Am. Proc.* **25**:506–510.

Massee, T. W., and F. H. Siddoway, 1969, Fall Chiseling for Annual Cropping of Spring Wheat in the Intermountain Dryland Region, *Agron. J.* **61**:177–182.

Miller, E. L., and W. D. Shrader, 1973, Effect of Level Terraces on Soil Moisture Content and Utilization by Corn, *Agron. J.* **65**:600–603.

Moldenhauer, W. C., and W. H. Wischmeier, 1960, Soil and Water Losses and Infiltration Rates on Ida Silt Loam as Influenced by Cropping Systems, Tillage Practices and Rainfall Characteristics, *Soil Sci. Soc. Am. Proc.* **24**:409–413.

Newman, J. E., 1970, Climate in the 1970's, *Crops Soils* **22**(4):9–11.

Robinson, E. E., O. D. McCoy, G. F. Worker, Jr., and W. F. Lehman, 1968, Sprinkler and Surface Irrigation of Vegetable and Field Crops in an Arid Environment, *Agron. J.* **60**:696–700.

Rycroft, D. W., and A. A. Thorburn, 1974, Water Stability Tests on Clay Soils in Relation to Mole Draining, *Soil Sci.* **117**:306–310.

Schwab, G. O., G. S. Taylor, J. L. Fouss, and E. Stibbe, 1966, Crop Response from Tile and Surface Drainage, *Soil Sci. Soc. Am. Proc.* **30**:634–637.

Thompson, L. M., 1969, Weather and Technology in the Production of Wheat in the United States, *J. Soil Water Conserv.* **24**:219–224.

Thompson, L. M., 1970, Weather and Technology in the Production of Soybeans in the Central United States, *Agron. J.* **62**:232–236.

Thompson, L. M., 1973, Cyclical Weather Patterns in the Middle Latitudes, *J. Soil Water Conserv.* **28**:87–89.

Unger, P. W., and J. J. Parker, Jr., 1968, Residue Placement Effects on Decomposition, Evaporation, and Soil Moisture Distribution, *Agron. J.* **60**:469–472.

Weakly, H. E., 1943, A Tree Ring Record of Precipitation in Western Nebraska, *J. Forestry* **41**:816–819.

Weakly, H. E., 1962, History of Drought in Nebraska, *J. Soil Water Conserv.* **17**:271–275.

Willett, H. C., 1964, Evidence of Solar Climatic Relationships, *CAED Rep.* 20, *Center Agr. Econ. Develop.,* Iowa State University, Ames.

Chapter 19

Soil Erosion
and Its Control

Landscapes are shaped by processes of erosion and deposition. Wind and water work relentlessly moving soil and rock fragments from one place to another. Gravity, too, is always present serving as the driving force for soil creep, landslides, and related phenomena. Ice occasionally adds a dramatic touch through the action of glaciers. All these are agents of geologic erosion.

The processes of geologic erosion may begin with a small gully and gradually enlarge it into a mighty chasm such as the Grand Canyon. Valleys continually enlarge, and the hills between are worn away. Even mighty mountain chains are worn down to stumps. Probably all land would have been eroded down to sea level if it had not been repeatedly lifted again by balancing forces. The most important of these forces is *isostasy*—the equalizing of pressures at depth in the earth's crust (Figure 19-1). The granitic rocks that occur very abundantly in mountains are less dense than basalt and must therefore protrude to high levels to balance the basaltic rocks that line the ocean basins.

A thin veneer of soil forms on the surface of rock that is exposed to the action of weathering. Erosion gradually removes the surface soil, but more rock from below is converted into soil. The supply of rock is renewed by isostasy. Thus the entire process goes on and on, though not always at the same rate. This constant renewal keeps most soils from becoming completely leached out, sterile, and unproductive.

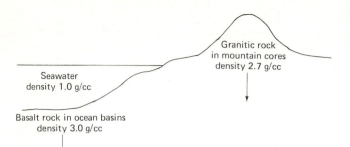

Figure 19-1 A schematic cross section illustrating isostasy. Thickness and density are balanced so that pressures beneath a mountain range and an ocean basin are equal at a depth of several tens of kilometers.

The processes of geologic erosion are vital to the renewal of soil fertility over very long periods of time.

Freshly exposed materials weather fastest because they receive the full impact of moisture and temperature changes. Later, soil covers the rock and insulates it, thus reducing the rate of formation of new soil. As the soil becomes deeper it provides more insulation, and finally slows the rate of soil formation to the same rate as soil erosion. The rate of geologic erosion must average the same in the long run as the rate of soil formation. Otherwise the soil gradually becomes either thicker or thinner until the rate of soil formation equals the rate of soil erosion.

Ignoring the materials that have been too recently exposed to weathering, a generality can be stated. The depth of the soil is inversely related to the long-term rate of erosion that has acted on it. Slow erosion permits a deep soil to form; rapid erosion keeps the soil shallow (Figure 19-2). This generality applies only to soils on erosional topography, but most topography is erosional. Refinements could be made for climate and rock resistance to weathering, but these refinements would only alter the scale factor; the inverse relation between soil depth and rate of erosion in any particular rock and environment would still hold.

The potency of geologic erosion as it acts over long periods of time is truly impressive. Its processes are natural, mostly unavoidable, and persistent. They are also slow. The best measurements and estimates of the normal average rates of geologic erosion are about four-tenths of a ton of soil per hectare each year. Ac-

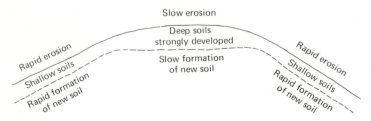
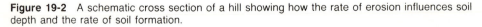

Figure 19-2 A schematic cross section of a hill showing how the rate of erosion influences soil depth and the rate of soil formation.

celerated erosion brought about by people's use of land is many times as fast as geologic erosion. Accepted tolerable rates of soil erosion are usually between 2 and 10 tons per hectare annually—probably 5 to 25 times the rates of geologic erosion.

The long-term effect of soil losses in excess of geologic erosion is certain to reduce the depth of most soils. Eventually soil depth will stabilize when the soil is thin enough to permit weathering that produces new soil as fast as the rate of soil erosion. Erosion at the maximum tolerable rates currently accepted will eventually result in soils about one-half to one-fourth as deep (estimated) as they would be under geologic erosion alone. Furthermore, many land users make no pretense of holding soil losses to any assigned limits. Erosion losses of several tens of tons per hectare are common; some annual losses are in excess of 200 tons per hectare (about 1.5 cm of soil depth).

WATER EROSION

Water erosion begins with raindrops. The devastating effect of raindrops striking bare soil (Figure 19-3) was long overlooked but is now recognized as the principal means of detachment of soil particles. Palmer (1965) reported that a thin film of water covering the soil surface adds "virtual mass" to raindrops and increases their impact forces. Maximum strain and maximum splash occur when the depth of water is about the same as the diameter of the raindrop.

Most raindrops are between 1 and 5 mm in diameter. Their terminal velocities are about 4.5 m/s for 1-mm raindrops and 9 m/s for 5-mm raindrops (any raindrop

Figure 19-3   Raindrop splash is the principal agent for detachment of soil particles. *(Soil Conservation Service, USDA.)*

falling from a height of 10 m or more is near terminal velocity). The larger raindrops carry much more energy and cause much more damage because of the combined mass and velocity differences.

A protective cover on the soil surface—usually growing plants or their residues —greatly reduces raindrop impact and erosion. The difference is illustrated in Figure 19-4 where pebbles have protected thin pedestals of soil from raindrop impact.

Soil particles are moved only a meter or two by raindrop splash. Even this movement is largely offset by particles splashed the other way (a small net movement occurs on slopes or when there is wind to direct the splash). "Splash erosion" is important as an agent of detachment of soil particles, not as an agent of transportation. However, transportation may be accomplished either immediately or later by flowing water or wind that might not have been able to detach the soil particles.

Sheet Erosion

A thin film of water forms on the soil surface when rain falls too fast to infiltrate. The water film soon becomes thick enough to flow, and a sheet of water moves downslope. The movement provides kinetic energy that can transport loose soil particles but cannot detach them. Continuing rain provides loose particles by raindrop splash and makes the flowing water turbulent, increasing its energy. Thus a thin sheet of soil is eroded from the surface.

Figure 19-4 Erosion pedestals capped by pebbles that offer protection from raindrops. The pedestal in the center of the photograph is 5 cm tall. The large, dark area is the inside of a rill.

Sheet erosion is the most widespread and the most insidious form of soil erosion. It causes the largest amount of soil loss and probably the largest economic loss of any type of erosion, yet it is the least noticeable form. The damage caused by sheet erosion is even greater, proportionally, than the tons of soil lost would indicate. Sheet erosion sorts the soil particles, leaves behind the larger particles, and carries away the silt, clay, and organic matter. Most of the soil fertility is associated with these fine particles.

The surface layer of soil is normally the most fertile part of the soil profile. Plant residues and fertilizer are added there and tend to remain more concentrated there than elsewhere. This concentration of plant nutrients near the surface greatly increases the loss resulting from sheet erosion. Barrows and Kilmer (1963) found soil erosion carrying away as much as 70 percent of applied phosphorus fertilizer. The concentrations of phosphorus and potassium in sediment from a fertilized field were 3.4 and 13 times as high, respectively, as their levels in the original soil. These nutrients are needed in the field but become pollutants when they reach a stream.

Channel Erosion

Water rarely flows very far as a uniform sheet; instead, it concentrates in low areas and forms a stream. The stream concentrates the erosive power of the water along a channel. The deeper the water becomes, the faster it flows and the higher its erosive power becomes. Erosion of unprotected soil usually becomes apparent at a water velocity somewhere between 0.5 and 1.0 m/s. Water flowing faster than this has enough energy to both detach and transport soil particles. Larger particles—including soil peds and clods—are carried away as the velocity increases. By far the largest part of the damage occurs when the flow is highest. The increased energy of the larger flow causes the erosion rate to be far out of proportion to the volume of flow.

The movement of any particular soil particle may or may not be continuous. The largest particles being transported are likely to move by rolling and frequently become lodged in depressions or behind still larger particles. Other particles somewhat smaller than those that roll may be picked up and moved only a short distance before they settle to the bottom again. Sand particles may be picked up and dropped any number of times. Smaller particles may remain suspended for long distances. Silt often moves for kilometers; colloidal clay may be carried all the way to the ocean.

Channel erosion is divided into three subtypes: rill, gully, and streambank erosion. Rills (Figure 19-5) are the smallest channels. They are small enough to be crossed and smoothed over during normal tillage operations. The end effect of rill erosion, including smoothing by tillage, is the same as that of sheet erosion. A layer of soil is removed from the field, and the soil profile is made thinner. Rills commonly form in the cultivated interrow spaces between sloping crop rows.

Gullies are too large and steep-banked to be crossed and smoothed by normal tillage operations. They commonly form where a concentration of water flows down a slope or over a bank, usually beginning at a low position in the landscape and eroding toward the source of the water. A stream of water that flows harmlessly across a gently sloping field becomes highly erosive as it pours over the steep bank at the upper end of an active gully. A waterfall causes a gully to erode deeply into

Figure 19-5 Rill erosion resulting from crop rows running up- and downhill. (*Soil Conservation Service, USDA.*)

a hillside or other slope that might otherwise resist erosion. Water coming to a gully from various sources may cause it to branch and develop into a gully system ravaging a sizable area of land. Islands of uneroded soil between gullies are often unusable because of inaccessibility.

Gullies erode away entire soil profiles and often cut into unconsolidated material beneath the soil. Their depth is limited either by tough underlying material such as bedrock or by the level of their outlet end. Gullies cutting into deep deposits of loess or other unconsolidated material usually become deeper as they erode into higher land.

Gullies between 1 and 5 m deep are considered medium-sized; those outside this range may be labeled small or large. Some large gullies are over 20 m deep and still growing. Gullies are also classified as active or inactive and U- or V-shaped. The V shapes are the most common except in loess. Loess has a tendency to stand in vertical banks and therefore to form U-shaped gullies. The U shape requires the most work to stabilize because the sides must be sloped before vegetation can be established on them.

Streambank erosion occurs where the main channel of an established stream impinges against its bank. Often this produces nearly vertical banks on the outside curves of a meandering stream. Streambank erosion gradually enlarges small valleys by cutting into nearby uplands.

WIND EROSION

Wind erosion, like sheet erosion, is often ignored. People often complain about dust in the air without any thought of its point of origin. Actually, many fields lose several tons of soil per hectare during a single windstorm. Exceptional cases have been noted where entire plow layers were removed overnight.

Much of the soil carried away by wind is first detached by raindrop splash. Rain on bare soil also has a smoothing effect that increases the susceptibility to wind erosion. Wet soil will not blow, but this does not prevent the wind from first drying and then eroding the soil. The rate of wind erosion after a day or two of wind is likely to be higher if the windy period was preceded by a rain than if it was not.

The process of *saltation* (Figure 19-6) is a significant factor in wind erosion. Particles and aggregates between 0.1 and 0.5 mm in diameter may saltate. Larger particles are too heavy, and smaller particles do not protrude enough above the soil surface. The significance of saltation is not so much in the meter or two of jumping movement of the saltating particles as in the impact of the falling particle when it strikes the surface. Small particles that can be transported long distances are knocked into the air by the saltating sand particles. The small particles are resistant to wind erosion unless saltation or some kind of traffic initiates their movement.

Sand moved by wind usually does not go very far. It may be deposited in a nearby fence row or a ditch that will have to be cleaned out before it can function properly again. Very commonly the sand blown from one field damages the crop in a nearby field, as shown in Figure 19-7. As it moves, it can have a sandblast effect that takes the paint off buildings, cars, or whatever else may be in its way.

Wind erosion is an especially serious hazard on cultivated organic soils. Occurring in low, level positions, these soils are virtually immune to water erosion. But cultivated, flat areas are exposed to the wind, and organic particles are easily transported because of their low densities.

PREDICTING AMOUNTS OF SOIL LOSS

The old saying "An ounce of prevention is worth a pound of cure" is very pertinent when dealing with erosion problems. It is much easier to limit erosion to tolerable rates than it is to repair eroded land. One of the most important keys to a successful

Figure 19-6 Saltation occurs as a fine- or medium-sand particle jumps to a height ranging from a few centimeters to a meter and falls again a short distance downwind. The jump occurs because the particle acquires a spin as it rolls onto the soil surface. A particle spinning in the wind has an airfoil effect that lifts it until the spin slows down.

Wind

Figure 19-7   This wheat crop is almost a total loss because it is being covered by soil blown from an unprotected field nearby. *(Soil Conversation Service, USDA.)*

erosion control program is to predict when and where excessive erosion is likely to occur so that something can be done in time to prevent it.

Many different factors interact to determine the amount of soil loss occurring at a particular time and place. The most important factors influencing sheet and rill erosion have been combined into the *universal soil loss equation* (Wischmeier and Smith, 1965). A similar equation has been developed to predict losses by wind erosion (Moldenhauer and Duncan, 1969), but only the water erosion equation will be discussed here. The universal soil loss equation is

$$A = RKLSCP$$

The average annual soil loss, *A,* in tons per hectare is estimated by multiplying together all the factors on the right-hand side of the equation. These factors are discussed in the following paragraphs.

The rainfall factor, *R,* evaluates the erosive potential of the rainfall. It is the average yearly sum of the products of the kinetic energy of each storm times the maximum 30-min rainfall intensity of the storm. This factor is the most accurate indicator yet developed for the erosive potential of rainfall on fallow land. Values of *R* for the central and eastern portions of the United States can be interpolated from the isoerodent lines shown in Figure 19-8. This map reveals why much of the southeastern United States has suffered more erosion than most of the rest of the nation. The high erosive potential coupled with the growing of row crops (cotton, corn, tobacco, etc.) in that region has stripped away the topsoil from much of the area and created many gullies.

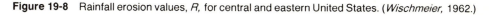

Isoerodents

Mean annual values of erosion index R

Figure 19-8 Rainfall erosion values, R, for central and eastern United States. (*Wischmeier*, 1962.)

The soil erodibility factor, K, indicates the relative resistance of different soils to erosion. The values of K are chosen so that $A = RK$ for unprotected fallow plots on 9 percent slopes 22-m long. These particular plot dimensions were used as data sources for an earlier soil loss equation developed by Browning, Parish, and Glass (1947) and were adopted by Wischmeier, Smith, and Uhland (1958) for the universal soil loss equation. Values of K for some selected soils are given in Table 19-1.

The factors for length and steepness of slope, L and S, can be combined into a single graph (Figure 19-9), constructed so that the LS value is 1.0 for the 9 percent slopes 22-m long used by Browning et al. (1947). Other slopes are scaled proportionally by comparison of their effects on erosion rates. Steeper slopes naturally suffer more erosion under otherwise comparable conditions. So do longer slopes, as shown by each of the curves in Figure 19-9. The lower part of a long slope receives accumulated runoff water from the area above it. The erosive effect of such accumulations is illustrated in Figure 19-10.

The cropping and management factor, C, is a composite of the effects of crops and crop sequence, tillage practices, and the interaction between these factors and the timing of rainfall through the year. The timing effect becomes especially important where the land is sometimes bare and sometimes well covered. The amount of erosion caused by a particular rainstorm on such land depends greatly on when it

Table 19-1 Soil Erodibility (K) Values and Soil Loss Tolerance (T) Values for Selected Soil Series (Most of the T Values Should Be Reduced by 2 tons per hectare in
Severely Eroded Areas)

Soil series and texture	K value	T value, tons/ha	Soil series and texture	K value	T value, tons/ha
Ashdale silt loam	0.72	9	Lindley loam	0.97	7
Ayr fine sandy loam	0.54	9	Lisbon silt loam	0.72	9
Barker silt loam	0.83	7	Manning fine sandy loam	0.54	9
Barnes loam	0.72	11	Marshall silt loam	0.72	11
Bates loam	0.72	7	Miami silt loam	0.83	7
Bayard fine sandy loam	0.45	11	Milton silt loam	0.83	7
Bloomfield fine sand	0.38	11	Monona silt loam	0.72	11
Boone fine sand	0.45	5	Moody silty clay loam	0.72	11
Clarinda silty clay loam	1.10	7	Muscatine silt loam	0.72	11
Clarion loam	0.72	9	Nicollet loam	0.72	9
Clinton silt loam	0.83	9	Norwalk silt loam	0.83	7
Crofton silt loam	0.72	11	Odell silt loam	0.72	7
Dawes silt loam	0.63	11	Pierre clay	1.10	9
Fargo clay	0.97	11	Primghar silty clay loam	0.72	11
Fayette silt loam	0.83	9	Rhoades loam	1.10	7
Fox silt loam	0.83	7	Seymour silt loam	0.97	7
Galva silty clay loam	0.72	11	Sharpsburg silty clay loam	0.83	9
Geary silt loam	0.72	11	Shelby clay loam	0.83	9
Gosport silt loam	1.10	7	Sogn silty clay loam	0.72	2
Gravity silt loam	0.83	11	Storden loam	0.72	9
Grundy silt loam	0.83	9	Tama silt loam	0.72	11
Hamburg silt loam	0.72	11	Tedrow loamy sand	0.63	9
Harmony silty clay loam	0.83	11	Terrill loam	0.72	11
Hastings silt loam	0.83	9	Ulysses silt loam	0.72	9
Hayden loam	0.83	9	Union silt loam	0.97	7
Hosmer silt loam	0.97	7	Varna silt loam	0.83	7
Hubbard loamy coarse sand	0.38	11	Volney silt loam	0.72	11
Ida silt loam	0.72	11	Warsaw loam	0.72	7
Kenyon silt loam	0.72	9	Weller silt loam	0.97	7
Lamont fine sandy loam	0.54	7	Wheeling silt loam	0.72	7

falls. This interaction factor is strong enough to require adjustment of the C factors
for specified cropping systems used in areas with different rainfall patterns.

The value of C is taken as 1.0 for land in continuous fallow plowed up- and
downslope. The C factor for a good growth of permanent pasture is about 0.004
(indicating that continuous fallow loses soil about 250 times as fast as permanent
pasture). Almost all other cropping systems have C values between these two
extremes. Some representative C values used in part of the corn belt are given in
Table 19-2.

The C factor for a crop rotation is a weighted average of the effects of all the
crops in the rotation. The amount of bare ground between plants is usually a good

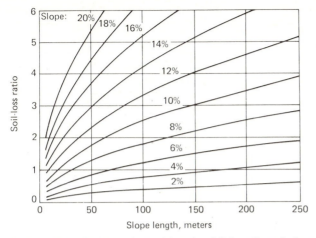

Figure 19-9 Chart for determining the *LS* (length and steepness of slope) factor for the soil equation. (*Adapted from Wischmeier and Smith*, 1965.)

guide to the influence of that plant or crop on erosion. A thick stand of grass has a *C* factor of only 0.004 because very little soil is exposed. Grain crops expose more soil and permit more erosion. Cultivated row crops leave large areas open, especially while the crop is young, and result in larger *C* factors. Clean-cultivated fallow land has the largest *C* factor of all because it is completely open.

The conservation practice factor, *P*, is assigned a value of 1.0 unless a special practice such as contouring or contour strip-cropping is used. The value of these

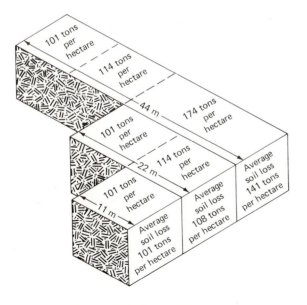

Figure 19-10 A schematic illustration of the effect of slope length on soil losses from continuous corn planted up- and downhill on Marshall silt loam. (*Data from Wilson*, 1947.)

Table 19-2 Crop Management (C) Factors for Wisconsin and Areas of Adjoining States. Ratio of Soil Loss from Cropping Systems to Loss from Continuous Fallow

Cropping	Residues left on surface until spring plowing			Residues plowed under in fall		Residues removed in fall; spring plowing		Residues left on surface; wheel-track planting	
Corn yield, q/ha	25–40	40–50	50+	40–50	50+	40–50	50+	40–50	50+
Hay yield, tons/ha	2–5	5–8	8–12	5–8	8–12	5–8	8–12	5–8	8–12
Continuous row crop	0.47	0.43	0.38	0.49	0.43	0.55	0.52	0.32	0.26
RRROx*	0.35	0.31	0.28	0.38	0.34	0.38	0.35	0.22	0.19
RROx	0.31	0.27	0.24	0.34	0.31	0.33	0.31	0.19	0.17
ROx	0.23	0.19	0.17	0.27	0.24	0.24	0.23	0.14	0.12
RRROM	0.25	0.20	0.175	0.25	0.21	0.28	0.26	0.14	0.12
RROM	0.19	0.15	0.13	0.20	0.15	0.21	0.19	0.10	0.09
RROMM	0.15	0.12	0.10	0.16	0.12	0.17	0.15	0.08	0.07
RROMMM	0.127	0.10	0.086	0.13	0.10	0.14	0.13	0.07	0.06
ROM	0.11	0.083	0.061	0.13	0.10	0.115	0.098	0.055	0.042
ROMM	0.083	0.063	0.047	0.098	0.074	0.088	0.075	0.042	0.032
ROMMM	0.067	0.051	0.038	0.080	0.060	0.071	0.061	0.035	0.026
ROMMMM	0.057	0.043	0.032	0.067	0.051	0.059	0.051	0.029	0.023

*R stands for row crop, O for oats, Ox for oats with green manure, and M for legume-grass meadow.

practices for reducing erosion on various slope gradients is shown in Table 19-3. The values in the table are the amount of soil lost *with* the practice divided by the amount lost *without* it. The most effective of these practices, contour strip-cropping with alternate meadow strips on slopes between 2 and 7 percent, reduces soil loss to about one-fourth of what it would be without the practice.

One other conservation practice, terracing, should be mentioned here. The estimated soil movement with terracing is calculated by using the spacing between terraces as the slope length and applying the P factor for contouring or strip-cropping where they are used. Most of this soil loss will accumulate in the terrace channel and may be considered more tolerable than soil loss that leaves the field.

The use of the universal soil loss equation may be illustrated by the example of a farmer in central Iowa. A field containing Clarion loam soil with 5 percent slopes 100-m long will be assumed. This farmer produces 60 q/ha of corn with a corn-corn-oats-meadow rotation and standard management practices without contouring or other similar practices. The value for P is therefore 1.0 and the values for RKLSC can be obtained from Figures 19-8 and 19-9 and Tables 19-1 and 19-2 to give

$$A = R \quad K \quad LS \quad C \quad P$$
$$170 \times 0.72 \times 1.00 \times 0.13 \times 1.0 = 15.9 \text{ tons per hectare}$$

The predicted average annual soil loss of 15.9 tons per hectare compared with the tolerable soil loss of 9 tons per hectare for Clarion loam (Table 19-1) indicates that

Table 19–3   Conservation Practice (P) Factor Values for Use in the Soil Loss Equation

| Percent slope | Contouring | Practice factor values | |
		Strip-cropping (with alternate meadow strips)	Strip-cropping (with alternate small-grain strips)
1.1– 2.0	0.60	0.30	0.45
2.1– 7.0	0.50	0.25	0.40
7.1–12.0	0.60	0.30	0.45
12.1–18.0	0.80	0.40	0.60
18.1–24.0	0.90	0.45	0.70

Source: Based on Wischmeier and Smith, 1965.

this farmer would have excessive soil loss under the assumed system. He could reduce his soil loss to below the tolerable limit in any of several ways. He could change to a corn-oats-meadow rotation with a C factor of 0.061 and reduce his soil loss to about 7.5 tons per hectare. He could retain the corn-corn-oats-meadow rotation and use contouring with a P factor of 0.5 to reduce his soil loss to 8.0 tons per hectare. Or he could use contour strip-cropping or terracing and increase the amount of row crop if he desired.

TECHNIQUES FOR CONSERVING SOIL

Erosion involves detachment and transportation of soil particles. Erosion control practices can be aimed at reducing either detachment or transportation, or both. For example, a vegetative cover or a mulch on the soil surface prevents detachment because it prevents raindrops from striking the soil surface. Rough contour plowing may permit detachment but prevent transportation because water is held in depressions until it infiltrates.

The most appropriate type of control measure depends on the type and severity of erosion hazard and on the nature of the soil. In general, the particles in sandy soils are easily detached but are difficult to transport. Erosion control practices that limit runoff to slow velocities are highly effective for conserving sandy soils. The particles in soils high in clay are usually difficult to detach but easy to transport. It is usually more practical to limit detachment of particles from clay soils than to prevent detached particles from being transported.

Aggregate stability is a major variable that is well correlated with soil erodibility. Soils containing significant amounts of clay form aggregates with varying degrees of stability. Unstable aggregates break apart under the impact of raindrops. Some silt and clay particles broken from aggregates are carried away in runoff water; other particles are carried into soil pores and tend to lodge there. The crust thus formed has a low permeability and results in greatly increased runoff and erosion. Stable aggregates resist raindrop impact for a much longer time and maintain the soil in a more permeable condition.

Most soil aggregates have some cohesion from one aggregate to another, which helps the soil resist erosion. Some aggregates, however, have strong internal cohesion but little or no external cohesion to other aggregates. Soils dominated by such aggregates behave like sandy soils so far as erosion is concerned even if they have high clay contents.

The usual relation between aggregate stability and soil erosion is illustrated in Figure 19-11. As shown in the figure, the percentage of water-stable aggregates usually decreases with successive years of row crop and results in increasing soil loss. This trend can be reversed by including grass-legume crops, cover crops, and green manures in the cropping system.

Aggregate stability is strongly influenced by the clay and organic-matter contents of the soil. The bonding mechanisms involved are discussed in Chapter 3 and will not be repeated here. It must be realized, however, that type of clay and the adsorbed cations are as important as the amount of clay present. Clays of the montmorillonitic type are generally more erosive than clays of the kaolinitic and oxide types. It is fortunate that the less erosive types of clay are commonly present in the soils of warm, humid climates. High-intensity rainfall is common in such climates and would undoubtedly cause very severe erosion if the montmorillonitic type of clay were dominant there.

The addition of plant residues to a soil results in a rapid increase in microbial activity. Aggregate stability increases markedly as the microbial activity increases, then decreases gradually as the microbial activity declines. McCalla (1946) showed that fungi are the most effective group of soil microorganisms in stabilizing aggregates. Aggregate stability depends more on the rate of microbial activity than on the amount of organic matter present in the soil.

Vegetative Methods of Erosion Control

Methods of controlling erosion are often divided into two groups—vegetative and mechanical. Both types are useful; sometimes one works best, sometimes the other. Sometimes both are needed.

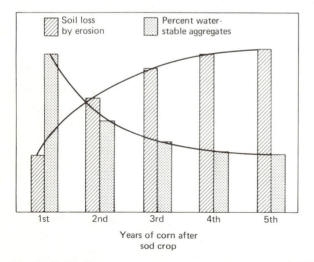

Figure 19-11   The inverse relation between aggregate stability and soil erosion. (*Based on data of Browning et al.,* 1948, *and Smith et al.,* 1948.)

Vegetative methods can be highly effective at controlling erosion. Natural prairie and forest vegetation are vegetative methods that normally limit erosion to less than half a ton of soil loss per hectare per year. Very few cropped fields have such low soil losses even if they are protected by intensive erosion control practices.

One of the first things to be considered is an appropriate vegetative cover for the land. In urban and suburban areas, along highways, etc., this usually means selection of the types of grass, flowers, shrubs, and trees to be grown. The principal erosion control problems encountered in such areas occur when the ground is bare before permanent vegetation is established. Gardens sometimes have erosion problems later and would have more if they were not usually on relatively flat ground. Areas that do need erosion control practices can be treated with some of the same methods that farmers use on cropland.

Selecting appropriate vegetation for cropland involves designation of a cropping system. Using the soil loss equation discussed earlier in this chapter is a good way to judge whether a particular cropping system is suitable for a designated piece of land. Systems that would result in too much erosion should be supplemented by special erosion control practices or replaced by other systems that provide better protection, usually through a thicker vegetative cover. Data showing the percent runoff and soil loss from several different crops are given in Table 19-4. These data

Table 19–4 Percent Runoff and Soil Loss under Different Management Systems and Crops at Temple, Tex. (Hill et al., 1944), Clarinda, Iowa (Browning et al., 1948), and Statesville, N.C. (Copley et al., 1944)

Location and crop	Runoff, %	Erosion, tons/hectare-year
Temple, Tex.:		
Continuous corn	13.6	46.2
Corn in rotation	14.5	43.9
Oats in rotation	3.3	4.7
Cotton in rotation	13.4	38.6
Clarinda, Iowa:		
Continuous corn on topsoil	18.7	85.8
Continuous corn on subsoil	20.2	115.6
Corn in rotation	12.6	41.2
Oats in rotation	9.9	22.6
Clover in rotation	3.8	12.1*
Continuous alfalfa	2.2	0.2
Continuous bluegrass	1.2	0.1
Statesville, N.C.:		
Corn in rotation	10.7	64.3
Wheat in rotation	13.5	12.6
Lespedeza in rotation	5.3	3.4
Cotton in rotation	10.4	48.9
Continuous cotton	12.4	70.0
Rotation average	10.0	32.3

*This loss resulted from failure of clover because of drought and grasshopper damage in 1937 and 1938. The annual report, published in March, 1950, shows no soil loss under rotation clover from 1943 to 1949.

come from three different states, Iowa, Texas, and North Carolina, but they tell a consistent story. Row crops such as corn and cotton result in the largest soil losses. The losses from small grains are smaller than those from row crops but larger than those from close-growing crops such as alfalfa or bluegrass.

A row crop grown as part of a rotation causes less soil loss than it does grown continuously, an effect that can be seen in Table 19-4. The carry-over effect of the previous crop shows clearly in Figure 19-12. Alfalfa, clover, or grass plowed to only a shallow depth is highly effective at reducing erosion from a row crop grown the following year. This favorable effect is reduced, however, where the forage is removed for hay or silage, leaving little or no residues to be returned to the soil. Excessive erosion can usually be avoided by increasing the frequency of close-growing crops on the less stable areas.

After deciding on the appropriate vegetative cover, the second thing to consider is how to make the chosen crops grow well. Slow growth after planting leaves the soil bare and erodible for too long. A poor stand leaves bare spots subject to erosion at any time. The crop should be given every advantage to grow well, with lime and fertilizer supplied in optimum amounts. The improved crop growth will not only produce more profit but will also reduce soil erosion. Selection of vigorous crop varieties and planting at the right time are also helpful for the same reasons.

A third consideration in growing crops is the plant population or number of plants per hectare. Appropriate varieties and high fertility may make it possible to grow more plants per hectare and thus reduce the open space between plants. Row crops may be grown with narrower row spacing that will permit them to protect more of the soil from erosion.

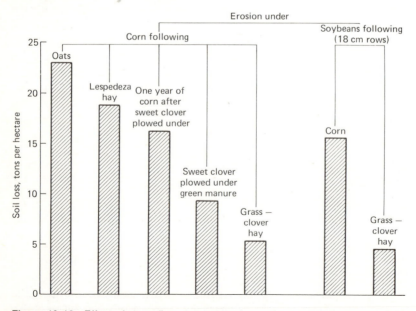

Figure 19-12 Effect of preceding crop on erosion under corn and soybeans. (*Based on data of Smith et al., 1948, for the growing season from April 27 to October 7.*)

Cover Crops   The principal crops in a rotation may leave the soil unprotected at times. For example, a crop may be planted in the spring and harvested in the fall. The erosion hazard is especially great if fall plowing is practiced with spring crops. Harvesting the entire plant also exposes the soil to both water and wind erosion over winter. A cover crop may be the answer to the problem if the weather and the supply of soil moisture permit one to be grown.

A cover crop must be fast-growing so that it can be established after harvest but before winter. It must be hardy enough to tolerate the winter of the area where it is grown. The amount of late fall, winter, and early spring rainfall has much to do with the practicality of a winter cover crop. If there is no water supply, it will neither grow nor be needed for controlling water erosion.

Small grains such as rye or wheat are often used as winter cover crops. Several legumes, including vetch and winter peas, are also used. The cover crop normally is not left long enough to mature but is plowed under as a green manure crop prior to the planting of the next crop in the rotation.

Companion Crops   Occasionally two crops are planted at the same time but harvested separately. These are commonly a fast-growing small grain such as oats or wheat and a slow-growing crop such as hay, pasture, or green manure seeding. The fast-growing crop has been called a "nurse crop," but this terminology is misleading. The two crops are in competition for water and nutrients—neither is likely to "nurse" the other. Care is usually taken not to provide a high level of nitrogen for companion crops because too much nitrogen would make the small grain grow too vigorously and result in a poor stand of its companion crop.

Properly managed companion crops have a very favorable effect on erosion control. The fast-growing crop provides early protection and helps to control weeds; the slow-growing crop provides a thick cover later. Grass-legume mixtures used as hay or pasture for 2 or more years, then incorporated into the soil for green manure, are highly beneficial to aggregate stability and erosion resistance for most soils.

The companion-crop approach is often useful in nonagricultural situations as well as to farmers. A highway right-of-way, for example, may well need the protection of a fast-growing crop during the time that its permanent vegetation is being established.

Special Vegetation in Problem Spots

Erosion potential is seldom uniform over an entire area. More water usually crosses the lower part of a slope than the upper part, often accumulating in swales where it may cause gullies to form. Sheet, rill, and gully erosion all become greater hazards on the lower part of long slopes. Wind erosion also shows a related tendency. A small open area may not suffer much from wind erosion either because the wind is deflected from it by a nearby obstacle or because it has few particles to saltate. The downwind portions of larger open areas are often subject to more severe wind erosion resulting from *avalanching.* Avalanching results from an accumulation of saltating particles—some from upwind areas and some from the local area, including many that would not have saltated had they not been knocked loose by others.

Most of the problems cited above can be overcome by placing special vegetation in appropriate areas. Sometimes the situation can be handled by planting the regular crops in special patterns as in strip-cropping. Some problems, however, call for permanent vegetation in certain places to avoid gullies, to keep silt out of ponds and reservoirs, or to serve as windbreaks.

Strip-cropping

A strip of close-growing vegetation placed perpendicular to the flow of water or wind can provide protection to an adjacent strip of row crop or of fallow land. Often a crop rotation is practiced by alternating the crops grown on the various strips. Several different applications of this practice are known as contour strip-cropping, border strip-cropping, buffer strip-cropping, field strip-cropping, and wind strip-cropping. Contour strip-cropping will be considered in some detail after a brief discussion of the other types.

Border strip-cropping involves the growing of strips of permanent vegetation placed around the border of a field, a farm, or a body of water. The border strip may be used to control either wind or water erosion. It is especially useful to filter the silt out of runoff water before it reaches a pond or stream. The protective vegetation can also utilize much of the plant nutrient supply of the runoff water and thus deter eutrophication of the body of water being protected. Border strips are very useful for this purpose when placed between feedlots and streams.

Buffer strip-cropping includes strips of permanent vegetation placed between some of the cropped strips in a field. The buffer strips are often irregular in width and perhaps discontinuous so the cropped strips may remain uniform in width across variable slopes. The buffer strips usually include any unproductive soils, steep slopes, or other problem spots that may be present in the field. Buffer strips are sometimes used for hay or pasture if they are large enough to make such use worthwhile.

Field strip-cropping is the practice of growing crops in strips across the general slope but parallel to a field boundary rather than on the contour. The strips are uniform in width and easy to cultivate and harvest, but the effectiveness of control of water erosion is reduced by the deviation from the contour. This method should be limited to fields where the chosen field boundary is nearly parallel to the contour lines or where the water erosion problem is not too serious.

Wind strip-cropping involves the planting of crop strips perpendicular to the prevailing winds irrespective of contour lines (Figure 19-13). The protection offered is partly that of a series of closely spaced miniature windbreaks. Probably more important, however, is the effectiveness of crop strips at catching saltating sand particles. This prevents the avalanching effect of saltation produced by wind blowing across large open areas.

Wind strip-cropping is useful in semiarid areas such as the Great Plains of the United States. The soils are frequently dry enough to blow in such areas because of the limited rainfall. The climate is generally too dry to grow trees for windbreaks. Furthermore, much fallowing is used in these areas in an effort to improve crop yields. The hazards of bare soil resulting from fallowing can be reduced by alternating fallow strips with crop strips. Sometimes the system is modified by using vegeta-

Figure 19-13   Strip-cropping to control wind erosion and conserve water in Nebraska. These are alternate strips of wheat and summer fallow. *(Soil Conservation Service, USDA.)*

tion or crops that are not a part of the rotation and may not be harvested. Sudan grass or grain sorghum is sometimes planted in very wide rows on land which will later be seeded to grass or small grain. Even strips of stubble left across fallow fields can help to control wind erosion. The same strips that help control erosion during the summer also help to catch snow during the winter. Holding snow on the fields improves the supply of water for plant growth the following year as well as reducing travelers' problems with snow drifts.

Contour Strip-Cropping

Contour strip-cropping runs rows around hills and alternates crops in the various strips (Figure 19-14). Erosion from fields where contour strip-cropping is best adapted and used is reduced to 25 percent or less of what it would be without the practice. Water moving across a strip of fallow or row crop picks up some soil and may even begin to concentrate in the swales and increase its velocity. But as it enters another strip with close-growing vegetation, the water is spread around the hill, its velocity is reduced, and much of the suspended silt is deposited. The water leaving the protective strip is no longer concentrated and moves more slowly than when it entered the strip. Grass waterways are often employed in major drainageways where there is too much water to be spread out again.

There is an upper limit to the length of slope that can be protected by contour strip-cropping alone. The reason is the gradual accumulation of water downslope. Protective strips spread and slow the water, but it still flows through them and is

Figure 19-14 Contour strip-cropping reduces erosion on hilly land.

added to the rainfall on the next cultivated strip. The amount of water is increased with each successive strip and eventually becomes too large to be handled in this manner. Maximum length of slope depends on the soil and climate and should be established locally. Limits of 300 m for well-drained soils and 150 m for poorly drained soils may be used if local guidelines are lacking. The effective length of longer slopes can be reduced by terracing the upper part of the slope or by placing diversions across the slope.

The minimum width of each protective strip should be enough to filter nearly all the silt from the runoff water (Figure 19-15). Wider protective strips are needed on steeper slopes because higher water velocities are involved. Suggested minimum widths of protective strips are shown in Table 19-5.

Maximum widths of clean-tilled strips depend on the soil, the slope, and the climate. Local guidelines should be established. Lacking these, the values shown in Table 19-5 may be used.

The first decision in planning the width of strips is the ratio of close-growing crops to clean-tilled crops in the rotation. Another important consideration is the width of the equipment to be used in the field. The strip width should be some multiple of equipment width. Suppose that a Minnesota farmer with 4 percent slopes 150-m long has calculated a probable soil loss of 20 tons per hectare if he uses a corn-corn-oats (with green manure) rotation. He notes that the P factor for corn with small-grain strips (Table 19-3) is 0.4 and would reduce his estimated soil loss to a tolerable value of 8 tons per hectare. Assume further that he has equipment

Figure 19-15 A small grain strip reducing soil loss by slowing the runoff water thereby causing soil deposition. *(Soil Conservation Service, USDA.)*

4.5-m wide (six 75-cm rows of corn). Three equipment widths would equal 13.5 m, which is more than the minimum width of the protective strip (Table 19-5). There are 2 years of corn in the rotation to 1 year of oats, and so the corn strips will be twice as wide or 27 m. This is approximately the width interpolated from Table 19-5 for 4 percent slopes and is equal to six times his equipment width. Both the soil loss equation and the strip width calculations indicate that this rotation would be too intense for slopes much steeper than 4 percent.

Contour strip-cropping is most effective where the amount of runoff is not excessive. Its application to soils of low permeability in regions that receive high-intensity rainstorms is limited to slopes of 1 or 2 percent, but permeable soils in areas receiving less intense rainfall may receive significant protection by contour strip-cropping on slopes up to 25 percent.

Contour strip-cropping reached a peak of popularity in several parts of the United States in the 1930s. It was probably overused at that time because it is an

Table 19–5 Suggested Minimum Widths of Protective Strips and Maximum Widths of Clean-tilled Strips for Contour Strip-cropping (to Be Replaced by Local Guidelines If Available)

Slope, %	Minimum width of protective strip, m	Maximum width of clean-tilled strip, m
1	7	50
3	9	30
6	12	20
9	15	15

inexpensive practice and little was known about its limitations. The popularity of strip-cropping has declined since then, partly because of larger farm equipment and partly because of border effects. Long, narrow crop strips give long, exposed edges where the crop may be damaged by hot winds, grasshoppers, etc. Still, it is an effective means of reducing soil erosion and is useful in many places.

Windbreaks The best protection against wind erosion is a permanent vegetative cover on all land. But since the cover cannot be permanent on cropland, some kind of protection is needed for areas that must be bare during times when strong winds may occur. Wind strip-cropping is a means of protecting narrow strips of bare land; windbreaks can be used to protect wider strips. The width protected is a function of the height and shape of the windbreak, the wind velocity, and the erodibility of the soil. An average width of effective protection is about 10 times the height of the windbreak. Within this zone the wind velocity is normally less than half of what it would be without the windbreak.

Maximum distance of protection is obtained by planting several rows of trees in the windbreak (Figure 19-16). The tallest trees are in the center and are flanked by medium-height trees including at least one row of evergreens for winter protection. Tall shrubs or short trees that have low branches are used on the outside so that wind will not blow under the windbreak. Such windbreaks provide maximum height, and their sloping sides cause maximum upward deflection of wind.

Figure 19-16 A seven-row windbreak in South Dakota.

The principal disadvantage of multirow windbreaks is that a large amount of land is required. Single-row windbreaks are therefore used in some places. Usually a medium-height tree that produces branches almost to the ground is chosen. It must be an evergreen if winter protection is needed. Spruce trees are among the best for this purpose. More windbreaks will be needed if a large area is to be protected because they are not as tall as the multirow windbreaks.

Single-row windbreaks leak more air between trees than do the multirow types, but this may be an advantage. A dense windbreak reduces wind to a low velocity near the windbreak, but the velocity increases markedly within a relatively short distance downwind. Porous barriers give less protection, but the protection they offer extends for greater distances than that from dense windbreaks. The leaked air is not moving fast enough to cause erosion unless there is a large hole (probably a missing tree) in the windbreak. Any tree that fails to grow in a single-row windbreak should be replaced promptly.

A possible problem arises at the end of a windbreak. The wind velocity there is usually higher than it would be without the windbreak. One good solution to this problem is to end the windbreak at a pasture or other area of permanent vegetation. Or, the windbreak may end like the one in Figure 19-16 in an area shielded by a hill.

Grass Waterways   A concentrated flow of water in swales is often unavoidable. Such areas are likely places for gullies to form, especially if cultivated crops are grown in the swales. Grass waterways are used where the risk of channel erosion would be excessive if the area were cultivated. The need for a grass waterway should be recognized and acted upon early because it is much easier to prevent a gully from forming than it is to repair the land later. The easiest way to establish a grass waterway is simply to leave that portion of the field unplowed when other crops follow hay or pasture. This will work, of course, only when close-growing perennial crops are included in the rotation and where the land is properly shaped for a waterway. Otherwise the land must be prepared and seeded, sodded, or planted with cuttings.

An ideal waterway is either concave or V-shaped. The V shape is preferred where the bottom of the waterway dries out slowly (it minimizes the area that stays wet). Either shape should be made broad enough to cross and must be large enough to carry the maximum probable runoff (Figure 19-17). Waterways that are too small may result in two gullies, one on each side of the grass strip.

Most pasture grasses can be used for grass waterways if the erosion hazard is not too severe. Kentucky bluegrass, for example, is often quite satisfactory. The worst places, however, call for an especially tough and aggressive grass, particularly where an active gully already exists and an attempt is being made to smooth and heal it. Reed canarygrass has proven to be the best grass to use in such situations in the northern United States. It is tough and aggressive enough to grow in an old gully or to keep a new gully from forming. Reed canarygrass is difficult to establish by seeding but easily established from sod cuttings.

Figure 19-17 A well-sodded grass waterway protecting a swale from channel erosion.

Protecting Grazing Lands from Erosion

Extensive areas of grazing lands are located in arid regions; smaller parcels are distributed almost everywhere. New vegetation is difficult to establish on much of this land because of dry weather, stony soils, steep slopes, etc., and so maintaining the vegetation already on the land is very important. Erosion control therefore consists largely of managing the livestock.

Grazing lands produce a maximum amount of usable forage if they are grazed moderately. Overgrazing weakens the plants and slows their growth; it also reduces the vegetative cover and increases the amount of erosion. A good rule for ranchers using semiarid rangeland to follow is to "take half and leave half" of the year's growth. This rule coupled with the rotational deferred-grazing system discussed in Chapter 17 provides a good general management plan. Additional care is needed to obtain uniform grazing over the entire area. Fences, water, and salt must be placed in appropriate locales to cause the livestock to cover the entire area. Salt blocks are especially useful for this purpose because they can easily be moved to an area that is not being grazed.

Mechanical Methods of Erosion Control

Vegetative methods can usually control erosion if they are applied soon enough, but areas that have already been seriously eroded may need mechanical methods of repair. A mechanical method may be preferred over a vegetative method also because of flexibility of land use. Vegetative methods automatically specify that certain areas must have certain vegetation. A mechanical method may permit growing a more profitable crop in these areas. Frequently, of course, vegetative and mechanical methods of conserving soil are combined to obtain maximum effectiveness. Contour strip-cropping is a good example of a practice that is partly vegetative and partly mechanical.

Contour Tillage Row-crop farmers have traditionally taken pride in the straightness of their rows. Teen-age boys learning to handle horses years ago were admonished "Don't look back," because looking back usually caused a crooked row that made cultivation difficult. The problem with long, straight rows is that most of them come to a hill sooner or later. Cultivating makes the area between the rows into a sloping channel lined with loose soil. Rill erosion is a very common result.

Contour tillage (Figure 19-18) is one of the simplest and least expensive soil conserving practices known. In fact, working around a hill rather than up and down it usually provides a saving in the form of reduced fuel consumption. Browning (1948) estimates an average fuel saving of 10 percent resulting from contour tillage because contouring eliminates pulling implements uphill. Contouring may also reduce the time requirement for tillage, but this saving occurs only where the slopes are relatively smooth and uniform. Contouring is less likely to be used on rolling topography because this combination produces many point rows, crooked rows, and an increased time requirement.

Figure 19-18 Contouring permits row crops to be grown on land that would erode excessively if the rows were straight.

Contour tillage forms small ridges across the slope. Water is stored behind the ridges, and therefore more water infiltrates and less runs off. Runoff that does occur is usually slower and less erosive than it would be otherwise. Contouring eliminates or greatly reduces erosion from storms of low to medium intensity but offers little or no protection from storms heavy enough to overflow most of the ridges formed by cultivation.

Contouring is most effective on gentle slopes (2 to 7 percent) of moderate length. On the average, contouring decreases soil loss by 50 percent on gentle slopes. Runoff from slopes less than 2 percent either is not a problem or is large in volume because of soil permeabilities much lower than rainfall rates. Contouring is therefore less beneficial on flatter slopes than it is on gentle slopes. Steeper slopes cause the storage depressions behind small ridges to be smaller, easier to overflow, and less effective for controlling runoff. The effectiveness of contouring on moderate slopes can be increased by using listers or other implements that produce very rough surfaces.

Tillage to Control Wind Erosion   Both timing and type of tillage have strong influence on wind erosion. Anything that reduces wind velocity at the soil surface will reduce wind erosion. A good cover of vegetation or of crop residues provides excellent protection. The duration of periods without such protection depends greatly on the timing of cultivation. Fall plowing for a spring-planted crop, for example, can result in greatly increased wind erosion.

A rough, cloddy surface is also good protection against wind erosion. Plowing may produce enough surface roughness for this purpose if the soil contains enough clay to produce stable clods. Often, however, clods produced by fall plowing are broken down by raindrop impact, wetting and drying, freezing and thawing, etc. Susceptibility of plowed land to wind erosion therefore tends to increase with time.

Emergency tillage is sometimes used to reduce wind erosion on fields that have become too smooth. The implement used should bring clods to the surface and make ridges across the path of the wind. Such tillage may sharply reduce wind erosion, but the effect is usually short-lived because windblown particles and raindrop action are likely to smooth the surface again within a few days or weeks.

Crop Residue Utilization   Most crops leave some residues behind that can be quite helpful in reducing soil erosion. The residues are present at the very time when protection is most lacking—during the season when no crop is growing. The benefit derived from these residues depends greatly on tillage practices. Their value for erosion control is nullified if they are plowed under long before the next crop is planted.

Plowing in or disking in residues gives much more protection than plowing them under. The part of the residues remaining above the soil gives protection from raindrop impact and reduces runoff velocities and wind velocities at the soil surface. Partially buried residues help to hold passages open for water penetration and thereby favor infiltration and reduce runoff.

The reduced-tillage practices discussed in Chapter 17 are excellent for erosion control because they leave crop residues on the soil surface. The effectiveness varies according to how much residue is left on the surface and whether all or part of the soil is covered. Some no-tillage systems are nearly as effective as permanent grass cover for controlling erosion. For example, Bennett, Mathias, and Lundberg (1973) noted no soil loss on hilly land where orchard grass was killed chemically prior to planting corn. Conventional tillage in the same experiment resulted in heavy soil loss. Several different grasses and small grains can be used as cover crops in no-tillage systems. In each case, an appropriate chemical is used so the cover crop is either killed or so retarded in growth that it will not interfere with the crop being planted (Robertson, Lundy, Prine, and Currey, 1976). Crown vetch is another recommended cover crop for no-tillage corn (Hartwig, 1974). Several different herbicides will retard crown vetch so it will not interfere with corn growth and yet will recover and protect the soil after the corn is harvested.

Mulching Crop residues or other materials spread on soil to protect it from erosion, help it absorb water, or control weeds are called *mulches.* Meyer, Wischmeier, and Foster (1970) reported that 1 ton of straw mulch per hectare reduced erosion on a 15 percent slope to less than one-third and 4 tons per hectare reduced erosion to less than 5 percent of the erosion rate on unmulched soil. Mulching is especially effective for controlling sheet erosion because it breaks the fall of raindrops. Runoff velocities are also reduced enough to reduce channel erosion but not always enough to stop rills from forming on steep slopes (Lattanzi, Meyer, and Baumgardner, 1974).

Mulching is as useful on roadbanks and other construction sites as it is on fields and gardens. Construction site erosion often exceeds 100 tons per hectare per year. Roadbank erosion of 46 tons per hectare in 6 weeks was reported by Meyer, Schoeneberger, and Huddleston (1975). Several different mulches were reported helpful by Dudeck, et al. (1970) with excelsior mat and jute netting being best for severe conditions.

Terracing Most people think first of terracing when they think of methods of conserving soil. Terracing is usually the most effective and the most expensive method available for general field use. Many terraces are as old as the Romans in Europe and the Incas in South America. These and other ancient peoples made many hillsides look like wide giant staircases (Figure 19-19) by terracing them from top to bottom. Many of these terraces completely eliminate the effect of slope by means of vertical stone walls. The exposed soil has a level surface and erodes no more than level bottomland erodes. Such terraces are known as *bench terraces.*

Many ancient bench terraces are still in use. Most of them were built by hand and are similar in style to the ones shown in Figure 19-19. Modern terraces, however, are built with machinery and designed to be farmed with machinery. Several different types of terraces have been developed in recent years.

Channel-type terraces were the first type to be built with machinery. They are formed by plowing around the hill and throwing soil from the channel downhill to form a ridge. The channel is given a slight gradient (0.1 to 0.4 percent) so that runoff water caught by the terrace is led slowly around the hill. Early experience with this

Figure 19-19 Bench terraces built by the Incas at Machu Picchu, Peru. These terraces average about 2 m high and 3 m wide and are held in place by stone retaining walls.

type of terrace proved the wisdom of constructing outlets first and terraces later. The concentration of water discharged from a terrace can readily cut a gully if the outlet is not protected. Sometimes the water can be spread out on a pasture. Frequently grass waterways are used to dispose of water from terraces. Some modern terraces dispose of water through surface inlets leading into buried pipelines.

Broad-base terraces were designed by Priestley H. Mangum and first built on his farm in North Carolina in 1885. Soil for this type of terrace is thrown from both uphill and downhill into a broad ridge (Figure 19-20). Broad-base terraces are made wide enough and smooth enough to be farmed with large machinery. The channel is also broad and shallow so that water spreads across a wide area. These and other terraces built in subhumid regions are often horizontal and closed at the ends so that no runoff is permitted. Water conservation and infiltration become as important as soil conservation in dry regions.

Seeded-backslope terraces are built on land that is too steep for the broad-base type. The downhill side of these terraces is made as steep as possible so that it will be quite narrow. It is then seeded to permanent vegetation to protect it from erosion.

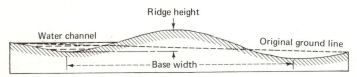

Figure 19-20 Diagram of the broad-base-type terrace designed by Priestley H. Mangum.

Figure 19-21 A schematic cross section of a slope showing two push-up terraces. The dashed lines show the original land surface.

These terraces are often built by the push-up technique in which a bulldozer pushes the soil uphill to form the ridge. This technique is advantageous because it reduces the slope below the terrace (Figure 19-21). The terrace channel is entirely above the original ground surface. Any silt caught in the terrace channel will raise its level toward the bottom level of the next higher terrace. The end result is a benching effect similar to that of the ancient terraces shown in Figure 19-19. Push-up terraces are practical for slopes up to about 8 percent. Terraces on slopes steeper than 8 percent are usually built from both above and below.

Terrace Systems Long slopes often require several terraces such as those shown in Figure 19-22. Each of the terraces in this figure follows its own specified gradient. As a consequence, the spacing between terraces varies wherever the slope changes. The inevitable result of farming such terraces is point rows. Many farmers refuse to build or farm terrace systems that have large numbers of point rows. Consequently, ways have been sought to make terraces run parallel to each other. Terraces across minor slope variations can usually be made parallel by allowing some variation in terrace gradients. Larger variations in slope may be handled by earth moving to form "cut-and-fill" terraces.

Figure 19-22 Terracing on a community basis near Temple, Tex. *(Soil Conservation Service, USDA.)*

The cut-and-fill technique is also used to make terraces much straighter than they otherwise could be. Some modern terraces disappear on ridges where water will not accumulate anyway. Where they do cross a high point they may consist only of a channel cut into the hill (Figure 19-23). This soil is used to make a fill go straight across a low place rather than following the contour around the swale. A pipe outlet is then provided to remove any water that would otherwise be trapped in the swale. This type of terrace tends to smooth the topography of the field because sediment washed from the high areas gradually fills the swales behind the terraces.

The amount of erosion suffered by sloping land between terraces can be estimated by the soil loss equation using the spacing between terraces as the slope length. Most of the eroded soil is caught in the terrace channels and never leaves the field. Laflen, Johnson, and Reeve (1972) measured the soil loss from four tile-outlet terrace systems in Iowa and found that over 95 percent of the eroded soil was caught in the terrace channels. Siltation in the channels requires that regular maintenance be performed to keep them from silting full and allowing water to overtop the terrace ridge.

Diversions Many situations call for diverting water away from a particular area. A very long slope may accumulate so much runoff water that it cannot be handled by contouring, strip-cropping, etc. A gully may be forming in a site where it can be controlled only if the water can be diverted away from it. A large amount of water may accumulate on a flat hilltop without causing damage, but if it runs down a sloping hillside the results may be disastrous. These and many other situations may be handled by means of diversions.

Diversions are a form of terrace designed to carry water away from an area where it might cause damage. Many diversions carry more water than most terraces and therefore need large channels. Diversions with cultivated fields above them should usually be protected with a border strip of permanent vegetation to keep excessive amounts of silt out of their channels.

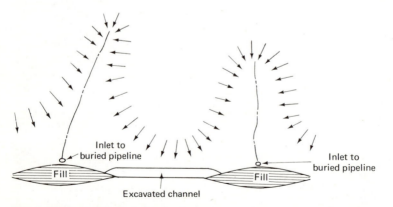

Figure 19-23 A plan view showing how excavated channels on high points and fills in low points can produce straight terraces. Ponded water is removed by buried pipelines. The arrows placed around the contour point downslope.

Controlling Gullies

It is best to avoid gullies by protecting critical areas with grass waterways, terraces, and other erosion control techniques. Once a gully is formed, however, every effort should be made to control it. This involves three important procedures: (1) control (often by diversion) of water above the gully, (2) reshaping the gully and adjoining area into an acceptable topography, and (3) establishing vegetation in the gully.

Reshaping the gully sides and the head end of the gully is especially important with U-shaped gullies. Vegetation usually will not grow on vertical gully banks. The vegetation finally established may be a tough pasture grass, a legume, or some kind of thick cover of vines or brush. The vegetation may have some value for wildlife or perhaps for limited grazing use, but this is of secondary importance. The primary requirement is that it stabilize the gully.

Gullies are sometimes stabilized by building a terrace or a pond across them. The water storage area in the pond or behind the terrace is likely to accumulate silt. Siltation has completely refilled many gullies and made them into usable land again.

Large gullies may be too costly to fill or even to reshape but should nevertheless be controlled. A diversion above the gully may or may not suffice. Sometimes the gully is located in the only place that can be used as an outlet. These situations require a concrete drop structure, a pipeline, or other device to prevent the water from eroding as it drops. Such structures need to be sealed tightly into the soil around the gully. Loose debris such as dead trees, stones, and old car bodies dumped into the gully usually make matters worse rather than better.

OBJECTIVES OF SOIL CONSERVATION

The basic reason for conserving soil is to maintain it as a permanent, useful resource. It is not necessary to retain every soil particle in its present location, but the rate of soil loss should be limited to a value that will not destroy the productivity of the soil. This rate varies from one soil to another according to the nature of the soil and the rate at which new soil can be formed.

Soil conservation is everybody's business. Every person relies on products that come from the soil for food, clothing, shelter, and other materials. An increasing population and an increasing standard of living result in an increased demand for these products. This increasing demand needs to be translated into careful use of soil and not into exploitation. Future needs for productive soil promise to be even greater than present needs.

REFERENCES

Armburst, D. V., 1968, Windblown Soil Abrasive Injury to Cotton Plants, *Agron. J.* **60**: 622–625.

Barrows, H. L., and V. J. Kilmer, 1963, Plant Nutrient Losses from Soils by Water Erosion, *Adv. Agron.* **15**:303–316.

Bennett, O. L., E. L. Mathias, and P. E. Lundberg, 1973, Crop Responses to No-till Management on Hilly Terrain, *Agron. J.* **65**:488–491.

Browning, G. M., 1948, Save That Soil, *Iowa Farm Sci.* **2**(8):3–5.

Browning, G. M., R. A. Norton, A. G. McCall, and F. G. Bell, 1948, Investigations in Erosion Control and the Reclamation of Eroded Land, Clarinda, Iowa, 1931–42, *USDA Tech. Bull.* 959.

Browning, G. M., C. L. Parish, and John Glass, 1947, A Method for Determining the Use and Limitations of Rotation and Conservation Practices in the Control of Soil Erosion in Iowa, *J. Amer. Soc. Agron.* **39**:65–73.

Copley, T. L., L. A. Forrest, A. G. McCall, and F. G. Bell, 1944, Investigations in Erosion Control and Reclamation of Eroded Land, Statesville, N.C., 1930–40, *USDA Tech. Bull.* 873.

Drullinger, R. H., and B. L. Schmidt, 1968, Wind Erosion Problems and Controls in the Great Lakes Region, *J. Soil Water Conserv.* **23**:58–59.

Dudeck, A. E., N. P. Swanson, L. N. Mielke, and A. R. Dedrick, 1970, Mulches for Grass Establishment on Fill Slopes, *Agron. J.* **62**:810–812.

Ellison, W. D., 1948, Erosion by Raindrop, *Sci. Amer.* **179**(5):40–45.

Hartwig, N. L., 1974, Crownvetch Makes a Good Sod for No-till Corn, *Crops Soils* **27**(2):16–17.

Hill, H. O., W. J. Peevy, A. G. McCall, and F. G. Bell, 1944, Investigations in Erosion Control and Reclamation of Eroded Land, Temple, Tex., 1931–41, *USDA Tech. Bull.* 859.

Jacobson, Paul, 1969, Soil Erosion Control Practices in Perspective, *J. Soil Water Conserv.* **24**:123–126.

Laflen, J. M., H. P. Johnson, and R. C. Reeve, 1972, Soil Loss from Tile-outlet Terraces, *J. Soil Water Conserv.* **27**:74–77.

Lattanzi, A. R., L. D. Meyer, M. F. Baumgardner, 1974, Influences of Mulch Rate and Slope Steepness on Interrill Erosion, *Soil Sci. Soc. Am. Proc.* **38**:946–950.

McCalla, T. M., 1946, Influence of Some Microbial Groups on Stabilizing Soil Structure against Falling Water Drops. *Soil Sci. Soc. Am. Proc.* **11**:260–263.

Mannering, J. V., L. D. Meyer, and C. B. Johnson, 1968, Effect of Cropping Intensity on Erosion and Infiltration, *Agron. J.* **60**:206-209.

Massie, L. R., and G. D. Bubenzer, 1974, Improving Roadbank Erosion Control, *J. Soil Water Conserv.* **29**:176–178.

Meyer, G. J., P. J. Schoeneberger, and J. H. Huddleston, 1975, Sediment Yields from Roadsides: An Application of the Universal Soil Loss Equation, *J. Soil Water Conserv.* **30**:289–291.

Meyer, L. D., W. H. Wischmeier, and G. R. Foster, 1970, Mulch Rates Required for Erosion Control on Steep Slopes, *Soil Sci. Soc. Am. Proc.* **34**:928–931.

Moldenhauer, W. C., and M. Amemiya, 1969, Tillage Practices for Controlling Cropland Erosion, *J. Soil Water Conserv.* **24**:19–21.

Moldenhauer, W. C., and E. R. Duncan, 1969, Principles and Methods of Wind-erosion Control in Iowa, *Iowa State Univ. Agr. Home Econ. Exp. Sta. Rept.* 62.

Nazarov, G. V., 1974, Permeability of the Soil as an Indicator of Its Resistance to Erosion, *Sov. Soil Sci.* **6**:336–339.

Palmer, R. S., 1965, Waterdrop Impact Forces, *Trans. ASAE* **8**(1):69–70, 72.

Richardson, C. W., 1973, Runoff, Erosion, and Tillage Efficiency on Graded-furrow and Terraced Watersheds, *J. Soil Water Conserv.* **28**:162–164.

Robertson, W. K., H. W. Lundy, G. M. Prine, and W. L. Currey, 1976, Planting Corn in Sod and Small Grain Residues with Minimum Tillage, *Agron. J.* **68**:271–274.

Siddoway, F. H., 1970. Barriers for Wind Erosion Control and Water Conservation, *J. Soil Water Conserv.* **25**:180–184.

Smith, D. D., D. M. Whitt, and M. F. Miller, 1948, Cropping Systems of Soil Conservation, *Missouri Agr. Exp. Sta. Bull.* 518.

Thompson, J. R., 1970, Soil Erosion in the Detroit Metropolitan Area, *J. Soil Water Conserv.* **25**:8–10.

Wilson, H. A., 1947, *Fifteen Years of Soil Losses, Run-off and Crop Yields in Page County, Iowa,* Agronomy Dept., Iowa State University, Ames.

Wischmeier, W. H., 1962, Rainfall Erosion Potential, *Agr. Eng.* **43**:212–215, 225.

Wischmeier, W. H., and J. V. Mannering, 1969, Relation of Soil Properties to Its Erodibility, *Soil Sci. Soc. Am. Proc.* **33**:131–137.

Wischmeier, W. H., and D. D. Smith, 1965, Predicting Rainfall-erosion Losses from Cropland East of the Rocky Mountains, *USDA Agr. Handbook* 282, 47 p.

Wischmeier, W. H., D. D. Smith, and R. E. Uhland, 1958, Evaluation of Factors in the Soil Loss Equation, *Agr. Eng.* **39**:458–462.

Woodruff, N. P., and F. H. Siddoway, 1965, A Wind Erosion Equation, *Soil Sci. Soc. Am. Proc.* **29**:602–608.

Chapter 20

Soil Pollution

People have long ignored pollution and its effects on the environment. Pollution has increased along with population until it can no longer be neglected. Much study has gone into pollution problems in recent years and a great deal of concern has been expressed for our environment. Pollution is such a complex and persistent problem that continued concern and action will be needed throughout the foreseeable future. Pollution is so far-reaching that everyone is affected, and united efforts are needed to control pollution.

A pollutant is something that degrades the quality of something else. Most pollutants either are or have been useful or they are by-products or residues from the production of something useful. Many pollutants are difficult to eliminate without losing the value that caused them to be produced. Pollution control therefore requires evaluation of advantages and disadvantages so that the best alternatives can be chosen.

Some pollutants are natural materials that either get in the wrong place or become too concentrated in some places. Consider soil as an example. Soil can become a pollutant by getting into water or air. Soil is also a part of the environment that receives pollutants. Soil can depollute some materials by decomposing them into harmless end products such as carbon dioxide and water. Useful plant nutrients are often released by decomposition of pollutants. But, there are some materials that are difficult or impossible to decompose. Accumulations of such materials can pollute the soil. Soil pollution therefore includes three different aspects: soil as a pollutant, soil as a depollutant, and polluted soil.

SOURCES OF POLLUTANTS

Pollutants come from all types of human activity. Agricultural, industrial, and residential land use all produce residues and waste products that pollute air, water, and soil. Wrappings, containers, leftover materials, items that have outlived their usefulness, and various remnants and products of decomposition can be pollutants (Figure 20-1). These pollutants may become a problem because of high concentrations in particular areas or because of high potency even in low concentrations.

A distinction is usually made between point sources and diffuse sources of pollution. Sewage effluent entering a stream through a pipe is a point source. Eroded soil entering all along the length of a stream is a diffuse source. The feasibility of remedies and the type of correction to be applied depend on whether the polluting material comes from a point source or a diffuse source.

Soil pollution often relates to diffuse sources. Leaching, volatilization, sediment, and dissolved material in runoff are all diffuse sources of pollution. Extensive soil conservation and other good management practices are the best means of limiting such diffuse pollution.

Pollutants reaching the soil are often diffuse as well. Spreading waste material across a wide area allows the soil to depollute the waste even though it would pollute a smaller volume of soil.

Figure 20-1   Discarded junk in a gully is a common type of soil pollution *(Soil Conservation Service, USDA)*.

SOIL AS A POLLUTANT

Soil is the source of dust in the air and sediment in water. On the basis of tonnage, soil is by far the largest pollutant. The 3.5 billion tons of soil erosion occurring each year in the United States is hundreds of times greater than the tonnage of sewage entering the country's streams. The largest fraction of the soil is inert material, but even inert material can fill a reservoir (Figure 20-2).

Eroded soil carries with it bacteria and virus, but disease problems from this source are unlikely. Other problems caused by pesticides and plant nutrients in the soil are much more common. Grass-killing herbicides carried into waterways with eroded soil have killed the grass and exposed many waterways to rill or gully erosion. Insecticides carried into fish ponds can build up into high enough levels to kill the fish. A problem can arise even if the fish are not killed because some insecticides are progressively concentrated through the food chain. The banning of DDT resulted from its tendency to become more and more concentrated as it moved from soil to worms or fish and then to birds and their eggs. Metcalf (1971) cites Lake Michigan as an example with 2 ppt (parts per trillion) DDT in the water, 14 ppb (parts per billion) in the bottom mud, 410 ppb in amphipods, 3 to 6 ppm (parts per million) in fish such as coho and lake trout, and as much as 99 ppm in herring gulls at the top of the food chain.

Figure 20-2 Reservoirs built for electric power, flood control, or recreation lose much of their capacity by siltation *(Soil Conservation Service, USDA)*.

Plant Nutrients as Pollutants

The plant nutrient content of soil is usually its most important characteristic as a pollutant. The growth of algae and other plant life in bodies of water is usually limited by the supply of phosphorus or, less often, of nitrogen. Soil contains both of these nutrients and can supply them to streams, ponds, and lakes through leaching and by runoff water carrying both dissolved material and sediment. The result is often eutrophication as shown in Figure 20-3. Sediment usually carries considerably more nutrients than the amounts dissolved in water. An especially large percentage of the phosphorus moving from soil to water is carried by sediment because most phosphorus compounds have low solubilities.

Phosphorus is the nutrient most likely to be deficient enough to limit algal growth in nature. Adding a phosphorus supply is therefore the most common way of stimulating unwanted growth. Phosphorus is also the nutrient most amenable to control when efforts are made to improve the quality of water. Erosion control practices such as contouring, terracing, and border strip-cropping can keep much sediment out of streams. Phosphorus and other nutrients associated with the sediment are also kept out of the streams. At least two other sources of phosphorus also need to be controlled to keep water clean. One is runoff from feedlots which are too often placed next to streams (Figure 20-4). The other is sewage that has had only primary and secondary treatment. The effluent from secondary treatment contains large amounts of phosphorus coming from detergents and various organic wastes. Table 20-1 shows the principal sources of phosphorus in surface waters in Wiscon-

Figure 20-3 Algae grow in water that has been eutrophied (enriched in nutrients).

Figure 20-4 Cattle feedlots next to a stream pollute the stream with nutrients from manure.

sin. These data show that sewage treatment must be included if the phosphorus supply is to be controlled.

Tillage and fertilization also have much to do with the amount of phosphorus moving from land to water. Reduced tillage reduces sediment but the reduction may be partly offset by increased soluble phosphorus in the runoff resulting from surface-applied fertilizer. Proper rates of fertilization are important in any cropping system because too much fertilizer results in excessive nutrient loss to streams. The right amount of fertilizer, however, can improve growth and reduce erosion enough to

Table 20–1 Sources of Phosphorus in Wisconsin Surface Waters

Source	Total phosphorus, %
Municipal, industrial, and private sewage systems	59
Runoff from urban land	10
Runoff from rural land	
Fertilized land	21
Other cropland	3
Noncropland	3
Other (groundwater, precipitation, etc.)	4
Total	100

Source: Data of R. B. Carey from Walsh and Keeney, 1970.

reduce the amount of sediment and nutrients carried to streams. Proper fertilization is therefore a good way to reduce pollution.

Soil loss is normally seasonal and sporadic. Most erosion occurs while the land is bare or else the crop is too young to provide good cover. Heavy rain at such times can carry large amounts of soil in the runoff. The effect on the phosphorus supply, however, lasts much longer than the runoff period. The phosphorus in the sediment establishes an equilibrium with that in the water and gradually replenishes the supply as it is used.

Soil loss from forest land is normally less than that from cropland. The greatest hazard comes when the forest is harvested. Even then, the nutrient loss need not be large, nor should forest fertilization cause much increase in nutrients carried by runoff. Forest fertilization rates are seldom very high and the runoff water passes through too much litter material to carry large concentrations of nutrients.

Soil as an Air Pollutant

Wind erosion contaminates the air with large amounts of dust that may be carried hundreds of miles from the source. The dust may be a nuisance or worse wherever it settles. Respiratory problems are aggravated by dust, clothing gets dirty, machines wear out, and experiments can go wrong as a result of dust contamination.

The smallest dust particles accumulate in the upper atmosphere, sometimes in large enough concentrations to intercept significant amounts of sunlight. The worst episodes of wind erosion occurred when dry climatic cycles struck cultivated land

Figure 20-5 Dust is a common air pollutant.

in semiarid areas. The desolate conditions thus produced have aptly been called *dust bowls*. Dust bowls in the 1930s and 1950s darkened the skies for weeks at a time even in states far removed from the principal sources. Smaller amounts of dust pollution such as those shown in Figure 20-5 can occur anywhere.

Soil also contributes to air pollution by releasing volatile compounds into the atmosphere. Nitrogen escapes by ammonia volatilization and by denitrification. Decomposition of organic materials in soil can release sulfur dioxide and other sulfur compounds. Most of the gases emanating from the soil eventually return to the soil as the atmosphere is cleansed by precipitation. Dust particles also return to earth, often as the nuclei of raindrops.

SOIL AS A DEPOLLUTANT

People have often assumed that anything deposited in the soil would decompose. This assumption is not quite true. Some plastics, metals, and other materials will remain unchanged because there are no soil microbes able to decompose them. But soil is still the closest thing there is to a universal depollutant. Most things will rot, rust, or otherwise decompose in soil if given enough time.

Using soil as a depollutant is an extension of natural recycling and can be a highly beneficial way to dispose of most organic wastes. Materials that would be pollutants anywhere else become energy sources for soil microbes and nutrient sources for plant growth when applied to the soil.

Disposal of Manure

The use of manure as a fertilizer is discussed in Chapter 9. Often, however, the fertilizer value of manure is secondary to the problem of waste disposal. Feedlot concentrations of tens of thousands of cattle and chicken concentrations of hundreds of thousands of birds in single operations produce tens or hundreds of tons of manure per day.

Feedlot operators seldom have enough land available to spread their manure at the rate of 10 to 20 tons per hectare, the rate often used for fertilizer applications. They may want to use rates 10 times that heavy so only one-tenth as much land is required. The manure concentration is therefore high enough to be a source of pollution both in the pens and on the land used for disposal.

Manure emits a variety of nitrogen and sulfur compounds that pollute the atmosphere. Some of these compounds have strong offensive odors. The best way to control these odors is to spread the manure at reasonable rates and incorporate it promptly into the soil.

Manure on the soil surface is also a source of water pollution. It yields soluble plant nutrients to runoff in addition to the nutrients carried in solid particles. Incorporation into the soil helps solve the water pollution problem as well as the air pollution problem. The possibility of leaching remains, but leaching losses are usually small compared to runoff losses.

Manure can be applied to soil in a variety of ways. Spreaders are available to distribute the solid material either while it is fresh or after it has been composted.

Or the manure can be fermented in a lagoon and then spread with a sprayer truck or an irrigation system. Any of these systems converts a potential pollutant into a valuable source of plant nutrients.

Other Organic Wastes

Most organic wastes can be spread on land as manure is spread. The use of land for tertiary treatment of sewage is discussed in Chapter 17. Larson, Gilley, and Linden (1975) estimate that sewage sludge contains nutrients equal to about 5 percent of the N, 12 percent of the P, and 1 percent of the K applied in commercial fertilizers in the United States. Additional nutrients that could be turned into fertilizer are contained in sewage effluent, garbage, and other organic wastes. For example, Gambrell and Peele (1973) discuss the disposal of peach cannery wastes on soil.

Not all soils are equally suited for waste disposal. Suitable soils should be reasonably permeable and well aerated so that oxidation can take place. Good drainage is essential, but a soil can be so coarse-textured that the soluble materials in the wastes pass through too rapidly to be decomposed. Other soils may be too fine-textured. Osborne (1975) describes a system used to purify campground wastes to protect a lake from eutrophication. Impermeable soil was removed, tile drains installed, and suitable gravel, sand, and peat materials were placed over the drains to receive the wastes applied through a sprinkler system.

Plants Aid Waste Disposal

A forage crop should be grown and harvested where land is used for waste disposal (Figure 20-6). The crop aids in erosion control, maintaining soil permeability, and

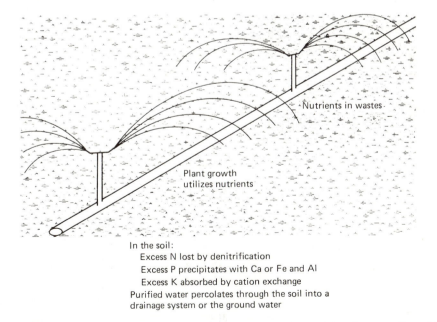

Nutrients in wastes

Plant growth utilizes nutrients

In the soil:
Excess N lost by denitrification
Excess P precipitates with Ca or Fe and Al
Excess K absorbed by cation exchange
Purified water percolates through the soil into a
drainage system or the ground water

Figure 20-6 Nutrients added by waste disposal are removed by plant growth and by various processes in the soil.

removal of soluble materials that would otherwise build up in the soil and groundwater. Harvesting the forage removes a maximum amount of nutrients and other ions from the area.

Wastes are sometimes applied at rates of 100 to 200 tons per hectare and may carry as much as 1,000 kg each of N, K, and Ca plus proportional amounts of other nutrients. Even forage crops are not likely to remove such large quantities of nutrients. Other means of nutrient removal are indicated in Figure 20-6. The loss of nitrogen by denitrification can be encouraged by having at least two areas to alternate where an irrigation system is used. Nitrification takes place during dry periods and denitrification during wet periods. Lance, Whisler, and Bouwer (1973) indicate that about 5 days were required to allow enough oxygen to enter their soil for nitrification of the ammonia contained in sewage effluent applied during the preceding 6 days. They suggest that more dry periods would be more beneficial than longer ones.

Maximum Rates of Waste Application

Relatively low rates of waste application would give maximum fertilizer benefit from the wastes, but high rates are likely to be used where waste disposal is the primary objective. Lance, Whisler, and Bouwer (1973) suggest that the maximum rate of waste application be determined by balancing the ammonium-N supply against the oxygen supply.

High rates of waste application increase the likelihood of polluting either the groundwater or the soil. Even with alternating wet and dry periods, it is possible to leach nitrogen from the soil. The ammonium form can be leached if the wet periods are too long. The nitrate form may be leached if the wet periods begin too suddenly for denitrification to lower the nitrate concentration.

Heavy applications of wastes increase the salinity of the groundwater under any conditions. Salts such as sodium chloride and calcium chloride are too soluble to be prevented from leaching. Usually the increased salinity is not a problem unless the water also receives large salt inputs from other sources.

Toxic Elements

Heavy metals and other potentially toxic elements are the most serious soil pollutants in sewage. Bradford et al. (1975) found toxic levels of boron and excesses of several heavy metals in some plants irrigated with sewage materials. The heavy metals are a matter of concern because they accumulate on cation-exchange sites and are very difficult to remove from the soil. Harvesting forage crops removes small quantities of the heavy metals as long as the concentrations are below toxic levels. Still smaller amounts of heavy metals are emitted to the atmosphere in particulate matter while plants are growing. Beauford, Barber, and Barringer (1977) reported zinc emissions of about 90 g/ha. The zinc in the emitted particles was 1.6 times as concentrated as in the rest of the plant and 5.3 times as concentrated as in the substrate.

Sewage sludge contains higher concentrations of heavy metals than most other wastes contain. The concentrations vary widely from one community to another and from time to time, but they are consistently high enough that heavy-metal concentrations should be monitored where repeated heavy applications of sludge occur. Larson, Gilley, and Linden (1975) recommend that the zinc equivalent not be raised above 5 to 10 percent of the soil's cation-exchange capacity. The zinc equivalent is defined as the zinc concentration plus twice the copper concentration plus 8 times the nickel concentration. The zinc equivalent will probably be reached when the tons of sewage applied per hectare totals between 25 and 75 times the cation-exchange capacity of the soil in milliequivalents per 100 grams.

Pesticides in the Soil

Pesticides can be subdivided into groups according to the types of pests they control as herbicides (weed killers), insecticides, fungicides, nematocides, etc. Many pesticides are chemically similar to several others and are therefore given family names such as chlorinated hydrocarbons or organic phosphates. Pesticides that can be decomposed readily are biodegradable whereas those that resist decomposition are called persistent.

Hundreds of different chemicals are used to kill weeds, insects, and other pests. Many are applied to the soil to kill soil-borne pests. Other pesticides reach the soil even though they were applied to plants, animals, or buildings. Soil microbes are able to detoxify most pesticides. Ideally the pesticides would be detoxified right after they have served their purpose. Practically, some decompose so rapidly that repeated applications are needed; others last so long that they become serious pollutants. An ideal pesticide would never cause a pollution problem because it would not harm anything but the pest it was designed to control.

Factors Affecting Pesticide Decomposition

Some pesticides need to last much longer than others. A pesticide that eradicates the pest problem in a short time may be adequate if it decomposes in a few days. Other pest problems may keep recurring so more persistent pesticides are needed. The rate any particular pesticide decomposes in soil depends on both the pesticide and the soil. Some biodegradable pesticides (such as the organic phosphates) were specially developed to reduce pollution by decomposing faster than the pesticides they replaced.

Many of the most persistent pesticides found in soil are insecticides. Nash and Harris (1973) reported that a sandy loam soil in Maryland still contained between 7 and 49 percent of the original concentrations of various chlorinated hydrocarbon insecticides 16 years after they were applied. Insecticides may need to last all summer to avoid the cost of repeated applications, but a half-life of 16 years is too long for chemicals that pose environmental hazards.

Half-life of Pesticides   The half-life concept (the time required for half of the material present to decompose) is often applied to herbicides but with less accuracy than in some other applications. Variable results come from different soil conditions, methods of measurement, and rates of application. The half-life of DDT, for example, was reported to be 10 years or longer in some of the early tests, partly because the rates were high enough to kill the microbes that would have decomposed the DDT. More realistic rates of application result in a half-life of 2 to 4 years for DDT (Table 20-2).

Complex chemicals like DDT go through several stages as they decompose. Some of the decomposition products are as toxic as the original material. Later stages of decomposition produce harmless products long before the carbon and hydrogen are all converted to CO_2 and H_2O. Different half-lives are indicated for disappearance of the original material, detoxification, or complete decomposition.

Soil Factors Influencing Decomposition   Soil microbes do most of the work of decomposing pesticides. The microbial population and factors such as energy supply, aeration, pH, and temperature that influence microbial activity are all important.

The presence of microbes able to decompose a particular pesticide depends partly on whether the pesticide or similar materials have been present in the soil before. It may take much longer to develop an initial population of a microbe than to rebuild the population later. A pesticide may therefore be more persistent when it is first introduced than it will be later. Even when microbes are present, they need an energy supply and enough nutrients to make them active. The addition of decomposable plant residues or wastes can therefore accelerate the decomposition of pesticides.

Aerobic microbes are able to decompose some materials that anaerobic microbes either cannot decompose or decompose more slowly. Aeration therefore has a greater effect on the decomposition rates of some pesticides than others. Lavy, Roeth, and Fenster (1973) reported rapid degradation of 2,4-D under aerobic conditions at depths of 15, 40, and 90 cm. The 2,4-D degraded too fast for full season weed control. In contrast, atrazine degradation was slower and influenced more by depth. The phytotoxicity of atrazine was gone in 5 months at a depth of 15 cm and in 17 months at a depth of 40 cm but some still remained after 41 months at a depth of 90 cm.

Microbial activity is often faster at a pH near neutral than under very acid or alkaline conditions. Smith (1972) adjusted soil pH to various values with sulfur and

Table 20–2   Half-lives of Pesticides in Soils

Type of pesticide	Approximate half-life (years)
Lead, arsenic, copper, mercury compounds	10–30
Dieldrin, BHC, DDT insecticides	2–4
Triazine herbicides	1–2
Benzoic acid herbicides	0.2–1
Urea herbicides	0.3–0.8
2,4-D and 2,4,5-T herbicides	0.1–0.4
Organophosphate insecticides	0.02–0.2
Carbamate insecticides	0.02–0.1

Source: Metcalf, 1971.

calcium hydroxide. He found that 2,4-D degradation was fastest when the pH was between 7.0 and 7.4 and that the rate dropped off more if the pH was on the alkaline side rather than on the acid side.

Adsorption of Pesticides

Pesticides are adsorbed to varying extents by organic matter and clay in the soil. Stevenson (1972) indicates that herbicide activity in soil is usually more closely related to organic-matter content than to any other soil factor. Snelling, Hobbs, and Powers (1969) related atrazine activity to specific surface and exchange capacities as well as to organic-matter content. All of these factors influence the adsorptive capacity of the soil and the mobility of anything that can be adsorbed.

Humus and montmorillonite are most effective for adsorbing pesticides because they have large surface areas and high exchange capacities. Pesticides that form either positively charged ions or strongly polar molecules are adsorbed to a greater extent than others. The chlorinated hydrocarbons, for example, are held so they are relatively immobile in soil (Pionke and Chesters, 1973). Negatively charged ions can also be adsorbed but there are fewer adsorption sites available to them. Acid herbicides are relatively mobile because their negative charges are not adsorbed on cation-exchange sites.

Adsorbed pesticides are less active in several ways. Their low mobility reduces loss by leaching, absorption by plants, and decomposition by microbes. The reduced leaching loss helps control water pollution but the reduced plant absorption often means that application rates will be higher. The recommended rates for many soil-applied herbicides are adjusted for the organic-matter content of the soil so the pesticide activity will be maintained at the desired level.

A pesticide is likely to be most persistent in a soil where it is strongly adsorbed. The increased persistence is the result of the higher rate of application normally used on such soils combined with the slower rate of decomposition. Slower decomposition results because the adsorbed pesticide is held tightly enough to reduce its accessibility for microbial action.

Removal of Pesticides from Soil

Pesticides can move from the soil into air, water, or plants. Herbicides are intended to be absorbed by plants so they can kill weed species. Insecticides, fungicides, etc, can also be absorbed and probably decomposed within plants. Plants thus provide a significant means for removal of persistent pesticides from soil.

Many pesticides are volatile enough to diffuse through the soil if they are not adsorbed by the soil colloids. The diffusion helps to distribute them through the soil, but it also helps them escape to the atmosphere. Sunlight becomes a factor when certain pesticides reach the soil surface. Some types of pesticides are subject to photodecomposition when exposed to sunlight.

Pesticides in Water

Pesticides are more likely to cause damage if they reach a body of water than if they remain in the soil. The ability of soil to adsorb certain pesticides on exchange sites is an important mechanism for controlling water pollution. But, some pesticides are not adsorbed and are therefore subject to leaching where significant amounts of water percolate through the soil. Leaching may not cause pollution if the pesticide is biodegradable, but persistent pesticides can reappear. They eventually reach the groundwater table and seep into streams or are pumped up in wells.

Soil erosion is a more serious source of water pollution than leaching. Erosion can carry adsorbed pesticides directly into streams with no time for decomposition. Herbicides washed into grass waterways may kill the grass. Insecticides are taken from the water by fish, then absorbed by anything that eats the fish.

POLLUTED SOIL

Soil can be overloaded in its role as a depollutant and become polluted itself. The productive potential of the soil may be reduced or eliminated, or plants growing on the soil may absorb toxic materials that cause problems at some point in the food chain. The damage may be either short-lived or long-lived. Many pollutants are so costly to remove that it becomes uneconomical to reclaim the land.

Materials such as the pesticides and wastes discussed earlier in this chapter should cause no problem if they are applied at appropriate rates. However, most of them can be applied at excessive rates and it is then that they cause damage. Excess amounts of some pesticides may kill or injure plants, insects, and animals that were not intended to be targets. The overload may make the pesticide more persistent than normal by killing the microbes that would otherwise decompose it. Sometimes the chemical applied is converted to a related form that is also detrimental.

Salinity Produced by Pollution

Excess quantities of wastes can cause a buildup of soluble salts in the soil. Part of the salts are carried in solution in the liquid portion of the wastes and others are formed as the solids decompose. Salts accumulate in the soil when the amounts added in wastes exceed the amounts removed by plant growth, leaching, and other means. Eventually a saline or a saline-sodic soil may be produced. The worst such condition occurs when leaching removes the excess salts but leaves enough sodium to form a sodic soil. The nature and treatment of saline and sodic soils are discussed in Chapter 8.

Saline and sodic soils are occasionally formed by other forms of pollution. For example, in places where salt is used to melt snow and ice, there may be enough salt entering the soil along roads and sidewalks to sterilize the soil. Usually the affected area is only a narrow strip and it may be only an occasional spot along the way. Salty water dumped after freezing ice cream may also kill vegetation and damage the soil. A similar problem arises when seawater floods land because of a tidal wave.

Heavy Metals Can Damage Soil

Soils normally contain small amounts of heavy metals like lead, zinc, copper, and mercury. In fact, copper, iron, manganese, molybdenum, and zinc are included in the list of essential elements. But, excessive amounts of these and other heavy metals are detrimental to life and can sterilize the soil. Heavy-metal pollution is serious because it can persist for many decades.

Heavy metals reach the soil in several ways. Small amounts cycling through plant growth and returning in residues are normal and cause no problem. Larger amounts of heavy metals contained, for example, in repeated applications of sewage sludge can be too much. The zinc equivalent defined earlier in this chapter is a means of estimating how much sewage can be applied. Some industrial wastes contain such high concentrations of heavy metals that they should neither be spread on land nor dumped in water.

Lead and other heavy metals used in fuels and lubricants find their way into the atmosphere and drift across areas near heavily travelled routes. Air, water, and soil may all suffer pollution from this source. MacLean and Langille (1973) reported substantial increases in the lead, copper, and molybdenum contents of soils and plants along roads.

Bordeaux mixture (copper sulfate and lime), various mercury compounds, and other preparations containing heavy metals were once the only kinds of fungicides and insecticides known. Applications had to be many times as heavy as modern pesticides and often had to be repeated every week or two. The soil beneath some fruit trees became heavily contaminated with heavy metals that dripped or washed from the trees after they were sprayed. Plant growth on some orchard sites is still affected even several decades after the trees were removed.

Heavy metals are difficult to remove from soils because they are strongly held on cation-exchange sites. The concentrations in solution are therefore low and leaching is relatively ineffective for removing them from the soil. Growing plants can remove significant amounts if the metal concentration is not high enough to prevent plant growth. Sometimes the activity of a heavy metal can be reduced by adding something that reacts with it to form an insoluble compound. For example, Kunishi and Taylor (1972) were able to reduce the extractability of strontium added to soil from 80 to 100 percent down to 2 percent. They added diammonium phosphate, lime, and ammonium fluoride to the soil so the strontium would be converted to insoluble apatite.

Some problems caused by heavy metals are actually deficiencies of other nutrients. For example, iron and manganese are antagonistic to each other because they enter roots by the same carrier molecules. Excess manganese can therefore cause an iron deficiency. A suitable iron source applied to the plants or the soil may solve such a problem. Also, excesses of iron and several other heavy metals form insoluble phosphorus compounds and can create a phosphorus deficiency. Phosphorus fertilizer and liming may solve the problem.

Junk

Significant areas of land have been covered with automobile junkyards such as that shown in Figure 20-7 and with various other types of debris and trash. Such deposits may not harm the soil but they do affect its use, and they constitute a form of sight pollution. It often becomes necessary at a later date to remove the junk so the land can be put to another use.

Figure 20-7 Auto junkyards cover extensive areas of land.

Gullies are favorite places to deposit junk. Scenes such as that shown in Figure 20-1 are much too common. The person who dumped the junk may have thought it would help control erosion. Actually, the stream is likely to wash around or through the junk and be as erosive as ever. Some of the junk contributes to water pollution. The junk usually has to be removed before a serious effort can be made to smooth and stabilize the gully so the land can be reclaimed.

Strip Mines

Strip mining is an economical way to remove coal and other deposits where they are located near the land surface. Often several meters or even tens of meters of soil and rock material are removed to reach the layer being mined (Figure 20-8). Most strip mines have left behind topography converted into a series of steep spoil piles with small ponds between them. The material is usually deposited in inverted sequence with that from the lowest layers being placed on top because it was dug out last. The result is a type of wasteland that has become a matter of environmental concern. Several states have passed legislation requiring that strip-mined areas be restored to a useful condition.

Many coal deposits contain large amounts of iron pyrites and other sulfides in and near the coal layer. Sulfides in the coal lead to air pollution when the coal is burned. Sulfides in the material above the coal lead to excess acidity in the spoil piles

Figure 20-8 A strip mine where several distinct layers of material are being removed and replaced as the coal is mined.

and adjoining ponds. The sulfur exists as sulfides because reducing conditions exist in the deeper layers. The presence of oxygen in the spoil piles allows bacteria to oxidize the insoluble sulfides into soluble sulfates including sulfuric acid. The pH drops until bacterial action ceases around pH 2 or 3. Plants cannot grow on such acid spoil and fish cannot live in the acid water that seeps into some of the ponds. The pattern is often spotty with acid barren areas interspersed with vegetated spoil piles that contained less sulfur.

Reclamation of acid spoil piles would require so many tons of lime that it is usually prohibitive to neutralize all of the acidity. Leveling the steep, partially vegetated spoil piles is also expensive. Consequently, most such areas have simply been abandoned. Leaching may eventually remove much of the acidity and erosion will smooth the topography, but nature may take thousands of years to reclaim the land.

Several experiments have been tried to find alternate ways of reclaiming acid spoil from strip mines and other sources. The most successful approach has been to apply large quantities of organic materials such as manure or sewage sludge. For example, Hortenstine and Rothwell (1972) applied composted sludge to nearly sterile sand tailings from a phosphate mine. The compost improved the cation-exchange capacity, water-holding capacity, organic-matter content, and plant nutrient supply of the tailings. Sorghum and oat yields were much improved, though still low by field standards. Compost and fertilizer together gave much better yields than fertilizer alone on the tailings.

The best way to overcome the acid spoil heap problem from strip mines is to keep the soil on top of the sulfide-bearing layers. There are pilot projects now in operation to develop the techniques and check the costs of restoring mined areas to productive cropland. Iowa, for example, has a project that divides the overburden into soil, parent material, and acid shales. Each material is replaced in its own proper layer as old areas are refilled with material excavated from new areas. The final topography is formed into terraces (Figure 20-9) with the soil on top so it should become better cropland than it was before the mining began.

SOIL POLLUTION—OLD AND NEW

Soil pollution is old as well as new. People have dropped things wherever they happened to be as far back as human history can be traced. In fact, the debris left behind is the archaeologist's treasure and the source of much of our history. We would not know nearly as much about departed civilizations if the people had been more tidy.

The significance of modern pollution is related to the large number of people living today and their manner of life. Pollutants are more abundant now and come in a wider variety than ever before. Recent decades have brought a new awareness

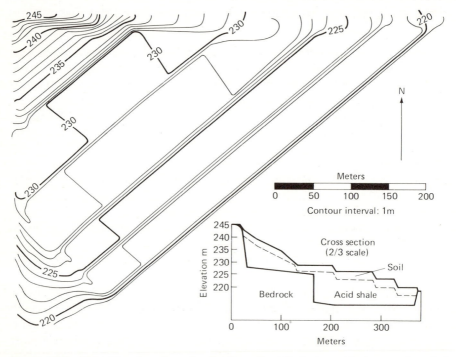

Figure 20-9 The planned contours for the Iowa Coal Project provide for a series of straight terraces in place of the original irregular topography.

of pollution problems and a desire to deal with them. The demands of a growing population are too great for polluted land to be abandoned in favor of new land.

Soil pollution is widespread and comes in many forms. The solutions are seldom simple but pollution can be controlled by concerted effort. It is possible for humanity to use soil productively without ruining it. Current efforts to learn how to use land wisely are most encouraging.

REFERENCES

Anonymous, 1976, The Aerial Photo—Water Quality Link, *Environ. Sci. Technol.* **10**:228–229.

Beauford, W., J. Barber, and A. R. Barringer, 1977, Release of Particles Containing Metals from Vegetation into the Atmosphere, *Science* **195**:571–573.

Bradford, G. R., A. L. Page, L. J. Lund, and W. Olmstead, 1975, Trace Element Concentrations of Sewage Treatment Plant Effluents and Sludges; Their Interactions with Soils and Uptake by Plants, *J. Environ. Qual.* **4**:123–127.

Burwell, R. E., D. R. Timmons, and R. F. Holt, 1975, Nutrient Transport in Surface Runoff as Influenced by Soil Cover and Seasonal Periods, *Soil Sci. Soc. Am. Proc.* **39**:523–528.

Gambrell, R. T., and T. C. Peele, 1973, Disposal of Peach Cannery Wastes by Application to Soil, *J. Environ. Qual.* **2**:100–104.

Graham, I. J., 1969, Diffusion of Organophosphorus Insecticides in Soils, *J. Sci. Food Agric.* **20**:489–494.

Hance, R. J., 1974, Soil Organic Matter and the Adsorption and Decomposition of the Herbicides Atrazine and Linuron, *Soil Biol. Biochem.* **6**:39–42.

Hinrichs, D. G., A. P. Mazurak, and N. P. Swanson, 1974, Effect of Effluent from Beef Feedlots on the Physical and Chemical Properties of Soil, *Soil Sci. Soc. Am. Proc.* **38**:661–663.

Hortenstine, C. C., and D. F. Rothwell, 1972, Use of Municipal Compost in Reclamation of Phosphate-Mining Sand Tailings, *J. Environ. Qual.* **1**:415–418.

Kunishi, H. M., and A. W. Taylor, 1972, Immobilization of Radiostrontium in Soil by Phosphate Addition, *Soil Sci.* **113**:1–6.

Lance, J. C., F. D. Whisler, and H. Bouwer, 1973, Oxygen Utilization in Soils Flooded with Sewage Water, *J. Environ. Qual.* **2**:345–350.

Larson, W. E., J. R. Gilley, and D. R. Linden, 1975, Consequences of Waste Disposal on Land, *J. Soil Water Conserv.* **30**:68–71.

Lavy, T. L., F. W. Roeth, and C. R. Fenster, 1973, Degradation of 2,4-D and Atrazine at Three Soil Depths in the Field, *J. Environ. Qual.* **2**:132–137.

MacLean, K. S., and W. M. Langille, 1973, Heavy Metal Studies of Crops and Soils in Nova Scotia, *Commun. Soil Sci. Plant Anal.* **4**:495–505.

Metcalf, R. L., 1971, Pesticides, *J. Soil Water Conserv.* **26**:57–60.

Nash, R. G., and W. G. Harris, 1973, Chlorinated Hydrocarbon Insecticide Residues in Crops and Soil, *J. Environ. Qual.* **2**:269–273.

Osborne, J. M., 1975, Tertiary Treatment of Campground Wastes Using a Native Minnesota Peat, *J. Soil Water Conserv.* **30**:235–236.

Pionke, H. B., and G. Chesters, 1973, Pesticide-Sediment-Water Interactions, *J. Environ. Qual.* **2**:29–45.

Romkens, M. J. M., D. W. Nelson, and J. V. Mannering, 1973, Nitrogen and Phosphorus Composition of Surface Runoff as Affected by Tillage Method, *J. Environ. Qual.* **2**:292–295.

Smith, A. E., 1972, Influence of Calcium Hydroxide and Sulfur on 2,4-D Degradation in Soil, *Soil Sci.* **113**:36–41.

Snelling, K. W., J. A. Hobbs, and W. L. Powers, 1969, Effects of Surface Area, Exchange Capacity, and Organic Matter Content on Miscible Displacement of Atrazine in Soils, *Agron. J.* **61**:875–878.

Sopper, W. E., 1975, Effects of Timber Harvesting and Related Management Practices on Water Quality in Forested Watersheds, *J. Environ. Qual.* **4**:24–29.

Stall, J. B., 1972, Effects of Sediment on Water Quality, *J. Environ. Qual.* **1**:353–360.

Stevenson, F. J., 1972, Organic Matter Reactions Involving Herbicides in Soil, *J. Environ. Qual.* **1**:333–343.

Syers, J. K., R. F. Harris, and D. E. Armstrong, 1973, Phosphate Chemistry in Lake Sediments, *J. Environ. Qual.* **2**:1–14.

Viets, F. G., Jr., 1971, The Mounting Problem of Cattle Feedlot Pollution, *Agric. Sci. Rev.* **9**(1):1–8.

Walker, K. C., and C. H. Wadleigh, 1968, Water Pollution from Land Runoff, *Plant Food Rev.* **14**(1):2–4.

Walsh, L. M., and D. R. Keeney, 1970, The Pollution Problem: Phosphorus in Surface Waters, *Hoard's Dairyman* **115**:870.

Warren, F. D., 1973, Action of Herbicides in the Soil—Affected by Organic Matter, *Weeds Today* **4**(2):10–11.

Weber, J. B., S. B. Weed, and T. J. Sheets, 1972, Pesticides—How They Move and React in the Soil, *Crops Soils* **25**(1):14–17.

Units of Measurement

Most of the units used in this book are part of the International System of units that is better known as the Metric System. Much literature, including previous editions of this book, contains a variety of units from the English system. The following factors are useful for converting data from one unit to another. The units are classified here according to their various functions.

Distance, length, thickness

kilometer, 1 km = 1,000 m = 0.621 mi
meter, 1 m = 39.36 in. = 1.094 yd
centimeter, 1 cm = 0.01 m = 0.394 in.
millimeter, 1 mm = 0.001 m = 0.0394 in.
Angstrom, 1 Å = 10^{-10} m = 10^{-7} mm

Area

square meter, 1 m^2 = 10.76 ft^2
hectare, 1 ha = 10,000 m^2 = 2.471 ac

Volume

cubic meter, 1 m³ = 1,000 l = 35.29 ft³
liter, 1 l = 1,000 cm³ = 1.057 quarts
cubic centimeter, 1 cm³ = 0.001 l = 0.061 in.³

Weight

metric ton, 1 ton = 1,000 kg = 2,205 lb = 1.102 tons (English)
quintal, 1 q = 100 kg = 220.5 lb
kilogram, 1 kg = 1,000 g = 2.205 lb
gram, 1 g = weight of 1 cm³ of H_2O = 0.035 ounces (Avdp.)
milligram, 1 mg = 0.001 g

Temperature

degrees Celsius, 1°C = 1.8°F (Fahrenheit)

°C

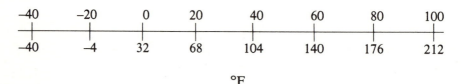

°F

Pressure

atmosphere, 1 atm = 1.013 bars = 1.033 kg/cm³
                 = 76 cm of mercury = 1,033 cm of water
                 = 33.9 ft of water = 14.7 lb/ft³

Density

density of water = 1 g/cm³ = 100 kg/m³ = 62.4 lb/ft³

Rates of application, yields

tons/hectare, 1 ton/ha = 10 q/ha = 1,000 kg/ha = 892 lb/ac
quintals/hectare, 1 q/ha = 100 kg/ha = 89.2 lb/ac
                       = 1.49 bu of wheat or 1.59 bu of corn or 2.79 bu of
                         oats/ac
kilograms/hectare, 1 kg/ha = 0.892 lb/ac

Proportions

depth basis, 1 cm/cm = 100%
concentrations, 100% = 1,000,000 ppm = 2,000,000 pp2m
map scales,   1:15,840 = 4 in./mile
              1:63,360 = 1 in./mile

Chemical units

mole, 1 mole = atomic, molecular, or ionic weight in g
equivalent weight, 1 eq wt = 1 mole/valence = 6×10^{23} reactive charges
molarity, 1 M = 1 mole/l
normality, 1 N = 1 eq wt/l
milliequivalent, 1 meq = 0.001 eq = 1 ml $\times$ 1 N

INDEX